W9-DER-605

EAST EUROPEAN MONOGRAPHS, NO. X

THE BOSNIAN CHURCH: A NEW INTERPRETATION

A Study of the Bosnian Church and Its Place in State and Society from the 13th to the 15th Centuries

JOHN V.A. FINE, JR.

EAST EUROPEAN QUARTERLY, BOULDER
DISTRIBUTED BY COLUMBIA UNIVERSITY PRESS
NEW YORK AND LONDON

1975

John V. A. Fine jr. is Associate Professor of History
at The University of Michigan

Library of Congress Catalog Card Number 75-6226
ISBN 0-914710-03-6

Printed in the United States of America

TO MY PARENTS

CONTENTS

Acknowledgments i

Terminology iv

Abbreviations vii

Map ix

INTRODUCTION 1
FOOTNOTES TO INTRODUCTION 6
CHAPTER ONE: RELIGION IN BOSNIA'S PEASANT SOCIETY 9
I: Peasants and Religion 9
II: Language, Experience & Thought 20
III: Family Structure 25
IV: Bases of Moral Order 27
V: Relations Between Rulers and Peasants 28
VI: Contact with and Spread of New Ideas 30
VII: Unrealistic Historical Models of Conversion 32
FOOTNOTES TO CHAPTER ONE 33
CHAPTER TWO: AN EXAMINATION OF THE SOURCES 41
Native South Slav Sources 42
Catholic Church Sources 46
Papal Sources 47
Inquisition Documents 54
Franciscan Sources 59
Chronicles 62
Pius II 63
Orbini and Pietro Livio of Verona 65
Luccari and Resti 73
Western 13th Century Chronicles 74
Franciscan Chronicles 75
Defters and Turkish Laws 77
The Gospel Manuscripts 81
The Apocryphal Tales 83
Traditions 84
Gravestones 88
Churches 93
FOOTNOTES TO CHAPTER II: 94

APPENDIX A FOR CHAPTER TWO: On Forged Documents 106
Footnotes to Appendix A 108
APPENDIX B FOR CHAPTER TWO: An Alleged Tract of Jacob de Marcia Against Dualists in Bosnia 110
Footnotes to Appendix B 111
CHAPTER III: BOSNIA FROM THE END OF THE TWELFTH CENTURY TO THE ACCESSION OF STJEPAN KOTROMANIĆ, ca 1315 113
I: Background: Catholicism and Orthodoxy in Bosnia and Environs 113
II: Heresy in the Balkans and Dualism 115
III: Ban Kulin and Catholicism 121
IV: Charges of Heresy and Foreign Interference 123
V: The Bilino polje Renunciation 126
VI: Religious Affairs from 1203 to the 1220's 134
VII: Charges of Heresy in the 1220's 135
VIII: Papal Action and the Hungarian Crusades in the 1230's 137
IX: Bosnian Catholic Church Subjected to Kalocsa 145
X: Establishment of the Bosnian Church 148
XI: Reference to Heretics from the 1280's to 1305 153
XII: Djed Miroslav of the Bosnian Church 155
FOOTNOTES TO CHAPTER THREE 157
CHAPTER FOUR: BOSNIA FROM Ca. 1315 to 1391 167
I: Ban Stjepan Kotromanić: Early Political Successes and Relations with the Different Faiths 167
II: Continued Papal Concern About Heresy in Bosnia 1325-1338 177
III: The Establishment of the Franciscan Bosnian Vicariat 180
IV: Tvrtko Establishes Himself in Power: Relations with the Different Faiths 187
V: Papal Letters About Bosnia in the 1370's 192
VI: Tvrtko's Involvement in the Hungarian Civil War 197
VII: Italian Documents about Dualists in Bosnia 199
VIII: Tvrtko, Catholicism and the Bosnian Church 200
FOOTNOTES TO CHAPTER FOUR 201
CHAPTER FIVE: BOSNIA FROM 1391 to 1443 211
I: Character of the Bosnian State 1391-1443 211
II: Three Religious Sources: The Serbian Anathemas, the Batalo Gospel, the Gospel of Hval Krstjanin 212
III: Political Events 1391-1421 and Role of the Bosnian Church 219
IV: Patarins in Secular Service and Relations Between Secular Leaders and the Various Faiths 238
V: Jacob de Marchia and the Franciscan Mission 244
VI: Bosnia and the Church Council at Basel 248
VII: Roman Catholic Gains in Bosnia 250
VIII: The Kosača Family: Church Building and Relations with the Orthodox Church 251
IX: Political Events from the late 1430's to 1443 and the Secular Role of Patarins 254
X: Patarin Hižas and Dubrovnik 256
XI: Gravestone Inscriptions Giving Information about the Bosnian Church 260

XII: Patarin Diplomats 264
XIII: Social and Political Position of the Bosnian Church 267
FOOTNOTES TO CHAPTER FIVE: 277
CHAPTER SIX: BOSNIA FROM 1443 to 1463 295
I: Increased Mention of Dualism in Sources from the 1440's 295
II: Civil War Between Stefan Tomaš and Stefan Vukčić 299
III: King Stefan Tomaš Accepts Catholicism 301
IV: The Bosnian Church: The Dragišić Charter and Radosav Ritual 303
V: Progress of Catholicism from 1446 305
VI: Papal Mention of Manichees 1447-1453 307
VII: Stefan Vukčić's Political Affairs (1448-1451) and Religious Associations 310
VIII: The Herceg's Economic Problems — War with Dubrovnik 312
IX: Peace Treaties of 1453-54 — Role of Gost Radin 317
X: The Pavlovići and Patarins in the 1450's 321
XI: The Herceg's Dealings with Catholicism and Orthodoxy 322
XII: The Question of Kudugers 324
XIII: Catholic-Orthodox Rivalry Begins in Bosnia 326
XIV: The Turkish Threat and Proposed Leagues to Meet it 328
XV: Kotruljic's Evidence 329
XVI: Catholic Progress in the Final Years of Stefan Tomaš 330
XVII: Persecution Launched Against the Bosnian Church in 1459 332
XVIII: Three Bosnians Abjure Fifty Manichee Errors in Rome 335
XIX: The Turkish Conquest of Bosnia and Part Played by Religious Differences 338
XX: The Position of the Bosnian Church in the Last Years of the Kingdom 341
XXI: Mention of Religion in the Hungarian Banate of Jajce 345
FOOTNOTES TO CHAPTER SIX: 346
APPENDIX A FOR CHAPTER SIX: The 50 Points Renounced Before Cardinal Torquemada and their Irrelevance for the Bosnian Church 355
CHAPTER SEVEN: HERCEGOVINA FROM 1463 to 1481 363
I: Patarins in Hercegovina after 1463 363
II: The Herceg's Relations with Different Faiths in his Last Years 364
III: The Testament of Gost Radin 365
IV: Patarins in Hercegovina After 1466 370
FOOTNOTES TO CHAPTER SEVEN 373
CHAPTER EIGHT: RELIGION IN BOSNIA AFTER THE TURKISH CONQUEST 375
I: Patarins and Other Heterodox Christians in the Turkish Period 375
II: Main Trends in Bosnian Religious History 1463-1600 377
FOOTNOTES TO CHAPTER EIGHT 387
BIBLIOGRAPHY 391
Information about the Organization of the Bibliography 391
Key to Abbreviations Used in the Bibliography 393
I: PRIMARY SOURCES 395
A: Written 395
B: Archaeological and Epigraphical 401
II: Secondary Works 406
REGISTER OF PERSONAL AND PLACE NAMES 435

ACKNOWLEDGEMENTS

The path along which an American of non-Balkan origin travels before reaching the decision to study a small medieval Balkan state could hardly have been direct; thus it is only fitting to acknowledge the influence of those who introduced me to this field. During my undergraduate years at Harvard four scholars had great influence on me: Professor Robert L. Wolff had the greatest impact in attracting me to History in general and to medieval Eastern Europe in particular. Professor Albert B. Lord introduced me to the culture and folklore of the South Slavs. Father Georges Florovsky introduced me to medieval Slavic Church History and the late Professor Clyde Kluckhohn opened up to me the fascinating world of Anthropology and the importance of applying its methods to the study of History. Upon graduation I received a Jugoslav Government Grant (through the Institute of International Education) to attend courses at the Filozofski fakultet in Beograd during the year 1961-1962. There the excellent lectures of Professor Sima Ćirković introduced me to the fascinating history of medieval Bosnia. With the above background and Professor Wolff's encouragement, when the time came for me to choose a topic for my doctoral thesis what could have been a more natural subject that heresy in medieval Bosnia? Once again I was fortunate to be able to spend a year in Jugoslavia (1966-67). This time Harvard University provided travel funds and the Jugoslav Government (through its Commission for Cultural Affairs with Foreigners) generously awarded me a stipend for a year's support. This time I was based at the Filozofski fakultet in Sarajevo. This institution helped me in every possible way; it provided me with an office, use of its library and contact with its stimulating faculty. There, I had the rewarding experience of working with the late Professor Ante Babić. He was a remarkable guide through the maze of sources; he not only made sure that I missed none but also that I should be aware of the innumerable complications connected with them. Through my many discussions with him I was saved from numerous pitfalls and faulty hypotheses. I also want to thank the following members of the Sarajevo History Department who all helped me in every way they could: Professors Milorad Ekmečić, Desanka Kovačević, Rade Petrović, Marko Šunjić, and Milan Vasić.

I am also grateful to the Zemaljski muzej of Sarajevo which made available to me all of its facilities — its library and exhibits (Archaeological and Folkloric). There I had the opportunity to spend many valuable hours with its stimulating scholars — both Anthropologists and Archaeologists — who warmly received me and shared with me experiences from their field work. In particular I acknowledge the help and advice of Doctors Mario Petrić and Vlajko Palavestra with whom I spent many hours of fruitful discussion and friendly debate. The staff of the Commission for the Preservation of Cultural Monuments was also helpful to me. There I am particularly grateful to Professor Zdravko Kajmaković with whom I spent many stimulating hours. At the Oriental Institute I am most appreciative of the long discussions I was able to have with the late Hazim Šabanović about Bosnia at the time of the Turkish conquest as well as about the Turkish sources for the period. I am also indebted to the Narodna Biblioteka in Sarajevo which allowed me to use its valuable collections.

The Jugoslav grant was generous enough to allow me to travel about Jugoslavia. Thus I had the opportunity to visit many villages and medieval sites in Bosnia and Hercegovina. I was also able to spend time in Beograd where Professor Ćirković was helpful to me. In addition I was able to visit and use the archives at Dubrovnik and Zadar. At Dubrovnik Professor Zdravko Šundrica was most helpful to me and at Zadar I am most appreciative of the assistance of Professors Foretić and Usmijani.

I returned to Harvard in the fall of 1967 and the thesis was written during that academic year. I am grateful to the Russian Research Center — which had previously (1964-66) given me fellowships — which then gave me a double fellowship for 1967-1968 so as to enable me to devote a full year (without having to teach) solely to writing the thesis which made it possible for me to finish the writing in one year. It is hard to find the words to express my debt and appreciation to Professor Wolff, whose advice and encouragement throughout was always helpful, but whose guidance through the writing process was instrumental. I am also indebted to Dr. Angeliki Laiou, the second reader of my thesis, who though tackling a work not in her field gave the text a most careful reading and who suggested many helpful improvements. Both Professor Wolff's and Professor Laiou's comments were valuable for the work's revision into its present form as were the comments of two friends who have read the complete manuscript: Professors Edward Keenan and especially Bariša Krekić whose careful reading spared me from a variety of embarrassing errors. Professor William Lockwood in Anthropology read and made helpful suggestions concerning my chapter on peasants and religion. Of course, none of the scholars and friends, whose help is ackowledged above are responsible for any errors of fact or of interpretation that may remain, and neither will they all necessarily agree with the conclusions I have advanced.

The text was revised for submission to the press in the course of 1968-1973. The Russian Research Center of Harvard awarded me a grant in 1968-69 that released part of my time from teaching for work on the manuscript. The Russian and East European Center at the University of Michigan financed the preparation of the manuscript for presentation to the press and the Horace Rackham Graduate School of the University of Michigan generously provided a subvention to enable the book's publication. To all three of these institutions I am most appreciative.

I am also greatly indebted to my wife Gena who not only put up with all the difficulties to be undergone when a spouse undertakes such a long project, but who helped me in every way she could and in particular with her firsthand knowledge of Serbian village life. And finally, I am most grateful to my parents, Professor John V.A. Fine and Elizabeth Bunting Fine, who have always encouraged and supported me in every way in my academic work. It is only fitting that this book be dedicated to them.

TERMINOLOGY USED IN THIS STUDY

The terminology for the history of religion in medieval Bosnia which we shall trace in this study is so complex that before we can begin we must define our terms.

Even though the medieval Bosnian might better have been termed *pagan* than *Christian,* we are going to call him a Christian, since the standard for a man's membership in a confession to be used in this study will not be based on his doctrinal beliefs, about which we know far too little, but on where his loyalty lay. This standard has been forced on me by the fact that, as we shall see, theology is fairly irrelevant to medieval Bosnia. Christianity, even among bishops and monks was so shallow that no one could possibly be called a Catholic if by the term is meant strict adherence to all the doctrines of Rome. Yet everyone appearing in the sources called himself, and therefore presumably felt himself to be, a Christian. Thus because everyone is found weak in doctrine, we must make loyalty our test of confession — loyalty to Rome made a man Catholic and loyalty to the Bosnian Church hierarchy made a man a Bosnian Christian.

The name *Bosnia* designates different areas at different times, and is derived from the River Bosna, which flows from Vrelo Bosne near Vrhbosna (modern Sarajevo) north into the Sava. The term then came to refer to the *župa* (county) of Bosnia, which included the area of Visoko-Zenica-Vrhbosna-Kreševo-Fojnica-Travnik and was the direct holding of the *Ban* (the ruler) of Bosnia.(1) However, at various times in the medieval period, the Ban of Bosnia also was suzerain over various territories to the north of his *banate* (state) as far as the Sava River. These lands (Sol, Usora, Glaž, Vrbanja, Zemljanik, and the Donji kraji) were frequently referred to as Bosnian lands. A magnate who held some or any of these territories frequently added the name Bosnia to his title. The Catholic diocese of Bosnia included the central core of the state and all the territory in the north as far as the Sava.

The title of *Bishop of Bosnia* refers to the Catholic Bishop of Bosnia who resided inside Bosnia until the 1230's and who then in the 1250's, after a period of dislocation, settled permanently in Djakovo in Slavonia. In theory he still ruled over this large Bosnian diocese, but his actual authority in the state of Bosnia for most of the period from 1250 to 1463 was non-existent. His activities were generally limited to affairs in Slavonia. In the middle of the fourteenth century the Franciscan mission to the South Slavs was established; the territory the Franciscans operated in was called the *Bosnian vicariat.* This vicariat was an enormous territory that included not only the whole diocese of Bosnia but also Serbia, Croatia, parts of Hungary, Srem, Bulgaria, Bohemia, Transylvania and Moldavia. Catholic sources may use the term "Bosnia" to refer to the state, the diocese, or the vicariat.

The term *Catholic* requires no definition. The term *Orthodox* refers to the Eastern Church and, after 1219 (unless otherwise stated) to the Serbian Orthodox Church. The term *orthodox* (with a lower-case "o") refers to the mainstream of Christian thought — Orthodoxy, Catholicism, or any other movement that had developed from them which, though schismatic, was not based on heretical doctrine. Of course, a body called orthodox, existing in the Bosnian setting, may well through ignorance have deviated considerably from the Catholic or Orthodox Church in certain of its beliefs and practices, and some of its deviations may even have been derived from the influence of heretics.

The term *Bosnian Church* (a translation of *Bosanska crkva,* the name used by the medieval Bosnians themselves) refers to the independent Bosnian Church established after the emigration to Slavonia of the Catholic hierarchy. First reference to this name in the sources is in the 1320's. The bishop of this church was called the *Djed* (literally "Grandfather"); a djed is first referred to in the sources in 1305.

The term *Patarin* has a complicated history. This Italian name was first used in the eleventh century to refer to a political party in Milan opposed to the Bishop of Milan. By the thirteenth century this term had come to designate Italian dualists who had links with the French Cathars. Papal and Hungarian letters from the beginning of the thirteenth to the middle of the fifteenth centuries on occasions used this term about Bosnia, always without defining it. From the end of the fourteenth century, Dalmatian (particularly Ragusan) sources consistently use the term Patarin to refer to members of the Bosnian Church, chiefly to its ordained clerics. However, on occasions Dalmatians used the term to designate lay members of the Church (e.g., Patarin slaves sold in Dubrovnik (Ragusa), or great noblemen who supported the Bosnian Church). I shall use the term Patarin in the Dalmatian sense to refer to ordained clerics of the Bosnian Church. However, since I shall cite sources in their own words, the term "Patarin" (in quotes) appears occasionally bearing the meaning — whatever it may have been — intended by the sources' authors.

When a source says *Christian,* it means the branch of Christianity to which the author of the source belonged. I shall use the term in a general sense to encompass all the different strains of Christianity found in Bosnia. The term *krstjanin* (literally "Christian") is a special Bosnian term for an ordained member of the Bosnian Church.(2) At times a Ragusan source will refer to such a cleric as "Christian so-and-so";

only in the Turkish *defters* (''cadasters'') does the term at times seem to acquire a broader meaning to include lay members of the church as well. I shall use the word *krstjanin* in the narrow Bosnian sense.

FOOTNOTES TO TERMINOLOGY

1. Many peasants today still use the term ''Bosnia'' to refer to the region of the medieval *župa* as opposed to the much larger area included in the modern Republic of Bosnia. (Oral communication from Wm. Lockwood. He notes that peasants from the vicinity of Bugojno speak about ''going to Bosnia'' when they go to towns such as Travnik.)

2. That this term *krstjanin* is used to refer only to ordained clerics in the Bosnian Church, will become apparent in the course of this study. On every charter witnessed by nobles and clerics, only the clerics are called *krstjani;* the lay figures, even if supporters of the Bosnian Church, never bear this title.

LIST OF ABBREVIATIONS

See bibliography for full titles of monographs cited with shortened title. Also see Bibliography when a writer's name is cited by itself (e.g., Orbini) for if a work is cited in that way (and does not appear on this list) it means that it is the only work of that particular author used.

AB, (E. Fermendžin ed., *Acta Bosnae potissimum ecclesiastica,* MSHSM, 23, Zagreb, 1892.)

CD, (T. Smičiklas, *Codex Diplomaticus regni Croatiae, Dalmatiae et Slavoniae,* II-XV, Zagreb, 1904-1934.)

Dinić, *Iz dubrovačkog arhiva,* III, (SAN, Zbornik za istoriju, jezik, i književnost srpskog naroda, III odeljenje, knj. XXII, Beograd, 1967).

GID, (*Godišnjak Istoriskog društva Bosne i Hercegovine,* which changed its title to *Godišnjak Društva istoričara Bosne i Hercegovine,* published in Sarajevo, I, 1949-)

GZMS, (*Glasnik Zemaljskog muzeja,* Sarajevo, I, 1889 - which began a new series in 1946 -)

JAZU, (Jugoslavenska akademija znanosti i umjetnosti, Zagreb; this academy has published a variety of important sources as well as two major periodicals, *Rad,* I, 1867-; and *Starine,* I, 1869-)

MSHSM, (Monumenta spectantia historiam Slavorum Meridionalium, series of sources published by JAZU, I, Zagreb, 1868-)

Prilozi, (*Prilozi za književnost, jezik, istoriju i folklor,* Beograd, I, 1920-)

Rad, (*Rad,* JAZU, Zagreb, I, 1867-)

Radovi NDBH, (Radovi Naučnog društva Bosne i Hercegovine, Sarajevo, I, 1954-)

SAN, (or prior to the revolution SKA, Srpska akademija nauka i umetnosti, Beograd; this academy has isssued a variety of important publications — a monograph series, posebna izdanja SAN, as well as a series of works of a variety of institutes, Zbornik radova . . . instituta . .)

Theiner, MH (A. Theiner, *Vetera monumenta historica Hungariam sacram illustrantia,* I-II, Rome, 1859-62)

Theiner, MSM (A. Theiner, *Vetera monumenta Slavorum Meridionalium historiam illustrantia,* I, Rome, 1863, II, Zagreb, 1875).

VZA. *Vjesnik Zemaljskog arkiva,* Zagreb, I-XXII, 1849-1920.

ZRVI, *Zbornik radova Vizantološkog Instituta,* SAN, Beograd, I, 1952-)

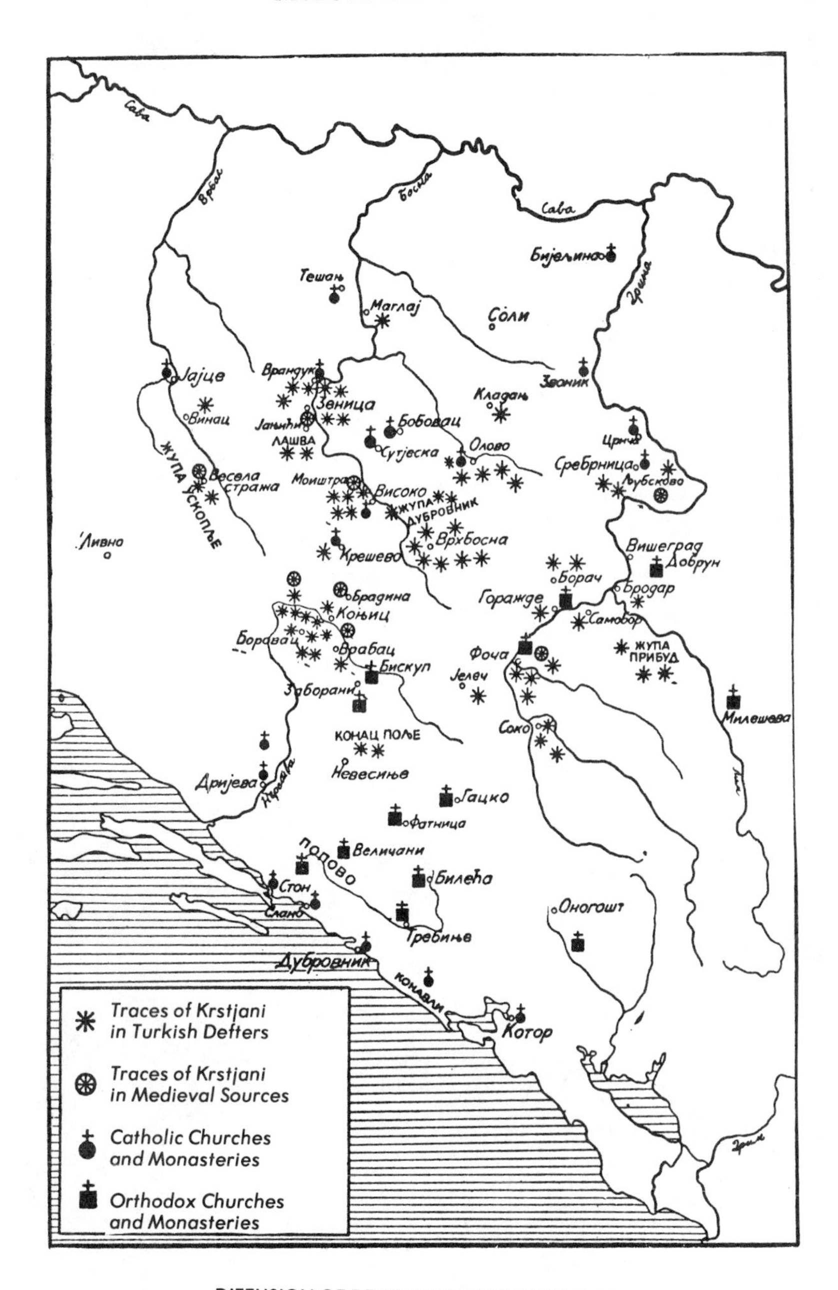

DIFFUSION OF DENOMINATIONS IN BOSNIA

(Reprinted by permission from S. Ćirković, *Istorija srednjovekovne bosanske države*. Beograd, 1964. Page 284)

INTRODUCTION

I

Many books and articles have dealt with the religious situation in medieval Bosnia. The sources speak of a Bosnian Church and of a heresy, and most scholars have assumed that the two were synonymous, and that a heretical Bosnian Church dominated religious life in Bosnia from the end of the twelfth century to the Turkish conquest (1448-1481). Yet there has been no agreement regarding the nature of this church or heresy, and some have argued that heresy is not an accurate term for this Bosnian movement. Two main views (one of which occurs in variant forms) about this movement have been predominant in the literature.

The first view asserts that the heresy was synonymous with the Bosnian Church and that its doctrine was dualist. Though called by other names, the Bosnian Church should be viewed as part of the same movement as the Bogomils in Bulgaria and the dualist heresies of northern Italy and southern France, and thus as one among many successors of the Manichees or the Gnostics. The essence of a dualist world view is that there are two principles — a good principle usually identified with the spirit and considered to be the creator of heaven and the soul, and an evil principle identified with matter and considered to be the creator of this world and the human body. Here on earth is waged one phase of an eternal battle in which human beings, with their good souls and evil bodies, play a role and at the same time strive to achieve salvation.(1)

Evidence supporting the contention that dualist doctrines akin to these existed in Bosnia has been drawn almost exclusively from papal and inquisitional documents. This material was utilized by the noted Renaissance humanist Enea Silvio de' Piccolomini (better known as Pope Pius II, 1458-1464).(2) His works, describing the heretics as Manichees, were incorporated into the writings of the late fifteenth-and early sixteenth century writers Marc Antonio Sabellico and Raphael Volaterrano,(3) and thereafter into a variety of later chronicles and histories. Whether accurate or not, therefore, Pius II is the source and founder of the school of historical thought that has continued through the eighteenth-century writers Farlati and Raynaldi down to the present.(4) This view has been the predominant one among scholars; it was most persuasively argued in *Bogomili i Patareni,* the major work of the nineteenth century scholar Franja Rački.(5) V. Ćorović, S. Ćirković, A. Babić, D. Kniewald, A. Solovjev, and other twentieth-century historians have more recently expounded and elaborated on it.(6)

The second view, and the one that appealed particularly to Serbian writers, maintains that the Bosnian Church, though often called heretical, really was not, but was, rather, essentially Orthodox in doctrine, differing from Serbian Orthodoxy only in that it was governed by its own Bosnian hierarchy. This view rests on abundant material in the sources. Scholars have found proof both in Bosnian domestic charters and in diplomatic records of Dubrovnik that Bosnians who were members of the Bosnian Church accepted a variety of standard Christian beliefs and practices that were foreign both to dualism and to the practices ascribed by Italian Catholic documents to the Bosnian heretics.(7) The most famous proponent of this view was V. Glušac, who advanced many sound reasons for scepticism in regard to Rački's interpretation, but whose general use of sources can only be called cavalier. Sources in support of his views were utilized, while those contradicting them were ignored or rejected for absurd reasons.(8) The vehemence with which he advanced his opinions and the tone of invective in academic replies to his work illustrates the degree of emotionalism that has characterized so many of the works discussing religion in medieval Bosnia.

The Orthodox theory also has its drawbacks. Not only is it contradicted by papal and inquisitional sources, but it is also refuted by various Serbian sources. Bosnian churchmen are anathematized in the Serbian sinodiks (texts drawn from the edicts of Church synods). In addition, two medieval Serbian biographies (one from the fourteenth and the other from the fifteenth century) refer to the Bosnians as heretics. Thus it is not surprising that we find serious attempts by scholars to propagate a variant of this second view which claims that the Bosnian Church was neither dualist nor Orthodox, but was more or less "orthodox" (with a lower-case "o")(9) in belief yet in schism with both East and West. This view was put forward and strongly defended by J. Šidak up until 1953(10) when he adopted a position in basic agreement with Rački. Since 1953 Šidak has written a number of excellent critical studies on specific aspects of the religious history of Bosnia.(11) In 1953 Fra L. P.(etrović) published a stimulating work, *Kršćani bosanske crkve* (Christians of the Bosnian Church),(12) which also takes this view and makes a devastating attack on the dualist viewpoint. His emphasis on the monastic character of the Bosnian Church also deserves serious consideration, although I believe he exaggerated connections between the Bosnian Church and earlier Catholic movements in Croatia.

Thus we are left with two basic theories: that the Bosnian Church was dualist, and that it was independent and schismatic but more or less orthodox in doctrine. Both views have a variety of sources to support them. One needs only to place side by side two documents of the 1460's to see this vividly; the fifty articles that Cardinal Torquemada drew up to be renounced by three Bosnian "Manichees" in 1461, and the Will of the Bosnian Church leader Gost Radin in 1466, which contradicts several of Torquemada's points and does not exude a Manichee spirit at all. Thus it is not at all surprising that serious scholars have advocated such opposing views and that the problem still has not been resolved.

To advocate one of the two basic views of the heresy, one needs to find legitimate reasons to reject a whole collection of sources. Because no one has yet been able to justify the discarding of such a large number of documents on the basis of solid arguments,(13) a totally different solution must be sought and a model created that takes into consideration these contradictions.

In 1601 when Gost Radin's will and many Bosnian charters were still buried in various archives, Mauro Orbini, writing his history of the Slavs, noted contradictions between the dualist tradition in the chronicles whose material can be traced back to Pius II, and the material contained in a now lost chronicle, most probably written in the first half of the sixteenth century, by Pietro Livio of Verona. Orbini took the natural way out of this dilemma and decided that there must have been two movements or heresies.(14) Three hundred years later, in 1892, an Orthodox cleric writing under the pseudonym of Atom wrote that historians had taken separate facts, (i.e., that in Bosnia there were heretics and that in Bosnia there was a Bosnian Church) and combined them to produce a Bosnian heretical church. Atom pointed out that the two need not refer to the same religious movement.(15) However, except for Orbini and Atom, no other historians have considered the possibility that there might have been two movements. And even Atom never followed through on his idea, but instead quickly fell in with the Orthodox point of view which he argued with considerable vehemence.

It is clear that Orbini did not clearly differentiate between the two movements and at times attributed to one what should have been attributed to the other. Yet his general idea does provide a means for resolving our dilemma. In fact, it is the only solution that does not require the discarding or ignoring of a large number of sources. Thus the possibility that Bosnia contained both a dualist heresy and a schismatic non-dualist Bosnian Church is seriously examined in this study.

The hope that heretofore unknown sources will yet be discovered is not totally bleak. Recently scholars have begun to utilize the Turkish archives, which contain *defters* (Turkish cadasters) describing Bosnia, most of them still unpublished. The type of material to be found in them is described by T. Okiç, who also presents considerable actual material drawn from the defters.(16) This material is tremendously important for the student of the Bosnian Church. It is to be hoped that the detailed and tedious work necessary to enable editions of these *defters* to see the light of day will be carried out in the near future. In addition to the Turkish, the Italian archives have not yet been exhausted by scholars working on Balkan problems, and from time to time they yield a previously unknown prize. Only recently, M. Šunjić found an interesting document about Gost Radin in the Venetian archives.(17) Whether the labor of going through these Italian materials, much of which has already been examined by Balkanists, will yield results proportional to the time spent is doubtful. But it is hoped that scholars using the Venetian, Veronese, and other Italian archives for other projects will look for material pertaining to Bosnia. Perhaps even a copy of Pietro Livio of Verona's lost chronicle about Bosnia still exists somewhere in Italy.

II

This study differs from previous studies on the Bosnian Church in several ways. First, its research has been carried out with no conscious preference for any of the conflicting viewpoints. And even as I write I am bothered by contradictions in the sources and must admit that I am not completely satisfied with the validity of any explanation. One thing, however, is very clear and that is that we should not seek a simple solution. It is highly unlikely that the dualist theory, the Orthodox theory (or its variants), or any other single-explanation theory is sufficient to explain all aspects of the religious turmoil in medieval Bosnia. We should, therefore, be willing to admit the possibility that there was more than one religious group in Bosnia which could be considered suspect in the eyes of Rome. This means that it is important that the sources be presented in their own terminology. One common fault in the work of many scholars is that, once they have decided to their own satisfaction that the heretics were of a particular character, they, when summarizing sources, substitute the label they have accepted (e.g. Bogomil) for the word "heretic" that actually appears in the sources. I shall in all cases faithfully follow the terminology of each document. It should be kept in mind, however, that medieval writers were not always careful about the way they used terminology, and as a result of lack of knowledge or for the express purpose of slandering an opponent, sometimes used the wrong term.

I intend in this study to utilize only sources concerning Bosnia. Many scholars, having decided that the Bosnian heretics were dualists, have filled in gaps, resulting from the dearth of source material, by describing the practices of dualists elsewhere. However, even if there was dualism in Bosnia it is necessary to remember that Bosnia was a backward and uneducated country; hence the practices of any religious confession there would be bound to acquire significant deviations from those of their co-religionists elsewhere. Thus a practice not attested in sources concerning Bosnia can in no way be assumed to have existed there. Such analogies tend to cloud the issue and bury the content of the few sources that do exist about Bosnia beneath a weighty mass of irrelevant material. The only way we can untangle Bosnia's complicated religious history is to rely on Bosnian sources, or foreign sources specifically about Bosnia. If these sources are insufficient to solve the problem, then we must reconcile ourselves to the fact that the problem cannot really be solved.

Moreover, many scholars have ignored the environment of Bosnia when they treat heresy there, and have assumed that heresy in Bosnia, Bulgaria, northern Italy, and southern France can be discussed in the same way, simply in terms of the ideas supposedly propagated. Bosnia differed from these other regions in many significant ways, and these differences must have had important effects upon religious beliefs and practices. For this reason, before discussing the history of the Bosnian Church, or the sources from which our picture of its history has been drawn, I shall first describe (drawing on some of the findings of Anthropology)(18) Bosnia's peasant society and assess the effect such a society would presumably have had on ideas with which it might have come into contact, as well as how its members would have regarded religious phenomena. In this way we shall have a realistic background against which to view the doctrines found in Bosnia and we shall not fall victim to the common tendency of treating ideas as if they had existences independent of the people who used them.

Since no religious movement can be considered in a vacuum, we shall discuss also other religious groups in Bosnia and shall take into consideration the limited material that exists about Catholic and Orthodox churches, priests, and populations in Bosnia and Hercegovina. It is also impossible to discuss any religious confession outside of the general historical framework. None of the beliefs and institutions of any confession existed independent of the social factors and political events of their time. Thus instead of skipping from document to document about the heresy and Bosnian Church with only passing references to historical figures and events, I shall in the narrative sections of this study present a brief but general historical survey of Bosnia (at times differing somewhat from standard accounts) into which I shall fit the documents about religious matters as each becomes relevant. This will make the study longer than I had originally intended, but since the heresy and Bosnian Church were Bosnian phenomena and not narrow religious issues, the benefits to be gained by this approach in terms of understanding will make the added length well worthwhile. This chronological method of presentation, though unfortunate in many ways, is necessitated by the fact that certain common generalizations in the historical literature hold true for only limited periods of time.

It also should probably be stated here that my study is not focused on theology. I am primarily interested in Bosnia and its society. And while we shall try as best we can to clarify the nature of the Bosnian Church's doctrines, our main interest will be the role of this church (and/or) heresy in the Bosnian state and how this "church" and all other religious organizations were related to political events and to society. In so doing, I hope to achieve a broader understanding of the nature of religion in medieval Bosnia. It is to be hoped that, by asking many new and different questions of the sources, we may be able to come up with some new insights into the religious history of Bosnia, as well as some new explanation for (or ways of looking at) some puzzling aspects of this history.

Finally, it should be stressed here that though much emphasis will be placed on religious syncretism and the weak grasp of Christianity among Bosnia's uneducated peasants, I am in no way claiming that rural Bosnia was more backward than other rural areas of medieval Europe. Much of what I say about the medieval Bosnian peasants holds true for peasants the world over well into the twentieth century.

FOOTNOTES TO INTRODUCTION

1. For a brief discussion on dualism, as well as for bibliography on the subject, see Chapter III.

2. Pius II, *Europa Pii Pontificis Maximi nostrorum temporum varias continens historias,* Basel, 1501, and his *Commentarii rerum memorabilium, quae temporibus suis contigerunt,* Rome 1584.

3. Marc Antonio Sabellico, *Rhapsodiae historiarum ab orbe condito Enneades,* 2 vols., Venice, 1494-1504, and Raphaelo Volaterrano, *Commentariorum urbanorum,* Paris, 1510.

4. D. Farlati, *Illyricum Sacrum,* 9 volumes (especially Vol. IV), Ve 1751-1819, and O. Raynaldi, *Annales Ecclesiastici.* Continuatio Baronii, 15 volumes, Lucca, 1747-56.

5. Rački's work was originally published serially in *Rad* (JAZU, in Zagreb), in 1869 and 1870. It was reissued in book form by SKA in Beograd in 1931. I shall cite the SKA edition.

6. V. Ćorović, *Historija Bosne,* SKA, Beograd, 1940; A. Babić, *Bosanski heretici,* Sarajevo, 1963; S. Ćirković, *Istorija srednjovekovne bosanske države,* Beograd, 1964 and his "Die Bosnische Kirche," *Accademia Nazionale dei Lincei, Problemi Attuali di Scienza e di Cultura,* Rome 1964, pp. 547-564; D. Kniewald, "Vjerodostojnost latinskih izvora o bosanskim krstjanima," *Rad,* (JAZU, Zagreb) 270, 1949, pp. 115-276. A. Solovjev has produced a large number of articles on the subject, see bibliography for details.

7. For example: fourteenth and fifteenth-century Bosnian Gospel manuscripts, some clearly copied by Bosnian Churchmen, but so Orthodox in form that they were in the seventeenth and eighteenth centuries used in Orthodox churches, and which contain religious pictures including the portrait of John the Baptist (who was supposedly hated by the Bogomils); plans for the participation in a Ragusan Catholic Church festival in 1426 by the prominent Bosnian nobleman and supporter of the Bosnian Church, Sandalj Hranić; the Last Will and Testament of the Bosnian Church leader, Gost Radin, written in 1466 which has a large cross (a symbol rejected by dualists) down the left margin and whose content strikes the reader as being quite orthodox and not at all dualist. These documents are all discussed in detail in the narrative chapters.

8. V. Glušac, "Srednjovekovna 'bosanska crkva,' " *Prilozi,* IV, 1924, pp. 1-55. Also his, *Istina o Bogomilima,* Beograd, 1945.

9. In this study "orthodox" with a lower-case "o" means the mainstream of Christian doctrine at the time, i.e. Catholicism and Orthodoxy, while Orthodox with a capital "O" will refer to the Eastern Orthodox Church, usually in reference to the Serbian Orthodox Church.

10. J. Šidak, "Problem 'bosanske crkve' u našoj historiografiji od Petranovića do Glušca," *Rad*(JAZU), 259, 1937, pp. 37-182.

11. See Bibliography.

12. Published as a book length article in the Sarajevo Franciscan journal, *Dobri Pastir,* 1953. The Sarajevo Franciscans also sell the off-print in book form.

13. In fact serious source study has been carried out only to show the validity of sources supporting a particular theory such as Kniewald's discussion of the Latin sources or Solovjev's of the Orthodox sources. D. Kniewald, "Vjerodostojnost . . . .," *Rad,* 270, 1949, pp. 115-276; A. Solovjev, "Svjedočanstvo pravoslavnih izvora o bogomilstvu na Balkanu," *GID,* V, 1953, pp. 1-103. Both works tried to establish the reliability of the sources, that showed the Bosnian heresy to be dualist. However, what is necessary is to scrutinize all sources, regardless of their content and point of view, collect what data we can about each document so that we can evaluate its authenticity and the reliability of its content, and only then on the completion of this task, try to come up with some conclusions about the heresy's nature. In this study we shall

try to do just that and in addition to the detailed chapter devoted to sources, we shall treat in the narrative chapters problems connected with the sources as we come to them. It is to be hoped that in the near future more scholars will turn to the necessary task of source study and the evaluation of the documents upon which our knowledge about medieval Bosnia must rest.

14. M. Orbini, *Il Regno de gli Slavi,* Pesaro, 1601. Having spoken of the Patarini as described by Pietro Livio and having described the conversion of many by the Franciscans, Orbini (p. 354) writes "Pellegrino . . . having converted the Patarin heretics, of which there were another sort in Bosnia, called Manichees. Which (according to Volaterrano and Sabellico) . . . ."

15. Atom, "Kakva je bila srednjevekova 'crkva bosanska,' 'vjera bosanska,' " *Bosanska vila,* 1892, p. 168.

16. T. Okiç, "Les Kristians (Bogomiles Parfaits) de Bosnie après des Documents Turcs inédits," *Südost-Forschungen,* 19, 1960, pp. 108-133. The material from the defters will be discussed in Chapter II.

17. M. Šunjić, "Jedan novi podatak o Gostu Radinu i njegovoj sekti," *GID,* XI, 1960, pp. 265-268. This document as well as the details of Gost Radin's colorful career are treated in Chapter VII.

18. The use made of Anthropology in this study consists in putting to the Bosnian sources some of the questions anthropologists put to the societies they study, making use of ethnological data on recent Bosnian peasant society and making use, for comparative purposes, of descriptions of the function and role of religion in various other societies. I make no attempt to apply concepts such as animism or totemism to which certain anthropologists have unconvincingly tried to give universal validity.

Since History tends to look at a society from the top down — emphasizing the ruling classes — it tends to present a picture of society that is too neat and orderly. Thus Anthropology which tends to think from the bottom up — from the level of the individual village on up to the local district — will be a helpful antidote. For after all, religious beliefs and practices in a peasant society are present on an individual, family and village level and are not dictated — at least not effectively — from the top down. And in fact, with the exception of certain taxes and possibly military service — probably as often rendered to the local noble as to the central state — most of our villagers probably had little contact with or feeling for the central state. Their ties would have been to the local noble who usually did all he could to assert his independence from outside interference and who presumably acted both toward the state and his peasants with as much license as he could get away with.

CHAPTER I

RELIGION IN BOSNIA'S PEASANT SOCIETY

(Great Tradition) religions are the product of urban and priestly classes. The extent of their acceptance among peasants is problematic. In general the existence of an undercurrent of belief distinct from that of the educated strata, often in direct opposition to it, characterizes peasant societies. Passed along by word of mouth from generation to generation, only fragments of this underground tradition are likely to find their way into historical records, and then very likely in a distorted form.(1)

The culture of a peasant community . . . is not autonomous. It is an aspect or dimension of the civilization of which it is a part. As a peasant society is a half-society, so the peasant culture is a half-culture. . . . To maintain itself peasant culture requires continual communication to the local community of thought originating outside of it. The intellectual and often the religious and moral life of a peasant village is perpetually incomplete; the student needs also to know something of what goes on in the minds of remote teachers (or) priests . . . whose thinking affects and perhaps is affected by the peasantry. Seen as a "synchronic" system, the peasant culture cannot be fully understood from what goes on in the minds of the villagers alone. Second, the peasant village invites us to attend to the long course of interaction between the community and centers of civilization. The peasant culture has an evident history; and the history is not local: it is the history of the civilization of which the village culture is one local expression.(2)

I: Peasants and Religion

One of the first things that should strike the student of heresy in Bosnia, but something that is generally ignored, is that Bosnia's entire religious history took place in a rural environment. This fact at once differentiates Bosnia's heresy from the heretical movements in southern France and northern Italy, which were essentially urban movements. In Bosnia, on the contrary, the towns, which began to develop only in the second half of the fourteenth and the early fifteenth centuries, became the centers of the Catholic restoration. This was largely because significant numbers of foreigners, the most important of whom were Catholic merchants from Dalmatia (especially Dubrovnik), established commercial colonies in those Bosnian centers most vital for their trade. They provided a solid nucleus of Catholics in these towns; they built churches and they attracted the Franciscans thither by offering them support and security. The Franciscans, in turn, had the major role in winning large segments of the Bosnian population to Catholicism.

But even after the development of these towns, the overwhelming majority of Bosnia's population remained rural, engaged in farming and herding. And this means that many standard methods for examining and studying heresies do not prove relevant when we apply them to Bosnia. Yet even so, such methods are usually attempted and the results are frequently quite meaningless. Thus it will be helpful to discuss a religious movement in a rural environment and explain what methods can and cannot be used. Some of the applicable methods will not be able to be successfully employed because the sources are so scanty, but if we are aware of them, we can at least avoid making erroneous generalizations.

This chapter focuses on how peasants — regardless of which confession they belonged to — regarded and reacted to religion. There will be considerable discussion of religious syncretism. Although syncretism may have been responsible for accusations of heresy against the Bosnians, and in fact may have resulted in practices that educated foreigners felt were heretical, it must be stressed that this syncretism was found among peasants of all faiths, including Catholics. Thus we should make a distinction between "heresy" arising spontaneously from a combination of old and new religious practices, occurring in differing patterns in different villages, and an actual heretical movement with defined doctrine such as neo-Manicheeism. Such distinctions frequently do not seem to have been made by the medieval churchmen who wrote our sources, and too often modern scholars have fallen into the trap of analyzing heresy in medieval Bosnia in the manner that the authors of our sources did.

Thus, one of the greatest mistakes often made in studies of the Bosnian heretics (or any of the other religious groups in Bosnia) is to examine in a sophisticated fashion religious ideas and to try to work these ideas into some kind of system. One clerical scholar expresses this dubious approach in general terms: "One of the more difficult tasks in attempting to assay the doctrines of a dissident religious group is to untangle emotional and popular opinions from the actual creed of the sect.''(3) This sort of approach is just what we must try to abandon. Those who think in these terms want to ignore the popular opinions which are real — and which therefore must be the subject matter that interests the historian — and to seek refuge in the unreal world of abstract theology and theory. Yet in an environment like that of Bosnia a consistent and constant abstract theological creed did not and could not exist. All that could exist would be a basic core of beliefs and practices associated with what people thought was Christianity. An illiterate society like Bosnia simply does not look upon its religion theologically. Practice is more important than belief, and much that is considered basic to a theologian is just not known, let alone understood.(4)

This will at once become apparent when we examine the case of a Catholic Bishop of Bosnia who was removed from office in 1233 for incompetence. This bishop did not know the baptismal formula, lived with a heretical brother, and did not seem to realize that his brother was heretical. Obviously theological analysis is irrelevant for a heresy in a country where such a situation could arise. This becomes even clearer when we stop to ask: If the bishop was an uneducated as that, what were the priests and common people like?(5)

Nor does this bishop seem to have been exceptional: ignorance of theology shows through in every detailed document that exists, from the 1203 renunciation of errors by the monks at Bilino polje, to Pope Gregory XI's replies to the questions submitted to him by Franciscans working in the Bosnian vicariat in 1373, to the reports of Catholic visitations or of travelers in the seventeenth, eighteenth, and nineteenth centuries. The pope's letter of 1373 refers to Orthodox Christians from somewhere in the Bosnian vicariat who were not aware of the existence of the schism between the Eastern and Western churches and who did not know the fundamentals of their own faith.(6) The level of understanding can be seen even more vividly in the report of a visitation to nearby Srem in 1612 by the Jesuit Bartholomaeus Kačić.(7) The Orthodox here also did not know of the schism and did not know the bases of the Christian faith. The writer lists as unknown: the articles of the Confession, the Ten Commandments, the words of any prayer, the Sacrament of the Eucharist, and even the meaning of the Cross. The local priests did not explain anything to the people, who knew no more than that they were *hrišćani.* Kačić found this level of ignorance among the Orthodox all over Srem.

The historian must therefore resist the temptation to put too much emphasis on theological issues, even though the inquisitional sources speak in these terms. He should remember that no Slavic source about Bosnia, be it Bosnian, Dalmatian, or Serbian, gives a detailed account of theology. This means that the only descriptions of the doctrines of the Bosnian "dualists" come from Catholic anti-heretical writings. A source written about a movement by its enemies is always a dangerous source, but in this case it is doubly so, for the Catholics who wrote the treatises were also foreigners and theologians. How well they understood this foreign culture and how much they knew about beliefs found in Bosnia is open to question. Also, much of the order and systematizing we find in their tracts may have been the result of the theologians trying to put order into the hodgepodge of their opponents' doctrines in order better to refute them. These theologians did not have the anthropological grasp to see that Bosnian beliefs were a haphazard mixture of many different cults, from paganism, Catholicism, and heretical ideas, to new beliefs arising from a mixture of the three, and from an unconscious drifting away from the three through ignorance and misunderstanding. Not only would heretics in one village hold concepts not present in another village, but contradictions (apparent to an intellectual) would have been common among the various beliefs simultaneously held even by a single individual.

Among most peasants, and certainly among Bosnian ones, practice is much more important than belief. Practices have continued through the centuries regardless of peasant acceptance of new religions; in fact, a new religion is quickly adapted to co-exist with old practices. In Bosnia the basic set of practices, regardless of what confession the peasant may nominally belong to, is the domestic cult. The most important domestic ceremony today among Serbs is the *slava*:(8) a family gathering to honor the family's patron saint on the saint's day, during which special rites are carried out which, by pleasing or appeasing the saint, will obtain his aid in guaranteeing the health and welfare of the family throughout the year. The central part of the *slava* and most other festivals is the ritual meal. This is clearly an ancient ceremony and evidence indicates it was important in medieval Bosnia.(9)

In addition to observing the *slava*, the peasants hold celebrations on other special days throughout the year, and on each occasion carry out special practices. On one day they perform a special ritual to insure the health of livestock, while on another day certain ceremonies guarantee the success of crops.(10) Each village has its own special holiday on which prescribed rituals are performed for the welfare of the whole village. If an epidemic strikes, it requires the usage of a special ritual. In medieval Bosnia, regardless of what confession a peasant belonged to, that confession was less important to him than the practices of the domestic cult. Of course, it should be stressed that the peasant set up no sort of opposition between domestic cult and Christianity, since by attaching the old domestic ritual to the person of a Christian saint, he had come to consider the *slava* and the practices going with it as the fundamentals of Christianity.

In addition to rituals for specific holidays, there are and were a vast number of magical practices and charms used for specific goals, good and evil. These magic practices also have not always been considered as part of a separate system from Christianity. A Ragusan chronicle discussing Ston, formerly the capital of Hum, for 1399 speaks of Serbian monks and priests there as schismatics and infidels "who did not believe in God or the Saints but in dreams, soothsayers and incantations."(11) Lead tablets unearthed near Mostar,(12) Foča(13) and Prozor(14) invoke the name of Christ and quote from the Gospel of John in hopes of protecting crops from assaults by nature and the Devil. These tablets reflect a practice that is still employed in Bosnia today: peasants go to a priest or hodja and buy from him a piece of paper on which he has written a prayer. The Christian prayers are generally addressed to particular saints. The prayer is then worn as an amulet (about the neck), which insures the health of the person from a particular disease or in some cases from illnesses in general. This means is commonly used by barren women who wish to become pregnant.

The wide variety of magic practices found in Bosnia can be seen in an Orthodox Church document from Bosnia, unfortunately anonymous, which attacks a wide variety of superstitions practiced and specifies the penalties to be imposed upon those who practiced them. Truhelka, who found the manuscript, dated it no later than the seventeenth century.(15) Clearly, most, if not all, the practices date back at least to the medieval period. The document condemns magic practices which weaken limbs, cause long illnesses, kill a man, make a man go crazy, separate a man from his wife and so on. Reference is made to various methods to prophesy the future, including the use of barley seeds by gypsies. Poisonous herbs are mentioned, as are special grasses worn as amulets to ward off the evil eye and other evils. Witches had rites to clear away clouds. Some people carried snakes in their bosoms for magic acts. Snake skin placed on the eye or ear seems to have been a way to cure diseases of these organs. The manuscript also refers to vampires and werewolves.

From all this it is clear that, in addition to regular annual rites to guarantee health and welfare, magic practices were employed on specific occasions. These magic acts (of course now considering only those we would call white-magic) are akin to the regular practices of the domestic cult because they all have the sole purpose of protecting the individual or family from evil and of coercing or soliciting from the dispensing powers one's physical or material welfare in "this world." The peasant's cult thus is primarily important for matters connected with "this world." Not surprisingly it consists almost entirely of practices and has almost no doctrine.(16)

In stressing practice, we should point out that though the procedure is fixed the meaning attached to it is vague, and certainly not particularly important. The explanation given for a practice frequently changes over the years. Different people in the same family may on the same occasion give two totally different meanings for a practice which is felt by everyone to be very important. Each hears the contradictory explanation but neither one seems concerned at the disparity in views, and no argument will ensue over it. It is simply not very important to understand a practice, what is important is that it be carried out. The peasants will protest, however, if they believe that the practice is being performed improperly. If the reader prefers to call such a cult as I have described "magical" rather than "religious," I certainly will not quarrel with him, but I might simply add that any theological approach to heresy will totally misunderstand the "this-world," practice-oriented Bosnian peasant mentality.

It is also clear that any religious doctrine that could have survived in this environment would have to have been able to meet, or at least adapt to and co-exist with, practices meeting these needs. This immediately makes one skeptical about the chances for success in this environment of the dualistic, anti-material, other worldly heresy that is often said to have thrived in Bosnia. To have acquired a following in Bosnia, dualism would have had to adapt itself considerably, or have given up trying to influence the daily lives of its believers and have been satisfied with simply initiating people on their death beds. Even so, it seems difficult to believe that it would have gained much of a following if it offered no benefits for the peasant in this world.(17) In addition, we should note that no source mentions a death-bed baptism in Bosnia.

In areas of Jugoslavia where there has not been Catholic pressure for church attendance, contact with churches, priests, and other aspects of formal religion are neither regular nor frequent. A peasant deals with the priest usually at times of a birth or a death. Village church attendance throughout the last two centuries has been sporadic; it is only in Catholic areas, and particularly in places near where the Franciscans have established themselves, that church attendance by some members of each family is regular.

We can be sure that the relationship between peasant and church was not much different in the Middle Ages. This certainty can be established since, throughout the Middle Ages and well into the Turkish period, the number of priests of any sort in Bosnia was minuscule. In the late twelfth and early thirteenth centuries if there were any priests (other than monks like those at Bilino polje in 1203), it is most likely that they were themselves peasants who were more or less illiterate and only doubled as priests on special occasions. At the end of the fourteenth century, the Franciscans (the only Catholic clergy present) numbered fewer than a hundred for the whole Bosnian vicariat, a region vastly larger than the Bosnian state.(18) Because they tended to live in communities, the Franciscans' influence would have been limited to the area around their monasteries; elsewhere, in regions farther away, people probably rarely or even never saw a Franciscan. There is no evidence that the Bosnian Church had many priests, either; their members also lived in monkish communities. Since these communities were few in number the influence of the Bosnian Church monks on the peasants was presumably restricted to those villagers who lived near their monasteries. We can assume that both religions sent out their members at specified times to visit more remote regions but such visits probably took place only once or twice a year. Therefore they would have had only limited effects on the peasantry, who would have gone on living as they always had, keeping their old customs and practices.

We have documentary evidence, confirmed by observations in more recent times, that in many rural places with churches — and even in some towns — it was rare for church services to be held every Sunday. Services by a visiting priest would be held on a limited number of Church holidays, perhaps no more than five or six a year.(19) At these times the priest would attend to such things as baptisms, confessions, and absolution. Today, in Orthodox villages scattered in the mountains of Bosnia and Hercegovina one comes across small wooden churches in appearance little different from sheds. A priest visits such churches about once a year to baptize children born since his last visit and to serve a general Mass for those who died during the year.

A church building is not always felt to be vital. Travelers, visiting Bosnia in Turkish times, frequently noted the absence of churches and the custom of holding Mass in private houses or under the open skies. Often a portable altar would be set up in a cemetery for the service.(20) Professor Kajmaković discovered in the region of Gacko a medieval church which he believes was simply composed of low walls enclosing the area of the church with an altar within, but never covered over with a roof.(21) Even where churches do exist it is common even today to find that for certain holidays Bosnians hold services outdoors, frequently in cemeteries.

Thus, there is seen a definite distance between the peasant and the formal religious apparatus. We also see that no strong feeling exists that the church building is of vital importance for worship.(22) A peasant's cult and religious loyalty center around his home and his village, and his religion is mainly expressed in terms of particular practices. His contact with church authorities is slight and is generally in regard to particular practices, namely those rites necessary to guarantee the health and welfare of a baby (baptism) or the rest and welfare of the dead, which the peasant believes are rites that must be performed by a priest. The majority of rites connected with one's life, however, are family rites and are carried out as prescribed by custom. For many occasions, the head of the household performs these rites, though certain rituals require specific actions by other categories of people (e.g., by twins, by young virgin girls, by children, etc.).

Certain specific problems, such as serious illness, or suspected witchcraft (milk is being drawn away from one's cow to that of a neighbor's, or one's fields do not produce properly, etc.) require a specialist. In Bosnia the specialist is simply a woman (often older) or, more rarely, a man, who knows more magic than the average man does, who is called in when the peasant's own practices cannot cope with a situation. She is of the same class of specialist as the midwife (who may also be a witch) or the peasant surgeon (who tends to be a man).

Many of the inadequacies found in Bosnian Christiantiy, that were to cause foreign clerics to condemn it, were surely not the results of heretical teaching but of deviations in customs and beliefs occurring on the individual or village level, where peasants tried to follow what they thought were Christian beliefs and practices which they did not understand very well. In time, without supervision by educated keepers of doctrine, a great part of any religious system will gradually "degenerate" into error. This will be particularly true in doctrinal matters about which peasants have never been particularly clear. Practices, if they seem important and if they have really been accepted, are likely to change less. However, practices which the peasants did not carry out themselves but left to a priest to perform, such as baptism, would not have made deep impressions on peasants in a region with few priests. In time, in the absence of priests, these rites could easily disappear or become bastardized.

As we have argued above, the bulk of Bosnian peasant practices were not Christian at all, but earlier rites. Some practices had taken on a veneer of Christianity — e.g., the *slava* rite becoming attached to a saint's day — but others — such as jumping over fires on St. John's Eve or rituals carried out by young maidens when rain was needed —

acquired no such veneer.

Many of these pagan practices have survived to the present day, and in the Middle Ages there were clearly more of them.(23) Jireček cites a 1452 glagolitic church document from the Croatian-Dalmatian area, that says that he who bows to the sun, the moon, or any created thing (stvorenije), and addresses prayers to them commits a mortal sin.(24) This document shows that in the Middle Ages some people still addressed prayers to the heavenly bodies. When we consider the number of superstitions and taboos about these bodies, especially the moon, that survive to the present day, the survival of a sun or moon cult in the Middle Ages should not surprise us at all. The Bosnian Orthodox Church document, condemning superstitions (dated by Truhelka as no later than the seventeenth century and already cited in connection with magic practices) also condemns following Hellenic (pagan) customs. It speaks of dancing in the squares, and refers to a special masked dance popular on the eve of the Day of the Assumption (V'znesenie) where men wear women's dress and women, men's dress. It also refers to belief in vilas (a species of nymph found only among the South Slavs).(25)

When we think of these and the countless other popular dances and ceremonies associated with various festivals, of the variety of magic practices and domestic rituals, of the number of taboos and superstitions, and of the folklore that still exists in extraordinary richness and variety, it is clear that Christian beliefs and practices had only a small, and often superficial, part in the cultural life of the Bosnian peasant. These old pagan rituals survived, most probably with the original meaning lost. We certainly cannot say that the pagan beliefs were retained. The performance of certain of the old seasonal rituals was probably felt to be important for village welfare, and others may not have been considered important in this way, but were retained because they were entertaining. We should always keep in mind that there is, and was, a difference in peasant attitudes toward diferent ceremonies. Some, we would say, the villager treated as religious ceremonies and other beliefs and customs were not so regarded.(26)

These earlier rituals combined very easily with newer rites and practices. Thus among the anathemas from the Council of 1211 held by the Bulgarian Tsar Boril to condemn Bulgarian Bogomils, is inserted: "anathema to all those in this heresy, to their practices, to their nocturnal reunions, their mysteries and . . . ." and later: "To those who on 24 June, the Nativity of John the Baptist, deliver themselves to magic operations and affect crops and to all infamous evils similar to the

Hellenic cult which they practice on this night — anathema!''(27) Here are described pagan survivals that were no part of Bogomil doctrine. Yet individuals who had accepted Bogomilism seem to have continued to practice these older customs, even though they would have been foreign to Bogomil thelogy. This shows both how mutually contradictory beliefs can co-exist in one person and also how well a religion (in this case, Bogomilism) can blend in and co-exist with the older beliefs. Unfortunately we do not know if the Bulgarian peasant thought of these rites as part of the same system as, or part of a separate system from, the Bogomil beliefs he held. In the 1430's somewhere in the Bosnian vicariat (most likely in Bohemia since that place is referred to later in the letter), according to the Franciscan Vicar (1435-39) Jacob de Marchia, people met in caves and cellars ar night and poured the blood of Jesus Christ (wine?) from certain hides lined with tar over themselves and their garments.(28) This sounds very much like the survival of a classical mystery cult — be it Mithraic or Dionysian. How much of the original meaning and ritual survived of course we do not know.

It is not surprising that educated Catholic foreigners looking at Bosnia and the surrounding regions, would have seen heretics everywhere, for almost no one, regardless of what confession he thought he belonged to, would have met the standards for being a good Catholic according to Rome's doctrine. Jacob de Marchia called those who practiced the rites in caves and cellars ''heretics.'' Thus, it seems safe to conclude that the following of such pagan rites and customs was grounds for being considered a heretic. If this conclusion is accurate then we have reason to believe that much of Bosnian ''heresy'' may well have been merely the following of customs that had survived from pagan times. We have seen above how Bogomilism and such practices co-existed in Bulgaria. And we can be sure that in Bosnia Christianity had blended with earlier beliefs and practices in a wide variety of different combinations and different ways. Anthropological literature has many accounts of what happens when a Great Tradition (such as Catholicism) meets Little Traditions (such as the peasant's domestic cult and rural festivals and rites). The two become mixed in a variety of ways; very rarely does either tradition remain pure and very rarely is the synthesis that results identical in two different places.(29)

A few illustrations may serve to show the way these syntheses occur in Bosnia. Unfortunately, owing to scanty sources, our examples must be post-medieval and thus will show syntheses between Christianity and Islam. The first is taken from a description of Turcisized Bosnians, called Poturs by Paul Rycaut in the 1660's. Solovjev tries to link Potur beliefs and practices with Bogomil survivals,(30) but this is dubious; all the beliefs and practices Rycaut mentions can easily be explained as Christian survivals mixed with Islam by converts who kept many of their old beliefs and did not wholly understand the new religion. The Christian beliefs could have been derived from any form of Christianity. The Poturs read the Gospel in Slavonic and were interested in learning the Koran; they drank wine in Ramadan but, to reduce the scandal, abstained from spices; they had charity and affection for Christians and tried to protect them from the Turks; they believed that Mohammed was the Holy Ghost promised by Christ; they abhorred images and the sign of the cross; and they circumcized themselves, using the authority of Christ's example for it. This is clearly a hodge-podge of Islam and Christianity; a deeper investigation of these people would surely have found much more left over from Christianity as well as other Islamic customs which were perverted.

Throughout Bosnia and Hercegovina Moslems still observe certain Christian holidays and in some cases participate in village celebrations along with Christian villagers. They particularly have maintained customs (many of which are pre-Christian) connected with the Days of St. George and John the Baptist — the latter is worth stressing since Dualists hated him for his use of a material substance (water) for baptism. Also observed are the Days of St. Procopius, St. Peter, St. Ilija, Good Friday, Easter, Christmas and in some cases the *slava;* some Moslems who have retained the *slava* even send flowers and oil to the village church on that day.(31) In certain villages of Hercegovina in the last century Moslem women did not veil themselves.

Also interesting is the phenomenon of double-faith. It appears in two forms: The first form is represented by people who are afraid to change their faith publicly yet do so secretly. In almost all cases in Bosnia, this concerned Moslems, who naturally feared the severe penalty that followed apostasy, but who wanted to be Christians. They secretly went to the Franciscans for baptism and the other sacraments.(32) These people then are really Christians, but secret ones. The second form of double-faith is that found among both Christians and Moslems who looked upon both faiths as religions whose magic worked. For example, a Moslem mother may bring a sick child to a church in hopes of a cure. This is still a common sight at the Orthodox monastery of Visoki Dečani near Peč. When St. Sava's relics were at Mileševo, the Turks, and we are told even the Jews, honored, and expected cures from the relics, to the same extent as the Christians. Examples also are found of Christian mothers taking sick children to the Hodja. A seventeenth century visitation reports that Bosnian Franciscans even frequently baptized

Turkish children; for the Moslems superstitiously believed that baptism brought good luck, success in battle, and was a protection against evil spirits.(33) The baptism of Moslems, who remain Moslems, for the purposes of healing and protection, is found throughout the Balkans in regions where conversion occurred.

A nineteenth century French traveller reports that Turks (i.e., Bosnian Moslems) prayed to the icon of the Virgin when ill and if they became sicker they would call a priest for baptism. They did not even hide this behavior although it was against Islamic law. Turkish officials in Travnik even allowed the priests to go to Moslem houses, "as doctors."(34) And it is very probable that the Turkish official regarded this baptism not as a religious act but simply as a medicinal remedy. Among Greek Moslems (prior to repatriation to Turkey in the 1920's) it was possible to find Moslem girls wearing Christian crosses for protection, and during epidemics, these girls frequently went to Christian monasteries to kiss a revered icon or to allow themselves to be sprinkled with Holy Water.(35) These last examples all fall under the category of practices and it is clear that they are believed to have magical effects of the same kind as the village healer's rites have. However, all this should cause us to hesitate before we make generalizations about people being of a particular faith or about their conversions from one faith to another.

II: LANGUAGE, EXPERIENCE AND THOUGHT

The vocabulary and the structure of a language have great influence upon the way a concept can be expressed, and a concept's expression exercises influence upon what other ideas may be related to it and therefore how that concept may develop.(36) The Bosnian peasant, illiterate and having little contact with the outside world, would not have been used to dealing with ideas and would naturally have a limited vocabulary. Many words, particularly abstract terms, simply would not have existed in his mind and if the ideas for which those terms stood existed at all, they would have been in an altered form brought about by the linguistic means at the peasant's disposal.

The peasant's language would tend to be concrete; his "speculation" would be confined to practical matters in his daily life and, to some degree, to the "why" stories — creation myths, etc. These myths would tend to be framed in concrete terms, and the actors, be they gods, men, or animals, would tend to act like people. Thus in village folklore we find a strong tendency to personify. Such things as diseases and days of the week take on human form. Here we have only attempted to call attention to the problem of language, which by its vocabulary and structure both may suggest new directions of thought via associations and also by these same structural features may restrict both comprehension and the development of other concepts.

Social structure, duties of kinship, type of life and work, daily experiences, and problems one regularly has to deal with all work together to determine the sort of things a man needs to be able to understand and the actions he must know how to carry out. The type of problems a man faces in his daily life has great influence on how he will reason, how he will relate ideas, what will interest him, what sort of thing he will remember, and how he will relate a given experience with other experiences. In effect then, a man's life experiences and education (based on similar experiences of the preceding generation) will impose limitations within which and will impose the manner with which he can observe existing reality and also what meanings he can attach to that reality. For ideas and opinions are never interpreted "as they are" but in light of the life situation and manner of reasoning of him who perceives and expresses them.(37) Thus it is impossible for us to take a statement of a belief or a given text used by a Bosnian Churchman (for example the beginning of the Gospel of John which seems to have been popular with them) and say this passage means such-and-such, therefore the Bosnian Church held such-and-such a doctrine. For us to know what the passage meant to a medieval Bosnian, would require us to know how he thought. And, unfortunately the process of thinking is not standard.

What ideas seem to bear relationships to each other or what objects may be linked somehow together vary the world over. To formulate knowledge it is necessary to organize one's perceptions into categories; and the way experiences and objects are classified varies from society to society; and it is vitally important to realize this, for one's logic, reasoning and whole thought-processes, and way of relating ideas, will depend upon how he classifies his knowledge. The Bosnian peasant's experiences are rooted in such factors as his village, his agriculture, and his limited contact with the outside world. His contact with ideas is clearly restricted. In addition, we know that practices of a magical nature were widespread. Magic requires belief in a connection between two objects, two events or an event and an object that does not necessarily correspond to our present day ideas of cause and effect or our ideas of logical relationships between ideas. Thus it is difficult to imagine how biblical texts or any religious idea would be interpreted. That the peasants' existence did not change much over the years, and that they rarely confronted problems of a different nature from those they had been dealing with for years suggest that the peasant would have little interest in speculation — or certainly speculation in new directions. He might well be willing to experiment with new home remedies. But it is unlikely that a peasant would ever think of or have interest in significant intellectual innovations. The fact that we are willing to think of innovation reflects a cultural attitude that we have, but is not a general world wide attitude. Most peasant societies that are left to themselves, without changes being forced upon them from without, reflect a certain stability and are unlikely to think of alternatives to their existing way of life, practices or ideas. And this is a matter worth emphasizing for one of the most impressive things, though not sufficiently emphasized, is the stability and constancy of beliefs, practices and attitudes toward life that the peasantry hangs on to despite changes of formal religion.

Thus we cannot really say that the peasants were converted to Christianity at a given time and that later some of them were converted to Islam. For conversion requires a significant change both in ideas held and ways of thinking. And what is impressive about the Bosnian peasants is that, in accepting a new faith, they accepted a few obvious and formal new practices, but basically continued to live and believe as they always had.(38)

In addition, because the peasants are not attracted to speculation and because they have had limited contact with ideas from the outside world, certain logical impossibilities, according to our way of thinking, frequently appeared in their thought, which cause great difficulty for us when we try to systematize their beliefs. Anthropologists have noted that primitive people also hold mutually contradictory beliefs without being troubled by the fact. And such contradictions regularly appear even in the thought of one individual.

Serbian peasants may believe a spirit is in two places at once. According to their Christianity, a dead spirit should immediately ascend to heaven (assuming it is not damned), yet for a full year the deceased's relatives dutifully carry out a series of grave-side practices for the benefit of a spirit felt by many to be present at the grave. When this inconsistency is pointed out to the practitioner, he looks puzzled and cannot explain how this can be reconciled. However, he does not suffer "disconfirmation" and will continue in the future to hold the two logically mutually exclusive beliefs. To him no contradiction really exists.

The situation is well expressed by Evans-Pritchard with regard to the Nuers:

..The great variety of meanings attached to the word kwoth (all

> *spirits) in different contexts . . . even in the same ceremony may bewilder us. The Nuers are not confused, because the difficulties which perplex us do not arise on the level of experience, but only when attempt is made to analyze and systematize Nuer religious thought. The Nuers themselves do not feel the need to do this. . . . Indeed, I never experienced, when living with them and thinking in their words and categories, any difficulty commensurate with that which confronts me now when I have to translate and interpret them. They move back and forth from the general to the particular never feeling there was any lack of co-ordination. Only when one tries to relate Nuer religious concepts to one another by abstract analysis do difficulties arise.(39)*

The relations the uneducated peasant makes between the general and particular will also bother us. But though in his "thought" and the way he relates ideas the peasant will trouble us with contradictions, in his actions, if we exclude certain magic rites and associations, we see that he is a practical and logical worker. The peasant, then, is basically a man of action. And magic and ritual practices are all action activities as opposed to being intellectual or speculative pursuits. Practices are done to make something happen (or prevent it from happening); why a particular practice works is not important, as long as it works. The peasant shows curiosity in how to work it, but not how it works. He is satisfied that there are things in this world that he does not understand, and sees no need to understand them. All he needs to know is how to deal with those aspects of the universe that affect his daily life. Things happen for that is the way God made them, and only let God give us . . . . Yet, we cannot really call this fatalism, since the peasant does try to alter matters by means of rituals, remedies, and magic.

When we come across abstract terms we must stop before making any assumptions. It is clear that the Bosnian Churchmen believed in the Trinity, but it is not clear how they conceived the Trinity. That they expressed belief in it in the same phrases and words as more educated Catholics and Orthodox churchmen does not give us a basis to assume that they necessarily conceived of it in the same manner. Every abstract term causes difficulty — justice, truth, soul, salvation, paradise. These words all have elusive qualities, for the meaning they have will vary not only among Bosnians, Catholic theologians, and ourselves, but also among different Bosnians in different social groups.

This problem of extracting meaning when dealing with words and phrases in texts is also encountered in our attempt to interpret the symbols that appear on the *stećci* (medieval gravestones). There is, of course, much dispute on what the symbols mean. Here I only want to point out that it is likely that several different systems of symbols were functioning at once. There may have been symbols of a religious nature, symbols of a purely decorative nature (which should not be considered symbols), and symbols connected with rites and practices the meaning of which was already lost in the Middle Ages but whose presence was important, since the rites with which they were connected were still practiced. In addition, many of the marks were probably juridical signs, substituting for writing in an illiterate society, giving name, family connection, or occupation of the deceased.

Also a given symbol need not have the same meaning each time it appears on a stone. A cross on one stone may be a symbol of resurrection, on another it may be a charm against evil — either for the benefit of the deceased in the after-life or for the benefit of the living, to keep the village safe from visits from the corpse by preventing him from leaving the grave — or it may be a charm to threaten and scare off a would-be grave robber.(40) Grave robbing seems to have been a common worry in connection with burial and we frequently find inscriptions which curse him who would disturb the bones resting there. A snake on one stone may have some cultic significance and on another simply mean that the deceased died as a result of snakebite. It is clear that the difficulty we have in reading the language of the pictures on the grave stones reflects our failure to stand in the shoes of the medieval Bosnian and view the world as he did and think about it in the terms that he used.

Before dropping the problem of meaning and patterns of thought, it may be fitting to quote a true story told by Dositej Obradović in the eighteenth century about a Montenegrin abbot. Obradović was gazing out of a window at a rainbow while conversing with Prince-Bishop Vasili.

> ..*And then there came riding up on an ass a certain abbot, a huge personage with an enormous beard. Prince-Bishop Vasili . . . caught sight of the abbot and said to me:*
> ..*'Deacon, I beg you, just ask that abbot what a rainbow is and why it has many colors.'*
> ..*I was overjoyed at the suggestion. My mischievous heart started to leap for joy at the thought of dumbfounding the big abbot. The door opened and in walked the abbot. I barely gave him time to bow to the bishop and sit down. Then I rushed in like a gamecock and blurted out:*
> ..*'Father Abbot, tell me please, what a rainbow is and why it has many colors.'*
> ..*I had already begun to feel amusement, anticipating that he would be disconcerted and would not know what to reply, and I could hardly keep from exploding with laughter. But if he had been disconcerted he would not have been a real Montenegrin. He seized his great bushy*

beard with one huge hand, looked at the rainbow for a moment, and then turned on me his large, terrible, black eyes: they would have scared Newton himself. And instead of answering me he inquired:
..'Do you see that jackass of mine?'
..'I see it,' I replied 'but I was not talking about it. What has a jackass to do with a rainbow?'
..'I understand your question,' retorted the abbot, 'but just let me tell you that that jackass of mine has a lot more sense than you have.'
..'I should like to know how you measure my sense and that of a jackass.'
..'Listen to me and I'll swear you will find out!' said the abbot. 'That jackass recognizes the chaff that's in front of him. If you don't believe it, look at him chewing. But despite all the years you've lived you don't yet know what a rainbow is! A rainbow's a rainbow, and not a hoop for a tub! But you ask why it has many colors. Why, confound you, how can it be a rainbow without having many colors? Did you ever see a black rainbow anywhere?'(41)

This story clearly shows how the abbot had no attraction toward speculation, how he regarded the world, and how his mind worked. And for us the problem of coming to grips with the way of thinking and how the peasant mind works is far more important than the matter of what ideas or beliefs were held. Understanding the peasant mind is the key to understanding our sources and the information they convey. No idea appearing in a source existed apart from the people involved. No examination of the sources about the Bosnian heretics can be successful if those sources are lifted outside of the context of Bosnia. A heretic is not everyman, and a Bosnian heretic was a Bosnian peasant with all that goes with that peasantness first, and a member of a religious organization second. The way he viewed that religious connection was from his rural, practical, worldly and unschooled point of view. His way of thought, his world-view, and his domestic cult and magical practices were far more important to him than any formal connection with any particular religious confession or church.

III: Family Structure

Families, be they noble or peasant, were patriarchal, and the father of a family had authority over the activities and behavior of all its members. On his death his possessions were not passed on according to primogeniture; the inheritance was shared equally by all his sons. If the family lived under one roof, one son (usually the eldest) became the family elder and assumed authority over the whole family and its property, though in theory the property was held jointly and the elder was expected to consult with his brothers on matters concerning it. This is confirmed by medieval Bosnian treaties which regularly were witnessed by "N. (the family head) and brothers." If the brothers lived under separate roofs, each received an equal portion of the inheritance. When several generations of a family live together under one roof, scholars often label the household a *zadruga*. However, it is hard to make a clear-cut distinction between the patriarchal family labeled a *zadruga* and any other patriarchal family. The main feature singled out to define a *zadruga* has usually been size. Yet while examples exist of extended families reaching 50 to 100 members, during the past three centuries (about which we have data) the average *zadruga* has always been under twenty, often nearer ten. At no time were enormous households standard. Unfortunately we have no real data on the size of Bosnian households prior to the eighteenth century. Many scholars assume medieval Bosnians lived together in large households, but others believe that it was the insecure conditions accompanying the Turkish conquest that caused people to band together in larger family units.

However, regardless of the size of households, medieval Bosnian families were patriarchal and the father (or grandfather) of the family made decisions on all matters of importance, be it family marriages or family economics. Thus regarding issues connected with religion, and any question of conversion, it is safe to assume that the father generally made the decision and the family followed.

It is inconceivable that a female family member would play a major role in a family's acceptance or rejection of a particular religious confession. In medieval Bosnia, marriages were a social matter arranged between two families and had no religiously sanctioned basis. The husband (or more often the family elder) possessed the right to break off the marriage arrangements if his wife "was not good." In the fourteenth century, the Franciscans complained that scarcely one in a hundred men remained throughout their lives with one wife.(42) Although this statement is surely exaggerated, it does illustrate the fact that women had few rights. The Franciscans thus encountered a major problem in their attempts to win the Bosnians to Catholicism, because Catholicism regarded marriage as a sacrament, and demanded that its adherents do likewise. Needless to say, the Bosnians strongly resisted this view of matrimony.

And in fact, the Bosnians continued, into the Turkish period, to believe that marriages were not a religious matter but a private legal relationship; both Orthodox and Catholics frequently were married by the Turkish *kadi* ("judge") instead of by their respective priests. The *kadis* were liberal in allowing mixed marriages, as well as in permitting divorces and re-marriages. That the Bosnians were willing to do what their churches considered sinful shows that the Bosnians believed that marriage was a private matter of no concern to the church. Thus in the Turkish period we continue to find Franciscans trying with little success to make the Bosnians accept the church's view on marriage; in 1600 a Catholic bishop complained that Bosnian men were simply chasing out their wives and taking new ones, that mixed marriages were being contracted, and that many marriages took place without any church ceremony. In 1618 there is even recorded a case of a Roman Catholic priest being married before a *kadi*.(43)

Scholars have at times linked this situation with neo-Manichee rejection of marriage and advanced it as evidence for dualism in medieval Bosnia. This is clearly an unmerited view, for the Bosnians had nothing against sex and marriage; it was simply that in their family-oriented culture they did not view marriage as a religious matter or a matter of concern for anyone other than the families involved. This view still can be found today in many parts of Jugoslavia, where families will send the bride back to her family if she proves unsatisfactory, be it a month or even a year or more after the wedding. Although it is clear that this concept of marriage was not tied to any religious belief existing in Bosnia, this does not mean that Catholic writers attacking the Bosnians did not regard this custom as "heresy." This, and not neo-Manicheen beliefs, could well have been the basis for the anti-heretical Catholic tracts' statements that the Bosnians "reject matrimony" or "the sacrament of matrimony." For most Bosnians did, in fact, reject marriage as a sacrament.

IV: Bases of Moral Order

In the West one of the most noticeable aspects of heresy was the element of protest. Too often students of heresies in other regions assume that this feature of Western heresy is a universal feature of heresies. Although it is not to be denied that church abuses and the immorality of individual clerics were a contributing factor to the Western heretical movements, possibly the positive sides of the heretics' doctrines should receive more emphasis than their criticisms of Catholicism. It is also certain that all the ignorance and moral laxity found in the West existed in Bosnia as well. However, there is no evidence of any Bosnian complaints about church or clergy or any sign that such moral protest was a factor in the Bosnian movements. Priests there were important for society only insofar as they played a certain part in ritual practices. Most rituals, to be effective, did not require any moral excellence on the part of the performer. Although Bosnian peasant society did have strict moral codes of behavior, these codes were not intimately connected with religion or religious sanctions.(44) The family and family honor were much nearer to being the bases for the moral order. In most cases we can assume that the Bosnian Catholic priest was a peasant himself and part of the village community, who lived the way everyone else did with no one expecting anything more of him.

Ideals about the church and priestly behavior do not seem to have existed in Bosnia. Since the Bosnian Church — as we shall see — was created out of the very organization of Catholic monks, whose morals and way of life had been criticized by the papal legate in 1203, it is apparent that there was no local opposition to the behavior of Bosnia's clergy. Thus it should be stressed that in Bosnia this protesting or dissenting aspect of heresy was absent. Later, after the violence and destruction caused by the Hungarian crusades, there surely was much hostility toward Catholicism. But this, of course, was due to external causes rather than to anything thought to be inherently wrong with Catholicism itself or with conditions in the Church in Bosnia. The hostility toward Hungary, however, because it was associated with the Catholic Church, probably did have the effect of driving some people into other churches or movements.

V: Relations Between Rulers and Peasants

Bosnia is not only a rural area with few towns, it is also a very mountainous country. Therefore it is not surprising that medieval Bosnia was divided up into many counties, most of which were marked by localisms and very little feeling for a Bosnian entity. Most counties were under particular families who were, to varying degrees, independent of the ruler (the ban). Thus throughout most of our period (except for roughly 1320-53 and 1368-91) Bosnia could most accurately be described as a series of local units loosely connected to the central state under the ban. The ban continually strove to bind these units to his authority and hold them together, and the local rulers equally actively strove to maintain their independence.

The mountainous geography contributed to the localism, and because of the terrain, communications were difficult both between various parts of Bosnia and between Bosnia and the outside world. For example, according to Ragusan records, fifteenth-century caravans from Dubrovnik took two days to reach Blagaj, four or five days to Visoko, five or six to Borač, seven or eight to Srebrnica, and ten to Zvornik.(45) Things were not much better in the beginning of the nineteenth century; a French traveller gives the following times by horse from Sarajevo: sixteen hours to Travnik, twenty-four hours to Srebrnica, thirty hours to Zvornik, seventeen hours to Višegrad and thirty hours to Mostar; and, of course, between Sarajevo and the outside world times were even longer: fifty-four hours to Dubrovnik and forty-eight to Novi Pazar.(46)

In most cases the peasants found themselves under the jurisdiction of local landlords. We have very limited source material on the status of the peasantry. In general, scholars believe that most peasants were more or less dependent and tied to the land, or, if they were shepherds, they were tied to the nobleman's flocks for which they were responsible. Life certainly was not easy; Bosnia is not a particularly fertile region.

It is often stated, as sort of a general rule about medieval societies, that the faith of the ruler defined that of his subjects. However, in the case of Bosnia, I wonder if we should not question this adage. As we shall see later, the nobles were tolerant about, if not downright indifferent to, religious matters. They often had relations with several different religions, at times played one off against another, and never seem to have let their attitudes toward a particular religion be influenced by any other-worldly consideration. Thus we get a picture of tolerance, not marred by any signs of fanaticism, among the medieval Bosnians. No ruler seems to have felt that his salvation or that of his people depended on his making his nation follow a particular faith. This indifference should be stressed, as it runs counter to the usual picture of the Middle Ages in Europe. It would be perfectly accurate to state that religion was not a major factor in internal Bosnian history, in spite of the tremendous attention that scholars have devoted to the religious history of Bosnia. Religion was a major factor only insofar as foreign powers used religion as an excuse to meddle in these issues which were of little interest to the Bosnians themselves. If pressure was put on a Bosnian noble to change his affiliation or if it seemed that he stood to gain something by doing so, he usually, as we shall see later, was quite willing to make the change.

The indifference of the rulers and nobles to the religion of their peasants is reflected in the ruins of medieval churches found on their estates. These churches are, without exception, small and appear to have been chapels erected only for the use of the noble's own family. In some cases they are clearly burial chapels. Their small size shows that they were not intended to serve general village needs, unless the peasant was only to attend ceremonies such as baptisms and funerals for his own family. But in any case, the fact that the noble built a tiny church meant that he accepted this customary behavior and had no interest in making regular churchgoers out of his peasants. It seems valid to conclude that the nobles did not particularly care what their peasants believed and left them to follow their own practices.

VI: Contact With and Spread of New Ideas

If a village lay near a town or trade route, it had some opportunity to meet new ideas. Other villages, and these were the majority, were quite isolated, having contact with a few other small villages and one or two markets. Most of the towns were also small and many of them were bypassed by trade routes which means that they, too, had little contact with the rest of the world. Their functions were that of local military defensive and administration centers which also provided a local market place and a place for a limited number of small craftsmen to live and work. Anyone who wants to obtain a vivid view of the isolation of these villages and the difficulties of communication between them and anywhere else need only take a flight from Beograd over the mountains of Bosnia to Split or Dubrovnik. The lack of frequent contact between villages is also apparent in the variations even among co-religionists in dress, songs, dances, etc., between villages in valleys on opposite sides of the same mountain.

The main source of contact between villages is the market, which is, and was, held at regular intervals. A town market usually would be held once a week. The most common pattern is that of overlapping market areas. This means that, although a village might have a near-by market which its villagers attend regularly, some of the villagers more or less regularly also go to one or two other markets. This pattern is the normal one, since it is rare to find villages where all the peasants go to only one market. In a region with a denser population, like central Bosnia, it is likely that villagers would attend as many as two to four different markets, depending on the proximity of their villages to the market places. When this situation exists, market days are staggered with one town having its market on Monday and the second town on a Tuesday and so on.(47) A market is a potential source for the spread of ideas and customs; and this overlap of market cycles facilitated their spread.

In recent times we have been able to observe the influence town markets have on surrounding villages in such matters as dress and the wide variety of household items which made their appearance first in towns. However, William Lockwood has demonstrated that markets are a poor medium for the spread of *ideas.* His field work in Bosnia has shown that the peasants coming to market are primarily concerned with business; they socialize only after business matters are completed and then they socialize in small groups with their own associates (e.g., relatives, in-laws, fellow-villagers, etc.). They tend not to meet new people and do not start conversations with strangers. At markets where people of different ethnic and religious groups buy and sell together, he found little or no mixing among the different peoples, other than in their economic relations.(48)

In our medieval Bosnian markets, then, it is unlikely that the Bosnian peasant would have spoken with strangers, including Dalmatians, unless he lived near one of the Dalmatian colonies in Bosnia and had friendly relations with them over a period of time. If a foreign missionary approached him, it is unlikely that he would have been receptive to new ideas from him.

Thus, most often, if the peasant encountered religious ideas at the market, he very likely would not have understood them and would have walked away unaffected or negatively affected. He already had his Christianity and the rites he connected with it; that was how his father and grandfather had done things and only let God give him health, and health to his family and animals, and let God grant weather conditions necessary for the prosperity of his crops, and that was all there was to the matter. In fact he probably would look upon religious speculation and new ideas as dangerous and likely to tempt the fates. On the other hand, if someone at the market had a new remedy or spell to cure rheumatism, well that was a matter worth looking into.

Thus markets were not an ideal place to preach a new gospel. However, they still provided the basic conduit through which ideas could be spread, and because market cycles overlapped, eventually most of the population would have heard of any significant idea that was able to attract the interest of peasant vendors. Whether a religious idea would captivate the interests of these vendors, however, is a debatable point.

These peasant attitudes and the limited communications between different regions must give us pause. And we should not make hasty generalizations about the diffusion of any religious doctrine across Bosnia. Despite the alarmist remarks in papal letters about the rapid spread of heresy, it seems probable that it — and any other religious ideas — circulated slowly; and this view is borne out by the relatively slow expansion of Islam in Bosnia after the Turkish conquest.

VII: Unrealistic Historical Models of Conversion

Too often the way people and societies react has been ignored by historians who deal with Bosnian history. For example, the "great man" theory, which has heresy brought to Bosnia by a particular individual, who was then instrumental in converting hordes of people in record time and in establishing some sort of church organization, is a popular one. Such a theory does not take into consideration the type of society found in Bosnia.(49) In a country of Bosnia's intellectual level and geographical conditions, ideas just do not spread as easily as that. In addition, historians who follow this approach do not stop to think about the process of conversion.(50) They believe it possible that masses of people, under no great pressure, will easily change religions in a short period of time. Having visualized a rapid spread of heresy in Bosnia, such historians then frequently tend to accept the inflated Franciscan claims of tens of thousands of converts during the first decades of the Franciscan mission in the middle of the fourteenth century and thus depict Bosnian society undergoing a rapid and large-scale return to Catholicism. No one denies that the Franciscans did have considerable success, (especially when we consider their limited numbers). However, no one should accept claims of so many conversions in so brief a time without solid evidence.

In addition, their success was not in "conversions," if we mean by conversion changing the belief of heretics. It is doubtful that the Franciscans did more than come into priestless areas full of Christians who held a wide variety of religious misconceptions and who had never been baptized. The populace was presumably relatively pleased to see the priests, was willingly baptized, and probably never felt that it had undergone any change in its religion.

The historian, who describes conversions in this rapid and wholesale manner ends up with a very unreal picture: There came a heretic and converted an area, then came Hungarian troops and many were converted to Catholicism but internal troubles then beset Hungary and many reverted to heresy, then came the Franciscans and a great many were converted again and then. . . . Conversion is a significant phenomenon; it does not occur as easily nor as often as some historians would have us believe. We need only look at the Turkish defters, which show the rates of conversion to Islam (from Christianity, confession unspecified) on a village level, to see what a gradual phenomenon this was. In the first years after the Turkish conquest, there were few converts; after a generation, the rate was increasing but still there were many Christians; and finally, after about sixty years, whole villages were converted. However, the rates of conversion varied from place to place, and in some villages conversion was rare, with the population remaining predominantly Christian. Thus in the case of Islam, a well-propagated religion which was also the religion of the conquering state, the rate of conversion was slow and occurred over a long period of time. If there is any question of conversion, rather than drift, in the Middle Ages when the cults were not being propagated by the state or any other apparatus with real authority, and when there were few preachers, it is certain that the rate of conversion from one Christian faith to another was much slower than that of conversion to Islam.

FOOTNOTES TO CHAPTER I

1. B. Moore, *Social Origins of Dictatorship and Democracy,* Boston, 1966, p. 456. The conflict between popular and official religion is also stressed by V. Lanternari, *The Religion of the Oppressed,* New York, 1963 (Mentor edition), p. 246. Lanternari describes this conflict as one between "the demands of the people and the needs of their society on one hand, and the action determined from 'above' by the clerical hierarchy on the other." See below, Note 29 of this chapter, for a discussion of more natural pressure-free syntheses between Great and Little Religious Traditions.

2. R. Redfield, *Peasant Society and Culture,* Chicago, 1956, pp. 40-41.

3. A. S. Shannon, in a book review in *Speculum,* XLI, 1966, p. 182.

4. B. Malinowski illustrates this point well: "When we say that 'Roman Catholics believe in the infallibility of the Pope' we are correct only in so far as we mean that this is the orthodox belief, enjoined on all members of the church. The Roman Catholic Polish peasant knows as much about this dogma as about the Infinitesimal Calculus." Malinowski stresses the need to distinguish between studying Christianity as a doctrine and as a sociological reality. (B. Malinowski, *Magic, Science and Religion and Other Essays,* (1948) Anchor book edition, p. 273.)

5. Unfortunately, we do not have any detailed descriptions of the intellectual level of medieval Bosnian peasants. However, we cannot expect them to be much more learned than their fellow tillers in Western Europe whose theological grasp is amusingly expressed by G. G. Coulton, *Medieval Village, Manor and Monastery,* Cambridge, 1925 (citation from Harper Torchbook edition, p. 257): "The peasant had scarcely the vaguest idea of the Mass service, except that at a certain moment, the priest 'made' the Body of God. Berthold of Regensburg finds it necessary to explain the different stages of the service to his congregation as though they were little children. They commonly came in late, and departed before the end; as soon as the priest had held up the consecrated Host they went clattering out (complains St. Bernardino) 'as if they had seen not God but the devil.' "

6. Text given in D. Kniewald, "Vjerodostojnost . . . ." *Rad,* 270, 1949, p. 157. For details on these and other passing references to Bosnian events, see my narrative chapters.

7. Text given by Dr. F. M. (Zagreb) editor, "Dva savremena izvještaja o Bosni iz prve polovine XVII stoljeća," *GZMS,* XVI, 1904, p. 261.

8. Now the *Slava* is found only among the Orthodox. However, Catholics of Bosnia and Hercegovina also practiced the *slava* well into the nineteenth century until a campaign carried out by the Catholic Church forced Catholics to give up this practice.

9. Though some scholars question its presence in medieval Bosnia it is probable that it was important then under the name of *Krsno ime.* See discussion, Chapter IV, note 20.

10. The connecting of a Christian holiday with agricultural needs is seen in a Christmas greeting from the vicinity of Glasinac in Bosnia, "Christ is born so that our grain grows. ("Hristos se rodi, da nam žito rodi.") M. Filipović, "Beleške o narodnom životu i običajima na Glasincu," *GZMS,* n.s. X (ist-etn), 1955, p. 125.

11. *Annales Ragusini Anonymi,* edited by S. Nodilo (*MSHSM* XIV, Scriptores I, Zagreb, 1883, p. 51.

12. Tablet from Hodbina. See *GZMS,* XVIII, 1906, pp. 540-41; and also M. Vego, *Zbornik srednjovjekovnih natpisa Bosne i Hercegovine,* I, Sarajevo, 1962, pp. 40-41. Attention should be drawn to the fact that the language and style of presenting those powers being called upon to defend the field against the devil is very similar to that of the curses against whosoever will violate the terms of the medieval Bosnian charters. Both types of documents reflect the same style of thinking and an expectation of heavenly action in the affairs of this world.

13. *GZMS,* XVIII, 1906, pp. 349-354.

14. *GZMS,* I, 1889, pp. 100-102.

15. *GZMS,* V, 1894, pp. 449-463.

16. There are various general beliefs in spirits, demons and strange personifications (such as diseases or days of the week) but these are usually vague, vary from place to place, and are clearly more of a folkloric nature. Basic practices are almost never tied to them though particular superstitions and taboos may be. Legends and superstitions about these spirits should not be considered part of any belief system which we might call religious.

That the function of religion in medieval Bosnia — besides contributing to village and family solidarity — was primarily to achieve goals in this world rather than to procure salvation in the hereafter is also seen in medieval gravestone inscriptions. References to the hereafter or to salvation are rare. Far more common than asking God to forgive one's sins is the request that the living do not disturb the deceased's bones. The content of the inscriptions tended to praise the worldly achievements of the deceased. The deeds noted were rarely of a religious nature. It is only after the Turkish conquest that God's forgiveness begins to be sought for with any frequency.

17. It, of course, is not to be ruled out that a dualist movement could have acquired a following in Bosnia on non-religious grounds — e.g. political or social. However, as we shall see later, there is no evidence that the Bosnian Church or the heretics as a group supported any specific political or social cause.

18. G. Čremošnik, "Ostaci arhiva franjevačke vikarije," *Radovi* (Naučno društvo BiH), III, Sarajevo, 1955, p. 19.

19. R. Grujić, "Konavli pod raznim gospodarima od XII-XV veka," *Spomenik,* SKA, LXVI, 1926, p. 120. In some places a priest may visit a church only once a year (e.g. on the Pivska planina, in Montenegro just across the Bosnian border).

20. M. A. Chaumette-des-Fossés, *Voyage en Bosnie dans les années 1807 et 1808,* Paris, 1816, p. 70; see also M. Šamić, *Les Voyageurs français en Bosnie à la fin du XVIII-e siècle et au debut du XIX-e et le pays tel qu'ils l'ont vu,* Paris, 1960, p. 242, and Dr. F. M. (Zagreb), "Dva savremena izvještaja . . .," *GZMS,* XVI, 1904, p. 255.

21. Z. Kajmaković, "Jedno starije kultno mjesto," *Naše starine,* XI, 1967, pp. 175-80.

22. In neighboring Montenegro, where we have no evidence of heresy in the Middle Ages, we find the same attitude toward priests and church buildings. A Russian envoy visiting Montenegro in 1760 describes the churches, which were so miserable as to be hardly recognizable as churches, as being empty. Inside them were to be found neither crosses, icons, nor books (cited by M. E. Durham, *Some Tribal Origins, Laws and Customs of the Balkans,* London, 1928, p. 77). Similar descriptions of medieval Bosnian churches have been taken by scholars, who have assumed the missing objects were rejected for theological reasons, to indicate the presence of dualism. The Montenegrin example shows that crosses and icons can be absent for reasons having nothing to do with dualism. The Montenegrins also did not respect their clergy; an Austrian envoy in 1788 heard a commoner tell the Vladika, 'Holy Bishop, you lie like a dog! I will cut your heart out on the point of my knife!" The envoy's comment was that except for keeping the fasts the Montenegrins have no religion (Durham, *op. cit.,* p. 77). The priests were so slovenly there that as late as 1855 it was necessary for Prince Danilo to issue a law requiring each priest to go to church every Sunday, and to keep the church clean and to punctually obey the canons of the church (Durham, *op. cit.,* p. 84).

23. Following V. Lanternari, *The Religions of the Oppressed,* p. 82, footnote 6: "By 'paganism' we mean native religious forms prior to the coming of the Christians. Paganism, in fact, has no significance other than, historically, that it is a form of religion that antecedes Christianity and to which Christianity is opposed. Naturally every form of paganism is, in turn, the crowning phase of a religious-historical process determined by culture; it should not be evaluated in a negative sense, as is so often the case among those who use the word 'paganism' to signify an inferior religious form." Unfortunately very little is known about South Slavic paganism; the bulk of our knowledge (and more often hypotheses) comes from a study of topographical names, folk tales, folk beliefs and folk customs. And of course the last three have only been recorded centuries after the demise of any and all forms of paganism as a living religious system.

The literatuere on these folk beliefs, superstitions, taboos and practices is vast. See for instance, E. Lilek, "Vjerske starine iz Bosne i Hercegovine," *GZMS,* VI, 1894, pp. 141-166, 259-281, 365-388, 631-674; in English see P. Kemp, *Healing Ritual, Studies in Technique and Tradition of the Southern Slavs,* London, 1935. Also scattered throughout the following journals is a mass of material about folk beliefs and practices, usually presented according to regions, but sometimes according to subject matter or type of practice (e.g. beliefs about the moon, funeral practices), *GZMS; SEZ; ZREI; ZNŽOJS;* (see key to abbreviations for Bibliography.)

24. K. Jireček, *Istorija Srba* (trans. by J. Radonić), I, pp. 93-94. In his chapter pp. 91-103 Jireček mentions a wide variety of other pagan survivals retained by South Slav Christians including various other beliefs and practices reflecting veneration of the heavenly bodies.

25. *GZMS,* VI, 1894, pp. 449-463.

26. To determine whether a belief or practice is taken seriously one should study emotional reactions. As Malinowski points out, "To describe the ideas of the natives concerning a ghost or a spirit is absolutely insufficient. Such objects of belief arouse pronounced emotional reactions, and one ought to look in the first place, for the objective facts corresponding to these emotional reactions." (Malinowski, *op. cit.*, p. 245) If we find that the peasants have little fear of or other emotional reaction to a spirit or to the importance of a certain ceremony we may conclude that it does not really fall into the category we would label religious. Thus we should not simply categorize peasant beliefs from the point of view of whether they are part of the Great of Little Tradition; for certain elements of both traditions are taken seriously while other elements of both traditions are either not understood or are considered trivial.

27. H. Puech and A. Vaillant, *Le Traité contre les Bogomiles de Cosmas le Prêtre,* Paris, 1945, p. 344. M. Popruzhenko, *B'lgarski Starini,* VIII, 1928.

28. G. Fejer, *Codex diplomaticus Hungariae Ecclesiasticus ac Civilis,* X, pt. 7, Budapest, 1834, pp. 812-813, 883-884.

29. It is pointed out that in Madras nowhere is there to be found a single unequivocal version of the Great Tradition but instead there are several overlapping and competing versions with varying degrees of admixture of regional and local traditions. (M. Singer, "The Great Tradition of Hinduism in the City of Madras," in C. Leslie, *Anthropology of Folk Religion,* New York, 1960, p. 110). India proves to be an ideal place to study great and little cults and their syntheses and variations. We find that rarely are all the major Hindu festivals observed in any one place; which ones are observed vary from place to place. The major festivals of every town include many regional and local festivals that have equal importance, in the eyes of the inhabitants, with the Great Tradition festivals. Even in the case of Great Tradition festivals, the bulk of ritual and practice accompanying their celebration has nothing to do with the Great Tradition but are local customs or a localization of elements in the Great Tradition. The form this localization takes varies from place to place. Both traditions evolve with each other; the two traditions at times mingle and at others remain separated yet are seen by villagers as being one tradition. (M. Marriott, "Little Communities in an Indigenous Civilization," in Leslie, *op. cit.*, pp. 169-214). This description has much in common with conditions in rural Jugoslavia. The *slava* has equal importance with major church holidays like Easter, many of the rites connected with Christian holidays such as Christmas have nothing to do with Christianity, which saints' days are particularly singled out and celebrated vary from area to area or village to village. Much of the practice connected with celebrating these days has nothing to do with Christianity but the peasant views both aspects as being part of one tradition. The Yucatan provides us with examples of the variety of ways Catholicism may mix with pre-Christian beliefs. Agains we find that how syntheses occur, varies from place to place. The Catholic and pagan elements have many different ways of mixing. In peripheral communities they tend to become more closely united and blend into a single religious system with

Catholic ritual being incorporated into pagan practices. When priests leave an area, those rites that it is believed can only be celebrated by them tend to disappear, while others felt not to be dependent on them are incorporated into popular ritual. In other places we find that the two cults remain separated, with both being considered equal parts of public worship, complementary ways of dealing with the supernatural. Each cult is equally respectable and equally powerful yet distinct, each with its own priests; but they are not seen as rivals to one another. (R. Redfield, *The Folk Culture of the Yucatan,* Chicago, 1941. The pertinent parts are given in Leslie, *op. cit.,* pp. 337-390.)

30. A. Solovjev, "Le témoignage de Paul Ricaut sur les restes du Bogomilisme en Bosnie," *Byzantion,* XXIII, 1953, pp. 73-86; P. Rycaut, *Present State of the Ottoman Empire,* Cologne, 1676, p. 131.

31. Much material on Christian customs kept by these Moslems can be found in the interesting study of T. Djordjević, "Preislamski ostaci medju Jugoslovenskim Muslimana" in his multi-volume work, *Naš Narodni život,* vol. VI, Beograd, 1932, pp. 26-56. Among the many observances carried out by Moslems on Christian holidays are: Moslems refusing to work on Good Friday, coloring eggs for Easter, decorating houses with flowers on St. George's Day. The largest number of observances are kept about St. John's Day; both Christians and Moslems of Banja Luka then bathe their cattle before sunrise to protect them from disease, while elsewhere people bathe themselves before sunrise for the same reason. The preceding evening is celebrated in particular by the young who jump over fires, make love charms, etc. Moslems also attend the dances held on hill tops on St. Ilija's Day. In addition many Moslems fear that if they fail to observe the Day of St. Procopius they may be struck by lightening or, as Hasluck observed, their crops may be ruined by the angry saint. (F. W. Hasluck, *Christianity and Islam under the Sultans,* vol. I, Oxford, 1929, p. 71).

32. K. Draganović, "Izvješće apostolskog vizitatora Petra Masarechija o prilikama katoličkih naroda u Bugarskoj, Srbiji, Srijemu, Slavoniji i Bosni g. 1623, 1624," *Starine,* XXXIX, 1938, p. 24.

33. On the interesting problem of the double-faith, see also M. Filipović, "Kršteni Muslimani," *Zbornik radova Etnografskog Instituta,* SAN, Beograd, II, 1951, pp. 119-28.

34. M. A. Chaumette-des-Fossés, *op. cit.,* p. 75.

35. M. Hardie, "Christian Survivals among Certain Moslem Subjects of Greece," *Contemporary Review* (London), CXXV, 1924, p. 229.

36. See E. Sapir, *Culture, Language and Personality,* Berkeley, 1964. (A collection of papers written in the 1920's and 1930's.) See especially "Language" (1933), pp. 1-44. Sapir, p. 36, stresses the importance of vocabulary as an index of culture.

37 K. Mannheim, *Ideology and Utopia* (1936), Harvest Book Edition, p. 56.

38. Malinowski's discussion of the influence of missionaries on native beliefs in the Trobriand Islands is useful for comparative purposes: "As to the danger of their (the natives') views being modified by missionary teaching, well, I can only say that I was amazed at the absolute impermeability of the native mind to those things. The very small amount of our creed and ideas they acquire remains in a watertight compartment of their mind." (Malinowski, *op. cit.,* p. 252) Lanternari agrees, ". . . so-called 'conversions' are more apparent than real, touching only the surface of native belief and never reaching into their true religious life . . . Bengt Sundkler, commenting on this situation, maintains that

it can be demonstrated that groups as well as individuals (in South Africa) have gone from the mission church to the Ethiopian, from this to the Zionist, and finally crossing the bridge of native Zionism, have returned to African animism, whence they had started out. Referring to Melanesia and therefore to an altogether different cultural environment, other anthropologists and missionaries have arrived at identical conclusions. Peter Elkins for instance remarks that the work of the missionaries is for the most part superficial as far as native life is concerned. With few exceptions, most of the natives preserve their fundamental beliefs and their traditional religions; their ancient rituals are performed or remembered in hiding, ready to be resumed in the open as soon as the opportunity arises . . . Fison a missionary in the Fiji Islands in 1884 . . . spoke of the nominal acceptance of Christianity by the natives. . . . The passing of the Wainimala tribe from paganism to Christianity he felt had been a nominal conversion and only a few charlatans could describe it as the 'conversion of the people' in a theological sense . . . and in recent times the situation on the islands is still such as to make it most improper to speak of the 'conversion of native communities.' " (Lanternari, *op. cit.*, pp. 250-51).

39. E. Evans-Pritchard, *Nuer Religion,* Oxford, 1956, p. 106.

40. It has been theorized that the reasons that such enormous gravestones were used in medieval Bosnia were either to prevent grave-robbing or to keep the corpse in the grave since there are various beliefs about stones having the ability to keep a corpse bound to his grave through the stone providing a house for the soul.

41. *The Life and Adventures of Dositej Obradović,* edited and translated by G. Noyes, Berkeley, 1953, pp. 311-312.

42. Text given by D. Kniewald, "Vjerodostojnost. . .," p. 158. It is to be noted that what a villager, acquiring a wife, seeks is to gain a hard-working woman to do her share of work in the household. If she does not fulfill this function well she is not "good" and thus ought to be replaced by someone else. The ruling classes were equally casual about matrimony though they married and divorced for considerations different from those of the peasantry.

43. The material about marriages in Bosnia under Turkish rule comes from M. Filipović, "Sklapanje i razvod hrišćanskih brakova pred kadijama u tursko doba," in *Radovi* (NDBH), 20, 1963, pp. 185-95.

44. The absence in South Slavic popular belief of a definite connection between ethics and religion should be stressed for it, too, should make one skeptical about the chances for success of a dualist religion among the Serbs, Croats and Bosnians. In their folk tales religious figures frequently appear with combinations of sympathetic and evil traits. Čajkanović sees the origin of such combinations in the tendency of primitive religions to make their gods not ethical figures but gods of specific functions (e.g., thunder). Thus we may find the Devil portrayed sympathetically in certain folk tales (e.g., A. Savić Rebac, "O narodnoj pesmi 'Car Duklijan i Krstitelj Jovan,' " *Zbornik radova* (SAN), vol. 10, Institut za proučavanje književnosti, knj. 1, 1951, pp. 266-67). On occasions the Devil is credited with teaching men certain skills and trades. S. Dučić ("Život i običaji plemena Kuča," *SEZ,* SKA, vol. 48, 1931, pp. 286-87) gives examples of the Devil returning favors, helping people and teaching them healing methods. See also, V. Čajkanović, *O srpskom vrhovnom bogu,* SKA, Beograd, 1941, especially pp. 105-110 but throughout the book. We also

find such popular expressions as "The Devil isn't as black as people say" (Čajkanović, *op. cit.*, p. 49). On the Devil see also, T. M. Bušetić, "Verovanja o Djavolu — v okrugu Moravskom," *SEZ*, SKA, vol. 32, 1925, pp. 400-405. On the other hand St. Sava (like such other saints as St. Procopius and St. Ilija) has a vindictive streak. He flies into quick rages and damns with slight cause. We find him damning a whole village for the sins of one man. He also takes revenge when people fail to observe his day: "You had better keep St. Sava's fast or the wolves will get your flock." That wolves are his weapon is derived from one of his less saintly attributes — being patron of wolves. (Čajkanović, *op. cit.*, p. 15). These beliefs reflect a society possessing a value system quite different from that of dualism. The family, not a religious creed, is the basis for the moral order. Moral excellence by the performer does not make rituals more effective. And folk tales about religious figures depict them as powerful supernatural beings who can either aid or harm you through a combination of their beneficial or vindictive traits, which in turn are derived from the functions attributed to those figures.

45. K. Jireček, *Trgovački drumovi i rudnici Srbije i Bosne u srednjem vijeku* (trans. Dj. Pejanović), Sarajevo, 1951, p. 96.

46. M. A. Chaumette-des-Fossés, *op cit.*, pp. 33-45.

47. A good description of markets and the role they play in peasant society can be found in W. Lockwood, "The Market Place as a Social Mechanism in Peasant Society," *Kroeber Anthropological Society Papers*, 1965, pp. 47-67. A revised and far more detailed presentation, with specific emphasis on Bosnia, can be found in his dissertation, *Čaršija i selo: The Market Place as an Integrated Mechanism in Bosnia, Yugoslavia*, Berkeley, California, 1970. I have also profited from discussing markets with Dr. Lockwood who has generously shared with me much still unpublished material.

48. Oral communication from W. Lockwood. This point is treated in detail and demonstrated in his dissertation (see note 47).

49. In a separate study I have discussed and argued against the role in spreading heresy attributed by several historians to two heretical goldsmiths from the coast. See, my "Aristodios and Rastudije — a re-examination of the Question," *GID*, XVI (1965), Sarajevo, 1967, pp. 223-229. The problem is also discussed in Chapter III of this study.

50. See above, footnote 38.

CHAPTER II

AN EXAMINATION OF SOURCES

We must devote considerable attention to the sources before we can attempt to draw any conclusions about the character of the Bosnian Church. Our problem is made immensely more difficult because of the unsatisfactory nature of these sources. We do not have a single narrative source about Bosnia from the medieval period. At the beginning of the seventeenth century three chronicles about medieval Bosnia still existed, at least two of which were written by foreigners: they are the Chronicle of Hrvoje Vukčić written by Emanuel the Greek,(1) and Bosnian chronicles by Pietro Livio Veronese(2) and by a man bearing the improbable name of Milich Velimiseglich.(3) Nothing is known about any of the three writers or about the character of their work. We do not even know in what centuries they lived, since all we have about them is the fragments of their work that the seventeenth century historians chose to excerpt.

From the excerpts, we know that Pietro Livio discussed religious matters in Bosnia fairly extensively: Orbini and a visitation report from 1623-1624 drew considerable material from him. We do not know if the other two chroniclers discussed religious matters, since surviving excerpts from them deal with secular affairs.

We do not have a single work written by the Bosnian Churchmen themselves, except for manuscript copies of the Gospels and notations of a non-doctrinal nature in them. Thus we must draw our material about this church from domestic charters and Dubrovnik records, mostly of a business and diplomatic nature, that mention members of the Bosnian Church; from papal letters about heresy, which generally simply reiterate old phrases and give amazingly few details; and from archaeological material.

In addition to the lack of firsthand chronicles, no archive in Bosnia (including those of the pre-Ottoman Franciscan monasteries) contains any documents from the period prior to the Turkish conquest of the 1460's. Thus we must be satisfied with documents (such as charters, reports of diplomatic missions, letters about internal and foreign affairs most often written by non-Bosnians) that have been preserved in archives outside Bosnia. The most important of these archives are in

Dubrovnik, the Vatican, Budapest, and Venice. In addition, material can be found in Zagreb, Vienna, Kotor, Split (only the Cathedral archive), and Zadar (where all the Dalmatian city archives as well as the archives of the Adriatic Islands, excluding Dubrovnik and cities lying south of Dubrovnik, have been collected). With the publication in 1967 of Professor Dinić's third volume of materials from the Dubrovnik archive,(4) scholars can be satisfied that all documentary material about the Bosnian Church existing in any archive in Jugoslavia has been published. However, we can always hope that new material will be turned up in the Italian or Turkish archives.

Native South Slav Sources

If we take documents (charters, letters, Dubrovnik registry material) as a whole, we find that they contain considerable information, though it is rare to find more than one or two items of value in any given document. We have a variety of Bosnian charters, a few of which are witnessed or guaranteed by members of the Bosnian Church. These provide information about the church's hierarchy and the position it held in the state. The Dubrovnik documents (chiefly minutes of the town councils and reports of its diplomats) tell a great deal about the diplomatic role of various individual members of the Bosnian Church; inform us about the role of the church vis-à-vis the Bosnian state; and also give many details about the locations and functions of the houses (monasteries) of the Bosnian Churchmen.

All of these documents, except certain Bosnian charters, are of unquestioned authenticity, though they may sometimes contain inaccuracies through ignorance or design. However, some forged charters exist, fabricated after the fall of the Bosnian state by emigré families trying to prove that their ancestors had been prominent nobility in the days of the Bosnian kingdom. Since the forgers had access to original charters they were able to make convincing imitations. Thus scholarly opinion is divided on the authenticity of a variety of charters. Particular care must be used in regard to any charter bearing an unfamiliar name which has not appeared in other charters. Unfortunately a register of names for Bosnia has never been drawn up, which makes the task of checking names formidable. Since many forgeries were made by members of families who regularly supplied Franciscans, and since the forgeries often turned up in the archives of Franciscan monasteries, documents found in these monastery archives and documents mentioning these family names should always be given special scrutiny.(5)

The theological references in all these Bosnian documents — be they charters or wills — reflect orthodox theology. This fact has been noted before; but various scholars, who describe the Bosnian Church as dualist, have argued that it is of little significance. They maintain on the basis of analogies with the behavior of dualists elsewhere that this was a deliberate attempt by Bosnians to hide their heresy behind orthodox phrases. However, unlike the French Cathars and Italian Patarins living in the midst of zealous intolerant Catholic societies, the Bosnian Church existed in the midst of a society indifferent to religious matters, in which the Catholics were a minority. There were no persecutions of the Bosnian Church by Bosnians until 1459 (4 years before the Turks put an end to the medieval kingdom) and this persecution may well have been limited to clerics of that church. Thus without persecutions against them there was no need whatsoever for church members to depict their religious beliefs and practices in a false way. And with no reason to do so, it would be absurd for religious leaders to put into documents religious beliefs and practices which they considered sinful. Thus it makes sense to take the content of these documents at face value and conclude that the theological references in these documents are there because church leaders believed in these things and wanted them there.

The most valuable Serbian sources on Bosnian heretics are several editions of anathemas written to be recited as part of the Church service during the week of Orthodoxy which condemn a variety of heresies, ancient and contemporary, and include anathemas of the Bosnian and Hum heretics. The anathemas give a limited amount of information about their practices, and single out various heretics by name.

Though most of the names cannot be identified, a few of those damned as Bosnian heretics are clearly members of the Bosnian Church. Thus the anathemas show that the Serbs anathematized as heretics members of the Bosnian Church, which is strong evidence against the Bosnian Church being basically a Serbian Church but with its own local hierarchy.(6) These anathemas, as will be shown later, were compiled in the late fourteenth century after Tvrtko was crowned at Mileševo in 1377 as King of Serbia as well as of Bosnia.

There are also three other Serbian sources that may or may not refer to heretics in Bosnia. The first is clearly about Bosnian heretics, though nothing is told about their beliefs or practices. The Life of Stefan Dragutin (King of Serbia 1276-1282 and subsequently Ban of Mačva and Usora, who died in 1316) written by his contemporary Archbishop Danilo mentions Dragutin's work in teaching and converting "heretics of the Bosnian land"; presumably he found them in his realm, which was in the north of Bosnia: Mačva, Usora, and probably Sol (Tuzla).

The second is a notation in a Serbian Gospel text found now on Mount Athos dated 1329 which tells of the rebuilding of a church of St. Nicholas in the Bishopric of D'bar and refers to godless *Babuny*.(7) In various Serbian texts about Serbia, *Babuny* refers to Bogomils. The question here is: where is D'bar? At times it has been thought to be near the Bosnian-Serbian border on the River Lim. Strong arguments have also been advanced that it is meant to refer to Debar in the vicinity of Ohrid in Macedonia. But be it the Lim or Macedonia, in 1329 neither was part of Bosnia. And in any case, nothing in the notation suggests these *Babuny* had any connection with the Bosnian Church.

The final Serbian source is a life of Despot Stefan Lazarević (1389-1427) by Konstantin Filozof, written in 1431-1432, which refers to a council held by the despot in the town of Srebrnica shortly before his death. The town, lying near the Drina, had been in Bosnian hands for much of that century, but at this moment was held by Serbs. There are several manuscript editions of the text and in one of them, and only in one, following the mention of the council in Srebrnica, is added the remark that the town's inhabitants are all of the Bogomil heresy. Excerpts from the manuscript were published in Russian translation in 1859 by a Russian, V. Grigorovič, and immediately attracted notice, since, if authentic, it is the only time the term "Bogomil" is used in a primary source for heretics anywhere in Bosnia.(8) The fact that the statement contains an obvious exaggeration — the impossible claim that all the inhabitants of a town with a large Ragusan colony and Franciscan monastery were Bogomils — need not disturb us too much. The problem lies in the fact that only one of the several manuscripts contains the phrase, and that this manuscript has subsequently disappeared. After Grigorovič's death, the great philologist Jagić examined his library, and among other things sought for this particular manuscript of the Life of Stefan Lazarević.(9) The text was not to be found. Grigorovič had never published the original text or even the full text in translation and no paleographer or skilled philologist had had the opportunity to study the manuscript before it vanished. Thus we do not know whether it was an early manuscript, whose copyist, for example, knew from personal experience that Srebrnica was full of Bogomils and added the phrase, or whether it was a late copy whose writer had no sound reason to add the phrase.(10) Of course I shall not be able to resolve the question here. However, it seems worthwhile to devote this space to it since the passage is frequently cited in such a manner that a reader would not suspect there were any difficulties about it at all.

A more useful chronicle for us is that of Thomas Archdeacon of Split:(11) Thomas died about 1260. His work, a useful source for events that took place on the coast, speaks of many heretics in Dalmatia; but he never says a word about the nature of their heresy. His lack of interest in doctrinal matters is surprising for a churchman in Split. His information will be discussed in Chapter III.

Most useful collections of Bosnian-Serbian-Dalmatian documents published prior to 1940 are listed by F. Šišić with brief comments about the type of material in the collection, the archives the documents came from, the degree of care of the editors, etc.(12) The following editions published since 1940 should be noted:

1. G. Čremošnik, "Ostaci arhiva bosanske franjevačke vikarije."(13) Though no documents dated prior to 1463 survived in any Bosnian monastery, some papal bulls to the Bosnian Franciscans have been preserved in other Jugoslav Franciscan archives. Here Čremošnik discusses the contents of these archives, gives the texts of bulls and documents that have not previously been published, and summarizes previously published texts, often adding relevant comments and occasionally supplying a new date.

2. M. Dinić, *Iz dubrovačkog arhiva,* knj. III.(14) This is the most valuable collection of sources that has appeared on the Bosnian Church since Šišić wrote. Dinić has excerpted, with context, passages from a variety of different types of documents preserved in Dubrovnik, ranging from the usual diplomatic documents, and minutes of council meetings, to wills, court cases, etc., under three headings: a)The sales of slaves. This has relevance to heretics, since by Ragusan law a Catholic could not be enslaved or sold as a slave; thus in theory most of these slaves (excluding slaves for debts and by birth) should either be heretics or people sold as heretics.(15) b) Material referring to Patarins. c) Acts of the councils referring to Patarins. In these sections Dinić republishes some previously published material, gives fuller texts in some cases for passages Jorga(16) printed in summary form, and publishes many new passages which give us a fuller picture of relations between Dubrovnik and the Bosnian Patarins.

3. M. Dinić, "Nekoliko ćiriličkih spomenika iz Dubrovnika."(17) Here Dinić publishes some Cyrillic documents about court cases over some of the fortune left by Gost Radin in his will. More material on these cases can be found in his *Iz dubrovačkog arhiva,* Knj. III.

4. V. Mošin, "Serbskaja redakcija sinodika v nedelju pravoslavija."(18) Though these texts have been printed before, the Dečanski text was printed only in the last century in a book that is a great rarity today. These are the most scholarly editions of the three sinodik texts whose value for the study of the Bosnian heresies has been noted above.

Besides the South Slav sources we also have a large number of letters written by the Hungarian king to the popes, South Slav rulers, or clerics working in South Slav lands. These letters, written by foreign rulers who hoped to subject Bosnia, do not reveal as intimate an acquaintanceship with what transpired in Bosnia as the Slavic sources' authors.

In addition, the Hungarian sources frequently are biased and reflect ulterior motives. However, they do contain interesting information on occasions. All editions of Hungarian letters are listed and discussed by Šišić.(19) In the Hungarian king's circle we also find the Catholic Bishop of Bosnia after ca. 1250 when that bishop moved to Djakovo in Slavonia. The letters of this bishop concern affairs in Slavonia and at the Hungarian court and have little bearing on Bosnia itself.(20)

Catholic Church Sources

Under this category will be discussed the papal correspondence about Bosnia, the letters concerning missions of papal legates, the documents connected with the work of the inquisition in Italy — insofar as the inquisitors concerned themselves with matters pertaining to Bosnia — and the Franciscan mission in Bosnia. Each one of the three categories of sources (papal chancellery, inquisition, and Franciscan) has sources of knowledge independent of the others, but at the same time, since their authors all corresponded with each other, they cannot be considered fully independent.

These Catholic sources are particularly important, for without them and the historical works of Pope Pius II — to be discussed under Chronicles — there would be little reason for the historian to suspect that there was any dualism in Bosnia. Almost all references to Manicheeism or to doctrines or practices that could be called dualist come from these Catholic documents. This fact has led many, who want to argue that the Bosnian Church was not dualist, to claim that since only one category of sources, and these from a foreign land (and an organization clearly hostile to the Bosnian Church) speaks of dualism, we should reject the evidence of these writings.(21)

However, this is too easy a way out and one that cannot really be justified. As noted above the Catholic sources include documents of several different categories written over a period of several centuries which renders any *a priori* rejection of them as a whole impossible. Each category of Catholic sources — and frequently each specific document — must be analyzed individually. Only in this way can we arrive at any valid conclusions about the reliability of these documents.

It should be pointed out here, however, that even in a church (be it Catholic or Orthodox) source, shown to be authentic, there is inevitable distortion resulting in an unbalanced picture of the heretics (or schismatics). This distortion arises from the fact that these sources only mention how their opponents deviate from orthodoxy and never state in what ways their opponents agree with orthodox belief. Thus if a group under attack were to have only a couple of ''errors'' in an otherwise entirely orthodox creed, we shall be given no hint of its basic orthodoxy but shall be shown only the deviations. Moreover, both the contemporary inquisitor and the later scholar tend to make these deviations into the bases of the heretics' creed (and even to build up from them whole systems) without stopping to think that these deviations may well be peripheral beliefs and practices, and not at all central to their acual creed.

It also should be mentioned that these sources frequently speak of heretics rejecting certain practices; this need not have meant total rejection, but could just as well have meant rejection of the practice as performed in a Catholic church. We of course have no way of learning how slight a divergence might be seen as a ''rejection'' by an inquisitor. Thus one must not conclude from accusations in these sources that the heretics necessarily rejected sacraments for given functions; instead they may have had different concepts of or different practices and prayers connected with them.

Finally we should note the tendency to link deviations with established heresies. This could easily have resulted in a Bosnian idea being extended or distorted so as to coincide with a classical heretical error. Such an identification would have not only given the Catholic Church more justification to condemn Bosnian ''errors,'' but also would have made available for polemics valuable and damning pejorative labels as well as the whole existing anti-heretical literature. And of course once having come up with such a label (e.g. making the Bosnians into Manichees), the theologian would have tended to bring into play a whole series of other ideas associated with dualism whether or not these ideas could be found in Bosnia.

Papal Sources

We have to take the papal sources seriously. Most are clearly authentic. There is no reason to suspect forgery in any of the papal correspondence we now have, though there is a handful of letters now apparently lost whose authenticity we must question. Yet because the papacy is obviously a foreign and hostile source, we must be on our guard in interpreting these sources. The papal documents throughout the medieval period speak of heresy in Bosnia.

A most important fact about the papal sources is that almost exclusively we have only the papal side of the correspondence. We very rarely have the incoming letters to which the popes are replying. This is particularly distressing, since the letters to the popes would presumably have described the situation in Bosnia and the nature of religious deviations. In the replies, the popes usually did little more than inform their agents of the means they may employ to wipe out heresy in Bosnia. Letter after letter from the papal chancellery refers simply to heretics. One begins to wonder how much the popes actually knew, since they were never more specific. Even if the papal agents and correspondents knew very well what the heresy's nature was, we would expect the popes at some time to refer to it by a name suggesting that nature (e.g., Cathar) instead of constantly saying only "heresy." We would expect them on occasions to give to their agents a specific suggestion about correcting a doctrinal error that might convey some specific information about the heresy. They never did. Yet we know that on at least one occasion, in 1351, Pope Clement VI went into some detail to describe to one of his legates the nature of the errors of schismatics in Albania, Hum, and Serbia. The deviations which he found it worthwhile to elaborate on were something as well known as the differences between the Eastern and Western Churches.(22) Did he never feel it necessary to inform a legate who was headed for Bosnia what errors he might expect to find there?

Thus we are left in a bit of a quandary. The popes clearly had several sources of information about Bosnia and the sources were independent of one another. If the popes were in doubt they could, and we know on occasions they did, send legates. The popes certainly were not always Hungarian dupes, as is seen from the fact that in 1248 Pope Innocent IV told the Hungarians to cease and desist from actions against Bosnia, after he had received a letter from the Bosnian Ban Ninoslav, assuring the pope of his Catholicism. The pope on this occasion decided to inquire into the matter and, clearly suspicious of Hungarian motives, selected two Dalmatians to look into it.(23) The results of their inquiry have not survived, but presumably the pope received a report. The popes also on occasion sent legates from Italy. In some cases the legates may not have grasped matters well because of language difficulties; it is doubtful that many at the Bosnian court, prior to the arrival of the Franciscans in the 1340's, knew Latin, and it is equally to be doubted that the Italians knew any Slavic language. However, on some occasions (such as Johannes de Casamaris' mission in 1203 to Kulin) the legates were accompanied by clerics from the coast (in 1203 by Marinus, an archdeacon from Dubrovnik) who clearly knew both languages.

However, despite the fact that some legates did send reports, a problem does exist about the quality of their information. Beyond the problem of communication arising out of language differences (and hence difficulties) is that of communication arising out of cultural differences. An educated Italian could well have taken for heresy the pagan practices which the Bosnian peasant mingled with his Christianity. But beyond this, since Bosnia was not a land seething with enthusiasm and interest in churches and formal religions, it is doubtful that the legate, if he wandered about in a town or village, ever saw anything heretical at all. It is certain that he did not come across heretics haranguing throngs of thousands. Thus if he wanted information about local beliefs and practices, he would have had to inquire. The only "heresy", then, that he was likely to have noted would have been that found in the answers to questions he asked. We can be sure that the peasant would have been confused by any question of a theological nature put to him. And if the peasant did not understand the question, one cannot put too much value in his reply. And since the legate presumably had a preconceived notion about what heresy existed in Bosnia, we could expect him to ask questions concerning beliefs connected with the heresy he thought existed there. Thus, by asking a peasant complicated theological questions about (e.g.) dualism a legate might well receive from that peasant, who could not make anything out of the question, affirmative but misleading answers.(24) The legate would then be satisfied that (e.g.) dualism and heresy did exist in Bosnia and make that report to the pope, and the "Christian" peasant would return to his ploughing and domestic practices.

This is pure hypothesis, but in this inquiry we have to go one step further than is generally done when one speaks about sources of information. It is not sufficient to say that the pope knew what was going on because various legates actually visited Bosnia. One must realize that visiting a country does not mean understanding it. The accuracy of a visitor's report depends on what he did when he was in that country; where he went, whom he talked to, what he talked about, what questions he asked, how he posed the questions. Unfortunately, we have no information of this nature about any of the legates. Thus we cannot say if they did or did not make reports which were worth anything. In addition, in some cases there is a question whether certain of the legates who described Bosnia ever visited the country; once the Franciscan vicariat of Bosnia, which encompassed half the Balkans, was established in the 1340's, it is impossible to say where a legate who went to "Bosnia" actually went, unless it is specified what courts he visited. However, we do know definitely that in every century at least a few legates visited Bosnia proper in person and presumably made reports. Since the texts of these reports have not survived, we cannot evaluate their quality.

Thus we have to conclude that the papacy had access to information of some sort from several independent sources (the Kings and high clergy of Hungary, coastal clerics, papal legates and, from the 1340's, the Franciscans) and thus should have been in a position to know something about what was going on. Then how can the vagueness of the letters be explained? The answer to this problem clearly lies in what happened at the Papal Chancellery. Where was the information received? By whom? What was done with the information? What officials were informed of its contents? Where was it filed? Who wrote the papal replies about Bosnia, and did the writers have access to all the material received? These are questions which I cannot answer but which would make worthwhile a full study on the procedure and functioning of the Papal Chancellery. Only on the completion of such a study would we be in a position to pass a really qualified judgment about the value of papal sources.(25)

In addition to the problem of information available to and utilized by the papacy is that of the way the popes presented the material. The papacy's manner of writing reflects more than just hostile bias; the tone and style with which papal letters frequently speak of Bosnian heretics bear many similarities to what Professor Hofstadter calls "the paranoid style."(26) The papacy in 1200, fighting strenuously against the large-scale heretical movement in southern France, projected upon Bosnia's peasant society this evil it was fighting elsewhere and as a result tended to describe Bosnia in terms of this projection. The popes, hearing of heretics going to, or being in, Bosnia, became alarmed that Bosnia would be another southern France; the shortage of information in Rome about Bosnia surely did nothing to lessen this worry. Thus we must take this into account when we note the hysterical tone of papal letters about heresy in Bosnia; and we probably are justified in concluding that there is great exaggeration in the early papal letters about the size, significance and speed of diffusion of heresy among the populace in Bosnia.

It also should be pointed out that even though it is likely that there were at least two different movements in Bosnia — the Bosnian Church and a small dualist heretical current — the papal correspondence — through ignorance, design or lack of interest — generally makes no distinctions and lumps all deviants together as "heretics." Thus usually it is impossible for the reader to determine whether a pope is referring to one or more heretical groups when he says "heretics" and even the same pope may use this term with varying degrees of breadth on different occasions. Thus the reader should not take anything for granted when he comes across the term "heretic" in a papal source.

A second very interesting fact about the papal correspondence is that it changes in character in the 1440's. Prior to that decade there are only a few references to suggest that the popes believed that there was dualism in Bosnia. Each of these references occurs in a document of unquestionable authenticity. 1) Innocent III in 1200 stated that he suspected the Bosnian heretics were Cathars. The term "Cathar" was never to be repeated about Bosnia again.(27) 2) The Bilino polje document of 1203 which the Pope surely saw and the 1319 letter of Pope John XXII refer to practices that might be associated with dualism.(28) 3) Gregory XI in 1373 mentioned the "adoration of heretics"(29), which could, perhaps, be associated with the Cathar custom of adoring the *perfecti;* as an isolated instance, though, it could also be attributed to some local practice. 4) Boniface IX wrote Sigismund, King of Hungary, in December 1391 about the dangers threatening the king's realm. The Turks were singled out, Manichees and other heretics were mentioned, and finally schismatics were stressed for Bosnia.(30) Exactly where in Sigismund's realm the Manichees were thought to have lived is not stated. The reference to Bosnia, however, specifically links that land with schismatics. 5) In 1435 at the Council of Basel, a "Terbipolensis" bishop, suggested that the Council take steps to convert the Kingdom of Bosnia which was infected with Manichee errors. He stated that when he visited Bosnia he had been so hospitably and reverently received by the inhabitants that it was all he could do to prevent them from kissing his feet.(31) Of these five, only the reference to Bilino polje clearly describes the Bosnian Church (i.e., the organization that would become that church). The other four could well be describing other movements.

Thus prior to 1440 we have only these few references in papal and conciliar sources to dualism in Bosnia. And one may well ask if the earliest references would have been remembered by fifteenth century popes. Besides these references there are also several inquisition documents from the thirteenth and also from the late fourteenth and early fifteenth century which speak of dualist Bosnian heretics. We shall examine the inquisition sources shortly, but in any analysis of the references to dualism in the papal documents, we must consider the possibility that the popes took this idea from the inquisition documents.

A second possible source for the papal belief could be that the popes were deceived by the meaning of the term "Patarin." In Italy this term had, by the thirteenth century, come to designate dualists. The Dalmatian sources, for some unknown reason, refer to the Bosnian Church as Patarin, even though, as we shall argue later, there is little evidence that the Bosnian Church was a dualist church.(32) Yet it is possible that the Dalmatians began to use the term ''Patarin'' to refer to the Bosnian Church because that church had certain practices similar to those of — and possibly even acquired from — dualist heretics. However, regardless of the origins of this Dalmatian label for the Bosnians, once it came to be used regularly there, it would have been natural for Italians (including the popes), upon finding the Bosnians called Patarins, to have assumed that the Bosnians were dualists. Since Dalmatian sources used the term ''Patarin'' frequently about the Bosnians only in the fifteenth century, and since the Church sources use the term ''Manichee'' in 1391, 1435 and then repeatedly after the 1440's, it is not farfetched to see a connection between the terminology in the two groups of sources. For example, we have a Ragusan letter written in 1433 to the Council of Basel which calls the Bosnian Church, ''Patarin.''(33) This letter could well have been the source for the Terbipolensis Bishop's use of the word ''Manichee'' at the Council. This foreign bishop, presumably not knowing Slavic, could not have discussed theology with the Bosnians when he had visited Bosnia. And finally, if, as we have reason to believe, there were dualists in Bosnia as well as a separate group of non-dualists called Patarins, it is not surprising that confusion between the two groups and confusion over terminology might arise in the minds of foreigners.

In any case, in the 1440's references to dualism in Bosnia began to appear with some regularity; and these references continued until the fall of the Bosnian state, after which the stories of dualism entered the chronicles and soon became the accepted view of what the Bosnian Church had been during the whole medieval period. Thus, the reason that people came to believe, rightly or wrongly, that it was dualistic is chiefly because of the contents of sources written in the final two decades of Bosnia's independent existence. The basic question we must pose is: What happened in the 1440's? Was there some change in the situation in Bosnia? Did the dualist movement there, mentioned occasionally in papal and inquisition sources and presumably up to then not significant in size, suddenly grow and attract a large following? Or take over the organization of one wing of the Bosnian Church? Or did some people at the papal court or connected with it, for motives of their own, decide to make a big issue out of dualism, or make a small sect appear to be a large movement, or make it seem that a schismatic church was linked with heresy? This problem will be discussed fully in Chapter VI. However, here in the discussion of the sources I want to call attention to the fact that at this time a shift in the message of some of the documents occurred.

It should be stressed, however, that, though from the 1440's we begin to find the term "Manichee" in some letters, the former general terminology of "heretics," "schismatics" and occasionally "Patarins" continued to be used exactly as it had been before and in fact is more frequently used, even in the final years of Bosnia, than the dualist label.

The papal letters which refer to Manicheeism in Bosnia fall into two categories: a) those letters which we still have which have appeared in scholarly collections of documents edited during the last century and a half. These letters are of unquestionable authenticity and we shall discuss them as we come to them in the narrative chapters of this study. b) those letters which have not been so published but which appeared in a variety of eighteenth century histories which extensively quote papal letters. Several letters, two of which have been frequently used by scholars, have appeared only in such works. It seems worthwhile to make a search of papal archives to see whether these letters can be found. Letters of questionable authenticity frequently appear either in Raynaldi's Ecclesiastical History or Farlati's *Illyricum Sacrum.* Both of these writers were clerics as well as historians and both believed the Bosnian Church to be the same as the Bosnian heresy and both to be dualist in nature. Raynaldi regularly in his narrative used terms specifying dualism, though the letters on which he based his narrative simply said "heresy." One cannot help wondering whether these two historians at times did not change a word here or there from "heretic" to "Manichee" in the texts they were copying to clarify things for their readers. After all, since they were convinced of the equivalence of the two terms, such a change for the benefit of clarification, could not have seemed a sin. Farlati, as we shall see, was also deceived by at least one forgery, and thus we have reason to suspect that all his papal letters may not be genuine. The letters whose authenticity is highly suspect are as follows:

1. A letter from Benedict Ovetarius, secretary to the King of Cyprus, to the Bishop of Padua dated 1 October 1442. It describes the arrival at the papal court of the Bishop of Ferrara with a legate from the Bosnian king, who in public and in the name of the king and the whole Bosnian kingdom renounced Manichee teachings and accepted the Roman faith.(34) Regardless of whether the document is authentic or not, the "Bishop of Ferrara" is surely an error for Thomas Bishop of Hvar (Farensis) in 1442 the papal legate to Bosnia. Several reasons lead me to suspect this document: Had such a public conversion been announced it seems unlikely that Nicholas V could have made the error five years later of calling Tvrtko II's successor, Stefan Tomaš, the first Bosnian king to accept Catholicism.(35) Other evidence also, to be discussed in detail in Chapter V, makes it apparent that Tvrtko II was already a Catholic prior to 1442. For example in 1428 the pope approved his marriage to a Catholic girl and in 1433 the pope released Tvrtko from Church fasts for reasons of health.(36)

2. Farlati, having told us that Stefan Tomaš had been a Manichee or Patarin (showing that he considered the two words synonyms), cites a long letter which he claims was written by Pope Eugene IV, 3 November 1445, which states that the Bosnian king complained that he could not take any action against the Manichees in his kingdom since they were too powerful, and in fact that he had to honor the leaders (primariis) of that sect owing to the strength of the sect; to do otherwise would endanger his kingdom. The letter then lists a variety of beliefs, which are clearly dualistic, supposedly held by the Bosnian heretics: they think that the Incarnation, Passion and Resurrection of Christ were apparent, that the devil is equal to God, that they believe in two principles, that they damn the Old Testament while mutiliating and corrupting the New, and that they condemned marriage and certain foods.(37) Since this letter is the only one which we have, in which a pope gives any details about heretical beliefs in Bosnia, this letter automatically separates itself from the rest of the papal correspondence. Then when we find it missing from all the standard collections of papal letters and collections of papal sources for the Slavic lands we must wonder about it, for surely editors of such collections would have wanted to include this letter if a text existed. In addition, other papal letters of this period do not utilize any of this significant information, nor even, except for one 1447 letter of Nicholas V(38) which refers to King Stefan Tomaš giving up Manichee errors, use the term "Manichee." In fact, the term "Manichee" appears in no other letter by Eugene. Further examination shows that this letter appears in the same general account in which Farlati presents a description of the bogus Council of Konjic. (I do not discuss this council's "edicts" here since Rački has earlier clearly shown the document to be a forgery.(39) That this letter of Eugene's should be dated within a year of the fraudulent council and appear in Farlati together with the description of the council is sufficient reason to be very sceptical of the authenticity of the letter, and to suggest that Farlati may well have found the letter in the same place he found the other forgery. The fact that the letter seems to be non-extant now can only increase these suspicions. Might not such a letter have been forged in order to provide more verisimilitude for the Konjic council forgery? A thorough study of Farlati and the documents he used is called for, and such a study would have to include a search for the documents he supposely had, which we no longer possess.(40)

These two documents, whose authenticity I doubt, will not be utilized in the narrative chapters of this study.

Inquisition Documents

From the thirteenth century we have the following three texts, all of them drawn up by inquisitors, which refer to heretical beliefs in connection with Sclavania, Sclavonia or Bosnia:

1. "De Heresi Catharorum in Lombardia" (ca. 1210).(41)
2. "Summa de Catharis et Leonistis" of Rayner Sacconi (ca. 1250).(42)
3. "Tractatus de hereticis" probably of Anselm of Alexandria (ca. 1270).(43)

These treatises all discuss primarily the Cathar heretics in northern Italy and presumably are, to a considerable extent, based upon testimony before the inquisition by Italian Cathars. In the case of Rayner's "Summa" we have the work of a former Cathar who converted to Catholicism and then became an inquisitor. All three documents are of unquestionable authenticity. Thus, we have little reason to doubt the information they provide about Italian Catharism — its doctrines and organization. The three documents are consistent in the story they tell and they link the northern Italian Cathars with various Slavic dualists. Nothing is said in them about the role of dualism in Slavic lands. Either the inquisition did not know much about that subject or had little interest in it. The documents make clear that besides Bulgaria and Dragovica, a dualist "church" existed in "Sclavonia." In Chapter III we shall argue that this church in the thirteenth century was centered in Dalmatia, but also had some following in Bosnia. Unfortunately, the inquisition documents do not define the geographical boundaries of Slavonia, or the size of this church. Our only cause to doubt the information these documents provide about Slavic dualism is owing to the likelihood that the inquisition would be ignorant about conditions in the Slavic lands (we have no evidence that the inquisition prior to 1461 ever interviewed a Slav), or owing to the fact that some of this information may well have been procured by torture, a device likely to produce what the inquisitor wanted to hear rather than what the victim knew to be true. The tracts discuss the beliefs of Italian churches but say nothing specific about the doctrine of Slavic churches. However, since these Italian churches were derived from Slavic ones, we can expect that they would have had various beliefs in common. However, a variety of Italian beliefs most probably did come from the Western (French, Lombard) Cathar milieu.

Rayner's "Summa," written by an ex-Cathar with direct knowledge of the movement, is of course a source of greater reliability than one based upon inquisitional testimony. For material about the Italians therefore Rayner's work is unsurpassable. Unfortunately he gives us little data on the Slavic churches; but what he does give is probably accurate. He lists sixteen Cathar churches and reports that together the five churches of Slavonia, Philadelphia, Greece, Bulgaria and "Dugmuthie" (Dragovica) total 500 finished Christians (i.e., initiated ordained clerics). He then states that all the Cathar churches were derived from either the Bulgarian or Dragovican church. Bosnia is never mentioned though presumably it is included in the Church of Slavonia. Rayner never tells us where the Church of Slavonia was centered, nor does he give us any notion as to how many of the 500 *perfecti* belonged to this Church.

A hundred years or so later we have three detailed documents of an inquisitorial nature. The first is the testimony of Jacob Bech of Chieri before the inquisition at Turin in 1387.(44) Second, we have a list of errors of the Bosnian Patarins (Isti sunt herrores, quos communiter patareni de Bosna credunt et tenent).(45) This list exists in two copies, with the title of the second referring to Bosnian "heretics" instead of Bosnian Patarins.(46) Third, we have a dialogue between a Bosnian Patarin and a Roman Catholic (Hic sunt omnia puncta principalia et auctoritates extracte de disputatione inter Christianum Romanum et Patarenum Bosnensem).(47) Both these tracts are found in the library of St. Mark in Venice. Rački, who published them, dated them both from the late fourteenth century. The second manuscript of "Isti sunt herrores . . ." is found in the Academy Library in Zagreb. Mošin, on the basis of the paper and watermarks, concludes that the Academy manuscript is from the 1370's.(48) The Academy list is found in a collection of manuscripts connected with the Franciscans. And Rački, in the introduction to his edition of the documents found in Venice, plausibly suggests that they were drawn up in Italy for the use of Franciscans being sent to work in the Bosnian vicariat. His view is supported by the fact that a marginal comment on one of them from 1421 refers to a Bosnian vicar.

The testimony of Jacob Bech and the two tracts clearly state that the "Bosnian Patarins" were dualists. These Patarins believed in two principles and docetism; they condemned the Old Testament, church buildings, baptism with water and various other "Christian" practices normally rejected by dualists.

When we note that Rački dates these tracts late in the fourteenth century, it immediately strikes us that these two documents were issued at roughly the same time(1387) that Jacob Bech and his cohorts were on trial.(49) Jacob's sect was clearly dualist; and the inquisition, among other things, called Italian members of it from Chieri "Bosnian heretics." Jacob described how he had learned his heresy from two Italians and a Slav, and how several heretics, whom he names, had gone to Bosnia to study doctrine. He then proceeded to give a list of his errors, most of which were akin to the theology in the two tracts. Jacob also gives many details found in neither of them, and they include points that Jacob did not make.

The coincidence in time between the trials in Piedmont and the two tracts, the obvious interest of the Catholic Church at the time in the heresy in Piedmont, dualists and "Bosnian heretics," and the fact that this was not a time of particular papal (or other Western) interest in heretics in Bosnia itself, leads me to conclude that these documents were very likely connected with one another. This would then suggest that the trials stirred up interest in the heretics of Bosnia, leading Franciscans in Italy, whose order was responsible for religious matters in Bosnia, to turn more seriously to the problem of dualists in Bosnia. And thus we could suggest that the two tracts were compiled for the purpose of educating Franciscans about to be sent off to Bosnia. Since the tracts seem to have been written in Italy, and their appearance in Venice and their similarity to other Western anti-Cathar tracts bears out this supposition, we may suggest that a leading source for their contents was the doctrines and practices of the "Bosnian heretics" in Piedmont, about whom investigations were being conducted at this time. And, since Jacob and his friends were Italians, we may wonder how many of the views they held were identical with views of dualists in Bosnia. Since there were many dualists in Italy, many of Jacob's errors could easily have come from Western Cathar sources. In addition Rački and many scholars who have written after him have noted the similarity of the two tracts in both content and form to Western anti-Cathar tracts. This suggests that one or more of these tracts about Italian or French dualists could have been used in the creation of our tracts. And if a major source for the tracts about "Bosnian heretics" was the beliefs of Italian heretics, then we cannot be sure that any belief or practice attributed to the heretics was also held by any Bosnian. Thus these two tracts may be much more accurate in describing the heresy of the Italians known as "Bosnian heretics" in Piedmont than that of dualists in Bosnia.(50) From this, it seems that we can use these tracts for little more than to demonstrate that there still existed a dualist current in Bosnia with which Italians had connections. Whether it was a significant or large movement, and whether it had any connection with the Bosnian Church is not stated in Jacob's testimony.(51)

In fact, the only reason to connect the "Bosnian heretics" of our tracts with the Bosnian Church is the fact that the two Italian manuscripts call them "Bosnian Patarins," and "Patarin" is a term used by Dalmatians and Hungarians to describe members of the Bosnian Church. However, since the tracts describe dualists and since the Italians generally called their dualists "Patarins," it would be quite natural for the authors of the tracts to say "Bosnian Patarins" when describing Bosnian dualists. Thus I do not think we can conclude from the term "Patarin" that Bosnian Church beliefs and practices were being described in these documents. In fact the Bosnian Church that emerges from our study of all sources about it is very different from the sect being described in the tracts. Thus I conclude that the Bosnians with whom these Italian ''Bosnian heretics'' had contact were Bosnian dualists who belonged to a movement separate from the Bosnian Church.(52) And since much of the content of our tracts seems to have come from a Western milieu — the Turin trial and anti-Cathar pamphlets from the files of Italian Catholics — we may well wonder whether our tracts are even reliable sources to describe the beliefs and practices of any dualists in Bosnia. Thus I shall use them in this study for little more than demonstrating the existence of some dualists somewhere in Bosnia in the late 14th century.

In 1461 three Bosnian noblemen were sent to Rome where they renounced fifty ''Manichee'' errors before Cardinal Johannes Torquemada.(53) Since prior to their arrival the three had been interviewed by presumably bilingual Franciscans in Bosnia as well as by the presumably bilingual Bishop of Nin, we can conclude that these Bosnians held dualist beliefs of some sort. However, since they spoke no Latin or Italian and since Torquemada certainly knew no Slavic, communications between the two sides must have taken place through an interpreter. Since the three were not theologians and since they certainly were scared out of their wits by the inquisition procedure in a foreign land, we can suspect that they would have willingly renounced anything put before them, simply to return home more promptly. The fifty points are similar in content to the two late fourteenth century tracts; thus we suspect that Torquemada, when he learned dualists were coming to abjure errors, turned to the archives (possibly using the two tracts among others) and drew up the fifty points prior to their arrival and simply presented this document to them to be renounced. Thus the ''Fifty points'' also is not a document we can use with certainty to demonstrate the existence in Bosnia of any particular beliefs. It can simply be used to show, since the three men had passed through the hands of bilingual theologians, after which they had been forced to renounce dualist doctrines, that most probably they really had held at least certain dualist beliefs. We can learn nothing about their specific beliefs or practices.

We do not know whether the three nobles were members of the Bosnian Church; but a comparison of the 50 points with what is known from Bosnian documents about the Bosnian Church shows that the 50 points have little or nothing in common with Bosnian Church beliefs. Such a comparison, which demonstrates the irrelevance of the inquisition documents for the Bosnian Church, is made in Appendix A of Chapter VI.

Franciscan Sources
(including all visitations)

As stated earlier, despite the fact that several of the Franciscan monasteries in Bosnia existed before the Turkish conquest, no documents have been preserved in any of them from the period prior to 1463. However, various documents (letters and reports) about Franciscans *in the Bosnian vicariat* have been preserved elsewhere. All of these documents have to be used with great care since they concern the Bosnian *vicariat* and not the Bosnian banate-kingdom. The vicariat, founded in the middle of the fourteenth century, included not only Bosnia and Hercegovina but also Slavonia, Serbia, Bulgaria, and parts of Croatia, Hungary, Moldavia, Bohemia, and Transylvania. In other words, the Bosnian vicariat was a label for the whole area of South Eastern Europe in which the Catholics were striving to convert schismatics and heretics.

Thus, the fascinating story, contained in two letters about Vicar Jacob de Marchia's work, about pagan-like rites during which heretics poured "the blood of our Lord" over themselves from leather sacks cannot be ascribed to Bosnia proper for we do not know where in the vicariat it occurred. In fact, if this story is related to the same area in which a subsequent event referred to in the first letter took place, it may concern Bohemia.(54) The second letter, written from the Ban Monastery in Srem, also provides no reason to connect the story with Bosnia.(55) And that it was not about Bosnia is most probable, since Jacob de Marchia, about whose activities we have many documents, devoted the bulk of his time in the vicariat to fighting the Hussite heresy, which never appeared in Bosnia proper. Our sources also show that he spent considerable time in Srem and Bohemia. The frequently cited letter of Pope Gregory XI to the Franciscans in 1373 also has to be used with care since it also concerns the vicariat. Thus, the interesting information about "rustic" priests and some sort of adoration of heretical leaders could concern any part of the vicariat.(56) And finally we probably should not make use of the interesting letter of Pope Eugene IV from 1446 which mentions many people of both sexes in many places in the territory of the vicariat who do not belong to any certified order but serve God with clean lives — some of them in the houses of noblemen and some in other places the bishop assigns.(57) This letter most probably does not refer to Bosnia proper since the last place mentioned in the letter, prior to this description, was the confines of the Kingdom of Hungary, Transylvania and Ruthenia.

The most noted figure in the history of the Bosnian vicariat prior to the arrival of the Turks was Fra Jacob de Marchia (special visitor to the vicariat 1432-33 and vicar 1435-39). In the documents about his mission in the vicariat we hear of many different sorts of heretics: Bosnian heretics, Bohemian heretics, Hussites, schismatics, and even pagans. Machichees, however, are never mentioned.(58) It is also interesting that even in his "Life," written later in the fifteenth century by Fra Venanzio da Fabriano, who claims to have heard much of the material in it from Jacob himself, Manichees are never spoken of.(59) It seems very strange, if dualism were a major current in Bosnia, that Jacob, an able theologian, did not notice it and make some sort of clear reference to it in one of the documents about his mission in the 1430's. Fabriano's "Life of Jacob" gives us few details about Bosnia. It never distinguishes between the different regions of the vicariat. It gives inflated figures on conversions which vary from manuscript to manuscript (50,000 to 100,000 in one case). Twice heretics called "Patarins" are referred to; although it is not specified where they lived, we can presume it was Bosnia proper since heretics by that name were not known elsewhere in the vicariat. This is noteworthy for Jacob's own letters never called the Bosnian heretics "Patarins."

In addition to the material about Jacob, the bulk of which can be found in Fermendžin and Fejer,(60) a limited number of other Franciscan documents are important for the history of religion in Bosnia. The "De Conformitate vitae Beati Francisci" of Fra Bartholomaeus of Pisa has a list of the monasteries of the Bosnian vicariat as of 1385 which shows that there were already four monasteries (Sutjeska, Visoko, Lašva, and Olovo) in Bosnia proper and several others in the adjacent regions.(61) A second valuable Franciscan source is the "Necrologium Bosnae Argentinae" a manuscript kept in the Monastery of Sutjeska and published by Jelenić.(62) It contains one important notation about the killing of five Franciscans of the monastery of St. Mary at Visoko in 1465 by Patarin heretics. Except for material from Hercegovina, particularly that concerning Gost Radin's will and relatives, this is the last document from Bosnia proper to speak about activity by Patarins. After 1465 no Franciscan monastery has a single document (excluding chronicle material referring back to earlier periods) that refers to Patarins or heretics in Bosnia. The sudden disappearance of the Patarins and heretics from the sources is one of the great mysteries surrounding the whole heretical movement. In the years that followed, the Franciscans' troubles with other Christians were to be exclusively with the Orthodox. It is also interesting to point out that even in 1465, in spite of the new terminology being used in Italy, the Bosnian Franciscans refer to the Bosnian heretics as Patarins and not as

Manichees.

Franciscan documents from the period after the Turkish conquest have some relevance to the medieval religious situation. The two major collections of these documents were compiled by D. Mandić and by J. Matasović.(63) These documents as well as the reports of episcopal visitations of Catholic regions of Bosnia and Hercegovina which began to take place in the closing decade of the sixteenth century and then continue more and more frequently over the following centuries supply data on the numbers and distributions of Catholics as well as on the locations of various Catholic churches and monasteries. Since the Turks had edicts against the erection of new churches (and since this rule was strictly applied to Catholics, though more laxly enforced toward Orthodox), references to Catholic churches existing in any document up to the end of the sixteenth century strongly suggests that a Catholic church had stood in that location prior to 1463.

The Franciscan documents tell us a great deal about Catholic-Turkish and Catholic-Orthodox relations in this period. The latter are particularly interesting since prior to the Turkish conquest, in most of Bosnia we have no evidence whatsoever of Orthodox churches or clergy; there were just Patarins (or Bosnian Church) and Catholics. But then in the last decades of the fifteenth and throughout the sixteenth century we find numerous references to the activities of Orthodox clerics and their relations with Catholic believers. Since, as we have stated before, after 1465 references to heretics and Patarins cease, the replacement in the sources of Patarins by Orthodox has been a major argument in favor of the theory that the Bosnian Church had really been Orthodox all along and that with the fall of the kingdom, the name "Bosnian Church" — or Patarin — died out and was replaced with the general term for its confession and ritual, Orthodoxy.

The visitations,(64) though unfortunately later in time, are a much more useful source than the documents for the history and anthropology of the area. First, they tend to be more detailed and give figures: figures on the numbers of believers in a town or village, figures on the number of monks at monasteries and, what is particularly interesting, though unfortunately not for our study, data on the size of *zadrugas* in the eighteenth century. In addition, the visitations often supply data on popular mentality, customs, and practices which are useful for a student investigating any aspect of religion in the area. For example, a 1612 visitation gives a detailed description of the abysmal ignorance, the superstitions, and the graveside practices of the Orthodox in Srem, who were not even aware that a schism existed between Orthodox and Catholic Churches. A 1623-24 visitation for Bosnia and Hercegovina stresses the incompetence of priests and the conversion of Catholics to Islam. One of the most interesting of these visitations is the detailed papal inquiry and correspondence connected with it in 1629 and the following years concerning the Bishopric of St. Stephan, a bishopric established on his own by a Catholic priest who had become disgusted by the conditions within the church and by the fact that the responsible bishop, the Bishop of Trebinje, resided in Dubrovnik and took no care of his flock. These documents clearly show the connection between lack of clergy (or the presence of a few indifferent and ignorant clerics) and apostatizing from the faith. They also show large numbers of conversions from Catholicism to Orthodoxy and to Islam at the end of the sixteenth and beginning of the seventeenth century.

In fact, the visitations are a fine source to show that conversions took place in every possible direction, and that any study of Islamization that does not take all these other shifts of religious allegiance into consideration must be regarded as an oversimplified one. These documents and also one 1640 visitation refer to Moslems who are secretly Catholics. A visitation in 1655 speaks of the absence of parochial churches in Bosnia, and the great distances people must walk to attend mass. It also speaks of masses being served in cemeteries and private houses, and states that, just as in medieval times, there were no Catholic clergy in Bosnia other than Franciscans.

And finally, in the eighteenth century we have in one visitation a final reference to Patarins. In 1703 we hear of Greek Patarins in the region of Trebinje, who are clearly simply Orthodox believers, since their errors as described are no more than those Orthodox practices objected to by Catholics.(65)

It is also worth noting that the history of the area often was a matter of interest to the visitor, and he sometimes made references to it, usually drawn from fairly standard Catholic sources. One of these historical descriptions, since it quoted at considerable length the now lost chronicle of Pietro Livio of Verona, will be discussed when we come to speak of chronicles.

Chronicles

No narrative source from Bosnia has survived, thus all such sources that we have are foreign (Italian or Dalmatian) narratives. In addition, with the exception of the works of Pius II and Thomas Archdeacon of Split, our chronicles are not contemporary. Basically these later chronicles when they discuss Bosnia draw on one or more of three sources: a) Pope Pius II's works, particularly his *Europa,* published in 1494 and again in 1501, a considerable time after his death in 1464. He believed the Bosnian heresy was Manicheean. The basis for his view will be discussed shortly. b) Pietro Livio of Verona's now lost chronicle of Bosnia. His chronicle, we shall argue, was written about the third decade of the sixteenth century and was still extant and utilized up to the first half of the seventeenth century. He does not seem to have utilized Pius, and thus provides an independent source. He did not view the heresy as dualistic and tried to link it with the original non-dualist Italian Patarini. c) The Dubrovnik archives. As far as we can tell, these were not utilized prior to the second half of the sixteenth century. In the seventeenth century they were still used only limitedly. More extensive work based on them appeared only in eighteenth century works, many of which used the older chronicle form. If these early historians copied accurately, we can be grateful to them, since some of this material is no longer to be found in the archives. Chronicles whose information is drawn chiefly from these archives are discussed, as their data becomes relevant, in the narrative chapters.

Pius II

Pius' *Europa* has a brief discussion on Bosnia.(66) Here I plan simply to list the items about Bosnia in *Europa* so that when they reappear in other chronicles we shall know where they came from. a) He reports that the King of Bosnia, Stefan Tomaš (1443-61) was baptized by Gioanni Cardinal di S. Angelo. The Cardinal had been a legate in the Hungarian-"Slavonian" region in the late 1450's and we know he visited the Bosnian court in 1457. (The Cardinal was thus another source of information for the papal court on the situation in Bosnia, though none of his letters about Bosnia have been preserved.) This reference in Pius was a source of confusion for later writers, since it came to conflict with the letter of Eugene IV of 1446 which stated that Stefan Tomaš had recently been converted to Catholicism by Thomas of Hvar in the 1440's, and with Nicholas V's letter of 1447 saying that Stefan Tomaš had adopted Catholicism. However, the cause for the confusion was that only part of Pius' account was taken. In *Europa* Pius stated that the king, although he had adhered to Christianity, had abstained from baptism until he accepted it from the Cardinal. Thus, it is clear that the two sources do not contradict each other and it was simply a case of the king being baptized some time subsequent to his acceptance of Christianity.

b) "In Bosnia were many heretics called Manichees, who believed in two principles, the one evil the other good, they rejected the primacy of Rome and denied that Christ was equal to and of the same substance as the Father.

c) He then says that the heretics lived in monasteries in distant mountain valleys, in which women, who served the holy men, also lived. This story is repeated in several later works.

Pius also left some state papers and diaries which unfortunately were later worked over by a papal secretary named Campanus (d. 1477) who was Bishop of Teramo. Later these papers were printed under the title of *Commentarii,* under the authorship of Gobellinus. These were first published in 1584.(67) Gobellinus was simply a copyist.(68)

Since all the stories in *Commentarii* will be discussed in detail in Chapter VI, we shall not discuss them here. All that need be mentioned is that in 1461 three Bosnian noblemen had been sent to Rome where they renounced fifty "Manichee" errors. Whether they were really dualists or not, does not concern us here; but what is important is that Pius clearly believed that the three men were dualists; and the fact that the three renounced the fifty articles, seemingly with no protest, surely confirmed the pope in this opinion. Pius II in *Commentarii* and his *Europa* speaks without hesitation of the Bosnians as Manichees. Since Pius' letters prior to 1461 do not use the term "Manichee" and his two books as well as his 1461 letter about the three nobles do, it seems probable that the abjuration had a large role in leading him to this opinion or at least in causing him to examine Bosnian affairs with more interest or through different eyes. Of course it is possible that this opinion of Pius may also have been derived from other sources unknown to us, or that his humanistic interests may have influenced his outlook on the heretics by making him seek in them a classical heresy. But whatever the reason, it is clear that Pius by 1461 had come to the conclusion that the Bosnian heretics were Manichees and described them as such in his *Europa,* where he itemized some of their supposed errors and began by noting a belief in two principles, one good and the other evil. Pius' description of the dualist heretics in Bosnia was then repeated in the chronicles and history writing of the centuries that followed and laid the foundations for the still prevalent view that the Bosnian Church was a dualist organization.

The material about Bosnia in two other early and widely read histories came entirely from Pius. These two works are Raphaelo Volaterrano's *Commentariorum urbanorum* (Paris, 1510) and Marc Antonio Sabellico's *Rhapsodiae historiarum ab orbe condito Enneades* (2 volumes, Venice, 1494-1504).

One strange fact about Pius' writings, as well as other papal documents about Bosnia that were written in the period immediately after its fall, is that neither he nor anyone else seems to have questioned or drawn any information from Bosnians who fled to Rome and other parts of Italy. Other than possibly a rumour or two (e.g., the one about the betrayal of Bobovac by Radak),(69) not a single statement in letters, chronicles and histories written at the Vatican or elsewhere in Italy contains information that we might have reason to think should be traced back to a Bosnian emigré.

Orbini and Pietro Livio of Verona

One of our basic chronicle sources ought to be the now lost chronicle of Bosnia of Pietro Livio of Verona. Our dating and discussion of its contents must come entirely from the passages from it used by later writers who copied from it. It is fortunate that Orbini (1601) quoted extensively from it.(70) Peter Masarechi's report about his apostolic visitation to Bosnia and Hercegovina in 1623-24 also made use of it.(71) Masarechi's account condenses much of the material and omits many details which Orbini gives; with one exception every passage from Pietro Livio used by Masarechi also appears in Orbini. Luccari in 1605 also used Pietro Livio but he drew many fewer items from him, and so greatly compressed that material that it at times seems somewhat garbled.(72)

Near the beginning of Orbini's narrative on heresy in Bosnia, he states, "as Pietro Livio Veronese wrote" and near the end of his narrative on the subject he cites Volaterrano and Sabellico. The question that concerns us first must be, how much of Orbini's material came from Pietro Livio? An examination of the relevant parts of Orbini's text suggests a solution. Orbini's acknowledgment of his debt to Pietro Livio begins with paragraph two below.
His text:

1. "At that time there were in Bosnia many heretics and especially the Patarins. Hence the Roman Pope, who was Clement VI, in 1349 [sic 1339] ordered to the realm of Bosnia some Franciscans, men of sainted life, among whom were Brother Pellegrino [sic Peregrino] and Brother Gioanni of Aragon to extirpate from their midst the disease of so much heresy. Those by whom it was first introduced into the realm, it would not be inappropriate to speak of here.

2. "The heresy then of the Patarins of Bosnia had its origin (as wrote Pietro Livio of Verona) in the 'Paterno Romano,' who was ex-

pelled from Rome with all his followers and subsequently from all of Italy. And thus, having been driven out and not finding any place [of refuge], nor being received, they passed on through Friuli to those regions of Bosnia where some of them settled and others penetrated further on to Thrace and settled along the Danube not far from Nicopolis.

3. "And they lived without sacraments, without the sacrifice [of the Mass] and without a priesthood, though they called themselves Christians. They fasted on Fridays, observed the days of Our Lord and all Christian celebrations [Feasts], especially the Ascension of Our Lord; they were not baptized,(73) abhorred the cross and were called 'Paulichiani.'

4. "They persisted in these errors to the beginning of the last war between the Empire [Austrians], the rulers of Transylvania [Zapolya] and the Turk [Suleiman]. But perceiving themselves attacked by the Christians and led into capitivity as if they were Turks, they therefore resolved to embrace the true cult of Christianity. There were in these places fourteen villages inhabited by these Paulichiani. Some Greeks believed that these had been followers of Paul of Samosata, [the Greeks being] deceived by the name. But the Greeks are in error. These Paulichiani are far from the errors of Samosata. And I believe that, as those of Bosnia are called Patarini [Peterini?] in allusion to the name of St. Peter, so those are called [Paulichiani] in allusion to St. Paul, the two apostles and patrons of Rome.

5. "Then to turn to our region of Bosnia, when the above-mentioned Franciscans came to this realm. The affair succeeded contrary to their expectations, because they had suspected that Ban Stjepan, who held the Greek rite and therefore did not give obedience to the pope would oppose them, but he did the exact opposite. Having received them with great kindness, he gave them permission to teach publicly against the aforesaid heretics and introduce the Roman faith, for it seemed to him that it was better to have in his realm men of the Roman Catholic faith, which differed little from the Greek rite, than the heretics who were opposed to both Greeks and Latins. In this the Franciscans were greatly aided by Domagna di Volzo Bobali, canon of Ragusa, man of letters, and of very exemplary life, who found himself near the ban, holding the office of chief secretary, and persuaded him to abandon the Greek superstition and embrace the Roman rite.

6. "And the ban being at war (as is said) with Emperor Stefan Nemagna [sic Stefan Dušan] and Nemagna, wanting to have the ban in [the palm of] his hands treated secretly with Domagna and with some other Bosnian barons, to whom he promised money and position [or

rank] in his kingdom and in that of Bosnia if they served him in this matter. But Domagna immediately refused all offers, saying that he was bound to stick by his lord by the love he bore for his country and the splendor of the Bobali family, into which he was born, and that forbade him to do that [i.e., betray his master]. Whereupon immediately he [Bobali] advised his lord of what Nemagna plotted. For this the ban was taken with great affection for him and granted him many privileges, that still are to be found in the house of Bobali in Ragusa. The ban never let Bobali depart from him during [the rest] of his life, frequently affirming in the presence of his barons (as is seen in the said privileges) that he owed his realm, nay even his life to him. This man was (as has been said) in great part the reason that the before-mentioned Franciscans obtained access into the realm of Bosnia.

7. ''Who with great fervor and spirit disputed with the heretics and thoroughly convinced them, and they attracted all of Bosnia to themselves. In Bosnia, with the help of those who embraced the Roman faith, they raised many monasteries and convents. They worked also in Usora, Hum and Ston with the permission and [good] will of the Ragusans who (as is said) had become overlords of that place [Ston]. And they converted and baptized many of these heretics.

8. ''Whence everywhere and by the fame of their goodness and of the works that they did, many people of sainted life flocked to the vicariat of Bosnia, as was called the main place, where they have stayed up to this time. The above-mentioned Fra Pellegrino was made Bishop of Bosnia after having converted the Patarin heretics; of whom there was another sort in Bosnia called Manichees (as Volaterrano and Sabellico relate it) which had monasteries, placed in valleys and in other remote places, to which the matrons having been cured from illnesses were accustomed to come and according to their vows to serve a certain predetermined period of time. Thus they stayed with the monks, or more accurately heretics. This lasted to the year 1520. The abbot of these monasteries was called ded, and the prior stroinik.

9. ''The priest, when he entered the church holding in his hand bread, would turn to the people and say in a loud voice, I shall bless it. And the people would respond, bless it, and then he says, I shall break it. And the people reply, break it, and when this was done the people receive communion.''(74)

Masarechi's one departure occurs at the place marked by footnote 73. Instead of saying that the Paulichiani were not baptized as Orbini does, Masarechi describes a baptismal rite. ''On the second day of Epiphany the priest, elected by the people without any ordination [orde] each year, sprinkles them with water going round about [among them] and by this act they think they are baptized, as the Pauliani of today who are not yet converted still do."(75)

Of Orbini's text, how much comes from Pietro Livio? The first paragraph (n.b. all paragraph divisions are mine) could be from him or from any number of Franciscan sources. Its information is not new. Paragraph two attributes itself to Pietro Livio. Since paragraphs three and four are a direct and natural continuation of two and since they contain completely original data, there is little doubt that Orbini drew them also from Pietro Livio.

Paragraph six, and most probably all of paragraph five (since it contains material not found elsewhere) come from the documents found in the Bobali archive in Dubrovnik. Unfortunately this material is no longer extant. We have no way of knowing whether Pietro Livio or Orbini himself had read the Bobali documents. Thus we cannot say whether the Livio material ends after paragraph four, or whether it continues through paragraph six. Unless the Bobali family had forged these documents to glorify its ancestor (who except for a last will dated 1348(76) is unknown in other sources), we can treat the material in these two paragraphs as evidence coming from a contemporary primary source.

Paragraph seven contains nothing not found in any number of different Franciscan chronicles. Since there is nothing original here it does not make much difference to us whether it came from Pietro Livio or not.

Paragraph eight finds Orbini referring to the Pius tradition of Manichee monasteries, which he attributed to Volaterrano and Sabellico. However, since Masarechi gives paragraph eight in the same words as Orbini, either Orbini drew the material from these two humanists through Livio as a middle-man, or else Masarechi besides using Livio also had Orbini before him and copied this paragraph from Orbini.

That Masarechi did have Livio in his hands and did not take everything from Orbini is shown by the story he tells about the Pauliani's rite of baptism which Orbini does not give. This description probably appeared in the original but was dropped by Orbini in the interests of consistency since a description of a baptismal rite by a priest contradicts the statement at the beginning of the paragraph about the Paulichiani having neither priests nor sacraments. Orbini, also, may have heard elsewhere that the Patarini did not baptize themselves.

Paragraph nine in Orbini probably also came from Livio, since the material contained in it is original and since in manner of presentation the description of the communion rite resembles the Livio story given by

Masarechi about baptism.

Now to date Livio. Livio in paragraph four says that the Paulichiani persisted in their errors until the beginning of this last war between the Emperor (Austria), Transylvania and the Turks. Since Transylvania was involved it seems probable to assume that Livio was thinking of the campaigns of the 1520's which came to a close with the truce of 1532. This dating is confirmed by the fact that Orbini later says the heresy (which of the two currents is not made clear) persisted until 1520. If we connect this date, which has puzzled historians for some time, with Livio's description of events surrounding this war, it would explain how Orbini could come up with such a specific date for something as impossible to be concrete about as the end of a heresy. And by the phrase "until the beginning of this last war" it is suggested that the war had occurred not too long ago. Therefore, we have some, granted not conclusive, evidence that Pietro Livio of Verona was writing at some time not long after the 1520's.

Having praised Pellegrino's work in converting Patarin heretics, Orbini goes on, "of whom there were another sort in Bosnia called Manichees," who according to Volaterrano and Sabellico lived in monasteries (and Orbini gives an abridged account of the item in Pius' *Europa* about these mixed monasteries in remote places). Then he states that this lasted until the year 1520. Either Orbini or Livio clearly had noticed the differences between the material presented up to this point and the Pius tradition; this led whichever author it was to try to distinguish a second sort of Patarin, the Manichee. Following Pius' story which was taken second hand from Volaterrano and Sabellico, Orbini dates the ending of the Manichees (at least such is the context) in 1520, presumably the date of the wars he spoke of earlier. But if the dating is based on the wars, as I think it is, then Orbini had already mixed up material pertaining to the second heretical current with that which pertains to the first. It is also notable that Orbini or Livio, whichever one was the first to cite the two humanists, omits the passages from Pius and Volaterrano which speak of the two principles and actual dualism, and takes only the word "Manichee" and the story about the monasteries which may have originally been drawn from the Patarin(non-dualist) hižas and which has nothing dualist about it. This suggests that the author of this passage, despite the Pius-Volaterrano statement, did not believe the Patarins were dualists. Orbini then continues with an interesting hodgepodge from which it is hard both to separate the two currents of heresy and also to determine his sources.

Before evaluating the material we have presented, it would be worthwhile to look at the other data Orbini's history gives about religious conditions in Bosnia, and the sources from which he obtained his data. a) Ban Stjepan died in 1357 (actually 1353) and was buried in the Franciscan church of St. Nicholas of "Milescevo." (p. 354) (Source unknown, the matter will be discussed in Chapter IV.) b) He later speaks about Stefan Tomaš, the king, and after a few biographical details, uses Pius' material, though crediting it to Volaterrano, about the king having been a Christian a long time, who abstained from baptism until finally he was baptized by Gioanni Cardinal di S. Angelo. He then adds that the king had previously been infected with the Manichee heresy. He then uses a Franciscan source to the effect that Jacob de Marchia had converted the king. (p. 368) This last item came from a legend recorded by Wadding about the conversion in the 1430's of Tvrtko II, and is only important to show that Orbini clearly did use Franciscan chronicles as well. Thus much of the material about the Franciscan mission given by Orbini could have come from Franciscan sources rather than from Livio. c) Orbini then repeats the story from *Commentarii* that in 1459 (date accurate, confirmed by Pius' letter) King Stefan Tomaš, to show his loyalty to the pope, gave the Manichees, of whom there were many in Bosnia, the choice of baptism or exile. Two thousand were baptized and forty, remaining persistent in their errors, went to Stefan, the Herceg of St. Sava, and a partisan (as some would want) of that sect. (p. 369) Orbini, thus, was not fully convinced about the question of the herceg's religious loyalties. He then speaks of the three leaders of the heresy (although he should have said three leading nobles who were also heretics, for there is no reason to look upon them as religious figures) who renounced their errors in 1461 and gives that story in the form that it appears in *Commentarii.* (p. 369) He also tells of the mission of Tomašević's legate to the pope in 1461 (p. 372) and of the betrayal by Radak Manichee, (p. 375) both of which were taken from *Commentarii.*

Finally, he concludes with some interesting data on the Bosnian Church leader Radin Gost (d. 1467) (since we can assume that it is really Radin whom Orbini is speaking of when he refers to Radivoj Gost and Rasi Gost). a) that Gost Radin built Mostar (p. 384) (possibly a popular tradition; the source is unknown); b) that he was major domo for Herceg Stefan. (p. 384) (This remark which reflects the important role that Radin occupied at the herceg's court could have come from a variety of sources); c) that the herceg was schismatic (p. 385) (which is true if we do not insist that this has to mean the herceg was Orthodox. Dubrovnik documents frequently refer to Herceg Stefan as schismatic); d) that Radin was a monk of the order of St. Basil. (p. 388) (This statement, if true, would have made Radin an Orthodox monk. The source is unknown, and may have been based on something that made Radin appear more Orthodox than heretical. There is no other evidence to make Radin Orthodox); e) that Radin was confessor to Herceg Stefan. (p. 388) The source for this is unknown, but from statements "c" and "d" it is apparent that Orbini thought that both the herceg and Radin were Orthodox. We may suggest that Orbini found in the Dubrovnik archive a copy of the herceg's will, which states that Radin played an important part in its composition. This fact may well have led Orbini to ascribe to Radin the role of "confessor" and since the Orthodox Metropolitan David from Mileševo was also a witness, Orbini may have inferred from his presence that the two were of the same faith and Radin was a Basilian monk. In fact, points "c," "d," and "e" could all have been inferred from Herceg Stefan's will.

With the possible exception of some of the material about the Franciscan mission (which is not original and could have been drawn from any number of sources), and the few remarks about Gost Radin, all of Orbini's information comes either from the Pius tradition *(Europa,* through the medium of Volaterrano and Sabellico, and *Commentarii),* from Pietro Livio of Verona, or from the Bobali tradition. The account of the Orthodoxy of Ban Stjepan Kotromanić, Bobali's role in the ban's conversion, and the ban's motives for allowing the Franciscans to come to Bosnia most probably are from the Bobali archives (possibly with Livio as intermediary). Unless the Bobali material was fabricated later to glorify Domagna for family interests, we should regard the Bobali material as a contemporary source. However, since we do not have sufficient evidence to reject the Bobali tradition we shall make use of it, though keeping in mind its reliability is not proven.

The story of the origin of the Patarini seems farfetched and probably is little more than a scholarly attempt to explain the name Patarin. What is important here though is that Pietro Livio did not link the term with the well-known Italian dualists called Patarini but traced the term back to its less well-known earlier non-dualist roots, which were actually in Milan rather than Rome as he states. Pietro Livio's avoidance of the more obvious solution suggests that he had some reason to believe that the Bosnian Patarins were not dualist, and for that reason made a particular effort to find an explanation for the name Patarin that was not associated with dualism. In addition, in his description, he mentions little, and most of that he contradicts later, that might be considered dualistic about them, and when he used the material from the Pius tradition he took the material about monasteries and left aside the material about two principles. Thus it seems as if Livio, writing (ca. 1530) at a time when Pius' Manichee view was generally accepted, did not believe that the

Bosnian Patarins were dualists and went out of his way to avoid saying that they were. We only wish we knew what the reasons were that led Livio to this conclusion. We can guess that he came to this conclusion because his Paulichiani were not dualists — but unfortunately we do not know why he linked them with the Bosnian Patarins.

What Livio says about the baptismal and bread-breaking rites is interesting, though we cannot confirm it and do not know what his source about these rites was. We may suspect that Livio observed the baptismal (and possibly also the bread-breaking) rite firsthand among the "Paulichiani" about whom he speaks as if he actually had some acquaintance with them. One may suspect that he had actually traveled among them, possibly as one of the Catholic clerics involved in effecting their conversion, and for that reason knew of the fourteen villages that he mentions (and hence could give such a specific figure) and knew something of their rites. However, since he links their conversion with a war involving Transylvania we must suppose that the Paulichiani lived somewhere in or near Transylvania, or possibly still around Nicopolis, which is near. Since the only links between the Paulichiani and Bosnian Patarins given by Livio is their alleged common origin, we must question the relevance of Livio's material on the Paulichiani for Bosnia. What evidence did Livio have that the two groups were of common origin? And even if they were, their alleged split had occurred two or three hundred years before Livio observed the Paulichiani early in the sixteenth century. It is quite likely that during those centuries the Paulichiani's practices and customs had changed considerably from what they once had been. These drawbacks combined with the fact that other sources show that the Bosnian Patarins had a priesthood and certainly by the fifteenth century did not abhor the cross, show that we cannot rely too heavily on Livio's material here.

Livio (or Orbini) claims that the Greeks (presumably the Orthodox) tried to associate these Paulicians with Paul of Samosata. This again smacks of scholarship; the noted Paulicians of Tephrice (some of whom were transplanted by the Byzantines to Bulgaria and Thrace in the ninth century) were also at times linked in tracts by their enemies with Paul of Samosata. We do know that small communities of non-dualists called Paulicians continued to exist in Bulgaria, (generally thought to be descendents of the Armenian and Byzantine Paulicians) who accepted Catholicism in the seventeenth century.(77) Could there have been more of them on the Danube? As interesting as the material that Livio gives us about this deviant group is, we cannot make use of it in our study since if a link existed between them and the Bosnian Patarins, Livio (or Orbini) fails to tell us what his source for that link is.(78)

Orbini (most probably not Livio) presents the view that the titles of the hierarchy of the Bosnian Church were derived from monastic offices. Though he misuses the titles in the example he gives, his general theory seems valid and is corroborated by other sources that speak of the church hierarchy. It is probable that Orbini found this material in the Dubrovnik archives.

Luccari and Resti

Luccari compiled a history of Dubrovnik that was published four years after Orbini's in 1605. He made use of three no longer extant chronicles about Bosnia. In addition to Pietro Livio, he made use of the Chronicle of Hrvoje Vukčić by Emanuel the Greek and a chronicle of Bosnia by Milich Velimiseglich.(79) It is only through Luccari that we know that these last two works existed. The citations Luccari gives from these two tell us nothing about religious matters. His extracts from Pietro Livio, far briefer than Orbini's, are highly condensed, often garbled, and at times contradict the longer straightforward versions presented by Orbini and Masarechi. Luccari claims that Ban Stjepan's land was infected with the Patarin heresy and that the Patarins were disciples of Paul of Samosata (which is just what Orbini says the Patarini were not) and believed in two principles (which is not in Orbini, but is in agreement with Pius' tradition) mixed with much from ancient philosophy and poetry (source?) believing that divine providence at the time of the creation of mankind ordained once and for all for everything, and after that nothing could be altered.(80) Luccari's source for attaching these views of strict predestination to anyone is unknown.

We know he examined the archives of Dubrovnik, for he cites, though inaccurately, from the 1433 letter Dubrovnik sent to the Council of Basel about the Patarin hierarchy. But in speaking of the letter which called the Bosnians "Patarins," he calls them Nestorians and claims erroneously that some of the Bosnian Church hierarchy tried to attend the Council of Basel. Whether Luccari misread the original document or had in his hands an inaccurate copy, we cannot say. Besides this material, Luccari also calls Gost Radin the Major Domo of Herceg Stefan. However, Luccari accurately calls him a Patarin, but adds that he was an enemy of the Roman Church. This judgment is not confirmed by the good relations Radin had with Catholic Dubrovnik. Luccari also has material from the Franciscan tradition: the murder by his son and brother of King Stefan Tomaš and his burial at the Franciscan Monastery at Sutjeska in 1461, and the legend that the Bosnian bishops in the twelfth century resided at Kreševo — a claim seen frequently in Franciscan writings of the seventeenth through the nineteenth century but which has no basis in the sources. Finally, Luccari provides an interesting description of the hair-cutting *kumstvo* (godfathership) when the town of Dubrovnik, to seal a peace treaty, stood as godfather for the Bosnian nobelman Radoslav Pavlović's son Ivaniš. This information presumably was drawn from the Dubrovnik archive. Since much of Luccari's data contains inaccuracies, it seems that he did not use his sources carefully. The modern scholar must use him with care.

An early eighteenth century history which contains much valuable data is the chronicle of Dubrovnik, by Junius Resti (1669-1735).(81) Most of his material seems to have come from the Dubrovnik archives. The majority of the documents that he relied on for the late fourteenth and fifteenth century still exist and we can see that he copied them with accuracy and did not invent facts or names. For Bosnian religious history Resti is most useful for the thirteenth century since some of his material (particularly about Bosnian bishops at the beginning of that century and their relations with Dubrovnik) can be found nowhere else. These descriptions, since they usually reflect ties between Bosnia and the Archbishops of Dubronik almost certainly also came from original Dubrovnik documents that existed in Resti's day but that have disappeared since then. Of course, we have no way of knowing what sort of documents he used for the thirteenth century, and whether they were reliable sources. But judging by his use of later documents, we can at least conclude Resti copied them accurately.

Resti's chronicle, which ends in 1451, was continued by the "Croniche ulteriori di Ragusa" which most probably was the work of Giovanni di Marino Gondola (which Nodilo includes in the same volume with Resti's text.) This chronicle, like Resti's, is based on material from the Dubrovnik archive.

Western 13th Century Chronicles

Three thirteenth century chronicles — two English (Roger of Wendover and Michael of Paris), giving one text, and one French (Theobald of Rouen) with a variant version — preserve a no-longer extant letter written in 1223 by Conrad, Papal legate to France.(82) This letter describes a Cathar anti-pope. In one chronicle the anti-pope appeared in France himself and in the other he sent a deputy. In either case, the man who came to France was called Bartholomaeus. He was received with great honor by the Western Cathars, and was allowed to settle their disputes and to install bishops. Whether the anti-pope came in person or dispatched a deputy, and what he did in France, is of little importance to our study. Yet the letter says that he came from "in finibus Bulgarorum, Croatiae et Dalmatiae, juxta Hungarorum nationem." Just where does this mean? Two theories have been advanced: a) Bulgaria since it is the first place named and was also the home of the Bogomils, and b) Bosnia, since Bosnia seems to lie in the midst of all the regions listed.

For several reasons I am sceptical that Bosnia could have been the home of a dualist anti-pope in 1223. We have no evidence in either foreign or Slavic sources that the Bosnian dualist movement had by 1223 grown to significant size or reached a level of real importance. In addition, Bosnia does not border on Bulgaria. Serbia and Macedonia (another dualist center) do, but then neither of those places borders on Dalmatia. Thus we must conclude that either our source had in mind a wandering heresiarch who had no permanent see and who had a following in this whole region, or else our source was confused about Balkan geography. If the second explanation is correct, we cannot find a basis in its confusion to come to any conclusion about what it may have meant.

Franciscan Chronicles

There are a large number of Franciscan chronicles, most of which borrow from other chronicles, and contain no original material of their own. Many of them were kept by particular monasteries, which began to keep records only in the seventeenth or eighteenth century; hence all information in them about earlier centuries is simply copied from other Franciscan histories. Besides the Franciscans' own material, we can find in their chronicles some letters and documents connected with their missions and papal letters to their order. The bulk of these letters we have elsewhere. And if we do not have the text elsewhere, we have reason to wonder whether the letter is authentic, and thus cannot freely make use of it. While some of their chronicles utilized works such as Pius' *Commentarii* or later histories, whose original texts have been preserved, none of them has material from the lost chronciles and histories. The only original materials to be found pertain to the order and its work; quite often, such material needs to be treated with reserve.

There are three general categories of Franciscan chronicles: a) Chronicles of particular Bosnian monasteries, such as Fojnica and Sutjeska.(83) These chronicles being later compilations are almost worthless as sources for the medieval period. They repeat the story of the murder of King Stefan Tomaš and his burial at Sutjeska. Fojnica's chronicle reprints a forged charter to Knez Radivoj and contains some information about the destruction of monasteries in the sixteenth century by the Turks.

A Bosnian chronicle of interest is that of Fra Nikola Lašvanin, another eighteenth century work.(84) Lašvanin takes older material and makes a readable yarn out of it. Under 1319 he utilizes the letter of Pope John to Mladen Šubić(85) and elaborates; he makes the "heretics" referred to in the letter into Manichees and Patarins, and claims that they lived in the mountains and that some Dalmatians revolted against them (source of this remark unknown). Then the pope wrote the ban to take action, and Lašvanin proceeds to describe the action against them by some German crusader knights to force baptism upon them (p. 37). Later Lašavanin takes a story from a renaissance history of Split by an anonymous author which describes hugh fires in Bosnia which melted the rocks and leveled mountains, after which in those regions it became possible for settlement and Bosnian Patarins, called Manichees, settled there and said that God burned the mountains for our use because he liked our faith. Lašvanin repeats the story leaving out the pro-Patarin moral.(86) Lašvanin also records the tale from Pius' *Commentarii* of the choice given the heretics by the king of conversion or exile. Finally, he mentions old believers *(starovir'ci)* in the Vrbas region in 1737 and 1739. (pp. 76, 78) Unfortunately nothing is said about what sort of beliefs they held.

b) The Chronicles of the Bosnian order: The most important text was published by Fermendžin,(87) whose valuable introduction shows what letters, chronicles, and documents were utilized to make this eighteenth-century compilation. The chronicle describes the foundation of the permanent Franciscan mission in Bosnia in 1339, and the founding of the vicariat, shortly thereafter; it names the various vicars and gives as accurately as possible their dates. It does not tell us about relations between the Franciscans and the Bosnian court and nobles or give details on where the Franciscans worked or on their relations with the populace. In this chronicle too the term "Bosnia" usually refers to the whole vicariat. It contains only one reference to Manichees, under 1411, where it refers to the thirty years of work against them in Bosnia by a Fra Ambrosius. It gives limited data on specific churches, but one must inquire at every turn whether the dates are based on solid documentation or on the traditions of a given church.

c) General Franciscan chronicles: The best collection is the multi-volume work L. Wadding, *Annales Minorum,* which began to be issued in the eighteenth century. Bosnia plays a very minor role in this large compilation, appearing only here and there in scattered references. A few letters whose texts have not been preserved elsewhere, which must be used with care, are found in it.

Defters and Turkish Laws

A very valuable series of sources, though of course written after the fall of Bosnia are the Turkish tax records called defters, recently discovered and mostly still unpublished. The earliest covers Ottoman-occupied parts of Bosnia in 1455. Then after the main conquest, follow defters of regions of Bosnia and Hercegovina from 1468, 1475, 1485, 1489, 1516, 1528-30, 1529, 1542, 1561, 1575. The originals are to be found in Ankara and Istanbul. The Oriental Institute in Sarajevo has microfilms of them all. A summary article of the type of material to be found in them about the Bosnian Church, with a great many specific references (though unfortunately not by any means a complete compilation of the relevant material) has been published by Okiç.(88) Until all the material is available, conclusions based on them should be delayed.

The Turkish defters have numerous references to "kristian" land, "kristian" villages, etc. Since the defters call the Catholics and Orthodox "*gebr*"or "*kafir,*"and since twice moreover *gosts* are referred to in connection with "kristian" lands, we take the designation "kristian" in the defters to mean members of the Bosnian Church. The defters were drawn up later in the fifteenth or even in the sixteenth century; so most of the kristians referred to were previous owners who no longer owned the lands when the surveys were made.

It is a matter of dispute whether the Turks used the term "kristian" in the defters to designate ordained clerics or simply believers in the church. Since the Turks were interested only in drawing up a tax census to facilitate tax collection, they presumably did not care whether an owner (or past owner whose name had been given to a piece of land) was a lay follower of the church or an ordained cleric; so we have no reason to expect any precision or consistency in their use of the term kristian.

It is apparent that when the defter lists the number of households and unmarried males in a "kristian" village, the defter is describing peasant households (married couples and unmarried sons of adult age) and thus is using the term for lay members (believers) in the Bosnian Church. In some cases the defters may have said "kristian" villages because the peasants lived on and worked land belonging to the Bosnian Church.

I believe that in most cases when a man is named with the term kristian as his title (e.g. Kristian Radoslav's land), it refers to a Patarin monk, as is the case in all Slavic documents. Presumably the Turkish census taker would have written down names and titles in the form given him by the local inhabitants; hence when we find the title kristian preceding a name, we can assume it followed local usage and means an ordained cleric. Such land could either have been the kristian's own, which he continued to farm after ordination, or else land which the Bosnian Church had assigned him to work after his ordination.

When the term *baština* (hereditary land) is associated with Krstjanin So-and-so,(89) we may assume that these were the personal estates of the krstjanin which he retained after ordination. Since some of these seem to have been extensive estates, it indicates that some well-to-do Bosnians became ordained clerics in the church. Unfortunately, we do not know what became of these lands after the krstjanin died; were they passed on to other members of his family or did they become estates of the Bosnian Church? Since the krstjanin's name is associated with such lands, rather than a later member of his family, we may suspect that frequently the lands went to the church — be it with his ordination or be it after his death — but we cannot prove this point.

In addition to its specific meaning described above, on certain occasions, "kristian", as a title, seems to have described laymen. Okiç cites several cases of "kristian families," seventeen from two villages in the 1469 register with no names given, and seven from the village of Rastoka in the Nahija of Samobor from the 1477 defter, who are named: Jelonja Kristian, Raduh Kristian, Radeljko Kristian, etc.(90) Here is a case, then, of men, bearing the title "krstjanin," heading families. The obvious explanation, and the one which I accept, is that these men were lay members of the Church who headed households; thus this would be an example of the term "krstjanin" being used as a title for a non-cleric.

It is, however, within the realm of possibility that the seven kristians were ordained but remained heads of households *(zadruga(s));* they might have been ordained late in life after establishing a family of which they still remained head, or possibly they, despite being celibate, were the oldest living males of families (composed of nephews with families) which they headed. One Ragusan document refers to Gos Patarin (probably Gost Gojisav) and his "family"; "family" may well have had a broader meaning — be it nephews, etc., one's retainers or servants, or even the monks under his care — than one's wife and direct descendants. Thus it is possible that these seven kristians were monks who had "families" in such a sense. However, since it is odd to find seven examples of what seems to be an exceptional situation existing in one village, I retain my belief that these kristians referred to are laymen. We hope that the publication of more material from the defters will enable us to resolve this problem.

When the defters speak of "kristian lands" or "kristian villages" in a general way with no individual's name given, I think that these were lands or villages which belonged to the Bosnian Church, which may have been farmed by the monks or by peasants who turned over a share of the produce to the monasteries. When a village is referred to as kristian and then the number of families and of unmarried males is given this would have been a village, belonging probably to the church, farmed by regular peasants. Of course since the defters are documents that refer to the situation at the time the census was taken, we cannot be sure that at some earlier date monks had not farmed that land or lived in the village. These defters then suggest that the Bosnian Church owned and farmed villages. We have no contemporary Slavic charters or other sources that state this, but since the Catholic and Orthodox Churches both owned villages, we have every reason to expect that the Bosnian Church, a continuation — as we shall see — of a Catholic monastic institution, would have maintained this custom.

Since there is a definite correlation between the locations of the kristian lands mentioned in the defters and the locations of the hižas we come across in Bosnian and Ragusan sources, we may also suggest that this church land was held by the individual hižas. It is also apparent that in many cases the ordained monks actually worked the lands themselves. For example, we hear of farms "cultivated by kristians."(91) This supposition is confirmed by a reference in Gost Radin's will to krstjani and krstjanica kmeti (Christian peasants of both sexes).(92) As a leader of the church, Radin clearly would have used the term "krstjanin" in the strict sense of an ordained monk. That Radin spoke of the ordained man as a "kmet" makes it clear that the monk also farmed.

The defters help us locate communities of krstjani (ordained or not), thus providing the basic material for a map of their settlements. The defters also give us a limited amount of data on the locations of churches. We also learn about the size of the communities and the extent of their land holdings. Thus we impatiently await the publication of all the defters.

Finally, the defters are a vital source on the conversion of Bosnians to Islam. By comparing the records of the same villages over a period of time we can see the numbers and rate of conversion.(93) Of course we must always keep in mind that migrations also occurred. Thus, if village "a" in 1489 had two Moslem houses and twenty-one Christian houses and then in 1516 twenty-five Moslem houses, we cannot be sure that the twenty-one Christian families had all converted. It is possible that they had fled, and that the Turks moved in an equal number of Moslem families to farm the land. In addition, one must be careful, since in one defter two villages will be joined together and in the next be separated or put into other combinations of villages. But overall we do get an idea of the rate with which an area was becoming Moslem, even though we cannot always say that it was a result of conversion. The defters do not show changing of confessions from one Christian group to a second. The defters will also be a valuable source on migrations. However, for this they will have to utilized with great care.

Since personal names are rarely given, figures remaining constant between two defters can either mean a stable population or a total changeover of population or anything between. However, when in 1468 we find many abandoned villages, and in 1485 we find these same villages populated, we can safely postulate a migration of population, though we rarely learn where the new arrivals came from. In the defter the term "house" means "married couple"; thus defters give no information about the size of the *zadrugas* at the time. The Turks were interested in collecting a basic tax from each married couple and that was all they recorded. They were not concerned with how many people lived under one roof.

In addition to the defters we have the Turkish law codes: *Kanuni*.(94) Here we have the laws which specify that Christians are not allowed to build churches where there had been no churches previously, and if a new one is built, it should be destroyed. This law is repeated in *kanuni* of 1516, 1530, 1539 and 1542.(95) In addition, though the laws do not distinguish between Catholic and Orthodox, we know that the Orthodox, presumably with Turkish permission, built several churches and monasteries in Bosnia during the period of the laws. That the laws were generally enforced is testified to by the Austrian envoy Kuripešić, who, referring in 1530 to the sad plight of Christians, says that they not only could not build churches, but they could not ever repair them.(96) However, from charters given to certain Franciscan monasteries in Bosnia we know that permission was given at times on request to repair or even to rebuilt an existing church, provided that its dimensions did not exceed those of the original church. From this we can use these laws to show that almost every Catholic church we find in visitations and other sources for the first century after the conquest, must necessarily have existed prior to 1463. This then is useful to show where Catholic churches, presumably having at least some believers, had existed in the medieval period.

The Gospel Manuscripts

Several Gospel texts from medieval Bosnia have survived. Three of them, on the basis of their dedications, can clearly be tied to the Bosnian Church: a) The fragment of a Gospel made on the order of Tepčija Batalo in 1393 for Starac (elder) Radin. This clearly is a Bosnian Church elder, since on a second sheet is a very important list of the hierarchy of the Bosnian Church up to that time.(97) b) The Hval Gospel, written by Krstjanin Hval for Hrvoje Vukčić in the days of the episcopacy of the Bosnian Church of Djed Radomer in 1404.(98) c) The surviving parts of the Gospel of Krstjanin Radosav from the reign of Stefan Tomaš (1443-61) which includes the Apocalypse and a ritual.(99) In addition, several other similar Gospel manuscripts have been preserved and may well also be tied to the Bosnian Church.

Much has been made of these Gospels by scholars, some of it convincing, but much that is, to say the least, farfetched. What is included in them was clearly the "canon" of the heretical church, but we do not know what the texts meant to the Bosnian Churchmen and how they interpreted them. The Gospels included the whole New Testament (generally with slight changes in order, the most common change being the placing of the Apocalypse immediately after the Gospels.) This shows that the Bosnian Church attributed great importance to this book, but so did the Catholic West. In addition, the Gospel texts often included from the Old Testament, Moses' Ten Commandments; in Hval's Gospel they were placed in the middle of the New Testament. The absence of the rest of the Old Testament probably had little significance. It was a common practice in the Middle Ages in the West, also, to prepare New Testaments and Gospels without the Old Testament, which played a very small role in services.

The Bosnian Gospels were so close to the texts of the Orthodox Church that in several cases they were actually used later in Orthodox churches. The fragment of Batalo's Gospel has a later notation from 1703 referring to its presence in the Orthodox monastery "Skr'batno." (100) In the margin of the Hval text the words "začelo" (beginning) and "kon'c" (end) appear beside passages at places corresponding to the readings of the Gospel throughout the year in the Orthodox Church.(101) Presumably this signifies either that it was later used in an Orthodox church, or that the Bosnian Church used the Gospel text in the same manner as the Orthodox did in their church services.

These Gospel texts refute much that has been written about the Bosnian Patarins in the inquisition sources. First, the New Testaments are complete; thus there is no sign of mutilation of the New Testament.

We find religious pictures in them, which shows the Bosnian Christians were not iconoclastic. We find flattering pictures of John the Baptist and Moses, both of whom according to the inquisition were supposedly condemned by the "Bosnian Patarins." By including the Ten Commandments, the texts show that the Bosnian Church did not condemn all the Old Testament or the Old Testament Patriarchs. None of this proves that there were not other heretics in Bosnia holding these iconoclastic anti-Old Testament views. But it does show that the ordained members of the Bosnian Church connected with these three texts were not of this opinion. It has been pointed out that the art work in these texts is closely copied from Western Catholic styles, but this does not mean that the heretical artist would blindly copy without thinking of what the subject of the picture was.

Various scholars, particularly Solovjev, have tried to use the vocabulary of certain Gospels to demonstrate the existence of dualism.(102) Šidak has shown it is impossible to do this.(103) Solovjev had pointed out that in the Nikolski Gospel (a fourteenth or fifteenth century Bosnian Gospel the name of whose copier is unknown) it is said that Jesus "izide" (came out) instead of "rodi se" (was born). Šidak, however, showed that the two words are synonyms and in fact "rodi se" does occur in the majority of Bosnian Gospels, including the Bosnian Church Hval text. Solovjev tried to argue that the most frequently used term to refer to Jesus' sonship was "iločedni" which he linked to the word "ini" (other) and would connect with the Bogomil belief that Jesus and Satan were both sons of God. Šidak, however, demonstrated that "ini" can mean "only" as well as "other," and pointed out that in the Nikolski text the more common Orthodox form of "edinocedni" is used. Thus, there is no consistent use of the variant terminology in the Bosnian Gospels and it cannot be demonstrated that the words have the sense that Solovjev wished to attribute to them. To try to argue about theological subtleties on the basis of translations of particular (and usually difficult) words, when we consider the general level of learning and ability in translating at the time is a very dangerous thing to do.

In addition to the Gospel copied for him by Krstjanin Hval in 1404, Hrvoje was given a beautifully illustrated missal by Pop Butko "according to the Roman Law of the Divine Church of Sts. Peter and Paul in Rome." Thus, the same man had a Catholic missal made as well as a Bosnian Gospel. We can also mention the Miroslav Gospel, also a beautiful text, made for Prince Miroslav of Hum in the end of the twelfth century. It is Orthodox in form, and since there is evidence from Miroslav's family background as well as his church building that he was

Orthodox, there is no reason to take this Gospel as anything but Orthodox.(104) The conclusion from this, then, is that all the faiths (at least when we are dealing with Slavic rite Catholics) read and used the same books of the Bible in the vernacular language.

Finally, the Radosav fragments include a ritual in Slavic, which has many similarities to the Latin Cathar ritual from Lyons. Solovjev presents the two rituals as parallel texts.(105) The Cathar ritual is in four parts. 1) A litany composed of five invocations and responses. 2) The Lord's Prayer. 3) A second litany of five invocations and responses similar to the first. 4) A reading of the Gospel of John, I, 1-17. The Radosav ritual is in three parts. It lacks the initial litany and begins with 1) the Lord's Prayer. 2) A litany, which, excluding repetitions in the Cathar ritual, is a Slavic translation of the Latin recitation, with one phrase added in the Slavic ("he is worthy and just") not found in the Latin. 3) The reading from John, I, 1-17.

The Radosav ritual thus seems to be an abbreviated version of the Lyons ritual. But as Šidak has pointed out there is nothing in the Radosav text (or in the Lyons text either for that matter) which is heretical.(106) However, the likelihood that the Radosav rite was derived from a ritual similar to that recorded in the Lyons manuscript is great. It is even possible that this ritual had been brought to Bosnia by dualists, and later Radosav, a member of the Bosnian Chruch, recorded it. Since there is nothing heterodox in its contents, the recording of this rite does not make Radosav a dualist. In addition, we do not know whether Radosav's church ever performed this ritual. However, since the service is clearly a simple one, and since Bosnia was a land with a shortage of churches which required services at times to be held in private homes and cemeteries, it would not be strange if an abbreviated service should become popular. Thus, it is quite possible that the Bosnian Church — whether regularly, only on certain occasions, or only in some places — did use a ritual which had originally been brought to Bosnia by dualist heretics.

The Apocryphal Tales

Much has been made of ties between the Bulgarian Bogomils and apocryphal tales about biblical figures. However, recently scholars have shown that most all of these tales are far older than the Bogomils and date back to the first centuries of Christianity.(107) Thus, to try to discover Bogomil doctrine from them is very difficult, for we have no idea which parts the Bogomils found vital nor do we know how they interpreted them, which is very important since we have been told that they tended to interpret the scriptures allegorically to suit their doctrines. That these tales reappeared in great numbers in the years the Bogomil Church was active in Bulgaria is true, and very likely there is some connection between the revival of the tales and the hersey. However, it should be pointed out that in medieval Europe from the twelfth century on these tales were popular among all groups including the Jews. They certainly were no monopoly of the dualists.

Only one work, a hodgepodge of several other apocryphal works, entitled "Interogatio Sanctis Johannis" is definitely associated with the dualists.(108) Both copies of it were found in the West in Latin translation. In one is the notation that it had been brought (in the twelfth century) from Bulgaria by Nazarius, Bishop of the Cathar Church of Concorrezo, a prominent figure about whom we have considerable material in the inquisition files. The other manuscript, now in Vienna, refers to Bosnia in a marginal note stating, "Hell fire is neither spirit nor anything living but a place just like Bosnia (Bossina), Lombardy, and Tuscany."(109) We cannot connect the manuscript itself with Bosnia, but at least it seems to suggest that at sometime, probably in the thirteenth century (although we cannot date the marginal comment), there were dualists in the three localities.

This study will not discuss the Apocryphal Tales since we have no medieval manuscript of any of them from Bosnia. It is true that we find certain elements of their content in Bosnian folk tales (such as the Tale of John the Baptist and the Devil)(110) but we have no way of telling when and how these possibly dualist elements appeared. These themes all remained popular throughout the Middle Ages, and long after, when there were no longer any dualist Christians left in the Balkans at all. Since most of their content does not contradict orthodox beliefs, the tales were popular with the Orthodox as well. Most of the stories told tales of other events in the lives of biblical figures and thus supplemented rather than contradicted canonical texts. As such the tales were probably told and repeated by any group of people having any amount of contact with them. Thus, the thematic material from these apocryphal tales found in Bosnia in the eighteenth century and later could have been brought to Bosnia just as well by Orthodox believers who migrated there, either during or after the Middle Ages.

Since we cannot date the appearance of apocryphal motifs in Bosnia, or even prove by whom they were brought and what use was made of them in the past, it seems best to ignore them altogether in this study.

Traditions

A source that has been utilized from time to time has been the popular nineteenth century traditions about people called Bogomils in various parts of Bosnia and Hercegovina.(111) These stories usually concern families in more remote places who "to this day" perform secret Bogomil rites or who have recently converted from Bogomilism to some other confession. Fra Martić in the 1870's reported that in Drežnica there were still families who, though supposedly Moslem, preserved secret Bogomil rites in their homes and never went to the mosque. Later Ibrahim Beg Bašagić sent a hodja there, for the villagers had no idea what Islam was about and were performing marriages by themselves. These reported deviations probably reflect ignorance and local custom rather than Bogomil survivals.

Fra Martić also reported that somewhere on the upper Neretva there were still sixteen Bogomil families, who were Christians but refused to accept Franciscans or Orthodox priests but administered to themselves. The Franciscan, however, was unable to find out which families these were. In 1887 Mehmedbeg Kapetanović Ljubušak claimed that around Rama and Neretva there were Moslem peasants who had retained within their family circles some prayers from Bogomil times which they knew by heart. The Orthodox Metropolitan Sava Kosanović reports that in his time, in the second half of the nineteenth century, there were near Kreševo a few families who held secret Bogomil traditions but went to church and made the sign of the cross. He could not discover which families these were. Finally, Bakula in 1867 stated that in Dubočani near Konjic the family Helež had been the last Bogomils, but that they had accepted Islam a few years ago.

Truhelka and various other scholars visited these areas, and were unable to turn up anything that could be connected with dualist beliefs. Professor Andjelić told me in 1967 that recently he had tried to track down the Helež family in the Konjic region. He had found no family by that name but discovered there had been a second family with a similar name, Elez, who had been prominent in the area in the eighteenth century because they had maintained an enormous extended family of over a hundred members. Then at the end of the eighteenth century an epidemic of plague had wiped the whole family out. That such a great catastrophe to an important family should have entered popular tradition is not strange, and that it should mingle with a second genre of tale within a period of over sixty years is also understandable.

Yet what should be pointed out at once is that, whether or not there were dualists akin to Bulgarian Bogomils in Bosnia and Hercegovina, the Bosnian heretics were never called Bogomils. The word is attested only once for the whole region of Bosnia, and that is about the border town of Srebrnica in a Serbian source (Konstantin Filozof's life of Despot Stefan Lazarević written in 1431); and the reference appears in only one of several manuscripts. This manuscript has been discussed under "Native South Slav sources" and is clearly a questionable one. Even if accurate it only shows that Serbs called the heretics in Srebrnica by that name, and not that the Bosnians there or elsewhere ever used it.

Clerics of the Bosnian Church called themselves *"Krstjani,"* and their neighbors and local opponents called them "Patarini." It is notable that neither the word "Krstjani" nor "Patarin" appears in any traditions, for if a tradition truly went back to medieval times we would expect it to retain one of these two terms. We conclude that the sources for all the popular traditions which use the term "Bogomil" are either the history books using the term from at least the late eighteenth century or the spoken word of priests and of the literate few who were able to read them. The villages considered to have Bogomil believers were usually off the beaten track and may well have had customs that in some way differed from the general customs of the area (such as villagers performing their own marriages). This may have given them an air of unorthodoxy and caused others to link their behavior to the evil heresy of the Bogomils the priests spoke about. Thus, we must conclude that any tradition bearing the term "Bogomil" in Bosnia is suspect and is probably derived at least indirectly from books.

To this must be added the seventeenth century poem of a Bosnian Moslem on the Evils of Tobacco. There is much dispute as to whether the poem actually uses the word Bogomil or "dear to God" since from the manuscript it is not clear if it is one or two words.(112) In addition, it is not clear if there is a period before the expression. Thus, we do not know if it says "we were in stench like the Bogomils," or "Like those dear to God, abandoning the pipe." However, even if the poem should be interpreted as saying Bogomils it is also certainly derived from schooling, for the Turkish sources also never used the term Bogomil for Bosnia and Hercegovina.

Folktales that seem to embody dualist ideas are known in Bosnia but since they are found in an area much wider than Bosnia, I shall not use them for the same reason that I am not using the Apocryphal tales. From time to time, however, I shall touch upon practices which seem to go back to the Middle Ages. Many of them such as the *slava* and *kumstvo* (God-fathership) are attested in medieval sources. Certain magic rites that seem to be old, and that appear in certain manuscripts from the Turkish period will also prove useful to us. In addition, certain specific stories of behavior or traditions remembered that are not tied to schooling may hold valuable information. Thus, much more valuable than a rumor that there were sixteen secret Bogomil families in Drežnica would be the following story recorded in 1928:(113)

In Krupa on the Vrbas an old-timer reported that he remembered the last "djed" (here, an elder) in that area. Near the old church was a grove of oaks, and the largest was called the Djed's oak. In the old days (he is referring to a nineteenth century tradition and how far it reached back into the past we do not know) one man from each of the better families, when he reached the age of sixty,would let his hair and beard grow and would put on long black dress and carry a "T" staff. These men were called djeds. Everyone respected them and showed them special honor. Women and youths when they met them would greet them by kissing their hand. At church all the djeds sat together under the Djed's oak and all the householders honored them by having their young women or girls bring them bread and raki. Then the old men would bless each woman loudly saying: "Thank you householder N— for so honoring us. Do this honor always and God will reward you." The last of these djeds died in the first years of the Austrian occupation (i.e., just after 1878).

At Imjani, in an old abandoned church a "T" staff was found which an old-timer also referred to as a "djedovski štap " and told of a tradition of a hundred years before (from 1928) when there had been an old man called a "djed" with a long beard and such a staff, who headed a *zadruga* of sixty-five members. He was very rich and had a large herd of animals, and according to the tale he moved the church to the Imjani location. Now there is a new church and the old one is used only on St. Ilija's Day. It is frequent, that when a new church is built and consecrated to a new saint, the old church, or even its ruins, will remain the site at which are celebrated services on the day of the saint to whom the old church had been dedicated.

We should not jump to any hasty conclusions from these traditions about djeds, since in addition to the former Bishop of the Bosnian Church, the term refers to the head (or elder) of a *zadruga,* and it was customary for the elder to carry a "T" staff. Thus when a staff is depicted on a medieval gravestone, regardless of what this motif signified when the stone was carved, peasants frequently call it "djedovski štap" (djed's staff) because they associate this motif with the *zadruga* elder's staff. And in the same way the terms "djedov kamen" (djed's stone) and "djedov krst" (djed's cross) as popular names for certain specific medieval stones probably are derived from traditions about such elders.

The djed at Imjani was a *zadruga* elder, and at Krupa the djeds were elders of important households. The elders were traditionally respected;

the blessing they offered is similar to that which a typical householder may bestow on a guest who on particular holidays brings cakes, sweets, or raki to his home. The exchange of such items on specific holidays is a widespread custom. Thus, I do not think the story reflects any continuity from djeds of the Bosnian Church, but rather it illustrates the ritual practices and enormous respect that center around the most important elders in this patriarchal society. That the leader of the Bosnian Church was also called a djed, however, reflects the honored position he had as one of the society's most important elders. It would not be at all strange if his contemporaries had shown him the same sort of ritual honor as the people of Krupa were later to bestow upon their family leaders.

Gravestones

In Bosnia and Hercegovina are to be found a large number of gravestones, about which a vast literature has grown up. The peasants call the stones by a variety of names, the most common of which is "Greek graves"; while popular books and schooled laymen refer to them as "Bogomil gravestones," and the scholarly literature calls them by the term *stećci.* Numbering in the thousands, these stones vary from being fairly large to massive; they come in three basic shapes — slabs, boxes, and sarcophagi. In addition, we find a number of crosses and a few obelisks, and as we get into the Turkish period we find the form transforming itself into the Turkish *nišan* style of gravestones, i.e., a column with a turban on top. The majority of stones though carefully shaped are uncarved; and thus appear today as totally unmarked. However, in the Middle Ages they may have been painted with motifs, which have washed away over the years. Today in parts of Serbia, particularly around Čačak, we find painted stones, and families of the deceased periodically renew the colors as they wash away. This would explain why we find expensive, beautifully shaped, massive but undecorated sarcophagi.

Of course what has interested scholars has been the surviving motifs and inscriptions on those stones that are carved. Over the years descriptions of sites and readings of inscriptions have regularly appeared in the publication of the Zemaljski museum in Sarajevo.(114) In recent years a corpus of inscriptions (not yet completed) is being edited by Marko Vego, of which four volumes have appeared.(115) On the motifs I shall only mention the monograph series of particular graveyards that has begun to be edited after the war, of which nine volumes have appeared.(116) Other major sites have been written up in the various journals.(117) In addition, an invaluable book that can be described as a catalogue of motifs, with maps showing the geographical distribution of each motif, has been compiled by Marian Wenzel.(118)

Besides this, there is an enormous literature about the meaning of the motifs. The theory that was popular with scholars, from the time of Arthur Evans through the Austrian period, was that these stones were Bogomil tombs, and this theory is still the one generally believed and the one to be found in Jugoslav tourist folders. In general, however, scholars have by now abandoned this view. There are many reasons for doing this.

First there are the motifs themselves. Only a few of them can be associated with dualist beliefs and often this is only by a stretch of the imagination and insistence on symbolic interpretations that cannot be proved. Secondly, crosses are common. This reason in itself is not conclusive since: a) If the Bosnian dualists were identical with the Bosnian Church — which, of course, is a doubtful assumption — we can find evidence such as the cross on Gost Radin's will that at least some members of the Church had nothing against crosses, b) In a superstitious society the cross could be put up for all sorts of magic reasons having nothing to do with the faith of the people — for reasons of protecting the deceased, imposing a curse to scare away grave robbers (a definite fear, since several inscriptions offer a curse to those touching their bones), or placing a charm to keep the corpse in his grave. The use of crosses as charms by people who were not Christian, such as Bosnian Moslems, has already been noted in the Introduction. Thirdly, from inscriptions we see that members of the Catholic, Orthodox and Bosnian Churches all erected *stećci.* Indeed some of the elaborate stones — including many at Radimlja the most famous of sites — belonged to Vlach families, known to be Orthodox in faith. Thus it is clear that the erection of tombstones was not limited to a particular confession. About dualists we just do not know, for not one inscription or motif clearly refers to a deceased with such beliefs.

Not only can we show no connection between the motifs and dualist beliefs but the view that the stones are dualist is belied by place and time. The stones are found in regions beyond the borders of Bosnia, such as Croatia, Montenegro, parts of the Dalmatian coast and western Serbia, particularly around Valjevo.(119) In addition, the majority of stones that can be dated seem to fall into the period from the mid to second half of the fourteenth century and continue into the sixteenth century. Thus they were erected in regions outside of Bosnia and Hercegovina at a time when there is no sign of any heresy in these regions, and began considerably later than the appearance of Bosnia's heresy.

The stones, however, do appear in Bosnia in the fourteenth century when that land was at the height of its economic prosperity, when its mines were active and its merchants were having lively trade with the West. At this time the Vlachs, who erected in the following century the most elaborate stones of all, began to take over the direction of caravans from the coast into and through Bosnia. By the next century the Vlachs had a monopoly on the caravans and had begun to amass large fortunes. The Bosnian nobles also profited from the trade and, with increased contacts with the West, they began to imitate the lives of the Western nobility. Luxurious and elaborate courts with actors and musicians appeared in Bosnia and Hercegovina in the fifteenth century. The *stećci* began to appear at the time the mines began to open and as wealth increased the *stećci* became more elaborate. Those who had wanted to live well when alive wanted elaborate monuments to mark their last resting places. Thus we may conclude that the *stećci* were set up by everyone rich enough to afford them. And since thousands of *stećci* were erected, it is clear that people of classes other than the nobility also erected them — clerics clearly did, and we may assume traders did, as well as artisans and well-to-do peasants and shepherds. Under these stones are to be found rich Catholics, rich Orthodox, and rich members of the Bosnian Church, both lay and ordained. And though here and there a motif may have a religious meaning, the custom of erecting a *stečak* itself is not tied to any religion and most of the motifs seem to be secular in nature.

It is unfortunate that we do not have more inscriptions but we are dealing with a society in which the vast majority of the people were illiterate. Thus, there was no point in writing; motifs had to serve the purpose. We must look for a variety of different reasons to explain what is to be found on the stones; no single theory can explain the *stećci* motifs. In some cases the motifs may represent customary juridical signs — and show the family sign, or the occupation or role in society of the deceased.(120) In Dalmatia and recent Serbia we frequently find on gravestones occupational signs — guns for soldiers, tools for a carpenter, etc. The swords, shields, bow and arrows, etc., which frequently appear on the medieval stones may well have been placed there for a similar reason. The deceased, we are perhaps being told, was a warrior, or of a noble family, or of a certain rank of nobility.(121) Many of the stones depict hunts, while others show two warriors on horseback facing one another as in a tournament. We know that the Bosnian nobility did participate in tournaments, for their prowess at a Hungarian court tournament at the beginning of the fifteenth century was noted in a Polish source.(122) Thus, it seems that scenes from daily life were frequently used, probably for the purpose of calling attention to the heroism of the deceased, which is in keeping with the worldly behavior these nobles exhibited in life. It is also possible that many of the motifs were meant to be simply decorative;(123) in this category plainly belong the borders (often imitating plants) carved around the edges of many stones. The decorative motive is also the logical explanation for the Roman stele placed on a medieval grave near Višegrad.(124) The use of the Roman stone suggests that the deceased's family found it, thought it attractive and for that reason used it to decorate the grave. Some motifs may have been connected with ancient burial rites; and even though the original meanings may have been long forgotten, people continued to use them since the practice of having the given signs on gravestones had become established. Among these more ancient motifs connected with death ritual may well be certain scenes that frequently appeared on Hercegovinian Vlach stones that Marian Wenzel has speculated may be taken from Vlach rituals which may be derived from classical mystery cults.(125) Often motifs of this type would have entered into the stock of motifs belonging to the carvers themselves.

We should avoid being categorical about the meaning of motifs and avoid trying to fit motifs into rigid systems. It is likely that a given motif can have more than one meaning, and meanings may well vary according to location, date, religion, social class, etc. For example, a snake on one tombstone may have connection with a snake cult or possess some symbolic meaning; at the same time it may signify that the deceased was killed by snakebite. It is interesting to note that the graveyard at Boljuni which has a relatively large number of snake motifs on its stones is also infested with snakes to the present day, some of which are poisonous. Therefore, we should use care before we try to find a single theory to cover all "snake motifs."

The only motif that I want to single out and discuss is that of the pastoral staff. Since it has been found on several stones which from the stones' inscriptions are clearly monuments of members of the Bosnian Church hierarchy, it is often assumed that wherever this motif is found it marks the grave of a member of that hierarchy. Even this is not a certain matter. From Kreševljani in Nevesinje we have a carved slab which presumably originally had been placed over the door of a medieval church on which is depicted the patron kneeling before the Virgin (to whom the church was probably dedicated), behind whom stands a bearded, robed figure with a pastoral staff.(126) Presumably the man with the staff was an Orthodox bishop. Thus, staffs were not limited to members of the Bosnian Church. And such staffs were used by both Orthodox and Catholic clergy in Serbia and Dalmatia. Thus, we may suggest that the staff could also mark the graves of Catholic and/or

Orthodox clerics, and can therefore be taken as a general sign for a cleric regardless of his denomination. And finally staffs on some stones (clearly not all, since we know in some cases staffs did mean Bosnian Churchmen) may denote *zadruga* or family elders, for the staff over at least the past two or three centuries is the standard symbol to denote this elder. It could well have had this meaning in the Middle Ages as well. We have seen this connection between the elder and the "djedov štap" in the story we recounted about the djedovi of Krupa.

One last item that ought to be mentioned about motifs is that it is probable that in a large number of cases the carver, and not the deceased or his family, chose the motifs. This is frequently the case today and this fact would explain why we find similar motifs on various examples of work which, we know by the inscription, were by the same carver. It is interesting to note the cases where the entire inscription simply states the carver's name and says nothing about the deceased. Since the carver at times signed his name by an inscription, we must keep in mind that on certain stones, the carver may have signed his name with a picture or a motif. Thus motifs could be symbolic, religious, magical, or convey information about the deceased or the carver.

Much more needs to be done about *stećci.* For example, it is necessary to excavate cemeteries. The few thorough excavations that have been carried out (such as those done in the summer of 1967 in the region of Bileća) show that many more bodies lay there than there were stones. This is not surprising, for in a village cemetery, many peasants who could not afford stones were probably buried under wooden markers that have since rotted away. Excavations have turned up few objects buried with the dead. None of these have had religious significance. Excavations, if extensively carried out, might well show us ties between particular motifs (or size and shape of stone) and the age and sex of the deceased. In the region around Tavan,for example, in the present day, the appearance of a half-circle at the end of the arms of a cross signifies that a woman is buried there.(127) If one were to excavate a large number of graves it is possible that one would find certain motifs were only found on the graves of females or of children. (It has often been speculated that tiny stones mark graves of children.) In addition, on occasions two people are buried under one stone. Possibly certain signs denote this.

V. Palavestra and M. Petrić recently excavated a grave at Žepe and found the skeleton of a man, who had obviously died as a result of a skull-smashing blow. On his stone were depicted a sword and shield, a bow and arrow, a man standing holding a bow and arrow, and an arm beckoning to a horse followed by a dog.(128) Possibly one of the motifs signified the cause of death. If we excavated enough graves, we might find certain links between particular motifs and skeletons that show signs of violent death.

Finally, it is the custom the European world over to bury bodies with the head to the west; however, the Catholic Church has the custom of distinguishing priests by placing their heads to the east. We know that at present in Bosnia this burial procedure is not strictly followed among the Franciscans, and it is probable that it had not been in the Middle Ages either. But it would be worth investigating. If in Catholic areas, in church cemeteries, a certain number of skeletons did lie with heads to the east, we might be able to link certain motifs on stones above these graves to the priesthood. However, all of this is impossible until a significant number of graves have been dug up. We can hope that funds for such a project can be found soon.

In this study, I shall chiefly use the stones for the data that can be drawn from their inscriptions. These inscriptions provide much important material about major figures from the Bosnian Church as well as from the nobility; Bosnian Church inscriptions will be analyzed individually in Section XI of Chapter Five. The inscriptions are also valuable in that they provide data about which faiths lived where; for it is evident that Bosnia and Hercegovina in the Middle Ages was not uniform in religion. Each confession had followings in only certain areas, and no faith could be found throughout the land. The data from inscriptions can aid one in compiling a map of where the different faiths did have adherents.

Churches

It is important for the history of religion in medieval Bosnia to find concrete evidence on where particular confessions lived. A basic source for this is data on the location of medieval churches.(129) A limited number are referred to in the sources. Besides these, archaeology and folk traditions provide evidence of many more. In fact contrary to general opinion it is evident that in medieval Bosnia there were a large number of churches. In various issues, particularly since the War, of the *Glasnik* of the *Zemaljski muzej* reports have appeared about the excavation of a variety of these churches. In addition to these many other foundations, often on plots of lands called "crkvica" or some other form of the word "church", have been found. In fact localities of this name, or places about which popular tradition retains tales of a former church, very frequently yield ruins of foundations when they are investigated. Many problems exist about these churches. Since inscriptions are rare, it is almost impossible to date them. Some from their shape, or from fragments of carving, are clearly prior to the end of the twelfth century. Others, particularly Orthodox churches, often (when documentary evidence can be found) turn out to date from the sixteenth century. But this can not explain the majority of these ruins, many of which must date from the Middle Ages. Since the ruins have no inscriptions, often we cannot learn the basic facts about them: When were they built? How long did they exist (though if medieval cemeteries are beside the church we can make some fairly accurate guesses about these matters)? What denomination was the church? (Again cemeteries are a help, as are data about the religious confession of the population from visitation reports, continuity between the location and or cemetery and a later church, and topographical data.) On the question of denomination we must keep an open mind to the possibility that the Bosnian Church also had church buildings and that many of these church ruins may have once been churches of that local denomination.

FOOTNOTES TO CHAPTER II

1. Luccari, p. 197. Not to be confused with "Manoilo Grk" who copied two gospels in the first half of the fourteenth century. On the latter see Z. Kajmaković, *Zidno slikarstvo u Bosni i Hercegovini,* Sarajevo, 1971, pp. 50-52.
2. Luccari, p. 30; Orbini, p. 353.
3. Luccari, p. 42.
4. M. Dinić, *Iz dubrovačkog arhiva,* knj. III, (SAN), Beograd, 1967.
5. See Appendix A for Chapter II for a discussion about specific charters of questionable authenticity.
6. V. Mošin, "Serbskaja redakcija sinodika v nedelju pravoslavija," *Vizantijskij vremennik,* XVII, 1960, pp. 278-353. See below Chapter V.
7. On this source see A. Solovjev, "Svedočanstva pravoslavnih izvora o bogomilstvu na Balkanu," *GID,* V, 1953, pp. 81-82.
8. V. Grigorovič, *O Serbii v eja otnošenijah k sosednim deržavam preimuščestvenno v XIV-XV stoletijah,* Kazan, 1859, p. 52.
9. V. Jagić, "Ein neu entdeckter urkundlicher Beitrag zur Erklärung des Bosnischen Patarenentums," *Archiv für slavische Philologie,* XXXIII, 1912, pp. 585-87.
10. Most scholars who have tackled the problem agree that the Grigorovič text is drawn from a later manuscript. See M. Dinić, *Za istoriju rudarstva v srednjevekovnoj Srbiji i Bosni,* I deo, Beograd, 1955, p. 94, and his footnote 8 on that page for further references.
11. Thomas Archdeacon of Split, *Historia Salonitana,* ed. F. Rački, MSHSM 26, Scriptores, 3, Zagreb, 1894.

12. F. Šišić, "Izvori bosanske povijesti," published in a collective work, *Poviest hrvatskih zemalja Bosne i Hercegovine,* Sarajevo, 1942, pp. 1-38. This book was published at unfortunate time and has a most unpleasant Fascist preface; however, on the book's behalf it must be said that most of the contributions had been prepared before the war broke out. As are many collective works, this volume is uneven and at times is marred by a Croatian bias. However, Sisic's chapter is perfectly straightforward. In addition since Šišić died early in 1940 he can bear no responsibility for the unpleasant format in which his article appeared. Unfortunately, however, despite the chapter's title, it discusses only documents and ignores all other types of sources.

13. In *Radovi,* (NDBH) III, 1955, pp. 5-56.

14. SAN, Beograd, 1967.

15. Though published prior to 1940, the following volume was not cited by Šišić and contains a great deal of material on the sale of slaves in Dubrovnik, some of which is pertinent to the Patarins: G. Čremošnik, *Istoriski spomenici dubrovačkog arhiva,* ser. III, sveska 1. (Kancelarski i notarski spisi 1278-1301) SKA, Beograd, 1932. A second volume with less material on slaves was subsequently published by Čremošnik under the title, *Spisi dubrovačke kancelarije,* knj. I (Zapisi notara Tomazina de Savere, 1278-82), JAZU, Zagreb, 1951.

16. N. Jorga, *Notes et Extraits pour Servir a l'histoire des Croisades au XVe siecle,* 2 series, Paris, 1899.

17. In *Prilozi,* XXIV, sv. 1-2, 1958, pp. 94-110.

18. Analysis of the texts, in *Vizantijskij vremennik,* XVI, 1959, pp. 317-394. The texts themselves are printed in XVII, 1960, pp. 278-353. For the anathematized names see pp. 302, 345, 348.

19. F. Šišić, *op. cit.,* pp. 1-38.

20. From ca. 1250, when the Bosnian bishop moved to Djakovo in Slavonia, until the fall of the Bosnian state, the Bosnian bishop had no role in Bosnia proper, except for Peregrin Saxon (bishop ca. 1349-55) who had been close to Ban Stjepan Kotromanić before he became bishop and who was able to continue this relationship after he assumed the post. The case of Peregrin will be discussed in Chapter IV. Otherwise the Bosnian bishop kept court outside of Bosnia. His main role (as far as we can tell from surviving documents) was witnessing charters for the Hungarian king or nobles living north of the Sava, and being a party to or judging disputes over land in Slavonia. A study of the documents of this bishopric gives no data about Bosnia at all. The main importance of the bishop in Djakovo was that,by his absence and lack of connection with Bosnian events, he left a vacuum in Bosnia that had to be filled by someone else. An interesting phenomenon which demonstrates the merely formal (and unimportant) role of the Bishop of Bosnia in Djakovo is the fact that on occasions the Bosnian bishop found witnessing Hungarian charters is already dead. For example, 18 May, 1364 we find Peregrino Bosnensis witnessing a royal grant to the Croatian family of the Babonići. The grant confirms an earlier one made in 1330. Peregrin had been Bishop of Bosnia from 1348 or 1349 until 1355. Innocent VI in 1356 named Peter to succeed him as bishop. Peregrin also was listed as a witness to the confirmation of a 1292 grant to Stjepan Frankopan also in 1364. This shows that the royal chancellery on occasions simply mechanically copied names of witnesses off some earlier charter. Thus in general we can be suspicious as to whether witnesses of Hungarian charters actually were present to witness the proceedings they were said to have witnessed.

Secondly, we see that the Bosnian bishop was of so little importance that the Hungarian clerk copying a list of names was not aware of who the bishop was in 1364, when Peregrin had been dead almost a decade. For these two charters see *CD,* XIII, pp. 371-72, and pp. 414-15.

21. This point of view can be found in fra L. P. (etrović), *Kršćani bosanske crkve,* Sarajevo, 1953, as well as in the works of V. Glusač, (see bibliography).

22. *CD,* XII, p. 33.

23. *CD,* IV, pp. 241-42.

24. Malinowski provides an interesting description of some of the problems a legate would be faced with in trying to draw conclusions about Bosnian beliefs on the basis of questions he might have posed. ". . . no 'natives' (in the plural) have ever any belief or any idea. Each one has his own ideas and his own beliefs. Moreover, the beliefs and ideas exist not only in the conscious and formulated opinions of the members of the community. They are embodied in social institutions and expressed by native behavior, from both of which they must be, so to speak extricated . . . .

"How do the natives imagine the return of the *baloma* (a type of ghost)? I have actually put this question, adequately formulated, to a series of informants. The answers were, in the first place, always fragmentary — a native will just tell you one aspect, very often an irrelevant one, according to what your question has suggested in his mind at the moment. Nor would an untrained 'civilized man' do anything else. Besides being fragmentary, which could be partially remedied by repeating the question and using each informant to fill up the gaps, the answers were at times hopelessly inadequate and contradictory. Inadequate because some informants were unable even to grasp the question, at any rate unable to describe such a complex fact as their own mental attitude, though others were astonishingly clever, and almost able to understand what the ethnological inquirer was driving at.

"What was I to do? To concoct a kind of 'average opinion?' The degree of arbitrariness seemed much too great. Moreover it was obvious that the opinions were only a small part of the information available. All the people, even those who were unable to state what they thought about the returning *baloma* and how they felt towards them, none the less behaved in a certain manner towards those *baloma,* conforming to certain customary rules and obeying certain canons of emotional reactions." (Malinowski, *op. cit.,* pp. 240-42). The problems he raises clearly pertain to any illiterate society.

25. A. C. Shannon, *The Popes and Heresy in the Thirteenth Century,* Villanova, Pa., 1949, pp. 3, 8-10 has called attention to the same problem. Having noted the common assumption among scholars that medieval popes had comprehensive knowledge of heresy and the individual tenets of particular sects, he states, "Such an assumption is based on the anachronism of attributing to the popes of the thirteenth century the same rapid and accurate facilities of communication and information that are taken for granted in the twentieth. Perhaps the inference is justified, but if the official correspondence of the popes is a fair criterion of their knowledge of the peculiarities of the various heretical groups, then only one of two conclusions is possible. Either the popes assume that the heretical doctrines were too well known to bear repetition — a supposition disallowed by the repeated requests of popes for additional information, or they possessed only a general knowledge of the prevalent heresies and contented themselves with combatting doctrinal dissidence as a whole. In either

case, if we were to rely on papal documents for information regarding the particular teachings of the different popular heresies, our knowledge would be extremely sketchy and unsatisfactory." He notes the tendency to call local sects by the names of major heresies without establishing either the identity of doctrine or their affiliation with one another. He stresses the fact that none of the pontiffs of the thirteenth century, though they continually condemned and excommunicated heretics ever presented the tenets of any particular group or revealed in their correspondence a precise acquaintanceship with the tenets of the dissident sects. Shannon concludes that Innocent III was more concerned with the anti-religious aspects of heretical teaching and the consequent contamination of the innocent than with carefully distinguishing individual groups and their peculiar teachings. He concludes, "A careful examination of the papal correspondence would seem to warrant the conclusion that the papacy was primarily concerned with the discovery and suppression of heresy in general with only subordinate interest in the intracacies of the exact errors involved." Shannon, of course, is writing about Western Europe, France in particular; but, all that he says above is even more relevant for Bosnia, a region on the periphery of Catholicism about which the popes had less interest and fewer sources of information.

26. R. Hofstadter, "The Paranoid Style in American Politics," pp. 3-40, in a collection of essays under the same title, New York, 1967 (Vintage paperback edition). Though Hofstadter stresses American examples, this "paranoia" is not at all a uniquely American or even modern malady.

27. *CD,* III, pp. 14-15. The content of this and the following documents will be discussed in the narrative chapters.

28. See Chapter III: *CD,* VIII, p. 535 for the 1319 letter.

29. D. Kniewald, "Vjerodostojnost . . . *Rad* (JAZU), 270, 1949, p. 162.

30. *Bullae Bonifacii IX,* pt. 1 (1389-96), Monumenta Vaticana Hungariae, Ser. I, Vol. 3, Budapest, 1888, pp. 178-79.

31. Monumenta conciliorum Generalium XV, Concilium Basileense, II, Vindobona, 1873, Joannis de Segovia, *Historia gestorum Generalis Synodi Basiliensis* (E. Birk ed.), lib, IX, Chap. 5, p. 750. There is question as to what See the Terbipolensis bishop represented. If it were Herbipolensis, he would be Bishop of Wurzburg, if Tervipolensis, Treviso.

32. M. Vego, *Povijest Humske zemlje,* Samobor, 1937, I, p. 145, suggests that the term Patarin was used for members of the Bosnian Church because they struggled against the authority of the Catholic Church and refused to accept reforms (rather than for any doctrinal error). He concludes that the term had come to be used for any rebel against Rome. We cannot demonstrate the accuracy of Vego's assertion, but we should keep it in mind throughout the narrative chapters. The use of the term "Patarin" on occasions to describe the Kačić pirates of Omiš gives weight to Vego's view; and we might add that rarely will the term's use in the sources suggest anything more specific than Vego's suggestion.

33. Dinić, *Iz dubrovačkog arhiva,* III, p. 193.

34. Martene-Durand, *Veterum scriptorum* . . . I, Paris, 1724, col. 1592.

35. Theiner, *MH,* Vol. II, p. 237.

36. See below, Chapter V.

37. Farlati, *IS,* vol. IV, pp. 256-57.

38. Theiner, *MH,* vol. II, p. 237.

39. Rački, *Bogomili i Patareni,* pp. 462-65.

40. The manner with which the enormous multi-volume work *Illyricum Sacrum* was compiled may lead us to have great admiration for the energy and labors of Riceputi, Farlati and Coleti — the three collectors and compilers who carried on the work one after the other between the 1730's and 1819 — but hardly for their critical acumen or ability to distinguish the authentic from the bogus. On the compilation of *Illyricum Sacrum,* see F. Šišić, "Hrvatska historiografija od XVI do XX stoljeća," *Jugoslovenski istoriski časopis,* II, 1936, pp. 39-42, and M. Faber, "Zur Entstehung von Farlati's 'Illyricum Sacrum,' " *Wissenschaftliche Mitteilungen aus B-H,* vol. 3, 1895, pp. 388-95. Several scholars should jointly undertake a thorough study, region by region, of Farlati's data and their sources. Such a study I am sure will show that many commonplaces in the historical literature will turn out to have no solid source base at all.

41. A Dondaine, "La Hierarchie Cathare en Italie," *Archivum Fratrum Praedicatorum,* XIX, 1949, pp. 280-313.

42. A. Dondaine, *Un Traité néo-manichéen du XIIIe siècle, Le Liber de duobus principiis,* Rome, 1939, pp. 64-78; and Kniewald, *op. cit.,* pp. 190-238.

43. A. Dondaine, "La Hierarchie Cathare . . .," *Archivum Fratrum Praedicatorum,* XX, 1950, pp. 234-324.

44. G. Amati, "Processus contra Waldenses in Lombardia Superiori a 1387," *Archivio Storico Italiano,* Firenze, 1865, ser. III, Vol. II, pt. 1, pp. 50-61.

45. F. Rački, "Prilozi za povjest bosanskih Patarena," *Starine,* I, Zagreb, 1869, pp. 138-140.

46. *Ibid.,* p. 138, note 1; and Kniewald, *op. cit.,* pp. 163-69.

47. Rački, *op. cit.,* pp. 109-138.

48. Kniewald, *op. cit.,* p. 146, note 63.

49. Mošin's dating the Zagreb manuscript slightly earlier than the 1387 trial of Jacob should not worry us too much. First he bases his dating on the paper on which the tracts were written. However, paper could always have been used a decade or so after it was manufactured. Secondly, even if the paper had been used immediately after its manufacture, interest in these Italian heretics was not new in 1387. Investigations had been going on about them for some time and Jacob was already in 1387 a relapsed heretic.

50. If it were not for the facts that the inquisition in Turin was carried out by Dominicans and the two tracts seem to have been written by Franciscans, we might well conclude that the heretics being described in the tracts were not intended to be Bosnians at all, but these Italian dualists in Piedmont who were called "Bosnian heretics." For this was not an ephemeral label for Italian heretics simply used during Jacob's trial. Over 20 years later in 1412 we have documents from Chieri referring to Chieri heretics by the name of "Bosnian heretics." Those called "Bosnian heretics" are named and are clearly Italians. (See, M. Esposito, "Un 'auto da fé' à Chieri en 1412," *Revue d'Histoire Ecclesiastique,* XLII, 1947, pp. 423-27. It is only the Franciscan connection that leads us to conclude that the documents were intended to describe Bosnians rather than these Italians.

51. Three things ought to be suggested about Jacob's testimony: first that he testified after being tortured which makes much of his testimony suspect;

second, since his doctrine had been linked to Bosnia (be it through his testimony or through fact), it is a reasonable hypothesis that he lied about trips thither to draw attention away from a heresiarch somewhere in north Italy; third, Jacob himself stated that according to the creed of the sect, it was no sin to lie before bishops and the inquisition, and in fact that it was a mortal sin to reveal facts about the heresy. Thus, if Jacob still held his heretical errors at the time of the trial, which is likely, since at the time he was already a relapsed heretic, it is possible that he believed it his duty to lie to cover up certain facts and to protect certain individuals. In fact, the whole idea of Italian city dwellers going to rural uneducated Bosnia to learn religious doctrine is very hard to imagine. What language did they communicate in? We may suggest that prior to the 1340's, when the first Italian mentioned by Jacob visited Bosnia, an Italian heresiarch had fled from Italy and sought asylum in Bosnia and that the Italians going to Bosnia went to visit that heresiarch and his successors. We should also note the matter of the intent lying behind the document. Charles Lea, pointing to the temptation to falsify records when the enemy to be struck down is strong, cites Friar Bernard Delicieus who, speaking for the whole Franciscan Order of Languedoc in a formal document, declared that not only were the inquisition's records not worthy of trust but they were generally believed to be so. (H. C. Lea, *A History of the Inquisition* (1887), New York, 1922 printing cited, Vol. I, p. 380) Lea goes on to point out that witnesses were encouraged to say everything that occurred to them; great weight was given to popular belief; no distinction was made between hearsay evidence, vague rumors, general impressions, idle gossip and solid information. Everything against an accused was taken down. He cites a case in Niort from 1240 where 108 witnesses, most of whom did not personally know the accused, gave testimony. (*Ibid.*, p. 431) He points to cases of testimony given by young children (p. 436) and the testimony of false witnesses (e.g., a case from 1323 at Pamiers, where six false witnesses, including two priests, were sentenced, p. 441). Thus, in addition to all the specific reasons given above for questioning particular inquisition documents referring to "Bosnians," we also have to be suspicious of the general character of inquisition records.

52. It should be stressed here that even if I have exaggerated the connection between these tracts and the inquisition in northern Italy and it should turn out that a considerable part of the material on the "Bosnian heretics" was drawn from no longer extant reports from Franciscans working in Bosnia, my basic point here is not refuted. For there is still nothing in the tracts to connect them with the Bosnian Church, and thus we would still be left with tracts describing the separate dualist current.

53. D. Kamber, "Kardinal Torquemada i tri bosanska bogomila (1461)," *Croatia Sacra*, III, 1932, pp. 27-93, and F. Rački, in "Dva nova priloga za poviest bosanskih Patarena," *Starine*, XIV, 1882, pp. 1-21. The 50 articles renounced are given as Appendix A for Chapter VI. This renunciation is the most detailed of these documents; thus I have chosen to use it for comparison with other sources on the Bosnian Church to demonstrate the irrelevance for that church of this and the other Italian tracts.

54. Fejer, *CDH*, X, pt. 7, Budapest, 1834, pp. 812-13.

55. *Ibid.*, X, pt. 7, pp. 883-84.

56. Kniewald, "Vjerodostojnost . . .," p. 157.

57. Theiner, *MSM*, Vol. I, pp. 392-94.

58. See Fejer, *op. cit.*, X, pt. 7; and *AB*, pp. 139-177.

59. *La Vita di S. Giacomo della Marca (1393-1476)* of fra Venanzio da Fabriano, edited by P. M. Sgattoni, Zadar, 1940. See Appendix B, Chapter II for a source on which is based a bogus claim that Jacob referred to dualists in Bosnia.

60. See two volumes noted above in note 58.

61. *De Conformitate Vitae Beati Francisci ad Vitam Domini Iesu,* of fra Bartholomaeus of Pisa, *Analecta Franciscana,* IV, 1906, pp. 555-56.

62. J. Jelenić, "Necrologium Bosnae Argentinae," *GZMS,* XXVIII, 1916, pp. 337-57.

63. D. Mandić, *Hercegovački spomenici franjevačkog reda iz turskog doba,* Mostar, 1934; and J. Matasović, *Fojnička Regesta,* in *Spomenik* SKA Vol. 67, Beograd, 1930, pp. 61-432. Mandić's collection despite its title has material on Bosnia as well.

64. Much visitation material can be found in the appendix of documents collected by B. Pandžić, *De Diocesi Tribuniensi et Mercanensi,* Rome, 1959. This work and its documents cover these dioceses during the Turkish period. Since the visitations have not generally been utilized by non-churchmen and since they are scattered throughout a variety of journals, I will list here all the ones that I have utilized from the earliest that I know of in the 1590's up to 1655. Since this date is roughly 200 years after the fall of Bosnia and since by that date the Turkish administration no longer effectively blocked Catholic church building, 1655 seems a good cut-off point:

a) K. Horvat, "Monumenta historica nova historiam Bosnae et Provinciarum Vicinarum Illustrantia," *GZMS,* 1909, pp. 1-104, 313-424 (index pp. 505-518). Contains visitations from the end of the sixteenth century, from 1600 and 1601.

b) Ambrosii Gučetić (Gozze), a report of 1610, taken from Farlati, see Pandžić, p. 109.

c) Dr. F. M. "Dva savremena izveštaja o Bosni iz prve polovine XVII stoljeća," *GZMS,* XVI, 1904, contains 1) reports of the Jesuit Bartholomaeus Kačić of 1612 and 1618 which chiefly discuss Srem and Slavonia, pp. 256-66, and 2) the report of Marijan Maravić from 1655.

d) Reports on the diocese "Stephanenses" from 1622, see Pandžić, pp. 109-112; from 1624, see Pandžić, pp. 112-116; and especially see K. Draganović, "Tobožnja 'stepanska biskupija' u Hercegovini," *Croatia Sacra,* VII, 1934, pp. 29-58. Draganović includes a valuable introduction which explains the chaotic situation the Catholic Church in Hercegovina was in at the time.

e) Visitation of A. Georgicea (Athanasius Georgijević) from 1626, in M. Batinić, "Njekoliko priloga k bosanskoj crkvenoj poviest," *Starine,* 17, 1885, pp. 116-36.

f) K. Draganović, "Izvješće fra Tome Ivkovića, biskupa skradinskog iz godine 1630," *Croatia Sacra,* VII, 1934, pp. 65-78.

g) K. Draganović, "Izvješće apostolskog vizitatora Petra Masarechija o prilikama katoličkih naroda u Bugarskoj, Srbiji, Srijemu, Slavoniji i Bosni g. 1623, 1624," *Starine,* 39, 1938, pp. 1-48.

h) Michael Restić in 1639, see Pandžić, pp. 116-119.

i) S. Zlatović, "Izvještaj o Bosni god 1640 od Pavla iz Rovinja," *Starine,* 23, 1890, pp. 1-38.

j) A final visitation in this period, to the best of my knowledge still unpublished, was that of Marijan Pavlović in 1623 whose contents are summarized by K. Draganović, "Katalog katoličkih župa: XVII vieka u Bosni i Hercegovini," *Croatia Sacra,* years 13-14, numbers 22-23, 1944, pp. 93-96, Draganović's article also contains summaries of several of the visitations noted above and contains much interesting material on the state of the Catholic Church in Bosnia and Hercegovina in the seventeenth century.

65. Report of Antonius Righus in 1703, in Pandžić, pp. 129-37, reference to Patarins, p. 132.

66. Pius II, *Europa Pii Pontificis Maximi nostrorum temporum varias continens historias,* Basel, 1501.

67. Pius II, *Commentarii rerum memorabilium, quae temporibus suis contingerunt,* Rome, 1584.

68. J. Matasović, "Tri humanista o Patarenima," *Godišnjak skopskog filozofskog fakulteta,* I, 1930, pp. 245-46.

69. See below Chapter VI.

70. Orbini, *Il regno de gli Slavi,* Pesaro, 1601, pp. 352-54.

71. K. Draganović, "Izvješće apostolskog vizitatora Petra Masarechija o prilikama katolickih naroda u Bugarskoj, Srbiji, Srijemu, Slavoniji i Bosni g. 1623 i 1624," *Starine,* 39, 1938, pp. 43-44.

72. G. Luccari, *Copioso ristretto degli anali de Ragusa,* first edition, Venice, 1605; citations in this study are from the Ragusa, 1790 edition, pp. 30, 91-92.

73. This note marks the place where Masarechi's one departure occurs. For its text see below, immediately following our translation from Orbini.

74. Orbini, pp. 352-54.

75. K. Draganović, *op. cit.,* p. 43.

76. Testament referred to in Ćirković's notes to the recent Serbo-Croatian translation of Orbini. M. Orbini, *Kraljevstvo Slovena,* Beograd, 1968, p. 343.

77. D. Obolensky, *The Bogomils,* Cambridge (England), 1948, p. 266. For further references to these latter day Paulicians see Obolensky's footnote 2 on p. 266.

78. The only other connection which I have been able to find between Bosnia and certain latter-day Paulicians concerned some in Bulgaria and is drawn from a 1640 visitation there by Bakšić. Having made use of Orbini's account, Bakšić says the Paulicians (he calls them Paulianisti) had come from Bosnia which explains why their books had all been written in Bosnia on parchment. He then briefly notes their conversion from error and then returning to their books, states that they were all in Slavic in Cyrillic. He had found certain book manuscripts written under King Tvrtko from which the Paulianisti knew that they had come from Bosnia and had brought with them this heresy of theirs. (E. Fermendžin (ed.), *Acta Bulgariae Ecclesiastica,* MSHSM, XVIII, Zagreb, 1887, pp. 79-80.) Unfortunately, he tells us nothing about the books' contents. I suspect that they were mainly, if not entirely, Gospel manuscripts. This information provides some confirmation for Orbini (or Pietro Livio) — though we must allow for the possibility that Bakšić's inquiries and interpretations were colored by Orbini's account. Unfortunately, just as with the Orbini-Livio material, the passage of considerable time makes it impossible to connect any practices of the 1640 Paulianisti with the Bosnian Patarins. There is considerable material about these seventeenth century Paulicians-Paulianisti in *Acta Bulgariae Ecclesiastica,* though unfortunately nothing further about possible connections with Bosnia. However, this interesting sect calls for a

thorough scholarly study. Then we would be in a far better position to discuss this group's relationship (if any really exists) with groups in Bosnia. And we may at least be able to tell whether the Paulianisti should trace their ancestry back to the Armenian-Byzantine Paulicians or whether they came to the eastern Balkans considerably later from the West — be it Italy or Bosnia — as Orbini (Livio) and Bakšić believe.

79. Luccari, pp. 30, 42, 197.

80. Luccari, p. 91.

81. Junius Resti, *Chronica Ragusina,* ed. Nodilo, Zagreb, 1893. MSHSM, JAZU, 25 Scriptores, 2.

82. Roger de Wendover (d. 1236), *Chronica sive flores historiarum,* IV, ed. H. Coxe, London, 1842, p. 87 ff. Matthaeus Parisiensis, *Chronica Majora,* ed. H. Luard, Rer. Brit. Medii aevi scriptores, III, 1876, p. 78 ff. contains the same text. The Bishop of Rouen Theobaldus provides a variant text in Martene-Durand, *Thesaurus novus anecdotorum,* I, Paris, 1717, pp. 901-903. The relevant passages about the anti-pope from the East are cited and discussed by J. Šidak, "O pitanju heretičkog 'pape' u Bosni 1223 i 1245," *Razprave V. Hauptmannov zbornik Slovenska akademija znanosti i umjetnosti, razred za zgodovinske in družbene vede,* Ljubljana, 1966, pp. 145-59.

83. "Fojnička kronika" (ed. Ć. Truhelka), *GZMS,* XXI, 1909, pp. 443-57; "Ljetopis franjevačkog samostana u Kr. Sutjesci" (ed. J. Jelenić), *GZMS,* XXXV, 1923; XXXVI, 1924.

84. *Letopis fra Nikole Lašvanina,* ed. J. Jelenić, Sarajevo, 1916. Unfortunately, Jelenić published this chronicle in a never utilized, idealized version of eighteenth century Slavic script which makes it infuriatingly difficult to read.

85. For the text of Pope John XXII's letter of 1319 to Mladen Šubić, *CD,* VIII, pp. 535 ff; it is discussed below in Chapter IV.

86. Anonymi Spalatensis, in *Scriptores rerum Hungaricum, Dalmaticarum, Croaticarum et Sclavonicarum veteres ac Genvini,* III, Vindobonensis, 1748, ed. I. Schwandtner, p. 659. According to the Split anonymous writer this wondrous fire occurred in 1356; Lašvanin dates it 1376 (p. 39). Perhaps the story of the huge fire melting rocks and leveling mountains was based on a major mine fire.

87. E. Fermendžin (ed.) "Chronicon observantis Provinciae Bosnae Argentinae Ordinis S. Francisci Seraphici," *Starine,* XXII, 1890, pp. 1-67.

88. M. Tayyib Okiç, "Les Kristians (Bogomiles Parfaits) de Bosnie d'après des Documents Turcs inédits," *Südost-Forschungen,* 19, 1960, pp. 108-133.

89. For example Okiç cites twenty-two cases of *baštinas* of named krstjani (Okiç, *op. cit.,* pp. 123-24.

90. *Ibid.,* pp. 122-23.

91. *Ibid.,* p. 125.

92. Text of Radin's Testament given in Lj. Stojanović, *Stare srpske povelje i pisma,* I, pt. 2, Beograd (SKA), 1934, pp. 153-56. On the krstjani kmeti, see pp. 153-54. The whole testament is analyzed below in detail in Chapter VII.

93. For an example of this sort of data, see H. Šabanović, "Lepenica u prvom stoljeću turske vladavine" in *Lepenica,* Sarajevo, 1963, pp. 193-207.

94. The most important of which were published as the first volume of a new series — *Monumenta Turcica historiam Slavorum Meridionalium illustrantia-*Kanuni i kanun name za Bosanski, etc., . . . Sandžak, Sarajevo, 1957.

95. *Ibid.,* pp. 31, 43, 56, 66.

96. B. Kuripešić, *Putopis kroz Bosnu, Srbiju, Bugarsku i Rumeliju, 1530,* trans. Dj. Pejanović, Sarajevo, 1950, p. 22.

97. Dj. Sp. Radojičić, "Odlomak bogomilskog jevandjelja bosanskog tepačije Batala iz 1393 godine," *B'lgarska Akademija na Naukite, izvestija na Instituta za Istorija,* 14-15, 1964, pp. 495-509. The hierarchy list is a very important source. It was without question written in 1393. It will be discussed in detail in Chapter V.

98. G. Daničić, "Hvalov rukopis," *Starine,* III, 1871, pp. 1-146, gives the text of the gospel and compares it with two other gospel texts. See also V. Djurić, "Minijature Hvalovog rukopisa," *Istoriski glasnik* (1-2), 1957, pp. 39-52.

99. F. Rački, "Dva nova priloga za poviest bosanskih Patarena," *Starine,* XIV, 1882, pp. 21-29; V. Jagić, "Analecta Romana," *Archiv für Slavische Philologie* 25, 1903, pp. 20-36.

100. Radojičić, *op. cit.,* p. 500

101. Daničić, *op. cit.,* p. 6.

102. A. Solovjev, "La Doctrine de l'église de Bosnie," *Acad. de Belgique,* Bulletins — lettres, 5e ser., 34, 1948, pp. 494, 496-97.

103. J. Šidak, "Kopitarovo bosansko evandjelje u sklopu pitanja 'crkve bosanske,' " *Slovo,* 4-5, 1955, pp. 47-63, and his "Marginalija uz jedan rukopis 'crkve bosanske' u mletačkoj Marciani," *Slovo,* 6-8, 1957, pp. 134-153.

104. Lj. Stojanović, *Miroslavljevo jevandjelje,* facsimile edition, Beograd, 1897. S. Kuljabakin *Paleografska i jezička ispitivanja o Miroslavljevom jevandjelju,* Srem. Karlovci, 1925. Dj. Stričević, "Majstori minijatura Miroslavljevog jevandjelja," *ZRVI,* I, 1952, pp. 181-203. R. Ljubinković, "Humsko eparhisko vlastelinstvo i crkva Svetog Petra u Bijelom polju," *Starinar,* IX-X, 1959, pp. 97-123.

105. A. Solovjev, "La Messe Cathare," *Cahiers d'études Cathares,* III, No. 12, 1951-52, pp. 199-206.

106. J. Šidak, "Problem 'bosanske crkve' u našoj historiografiji od Petranovića do Glušca," *Rad,* 259, 1937, pp. 116-122.

107. See especially E. Turdeanu, "Apocryphes bogomiles et apocryphes pseudo-bogomiles," *Revue de l'histoire des Religions,* CXXXVIII, 1950, pp. 22-52, 176-218, and A. Vaillant, "Un Apocryphe pseudo-bogomile: La Vision d'Isaie," *Revue des Etudes Slaves,* XLII, (1-4), 1963, pp. 109-121.

108. J. Ivanov, *Bogomilski knigi i legendi,* BAN, Sofia, 1925, pp. 73-87.

109. I. Döllinger, *Beiträge zur Sektengeschichte des Mittelalters,* II, p. 86. This note tells us nothing about the history of the manuscript where this scholium appears, nor is it clear whether heretics or later inquisitors wrote it. However, since this manuscript was popular with the dualists and since dualists were active in Lombardy and Tuscany, we may take this gloss as a suggestion that there also were dualists in Bosnia.

110. A. Savic Rebac, "O narodnoj pesmi 'car Duklijan i Krstitelj Jovan,' " SAN, *Zbornik radova,* X, Institut za proučavanje književnosti, 1, Beograd. 1951, pp. 253-271.

111. These various rumors have been collected and published by M. Hadžijahić, "O islamizaciji bosanskih krstjana," *Obzor,* 31 December 1937.

112. M. Hadžijahić, "Kaimija o Bogomilima," *Život,* 1952, No. 5, pp. 125-26. Dj. Popović, "Reč 'Bogomil' u Bosni" *Život,* 1952, No. 7, pp. 273-

75. Dj. Popović, "Da li su Bosanci u XVII veku znali za Bogomile," *Život,* 1952, No. 3, pp. 187-89, M. Traljic, "Da li su Bosanci u XVIII vijeku znali za Bogomile," *Historijski zbornik,* V. (3-4), 1952, pp. 409-10.

113. M. Karanović, "Jedan zanimljiv mramor kod Skender-Vakufa," *GZMS,* XL, 1928, pp. 137-38.

114. *Glasnik Zemaljskog muzeja,* Sarajevo, *(GZMS).*

115. M. Vego, *Zbornik srednjovjekovnih natpisa Bosne i Hercegovine,* Sarajevo, vol. I, 1962, vols. II, III, 1964, IV, 1970. Various corrections (or other readings) can be found in E. Hamp "Notes on Medieval Inscriptions of Bosnia and Hercegovina," *Zbornik za filologiju i lingvistiku* (Novi Sad), XII, 1969, pp. 83-91.

116. *Srednjevjekovni nadgrobni spomenici B i H,* published by the Zemaljski muzej in Sarajevo. 1, A. Benac, *Radimlja,* 1950; 2, A. Benac, *Olovo,* (Beograd, 1951); 3, A. Benac, *Široki brijeg,* 1952; 4, D. Sergejevski, *Ludmer,* 1952; 5, Š. Bešlagić, *Kupres,* 1954; 6, M. Vego, *Ljubuški,* 1954; 7, Š. Bešlagić, *Kalinovik,* 1962; 8, Š. Bešlagić, *Popovo,* 1966; 9, Š. Bešlagić, *Stećci centralne Bosne,* 1967.

117. See especially *GZMS, Naše starine (NS)* and *Starinar.* The number of sites listed in these journals runs into the hundreds, thus here I shall just list a handful of the most important sites. All are announced by Š. Bešlagić; "Stećci u Opličićima," *NS,* VII, 1960, pp. 145-54; *Stećci na Blidinju,* published as a monograph by JAZU, Zagreb, 1959; "Stećci u Ziemlju," *Starinar,* n.s. 15-16, 1964-65, pp. 279-92; "Boljuni — srednjovjekovni nadgrobni spomenici," *Starinar,* n.s. XII, 1961, pp. 175-205; "Ljubinje — srednjovjekovni nadgrobni spomenici," *NS,* X, 1965, pp. 113-163; "Stećci na Nekuku kod Stoca," *Glasnik Etnografskog muzeja u Beogradu,* XXI, 1958, pp. 155-75.

118. M. Wenzel, *Ukrasni motivi na stećcima,* Sarajevo, 1965.

119. For example, see M. Ćorović-Ljubinković, "Nekropole i grobni belezi" in *Gradja* IX, Arheološki institut, knj. II, Arheološki spomenici i nalazišta u Srbiji, I. Zapadna Srbija, Beograd, 1953, pp. 169-98; A. Horvat "O stećcima na području Hrvatske," *Historijski zbornik,* IV, 1951, 1-4, pp. 157-62; A. Benac, "Srednjevekovni stećci od Slivna do Čepikuća," *Anali Historijskog instituta u Dubrovniku,* II, 1953, pp. 59-82.

120. On this subject see the interesting article of R. Vulcanescu, "Les Signes Juridiques dans la region Carpato-Balkanique," *Revue des Études Sud-Est Européennes* (Bucharest), II, 1-2, 1964, pp. 17-69.

121. Since the stones with coats of arms are almost entirely found far away from the major medieval castles and since none of the known coats of arms of leading families are found on *stećci,* Marian Wenzel — who has shown the connection between various of these stones and the Vlachs — has suggested that some of these military motifs may have a tribal basis and reflect the family, clan or *katun* to which the deceased belong. M. Wenzel, "Štitovi i grbovi na stećcima," *Vesnik — Vojni muzej* (Beograd), 11-12, 1966, pp. 89-109.

122. J. Dlugosz, *Historiae Polonicae,* XI, in his *Opera Omnia,* XIII, Cracow, 1877, p. 141. The Polish chronicler who describes the games erroneously refers to Hrvoje Vukčić as the King of Bosnia. The games took place in 1412. The chronicler died in 1480.

123. That decorative reasons could be important is illustrated by the story of an Albanian stone, decorated with a sun and moon, erected at the beginning of this century. Miss Durham asked the daughter of the woman buried beneath the

stone the meaning of the markings. The girl replied, "We are poor and can't write. We put it up to remember Mother. The sun and moon are there to look pretty." (M. E. Durham, *Some Tribal Origins, Laws and Customs of the Balkans,* London, 1928, p. 123).

124. I. Bojanovski, "Zaštita spomenika kulture i prirode u području buduće akumulacij na Perućcu na Drini," unpublished manuscript, p. 4 of typescript.

125. See M. Wenzel, "A Medieval Mystery Cult in Bosnia and Hercegovina," *Journal of the Warburg and Courtauld Institutes,* 24, 1961, pp. 89-107; and her "Bosnian and Hercegovinian Tombstones — who made them and why?," *Südost-Forschungen,* XXI, 1962, pp. 102-43; and her "The Dioscuri in the Balkans," *Slavic Review,* XXVI, no. 3, 1967, pp. 363-81.

126. D. Sergejevski, "Putne bilješke sa Nevesinjskog polja," *GZMS,* ns III, 1948, p. 53.

127. M. Dragić, "Oboljenje, smrt, i pogrebni običaji u okolini Tavne," SAN, *Zbornik radova,* XIV, Etnografski institut, 2, Beograd, 1951, p. 134.

128. V. Palavestra & M. Petrić, "Srednjovjekovni nadgrobni spomenici u Žepi," *Radovi,* (NDBH), XXIV, Sarajevo, 1964, pp. 139-79, especially pp. 174-76.

129. Sources on churches: I) Archaeological: Of particular value are the issues of *GZMS* from its first issue in 1889 to the present, *Starinar,* and the manuscript of *Arheološki leksikon,* compiled at the end of the last century by V. Radimsky which exists in a German handwritten edition and a typed Serbian translation at the Zavod za Zaštitu spomenika kulture in Sarajevo. I am indebted to Professor Bojanovski who was kind enough to lend me this manuscript. II) Ethnographical: the traditions about churches and legends concerning them have been preserved in ethnographic works. Again the best sources are the ethnographic articles in *GZMS.* Also see the volumes of *Etnografski zbornik,* poreklo stanovništva, SAN, Beograd, (J. Cvijić, general editor) which pertain to Bosnia and Hercegovina; these volumes contain many traditions about churches.

APPENDIX A FOR CHAPTER II: ON FORGED DOCUMENTS

A typical forgery is the document which connects August and Thomas Knezović (alleged predecessors of a prominent eighteenth century Franciscan family) with the foundation in about 1398 of the Franciscan monastery at Kreševo.(1) The Knezovići, not known from any fourteenth or fifteenth century document, are credited with restoring a church of St. Catharine there and building a chapel of St. Mary's. In 1385, no Franciscan church existed at Kreševo, as we know from the list of monasteries in the Bosnian vicariat written by the vicar at the time, Bartholomaeus of Pisa.(2) It is hard to believe that a church would have been built there and ruined within thirteen years subsequent to 1385. In addition, since the wife of King Stefan Tomaš (1443-61) was named Katarina, and since both king and queen did build churches including a St. Catharine's at Jajce, it seems probable to attribute the building of the Kreševo church to that royal pair.

A second questionable charter which refers to the hierarchy of the Bosnian Church is an undated one issued, according to Šurmin, its editor, around 1370 by Tvrtko I (1353-91) to Stjepan Rajković.(3) We do not have the original; all that exists is an eighteenth century copy in which the Cyrillic letters have been transliterated into Latin ones. The grantee, Rajković, is not known from other charters, and considering the size of the grant (Lašva, Brod and Klopče near Zenica, Čukle (?), and two villages in Usora) and the important role the grantee had played, allegedly saving the key fortress of Bobovac from a Hungarian attack (presumably in 1363 — the only known Hungarian attack in this period), his anonymity is odd. Furthermore, we have no evidence that the Hungarians penetrated that far into Bosnia in 1363; other evidence suggests that they attacked in two waves, each of which was stopped to the north of Bobovac — once at Soko in the Plivska župa of the Donji kraji and the second time at Srebrnik in Usora. The document has Tvrtko calling himself "Lord Ban Tvrtko, by the Grace of God, Lord of many Bosnian lands, Bosnia and Serbia, Sol, the Podrina and many Bosnian lands."(4) In no other charter in which Tvrtko is entitled ban does Serbia appear in his title. Tvrtko only added Serbia to his title after his 1377 coronation as king at Mileševo, after which he almost always placed Serbia before Bosnia in the title.(5) The repetition of "many Bosnian lands" in the title also is odd. The grant was issued by Tvrtko and his younger brother Vuk. Vuk revolted against Tvrtko in 1366 and we know that in 1369 Vuk was in exile, still in revolt, against his brother. In 1374 on the occasion of Tvrtko's wedding we find Vuk back in favor. It is unknown when between 1369 and 1374 this reconciliation had occurred, but the occasion of the wedding does seem a likely time. Since the action of defending Bobovac against the Hungarians would have had to have taken place in 1363, we would expect the charter, if authentic, to have been issued not long after that date. We know that Tvrtko awarded Vukac Hrvatinić, defender of the key Soko fortress in the Plivska župa, with a large land grant in 1366.(6) The presence of Vuk with Tvrtko means that the Rajković charter would have to have been issued between 1363 and 1366, or sometime after 1370 (at whatever time Tvrtko and Vuk were reconciled). If the charter supposedly dates from before 1366, then the reference to Serbia in the title is anachronistic. If the charter was supposedly issued after 1370, then it is odd that Tvrtko should have waited 7 or more years to reward this very loyal servitor.

The wording of the Bosnian Church's guarantee is also unlike that on other charters. The charter is valid as long as "the church of God has root in Bosnia," a most unusual phrase, that seems to look forward to a time when the church will not exist there; something that we would not expect the bishop of a church to do. A joint court of djed (Bishop of the Bosnian Church) and two stroiniks and three nobles will judge charter violations. Such a court is not an impossibility, but we know nothing of it from other documents, and most charters are not explicit about the procedure that will take place if a charter is violated. When we consider all of the above and add to it the fact that the Bosnian Church played no role in any other surviving charter issued by Tvrtko, we must conclude that most probably this charter is a forgery.(7) Thus I make no use of its contents in this study.

There is one more charter which allegedly mentions a djed. This one, published in 1770 by B.A. Krčelić(8), was one of six documents published by Fra Ivan Tomko Mrnavić (1580-1637) in a 1632 work about the history of the Mrnavić family. This document purports to be a grant by Tvrtko II to Ivan Mrnavić in 1427 for services rendered as a diplomat to the Turks. It was witnessed by Abbot Mirohna. If such a grant had been made, the charter would have been in Slavic; all that Mrnavić passed on in 1632 was a Latin translation. F. Šišić in 1901 demonstrated convincingly that Fra Ivan Tomko Mrnavić was involved in a variety of forgeries, that none of the six charters are known otherwise, that all of them are at least highly suspect, and that no Mrnavić is known from any other medieval documents.(9) I am convinced that Šišić has completely demonstrated the forged nature of this 1427 charter. Recently Šidak has tried to rehabilitate the charter and has asserted that "abbot" meant "djed."(10) However, Šidak fails to come to grips with Šišić's arguments; thus I shall not make use of this charter in the study.

The most famous of the forged documents is one which describes a grand council, which never took place, of ecclesiastical grandees at Konjic in 1446 which issued various edicts against the Manichees. This document has been utilized by a variety of historians, including Runciman, even though Rački long ago proved it to be a forgery.(11) In the same vein is a forgery, dated 1450, in which the King of Bosnia orders Radivoj Vladimirović to persecute the Patarins of the Neretva and to defend the faith received from the Franciscans (an odd way for anyone but a Franciscan to describe Catholicism). The document also refers to the Council of Konjic, which cinches its forged character.(12) The Vladimirović family also was closely tied in the later period to the Franciscans. A variety of forgeries also turned up concerning the Ohmučević family.(13)

FOOTNOTES TO APPENDIX A

1. Published by M. Strukić, *Povjestničke crtice Kreševa i franjevačkoga samostana,* Sarajevo, 1899, pp. 38-39.

2. Bartholomaeus de Pisa, "De Conformitate Vitae . . .," *Analecta Franciscana,* IV, 1906, pp. 555-56.

3. Dj. Šurmin, *Hrvatski spomenici,* I, pp. 85-86.

4. V. Ćorović, *Kralj Tvrtko I Kotromanić,* Beograd, 1925, pp. 97-100, who also questions the authenticity of this charter — refusing to utilize its contents but also refusing to reject them entirely — believes that Tvrtko's title as given in the charter is clearly a combination of two different sets of titles. Whether the error was made by a forger or a careless later copier is of course not known.

5. I am aware of only one example (other than the questionable Rajković charter) in which Tvrtko placed Bosnia before Serbia in his title; this exception occurs in a charter to Dubrovnik in 1387. (Lj. Stojanović, *Stare srpske povelje i pisma,* I, pt. 1, p. 86.) Tvrtko's seal, significantly, places Serbia before Bosnia. (*Ibid.,* I, pt. 1, pp. 83, 85.) Šidak ("O vjerodostojnosti isprave bosanskog bana Tvrtka Stjepanu Rajkoviću," *Zbornik radova Filozofskog fakulteta,* Zagreb, II, 1954, p. 42) ingeniously suggests in place of "Serbia" we should read "Usora" which was frequently part of the ban's title. This would necessitate, however, a major error by the copier since the two words when written are not similar. But even if Šidak should be correct on this suggestion, there are enough other strange things about the charter — which I point out here — to make us still doubt its authenticity.

6. Šurmin, *Hrvatski spomenici,* I, pp. 83-84.

7. Also odd is the inclusion of Lašva among the lands given to Rajković. In 1380 when Tvrtko made Hrvoje Vukčić Great Vojvoda to succeed his deceased father, he awarded Hrvoje three villages in Lašva (Tribouša, Lupnica and Bilo) as hereditary property. To have made such a grant to Hrvoje would mean that

had Tvrtko really granted Lašva to Rajković around 1370 he would have had to take part of it back within the decade. In addition in the 1390's Lašva was clearly part of the family estates of Tepčija Batalo (Dj. Radojicic, "Odlomak bogomilskog jevandejelja bosanskog tepačije Batala iz 1393 godine," BAN, *Izvestija na Institut za istorija,* XIV-XV, 1964, p. 504. In the inscription in this Gospel, written in 1393, it is stated that "Batalo, who was greatly renowned, who held Toričan and Lašva . . .") Batalo is known only from the 1390's but for him to have been given the high court rank of tepčija suggests that he had previously held significant territory. Hence it is probable that his family held most of Lašva for some time. His holding all of Lašva (as the inscription states) in 1393 in no way contradicts the 1380 grant to Hrvoje, for Batalo had married Hrvoje's sister; thus it is likely that he had received Hrvoje's three Lašvan villages as part of her dowry. Thus we may suggest that Batalo's original territorial base had included the greater part of the župa of Lašva; in this case, it is highly unlikely that Tvrtko would have been granting Lašva to anyone around 1370.

8. B. Kerchelich, *De regnis Dalmatiae, Croatiae et Slavoniae notitiae praeliminares,* 1770.

9. F. Šišić, "Kako je car Justinijan postao Slaven," *Nastavni vjesnik,* IX, 1901, pp. 390-415 (esp. 413-14).

10. J. Šidak, "O autentičnosti i značenju jedne isprave bosanskog 'djeda' (1427)," *Slovo,* XV-XVI, 1965, pp. 281-91.

11. J. Jelenić (ed.), *Letopis fra Nikole Lašvanina,* Sarajevo, 1916. Along with the chronicle, Jelenić published a variety of documents. The edicts of the "Council of Konjic" are published here pp. 108 ff. For Runciman's use of the article banning the "Manichees" from building any more churches, see his *The Medieval Manichee,* Compass book edition, p. 113. Rački demonstrated that not only were the Council's articles forged but there never was such a council, see *Bogomili i Patareni,* SAN edition, pp. 462-65.

12. J. Jelenić, *Kraljevsko Visoko i samostan Sv. Nikole,* Sarajevo, 1906. The forged order to Radivoj Vladimirović is included in the Appendix.

13. On these and other forgeries see A. Solovjev, "Vlasteoske povelje bosanskih vladara," *Istorisko-Pravni zbornik,* (Sarajevo Pravni fakultet), I, No. 1, 1949, esp. pp. 82-84.

APPENDIX B FOR CHAPTER II: AN ALLEGED TRACT OF JACOB DE MARCHIA AGAINST DUALISTS IN BOSNIA

Presumably to make up for the lack of mention of dualism in sources connected with Jacob de Marchia's mission, the contemporary Croatian Hercegovinian Franciscan emigré D. Mandić.(1) reverts, it seems, to methods akin to those employed by some of the Bosnian Franciscan families of the sixteenth-eighteenth centuries. Mandić's work provides a good illustration of the pseudo-scholarship utilized to support preconceived beliefs, common in works about the Bosnian heretics.

Mandić refers to a work of Jacob which he calls "Dialogus contra manicheos in Bosna," and claims it was referred to in the 1697 investigation of Jacob's qualifications for sainthood. The book or tract, written by Jacob, still existed in 1697 but is now lost; but the chapter headings of it as recorded in 1697 have been preserved, and Mandić gives us these chapter headings as an appendix at the end of his book. As a title for these chapters Mandić writes,"Dialogus contra manicheos in Bosna." Since Mandić's book is in Croatian, one would naturally assume that this Latin title was in fact taken from the title of Jacob's tract as recorded in 1697. The chapter headings, among other things, do refer to items such as "on material churches" and twice refer to Patarins (though never specifying Bosnia); so the reader accurately concludes that a certain number of chapters truly discussed dualism. However, in the 1697 list of chapter titles the word Bosnia never appears. Puzzled, I looked at the very bad photographic reproductions of this 1697 list which Mandić publishes and nowhere could I find the title of the work as given by Mandić, which was the only link between the chapters and Bosnia. In other words, the title did not come from a document of the 1697 investigation but was a creation of Mandić's. The chapters on dualism and Patarins could easily have been written about Italians or heretics anywhere and be based on inquisitional material; Mandić simply decided to supply them with a title to connect them with Bosnia. Where did Mandić find this title? In his narrative Mandić cites a Franciscan chronicle written by a "younger contemporary and fellow countryman Fra Marianus of Florence," that Jacob when in Bosnia wrote "Dialogus contra manicheos in Bosnia."(2) Again Mandić goes to great pains to give the complete title as if he had a source saying that Jacob had written a work with such a title. In a footnote, though, we see that the chronicle does not give such a title but simply lists various of Jacob's writings including, ". . . Item alium contra manicheos de Boemia, et alium contra Manicheos in Bosna . . . ." The authority of Fra Marianus on anything to do with theology or Jacob has to be questioned. For how could anyone who knew anything about Jacob (or even theology or contemporary history) mix up the most prominent heresy of his time, the Hussites in Bohemia, against whom Jacob worked so unflaggingly, with Manichees about whose existence in Bohemia we have no evidence? And anyone who can make such a glaring error about Bohemia cannot be trusted for his data on Bosnia, where Jacob spent far less time. This is especially true if we consider the fact that Fra Marianus, as a younger man, was probably writing after Pius II and the establishment of the Manichee tradition, Had Jacob written a sermon about the Bosnians he probably would have entitled it "Against Bosnian heretics" since that is what he called the heretics in Bosnia in his letters. And it should be emphasized that Fra Marianus' chronicle does not provide the title which Mandić's narrative would have us believe it does. Thus, on the basis of this flimsy reference in the chronicle, Mandić makes up a title which he places in Latin as a heading over a document in his appendix to lead the unsuspecting reader to think it was the title of the no-longer extant work whose chapter headings he publishes.

FOOTNOTES TO APPENDIX B

1. D. Mandić, *Bogomilska crkva bosanskih krstjana,* Chicago, 1962, Chapter headings as listed in 1697 published pp. 442-44. Photostat of the original of these chapter headings, pp. 463-66.

2. *Ibid.,* p. 33. Wadding, *Annales Minorum,* vol. X, original pagination, p. 233 states, "Here (i.e., in the Bosnian vicariat) preserved in the mountains and protected places flourished the sect of Manichees against whom this servant of God (Jacob de Marchia) published a tract." Wadding, of course, put together his chronicle in the eighteenth century and thus is no source. Presumably he drew his remark about "Manichees in mountains and other protected places" from Pius II, and the remark about Jacob's tract from Fra Marianus' chronicle. Wadding, however, also provides no title for the tract.

CHAPTER III

BOSNIA FROM THE END OF THE TWELFTH CENTURY TO THE ACCESSION OF STJEPAN KOTROMANIĆ CA. 1315

I: Background: Catholicism and Orthodoxy in Bosnia and Environs

Bosnia's geographical location has made it a meeting ground between the cultures of Eastern and Western Europe. However, its mountainous terrain has limited the degree of penetration of both these cultures and has left Bosnia relatively isolated. Prior to the Slavic invasions, Christianity had, to some degree, spread through Bosnia in the fifth and sixth centuries, as evidenced by various ruined basilicas, and by the participation of a bishop from Bistue (from Bistue Nova near Zenica, at whose site are ruins of a large basilica) at two church councils in Salona (near Split) in 530 and 532.(1) Christianity seems to have been of two types — Nicene, which spread inland from Dalmatia, and Arian, which was the faith of the Visigoths, who were in the area from roughly 490 to 535. In some locations, two churches both dating from this period are found in close proximity to one another, which, it has been argued, shows the co-existence of the two faiths and considerable tolerance by the respective congregations.(2) Late in the sixth century, the waves of the Slavic invasion began. Not yet Christianized, the Slavs brought their own religious beliefs with them. Unfortunately very little concrete information about those beliefs has survived. In time, probably by the tenth century, most of the Slavs were nominally Christian. Once again, Christianity had spread inland from the coast. During this period there was a language dispute among the Dalmatian churches between the partisans of the Slavic and Latin liturgies. There is no evidence that the dispute ever reached Bosnia, however, nor is there evidence that Latin even penetrated into central Bosnia during our period until the Franciscans came in the fourteenth century.

We have almost no sources for the region of Bosnia prior to the reign of Ban Kulin (1180-1204). Except for a brief period of Byzantine overlordship in the third quarter of the twelfth century, the state was governed by a ban who ruled under nominal Hungarian suzerainty. Extant sources suggest that, at least from Kulin's time, the ban was generally able to act as an independent ruler. Bosnia was divided into many *župas* ("counties"), and presumably the tendency toward local units seen throughout Bosnia's subsequent history was already in existence. It is hard to believe that Kulin had much control over those regions away from the center of his state (the Visoko-Zenica-Vrhbosna-Kreševo area), or that he even had much knowledge of what was occurring in more distant parts of his banate.

The region of Hum, south of Bosnia, was less isolated than Bosnia from the coast and thus received more outside influences. Christianity had penetrated parts of Hum earlier and deeper. From 1168 to 1326 Hum was ruled by princes of the dynasty that ruled in Serbia under which Orthodoxy spread. An Orthodox bishopric was established for Hum in 1219 with its seat in the coastal town of Ston. In the coastal region of Hum, there already existed a Catholic population. The Serbian rulers brought into the area and supported Orthodox priests, who converted a considerable portion of the population. This caused a certain amount of conflict between the two churches, and in turn contributed to and exacerbated the quarrel between Prince Miroslav of Hum (ca. 1170-ca. 1198) and the papacy which broke out after Rainer, Archbishop of Split, had been murdered and robbed in "Neretva."(3) Pope Alexander III expected Miroslav to repay the money and punish the murderers. In the course of the dispute, Miroslav drove the Catholic Bishop of Ston from that town and refused to let the vacancy be filled. The pope then excommunicated Miroslav, who in turn allowed Orthodox priests to take over many of the Catholic churches in the vicinity of Ston.(4) Miroslav, brother of the Serbian ruler Stefan Nemanja, was clearly Orthodox. Not only did he favor the Orthodox clergy, but he also built the church of St. Peter and St. Paul on the Lim River to which he gave over twenty villages for its support.(5)

Whereas Hum thus had a mixed population of Catholics and Orthodox in the coastal regions and a predominantly Orthodox population inland, Kulin's Bosnia (ca. 1200) was nominally Catholic. All of the Bosnian state, as well as the territory north of it as far as the Sava, was included in one diocese under the Bishop of Bosnia. The bishop was consecrated by the Archbishop of Dubrovnik (Ragusa), whose suzerainty over Bosnia had been recognized in papal confirmations throughout the twelfth century up to 1187 and 1188.(6) All the Bosnian bishops mentioned in sources up to 1233 were native Bosnians.

If our sources are accurate, the Latin rite was not used in Bosnia. Writing early in the eighteenth century on the basis of archival material then in Dubrovnik, Resti reports that in 1189 the Archbishop of Dubrovnik consecrated Radigost as Bishop of Bosnia. Radigost "knew no Latin, nor other language, except the Slavic, so when he swore his oath of faith and obedience to his Metropolitan, he swore it in the Slavic language."(7) In addition, the surviving medieval Gospel manuscripts and church inscriptions from Bosnia are all in Slavic. Thus we can assume that a Slavic rite was used in churches.

II: Heresy in the Balkans and Dualism

Actual heretical movements also existed in the vicinity, but we do not know how deeply they penetrated into Bosnia. The most significant heretical movement of the twelfth and early thirteenth centuries was dualism, which in various forms and under various names existed all over southern Europe as well as in Anatolia. Dualism which believes that two principles were engaged in struggle — usually good against evil, spirit against matter — can be traced back to ancient Persia. Between 100 B.C. and 300 A.D. dualism was the dominant and common aspect of many different Gnostic sects, appearing in the Near East, some of which combined with Christianity.(8) These Christian Gnostic sects generally identified the New Testament God with the good principle who created the spiritual world and opposed him with Satan (often identified with Jahweh of the Old Testament), the evil principle who created matter, this world, and man's body. The most famous of the dualist "heresies" was Manicheeism, founded by a Persian named Mani (born in ca. 216, crucified in 275). Manichee ideas spread rapidly in North Africa and southern Europe but were successfully suppressed after a long and arduous battle by the Catholic Church.

Most scholars agree Manicheeism had been destroyed in Europe by the early Middle Ages. It survived as a small current in Persia and in some lands farther east and may have had some influence on the Paulicians who had appeared in eastern Anatolia by the early ninth century.(9) The Paulicians were hardy and brave warriors, and the Byzantine empire waged a long and bloody war against them before finally destroying their fortified capital of Tephrice in 872. In the years that followed, many of the captured Paulicians were transferred to the Balkans and settled in Thrace and Bulgaria, especially in the region of Philippopolis (modern Plovdiv).

During the reign of Tsar Peter of Bulgaria (927-969), a dualist movement arose in that kingdom. The Bulgarian movement called Bogomil, whose name means the beloved by God, had both religious and social aspects.(10) The religious doctrine was dualist. The good spiritual God created a spiritual universe, the angels, and human souls. One of the great angels, Satanel, sinned or revolted in heaven from which he was expelled. Having thus been exiled, Satan (who, as a result of his expulsion, had lost the "el" from his name) set about making his own world and created the visible and material world and all material things in it. He created man from clay, but in order to give him life he (stories vary) either induced God to blow upon the clay or else captured an angel and sealed him up in the clay body. Through sex and reproduction the human race of souls — angels imprisoned in human bodies — has been perpetuated. Finally, taking pity on man, God sent a spiritual being — Christ — to earth to bring his message of salvation. The human aspects of Christ's life were only apparent and after his apparent death he returned to God in heaven. The way to salvation was basically through an ascetic life, renouncing material things.

Society was divided into two orders: the layman and the initiated perfected Christian. The latter, having undergone a spiritual initiation, became possessed of the Holy Spirit. If he did not sin thereafter, upon his death, his soul escaped from matter forever and returned to heaven. The uninitiated mortal who died without this initiation had to endure another worldly existence with his soul once again imprisoned in another body. The perfected Christians led ascetic lives, worshipped (for the Holy Spirit within them) by an act of adoration by laymen. They rejected as many material things as possible, renounced family ties, recited the Lord's Prayer a prescribed number of times a day, did not work, lived on the charity of believers, kept long fasts (never eating meat or drinking alcohol), and rejected marriage and sex. Linking the Old Testament God with Satan, they rejected that book and all the patriarchs and prophets in it, since they believed those men served Satan. In practicing their religion, they rejected all material aspects of the service (e.g., church buildings, icons, crosses, baptism with water — hence their hatred of John the Baptist — and the bread and wine of the Eucharist).They had a very simple service held in private homes which consisted of the Lord's Prayer, the ceremonial breaking of bread, and the exchange of benedictions with the believers.

Bogomilism soon acquired the character of a social movement. Adherents believed they should not obey their masters, pay taxes, or fight wars. Such teachings must have been attractive to the Bulgarian peasantry, who had suffered greatly during Tsar Symeon's (893-927) long wars with Byzantium. Since the Bogomil movement seems to have been quite widespread in Bulgaria and Macedonia, it seems probable that these social aspects contributed to its successes far more than its theology. It is hard to conceive that the Bulgarian peasants would have been any more likely to be attracted to a theologically speculative doctrine (particularly one that rejected this world) than the Bosnian peasants whom we discussed in the first chapter.

Bogomilism spread to Constantinople, where its philosophical ideas attracted a following among the intellectuals and pseudo-intellectuals of the upper classes. From there its doctrines spread into Anatolia and also, through merchants and crusaders, to the West. It quickly attracted a sizable following in northern Italy and southern France, where a variety of dualist churches were formed.(11) In France the dualists were called Cathars (the Pure Ones) or Albigensians (after the town of Albi, one of their centers). The crusade sent against them, which eventually destroyed them (1208-1228), is called the Albigensian Crusade. In northern Italy, the dualists were often called Patarins.

These Western Europeans kept up their contacts with their co-religionists in the East. In 1167 a bishop named Niketas from Constantinople appeared in Saint-Felix-de-Caraman near Toulouse to attend (and dominate) a Cathar church council.(12) Later in the century, a Lombard Cathar bishop named Nazarius went to Bulgaria for baptism and brought some sacred Bogomil texts back to Italy. Inquisition sources show that all the Lombard churches derived their baptism and doctrine from one of three Slavic centers — Bulgaria, Dragovica (near Philippopolis)(13) or "Sclavania."

Bulgarian Bogomil influence spread west chiefly via the heretical church founded in Constantinople, which had extensive contacts with the Western world. Bulgarian dualism also spread directly from Bulgaria to the rest of the South Slavs in the eastern Balkans. The movement is attested in Macedonia, Thrace, and possibly Serbia in the twelfth century, although we have no evidence of it this early in Bosnia, Hum or Dalmatia. Late in the 1170's, Stefan Nemanja held a church council which exiled some heretics from Serbia.(14) Even though the description of the council in his *Life* gives no details that would help identify the nature of their heresy, most scholars assume that the heretics were Bogomils. This source also does not state where the exiled heretics went. They may well have fled to Macedonia or Bulgaria; however, it is also possible that some fled into Hum and Bosnia. When we begin to hear of heresy in Bosnia in the thirteenth century, only Anselm of Alexandria expressly links that heresy with heretics from any region east of Bosnia.(15) However, most of our material about Bosnia comes from Western sources, which lacked information about Bosnia's ties with the East.

The sources for the late twelfth and early thirteenth centuries also mention heretics in Dalmatia, but neither the papal agents who visited Dalmatia nor the Dalmatian clerics who wrote about local heretics give any details about their doctrines.

Thomas the Archdeacon of Split, whose chronicle of his city was written prior to 1268, tells a story of two brothers, Aristodios and Matheus, born in Apulia and brought up in Zadar. They were goldsmiths by trade, and frequently went to Bosnia in connection with their work. They knew Slavic and Latin. They fell into heresy and began to teach it with much success. We are not told where they acquired their heresy or what the nature of their heresy was. Archbishop Bernard of Split tried to convince them to give up their heresy but failed; so he confiscated their property and exiled them from the city. We are not told where they went. They may well have returned to Zadar, which Thomas elsewhere tells us was then a hotbed of heresy (without stating what sort). Smarting under their heavy punishments the two goldsmiths recanted, accepted Catholicism again, and recovered their lost property; so did all their ''misguided'' victims.(16) Since we find Matheus witnessing a decision to award some village lands to the monastery of sv. Krševen (St. Chrysogonus) in 1198,(17) we may suspect that their renunciation of heresy occurred prior to 1198. Thomas never hints that the brothers relapsed into heresy again after their reconciliation with the church. Much has been made of this story, but it really tells us only that two cultivated and persuasive artisans were heretics for a time and often went to Bosnia. They may well have been in Bosnia at some point that coincided with the time they were heretics. Since the brothers seem to have returned to the fold before 1198, we probably cannot link them with the Dalmatian heretics, whom we shall meet shortly, who fled to Kulin's court from Split in 1200.

Whether the two brothers were dualists or not is unknown. However, it does seem likely that there were some dualists in Dalmatia. Inquisition sources speak of dualists in ''Sclavania.'' The earliest of these inquisition tracts is ''De Heresi Catharorum in Lombardia,'' written about 1210, which describes the chaotic state of the Cathar church in Lombardy during the preceding decades when it splintered into six distinct churches. For ordination and establishment, each of these churches turned toward the Slavic world. The two largest Lombard churches turned to the churches of Dragovica and Bulgaria. Two others had bishops who received their ordinations in ''Sclavania'': Caloiannes, Bishop of Mantua (a church later known as Bagnolo), and Nicola, Bishop of Vicenza. We shall concern ourselves only with these last two Cathar churches. ''Sclavania'' is a vague term referring to the whole South Slavic world; but since Bulgaria and Dragovica had previously been distinguished, most probably it refers to the Slavic world excluding Bulgaria and Dragovica (i.e., the general area of Dalmatia, Croatia, Bosnia, Zeta, and Raška, or some specific place within that area).

The "De Heresi Catharorum in Lombardia" gives a fairly detailed description of some of the sophisticated dualist beliefs held by Caloiannes, which were also shared by the church that derived its doctrine from Bulgaria.(18)

Bosnia is not mentioned in this document. Since elaborate doctrines could only have arisen among at least semi-educated people having some acquaintance with theology, it seems plausible to place the center of this church of the Sclavani in Dalmatia, where intellectual life was considerably more developed, rather than in rural Bosnia. In addition, Italian was widely spoken in Dalmatia, which would have facilitated communication between Italians and Dalmatians. In fact, Dalmatia had contacts, through commerce, with both Italy and Constantinople (where there were Bogomils). It is quite possible, however, that during persecutions on the coast the leadership of this Dalmatian current had sought refuge in Bosnia and may even have attracted some Bosnians to its beliefs. It must be stressed that it is probable that some, or even many, of the beliefs held by the Italian churches and described in the document, regardless of where in the Slavic lands they derived their baptism, came from their own milieu — namely French and other Italian Cathars.

A second inquisition treatise, "Tractatus de Hereticis," which A. Dondaine attributes to Anselm of Alexandria and dates about 1270, also describes the period from the late twelfth century. This document contains much detail on the splintering and factionalism of the Lombard Cathars at the end of the twelfth and the beginning of the thirteenth centuries, and gives details on their hierarchies up to the date of writing. The most interesting part of this tract for us is the very beginning, where the author traces the origins of the movement. This is the first reference to Bosnia in the inquisition literature as well as an explanation about the origin of the Bosnian heresy.

Anslem starts his tract with Mani in Persia and his doctrine of two principles.

> And he (Mani!) taught in the region of Drugontie (Dragovica) and Bulgaria and Philadelphia; and heresy was spread there so that they founded three bishoprics: Drugontie, Bulgaria, Philadelphia. Afterwards the Greeks of Constantinople, who are three days from the borders of Bulgaria, came thither on account of commerce, and returned to their land and after having spread heresy there, founded there a bishop called the Bishop of the Greeks. After that the Franks

> came to Constantinople in order to subjugate the land and came into contact with that sect, and having (its doctrines) spread among them, they made their own bishop, who was called Bishop of the Latins. After this certain people from Sclavonia, that is from the land called Bossona (Bosnia), came to Constantinople on account of trade. They returned to their land and taught (heresy) and it spread and they constituted a Bishop who is called the Bishop of Sclavonia, that is of Bosnia.(19)

The author then goes on to speak of the spread of heresy to France and Italy.

This document written in about 1270 is the earliest extant reference to Bosnia in the inquisition literature. Anselm clearly equated Sclavonia with Bosnia and in fact states that the church of Sclavonia was, from the beginning, Bosnian. Dondaine dates the Franks' hope to conquer Constantinople to the crusade of 1147, but it is equally plausible to think Anslem was referring to the Latin expedition that really took the city in 1204. After the arrival of whichever crusaders Anselm had in mind, Slavic merchants came. Thus he would date the formation of the Church of Slavonia after 1147 or after 1204.

It is unfortunate that we do not know the source upon which Anselm relied for this information. Writing at least sixty years (if not over a century) after the event, Anselm is not a firsthand authority. I find it hard to believe that heresy could have arrived in "Slavonia" in just the way he described. There is no evidence other than this one work of Anselm (a foreigner who probably never visited the South Slav lands) that at any time in the twelfth or thirteenth centuries any Bosnian merchants went to Constantinople. However, Dalmatian merchants (from whom Bosnians bought goods) *did,* so it is more probable that, if heresy came to "Sclavonia" in this way, Dalmatian merchants, trading in the East, had brought it. In fact, this is a very likely way for dualist ideas to have come to Dalmatia and then, with Dalmatians as middlemen, to have spread inland from Dalmatia to Bosnia.

Anselm, however, is a source that clearly states that there were dualists in Bosnia in the thirteenth century. Since he links these dualists with merchants and the general term "Sclavonia" we can conjecture that the Bosnian dualists and the Dalmatian dualists were part of one current. Because of Anselm's emphasis on the Bosnians, we can conjecture that by 1270, when Anselm was writing, the Bosnians were thought to be more important or numerous in this current than any other people who lived in the region of Sclavonia. Though it is a likely conjecture, we cannot be certain from the above, however, that all or even most of the Dalmatian heretics referred to by Thomas the Archdeacon of Split and other sources were necessarily dualists. And we cannot assume that dualism was necessarily a sizable movement in Bosnia.

Confirmation that the Church of Sclavonia (Ecclesia Sclavoniae) included both Bosnia and Dalmatia is found in a Dominican source written in 1259, shortly before the composition of Anselm's "Tractatus." In his "Commentary on the Founding of the Hungarian Dominican Province," Suibert (Peter Patak), a prior of a Hungarian monastery, wrote of the Dominicans being sent at the time of Coloman — i.e., during the Hungarian campaigns against Bosnia and environs in the 1230's — to Bosnia and Dalmatia, where the Church of Sclavonia (Ecclesia Sclavoniae) was to be found and where many souls perished from heretical errors.(20) The author never elaborated on the nature of the errors of the Sclavonian church. But because the Dominicans were closely connected with the inquisition which called a dualist current "the Church of Sclavonia," it is reasonable to connect Suibert's term with dualists called by that name. This would indicate that there were some dualists in Bosnia, and that, as we have just suggested, Bosnian and Dalmatian dualists were part of the same church.

III: Ban Kulin and Catholicism

From 1180, when Ban Kulin is first encountered in the sources, until 1199, there is no reason to view him as anything other than a Catholic. In 1180 he received a friendly letter from the pope treating him as a dutiful son of the church.(21) Kulin's charter to the merchants of Dubrovnik in 1189 is couched in typical Catholic style, beginning with an invocation to the Trinity and ending with the date, given as the Day of John the Baptist.(22) We also have inscriptions from two churches erected during Kulin's reign, one by the ban himself and the other by one of his judges (sudija).

The inscription from Kulin's church appears on a stone plaque (now in the Zemaljski muzej in Sarajevo) found not far from Visoko at the village of Muhašinovići.(23) The plaque has on it six artistically carved crosses within circles and bears the following inscription: "Ban Kulin built this church when . . . Kučev'sko Zagorie and found for it stones (?) at Podgorie Sljepičist' and placed his image (portrait) over the threshold. God, give health to Ban Kulin and to (his wife) Banica Vojslava."(24) Thus the ban built a church and put up a plaque with six crosses over the entrance, and somewhere (presumably a fresco inside above the door) placed his own portrait. Since, as we shall see shortly, Orbini describes two Bosnian Catholic churches being dedicated in

1194, most scholars have attributed Kulin's church to that date. The "this-world" orientation of the Bosnian is reflected in the prayer which asks God to give the ban and his wife health, rather than to save their souls.

There is further writing on the plaque, a carved figure and some scratched symbols. Some historians have argued that these items were part of the original plaque. However, even a superficial examination of the plaque shows that this is impossible. The six crosses and Kulin's inscription are carefully and artistically placed and carved. The other markings, on the other hand, are crudely scratched across the face of the plaque and clearly deface it.(25) Professor Andjelić has succeeded in deciphering as much of this writing as is legible and has proved that it is subsequent to Ban Kulin since it states in part "Desjen' Rat'n'cevit writes in the days of Ban Stjepan and may God give him health and long life."(26) Because of the style of writing, Andjelić dates the second inscription as fourteenth century and places it in the reign of Ban Stjepan Kotromanić (ca. 1318-1353), though he points out the possibility of it being from the time of that ban's father, Stjepan Kotroman or, if his analysis of the writing style be incorrect, even from the reign of Kulin's son, Stjepan.(27)

The second church inscription was found in 1964 in the village of Podbrežja, three kilometers to the west of Zenica. It reads: "In the days of the Great Ban Kulin, Gra(d)ješa was a (or the) Great judge in it and (he) built (the church of) St. George. And he lies (was buried) in it and his wife said, 'put me in it' (also, when I die). Draže O(h?)mučanin (the master who carried out the building) built (it)."(28) The inscription continues with the wish that God appreciate the church (or the builder) and then owing to lacunae becomes unintelligible.

In 1194, Orbini tells us Archbishop Bernard of Dubrovnik consecrated a church in Hum, after which he went to the realm of Bosnia, to which he had been invited by Kulin Ban. There he was received regally by Kulin, consecrated two churches, and then returned home.(29) The same story about the consecration of the two churches is given in the Ragusan chronicle of Nicolo di Ragnina, compiled in the second half of the sixteenth century; Ragnina, however, dates Bernard's visit in 1190.(30) It is possible that the two churches Bernard consecrated were these two from which we have inscriptions. Because Kulin invited the archbishop to visit his realm, he could hardly have thought anything was amiss there from a religious standpoint. The archbishop, having carried out the consecrations, apparently went home without further ado, which suggests that everything in Bosnia appeared to be in order to him also.

Further evidence of Kulin's loyalty and good will toward Catholicism and the pope is contained in an unknown source used by both Orbini and Resti. Both wrote that when a certain Radigost arrived in Dubrovnik to be consecrated as Bishop of Bosnia by the archbishop, he brought presents for the pope sent by Kulin.(31) Resti's date of 1189 should be accepted in preference to the 1171 date given by Orbini, because we have no evidence that Kulin had become ban by 1171 and because the earliest chronicle compilation which discusses Radigost (i.e., that of Nicolo di Ragnina) states that Bernard, who became archbishop in 1185, consecrated Radigost.

In describing Kulin, Orbini says the ban "was a pious man, very religious, with affection for the Roman Pontiff."(32) Perhaps he drew this statement from another source; perhaps he came to that conclusion himself, on the basis of Kulin's gifts to the pope and the ban's invitation to the archbishop to consecrate two churches. In any case, Orbini's conclusion seems in agreement with everything else known about the ban prior to 1199.

IV: Charges of Heresy and Foreign Interference

The first act of outside interference which may have had an important effect on the course of future developments occurred in 1192, when the pope transferred Bosnia from the jurisdiction of the Archbishop of Dubrovnik to that of the Archbishop of Split.(33) This act is usually, and probably correctly, attributed to Hungarian machinations. Hungary, having close ties with Split, probably sought to assert, by means of the church, its authority over Bosnia. Bosnia seems to have had friendly ties with Dubrovnik, and there is no evidence that Kulin ever accepted the new arrangement, corresponded with the Archbishop of Split, or even let him consecrate any Bosnian bishop. If Orbini is correct in dating Archbishop Bernard of Dubrovnik's consecration of the Bosnian churches in 1194, this may give us reason to assume that Kulin simply ignored the new arrangement and continued to deal with Dubrovnik.

However, it is dangerous to rely on Orbini for dates about this period. His dating of Radigost's consecration as bishop was incorrect by eighteen years. In addition, the sixteenth-century chronicle of Nicolo di Ragnina gives 1190 for Bernard's visit, a perfectly reasonable date, and one that would place the archbishop's trip at a time when Bosnia was still officially under Dubrovnik. Unfortunately, then, we must conclude that we cannot ascertain with certainty the effects of and the reactions to this alteration in ecclesiastical jurisdiction. We can only suspect that the Bosnian reaction was unfavorable and that Kulin may have protested or ignored the change. And these suspicions are confirmed by the fact that we find Kulin dealing with Dubrovnik, and not with Split, in 1202 and 1203.

We do not have long to wait for mention of heresy in Bosnia. In 1199, Vukan the ruler of ''Dalmatia and Dioclea'' (Dioclea being an earlier name for Zeta and Montenegro) informed the pope that ''significant heresy is seen to sprout in the land of the King of Hungary, namely Bosnia, in such numbers that Ban Kulin himself, his wife, and his sister, the widow of Miroslav of Hum,as well as many of his relatives, have been seduced by the sin, and he has led more than 10,000 Christians into this same heresy.''(34) Vukan's emphasis on the adherence of Ban Kulin and the ban's relatives to the heresy makes one wonder if Vukan did not have ulterior motives when he called on the pope to take action against heresy in Bosnia. Ćirković points out that considerable tensions existed among the different Balkan rulers at this time, particularly between Vukan and his brother Stefan who were struggling for the Serbian throne. In 1202 Kulin was to attack some lands subjected to the Hungarian king; Ćirković suggests these were Vukan's lands. It is possible that the causes of friction that led to this attack already existed in 1199. We also know that, in his war against his brother, Vukan was supported by the King of Hungary, Imre, who also had designs on Bosnia.(35) The figure 10,000, like all such figures in medieval sources cannot be taken literally.

In October 1200, Pope Innocent wrote King Imre of Hungary asking him to take action against the heresy. The pope stated that he had heard ''that when recently our brother (Bernard) the Archbishop of Split expelled not a few Patarins from the cities of Split and Trogir, the noble man Ban Kulin of Bosnia not only gave asylum to their evil but bestowed upon them open help and (let them) display their evil to himself and his land and honored them the same as Catholics, nay even more than the Catholics, calling them by the name they called themselves Christians.''(36) When we turn to the Chronicle of Thomas Archdeacon of Split (written in the middle of the thirteenth century), we find much material about Archbishop Bernard's activities against the heretics of Split, but nothing on the nature of their errors. All that has been given us is the label ''Patarin'' that the pope attached to the heretics.

But if we consider the way this term was subsequently to be used in Bosnia, we see that we cannot assume that it necessarily referred to doctrines akin to those held by the sect of that name in northern Italy.

That Innocent may well have had this association in mind (or even called the heretics "Patarins" for this reason), however, will be seen in his letter of 1202. And since we have found reference in inquisition documents to dualism in Dalmatia (Sclavonia), making it likely that many of the Dalmatian heretics were dualists, and since we shall find references to dualism in Bosnia in various later inquisition documents, it is quite likely that these Dalmatians seeking refuge in Bosnia were dualists. Thus, this may well have been the beginning of a dualist current in Bosnia. All we know for certain, however, is that Innocent called these Dalmatian heretics Patarins and said that they called themselves Christians, and that not a few of them seem to have found asylum in Bosnia.

In November 1202, Pope Innocent wrote Archbishop Bernard of Split that "in the land of the noble man Kulin Ban there are a multitude of certain people who are strongly suspected of (belonging to) the condemned Cathar heresy . . . But he (Kulin) in his own justification answered that he believed them not to be heretics but Catholics, and that he was prepared to send some of them on behalf of all to the Apostolic See, in order that they demonstrate to us their faith and way of life, and that by our judgment they be confirmed in what is good or be converted from (what is) evil . . ." Kulin had recently sent to the pope an embassy including the Archbishop of Dubrovnik, Marinus the Archdeacon of Dubrovnik, and some of the above-mentioned alleged Cathars. He requested that the pope send a qualified legate to his realm to investigate matters. The pope decided to send Johannes de Casamaris thither to investigate and asked for the cooperation of the Archbishop of Split.(37) The sources say nothing about the beliefs and practices of the Bosnians Kulin dispatched to Rome. The pope clearly suspected them of Catharism, yet seems to have had some doubts as to whether they really were Cathars. We do not know whether they were Dalmatians who had fled earlier to Kulin's court, or Bosnians who were or were not linked with the Dalmatians. If Bosnians had been sent, if we consider their lack of theological education, it would not have been at all surprising that Innocent could not be certain of whether they were dualists or not. Part of the pope's concern that there may have been Cathars in Bosnia could well be owing to his preoccupation with that movement elsewhere (particularly in southern France) and his knowledge that Slavic dualists were active in the vicinity of Bosnia. The sources never again use the word Cathar in connection with Bosnia. Even in his letter to Bernard, Innocent said only that the heretics in Bosnia were *suspected* of being Cathars. Although there may well have been some dualists in Bosnia and although segments of the population may have acquired certain views or practices from dualism, we shall see that the main current of beliefs and practices in Bosnia was not Cathar.

Kulin's excuse that he thought the accused heretics were good Catholics is often interpreted as clever statesmanship in an attempt to buy time. However, because everything else we know about Kulin (excluding Vukan's tendentious letter) suggests that he was loyal to Rome, and because it is unlikely that the ban had a profound grasp of theological matters, I find it believeable that Kulin, if he really did have contact with the Dalmatian emigrés, actually accepted these more educated people from the coast as good Christians. His good faith is also seen by the fact that he sent some of the people, about whose faith the Catholics in Dalmatia had doubts, to Rome for the pope to examine.(38)

The possibility that ulterior motives were lurking behind the accusations against Bosnia about heresy should always be kept in mind. The Hungarians may well have hoped to use a charge of heresy as an excuse for military and political intervention, even for conquest. The Archbishop of Split may also have been deliberately maligning Bosnia. Having been awarded jurisdiction over Bosnia in 1192, he seems to have had no chance to assert his authority, since the Bosnians continued to maintain relations with his rival in Dubrovnik. If he was offended, he may well have believed that to show Bosnia as heretical would demonstrate Dubrovnik's inability to handle Bosnia's religious affairs and might lead to papal and Hungarian intervention, which could force Bosnia to submit to the Archbishop of Split. And we notice that much of the complaint about heresy in Bosnia that reached the papacy appears in correspondence with (or concerning the work of) the Archbishop of Split.

V: The Bilino Polje Renunciation

The legate, Innocent III's own chaplain, Johannes de Casamaris, was duly sent to Bosnia. Before him at Bilino polje on the sixth of April 1203 there was enacted a renunciation of errors. This abjuration provides us with some actual details about religious practices in Bosnia. The surviving document is generally thought to be a contemporary Latin translation of a now-lost Slavic original. Prior to its proclamation, Johannes de Casamaris had spent several months at Kulin's court and thus presumably had some notion of the state of affairs there. Marinus, Archdeacon of Dubrovnik, also witnessed the abjuration, and, knowing both Slavic and Latin, would have been able to insure some degree of communication between the legate and the Bosnians.

At Bilino polje were gathered "the priors of those men who up to now have alone been called by the prerogative of the Christian name in the territory of Bosnia'' on behalf of all those of their brotherhoods — i.e., the priors of several monasteries representing only their order and its membership. Since, in addition to the names of Kulin and Marinus, seven other names were appended to the document, there were seven priors, and therefore at least seven monasteries in Bosnia at the time. Besides the priors, present were: Johannes de Casamaris, the Archdeacon Marinus, and Ban Kulin, whom the priors called ''patron Ban Kulin lord of Bosnia.''

The priors promised to accept the rites and commandments of the Roman Church and to live according to them. Then they pledged themselves and their property (landed and movable), on behalf of their brothers, that at no future time would they ever pursue the depravity of heresy. There follows a series of promises to follow certain particular Catholic practices properly in the future. Obviously, the monks had not previously followed these practices properly, but we do not know exactly how or why they had erred.

They ended their schism with Rome and recognized her as the mother church. They promised that in places where they had convents, they would hold services where the brothers would participate morning, day and night, and would chant the hours. Henceforth they would have altars and crosses in their churches, and they would read, as the Catholic Church does, from the books of the Old and the New Testament. In all their convents they would have priests who at least on Sundays and holy days would serve the mass according to church prescriptions. These priests would hear confessions and administer penance. Beside their churches there would be established graveyards where they would bury brothers and also travelers who happened to die at the monasteries. At least seven times a year they would receive communion from a priest — namely at Christmas, Easter, Pentecost, St. Peter's and St. Paul's Day (a single festival for Catholics), the Day of the Virgin's Assumption, the Day of her Nativity, and All Saints. They promised to observe fasts and church festivals as ordered by the Church. Women of their order would be separated from the men both in sleeping and in eating arrangements, and no brother would talk to one of the women alone lest it give rise to evil suspicions. They would not accept married people into their order unless both parties agreed. In the future they would not accept into their order anyone they knew to be a Manichee or any other sort of heretic. They promised to wear particular dress, a habit without colors reaching to the ankles, to distinguish themselves from laymen.

They promised, that from now on, they would call themselves not ''Christians,'' as they had done up to then, but ''brothers,'' so that by monopolizing the name (Christian) injury is not done to other Christians (i.e., people other than themselves, who as Christians would want to use that name). On the death of each magister (presumably the head of the order), the priors with the consent of the other brothers would henceforth select a new magister to be confirmed by the pope. Whatever the Roman Church might want to add to or subtract from this document would be acceptable to them and they would accept it with devotion. The document was then signed by Dragice (Dragite), Lubin, Drageta (Brageta, Bergela), Pribis, Luben, Rados, Bladostus (Bladosius), presumably the priors of seven monasteries,(39) Ban Kulin, and Marinus Archdeacon from Dubrovnik. As a postscript to the document is the repetition of the renunciation by two of the priors before the King of Hungary and the Archbishop of Kalocsa. The visit to the Hungarian king will be discussed later.

The fact that Marinus of Dubrovnik was present, instead of someone from Split, suggests that Bosnia had kept up its connection with Dubrovnik and had not submitted to Split. That the renunciation was repeated before the Hungarian king reflects the status of the king as overlord of Bosnia.

The promise in the document that the priors would not pursue heresy in the future suggests one of two things: 1) that until 1203, the priors had had certain heretical practices (possibly dualistic or possibly errors through ignorance) which the legate noted, pointed out to them, and, possibly owing to extenuating circumstances, decided not to charge them with. However, even if they had had heretical practices, it is probable that they had not been full fledged heretics (i.e., following complete heretical rituals and doctrines); for if they had been, Casamaris surely would have made them renounce heresy in general as well as their specific errors in doctrine and practice. It is almost impossible to believe that a legate of Innocent III would have excused members of a determined heresy from renouncing heresy, or (2) that the priors were not implicated in heretical practices, but Casamaris, knowing that heretics were in the area and that these monks were ignorant of theology, feared the monks might fall under the influence of heretics and thus by this article tried to anticipate and prevent this.

There then follows a renunciation of schism. What was the schism about? Had Rome broken off relations pending the investigation? Or had Split, angry at being ignored, broken off relations with Bosnia? The answers to these questions are unknown.

The men renouncing schism and possibly heresy were priors of monasteries who acted only on behalf of their monasteries. Here, then, was a group of Catholic monks who were abjuring errors or bad practices which they had acquired. No mention is made in the document of Bosnians in general, the supposed 10,000 heretics, heretics from Dalmatia or heretics in Kulin's family. Although Kulin witnessed the proceedings, we do not know how closely he was associated with the heresy or schism of the monks. They called him "patron", but whether this meant that Kulin had actually aided their order or had given it gifts or whether this term merely reflected Kulin's status as ruler or patron of his people is unknown. In fact, since the document is most probably a translation from the Slavic, we do not know what Slavic word, the Latin word "patronus" was meant to render. I can think of no word found in titles or in Bosnian documents that bears the connotation of "patron"; thus it is possible that "patronus" is a poor translation of some word reflecting Kulin's status as Lord of Bosnia. And even if Kulin had patronized them, he could easily have done so believing these men to be good Catholics.

There is no evidence that the leaders of any heretical movement were present at Bilino polje. This all suggests that Casamaris' personal investigation had turned up no hard-core heretics. This, of course, only means that the legate did not find them. It does not mean that such heretics were not to be found somewhere in Bosnia. After all, Bosnia was a large region with rugged mountains and poor communications. To cross the country was then a trip of at least ten days.(40) However, had Casamaris heard about the existence of heresy anywhere in Bosnia, it is inconceivable that he would not have at least suggested prolonging his mission or sending further missions. But he did not. Casamaris seems only to have found deep ignorance among the monks and tried to correct that, presumably with the hope that they, in turn, would be able to provide proper church services and thereby straighten out the errors that must have existed among the general lay public.

There is almost nothing in the document that points to dualism with certainty. The absence of a cross in a church may well have been interpreted by the pope and his agent as a dualist practice and may have led them to suspect dualism. It may even have been that this lack was owing to the influence of dualists (directly or indirectly) upon the Catholic monks. However, we could also explain the absence of crosses on the basis of ignorance of practice; the monks may not have realized that it was necessary to have crosses in their churches. Thus the absence of a cross need not indicate a conscious rejection of it because of dualist theology.(41) We also do not know how prevalent this "error" was; it is possible that some churches had crosses and others did not and Casamaris wanted simply to make them universal. It seems to me that the absence of various holy items from churches can most easily be explained by the monks' abysmal ignorance about how churches should be furnished. In no case is it certain that any of these items were generally lacking in all churches. It is probable that the errors of one monastery would frequently have differed from those of a second, and that in the abjuration, Casamaris had simply lumped together in one general condemnation all the deviations he had found during his investigation.

The statement in the abjuration about graveyards has been interpreted by scholars to mean that at the time the Bosnians had no cemeteries; and this lack has been attributed to the dualist rejection of material things and the denial of bodily resurrection. However, the text of the abjuration merely says that henceforth the brothers would bury their dead beside churches. Thus Casamaris was trying to enforce the church canons requiring church burial for clergy. We find numerous fourteenth- and particularly fifteenth-century gravestone inscriptions in Bosnia and particularly in Hercegovina, referring to nobles, being buried on their own land. This suggests that there was no local custom against burial of the dead. We can conjecture that the monks prior to 1203 had either been buried on their own family estates or else in unconsecrated village cemeteries. The aim of this article, then, was only to make the monks follow the church canons, laymen presumably could continue to treat corpses according to their customary practices, whatever they were.

The document's statment about Manichees says simply that the priors would not accept them or any known heretics into their order. It does not say that the order had been accepting Manichees. And one could argue that the reason Manichees were singled out and named was because the papacy was preoccupied with dualists elsewhere in Europe at the time. However, that Manichees were singled out, it can be soundly argued, implies that at some time in the past the order had accepted into its ranks people who were Manichees, or who Casamaris thought were Manichees. Casamaris may have been in error and have assumed that certain practices (e.g., absences of crosses) resulted from Manichee influence. In such a case we can be certain that the uneducated Bosnians, who probably had no idea what a Manichee was, would not have been able to understand the accusation sufficiently to prove to the legate that they were innocent of the charge. Yet we do know that there were dualists in the vicinity of Bosnia, and very probably within its borders by 1203 as well. The order may well have unwittingly accepted some dualists, or people influenced by them, into its ranks, not realizing that they were heretics. However, this does not make the Bilino polje monks dualists, and their abjuration strongly implies that they were not.

Surely, if Casamaris had considered the monks dualists, they would have been forced to renounce Manicheeism itself, as well as belief in two principles and a whole series of beliefs and practices connected with that creed.

Thus our document is not a renunciation of dualism. Through ignorance the Bosnian monks had fallen into grave errors in practice and in discipline, but there is no reason to believe that any of their errors were made consciously or were owing to intentional rejections of practices on theological grounds. In fact, the document never touches on matters of belief at all.

Nor is there evidence that the Dalmatian heretics had any connection with or influence upon the monks at Bilino polje. We certainly do not need to postulate ties between any heretics and the monks to explain the situation that was being rectified at Bilino polje. The whole document therefore reflects only a reform of discipline and practice in the spirit of the Catholic Reform movement.

Finally, we may note that the monastic order was composed of monks, calling themselves "Christians," who lived in monasteries, each of which was under a prior. The order itself was headed by a "magister" (we would like to know his Slavic title), who henceforth was to be elected by the priors with the consent of the brothers and confirmed by the pope. The absence of the magister from the important proceedings at Bilino polje is noteworthy and gives us reason to believe that the position was vacant at the time. The situation parallels (or may even be identical to) that of the fourteenth century when we shall find the Bosnian Church based upon a monastic organization, whose clergy bear the Slavic rendering of the word Christian, krstjanin. Thus we may suspect that by 1203 the organization of the future Bosnian Church already existed. This monastic order of 1203 may well have started out a century or so earlier as a branch of a certified Catholic order, such as the Benedictines, but have subsequently lost contact with its original order and the rest of the Catholic world and have come to be an autonomous, self-perpetuating "Catholic" body, under its own locally elected priors and magister. Quite possibly it was simply this autonomy which constituted its schism. As we have seen, despite certain errors, some of which could well have been owing to heretical influences, there is no reason to consider the order heretical.

The monks, in calling themselves "Christians", seem to have claimed that they alone among all the people of Bosnia had the right to use that name. This obviously bothered Casamaris since their monopolizing the name, would imply that other Christians were not really Christians. Since the priors stated that they claimed this prerogative only in the territory of Bosnia, we suspect that they were slandering, not Catholic clerics and believers beyond the Bosnian borders, but laymen who were Catholics in Bosnia. Since in the fourteenth century the term "krstjanin" was to designate only ordained monks in the Bosnian Church, we must speculate that, already in 1203, the Bosnian monks believed that a man had to be ordained as a monk to be a Christian.

Another reason that may have caused them to insist on their exclusive right to the name "Christian" is the fact that all groups in Bosnia, claiming to be Christians, called themselves by this name — the Dalmatian heretics who fled to Bosnia in 1200 were self-styled Christians, the Bilino-polje monks were Christians, shortly we shall find the Bosnian Church clerics calling themselves krstjani, and in fifteenth-century Slavic sources we shall find Catholics calling themselves karsteni (or what was to become the standard form, kršćani) and the Orthodox, Hrišćani. Thus any group in Bosnia, faced with Christian rivals using the same name, would have had to insist on its exclusive right to this name.

Most scholars have failed to realize that the Latin documents of 1200-1203, which we have examined, may well deal with two different phenomena. In 1200, Innocent was complaining about Dalmatian heretics being received at the Bosnian court. The 1203 renunciation clearly was made by Bosnians. It has usually been assumed that the Bosnians had acquired errors from these Dalmatians and that Casamaris, the papal legate, had been dealing with a single Bosnian-Dalmatian movement. However, if I am correct in arguing that the Bilino polje renunciation shows only that the monks, through ignorance, were abusing church practices and says so little about heresy, we may well conclude that there had been no connection between these monks and the Dalmatians. Thus the abjuration sheds no light on the nature of the Dalmatians' heresy. The few Dalmatian heretics, of unknown doctrine (very likely dualist), who had come to Bosnia and taken refuge at Kulin's court, must be considered a separate phenomenon. There is nothing in the legate's reports to the pope about Dalmatians. Were they still in Bosnia and kept out of sight of the legate? Had Kulin, on learning they were heretics, exiled them from his court or even from his land? In any case, the Dalmatians are never heard of in Bosnia again. If they had been dualists, since from time to time we shall hear of dualism in Bosnia, we may conclude that their teaching had borne some fruit and had attracted the allegiance of various Bosnians.

But because our local sources do not hint of dualism in Bosnia until the fifteenth century, we probably are justified in concluding that dualism never acquired a large following in Bosnia. This does not mean, however, that the pope did not think it had or was not afraid that it might; thus, many of the papal references to "heretics" throughout the thirteenth and following centuries may well be to the dualists. He also may well have believed, or been led to believe, that other religious bodies in Bosnia, which were actually quite independent of the dualists had been infected with their heresy. The Hungarians, who sought an excuse to reassert their authority over Bosnia, may well have tried to convince the pope that dualism was widespread in Bosnia, and that institutions which were actually orthodox had fallen into this heresy. Such a view of Hungarian intentions is supported by Suibert's "commentary" (written in 1259),(42) which suggests that the Hungarian crusade of the 1230's was sent against the "Church of Sclavonia" (i.e., dualists).

However, there is no evidence that our monks at Bilino polje in 1203, or even later when they were to break away from Rome and form their own independent Bosnian Church, were influenced to any extent by, or even had any contact with, the dualists. Thus the Dalmatians (probably dualists) should be regarded as separate from the movement of well-intentioned but misguided Bosnian monks who were the mainstream of Bosnian formal religion. In May, two of the priors, Lubin and Drageta, visited the King of Hungary and repeated the promises made at Bilino polje.(43) Later that year King Imre wrote the pope about a visit to his court made by Kulin's son. His letter is undated but it clearly was subsequent to the visit of the two priors. The king refers to Casamaris having brought before him two leaders who in the land of Kulin Ban had been instigating the condemned sect of heretics. Kulin's son took on the obligation neither to defend nor support in his lands the above-mentioned or other men in heresy lest he have to pay a thousand silver marks.(44) It is clear Hungary wanted to depict the monks as heretics. We do not know how faithfully the monks fulfilled their pledge. However, because the sources from the 1220's and 1230's show little change, things probably continued much as they had been.

Shortly after the abjuration Casamaris wrote the pope about his mission. Only part of the letter is extant and we find the legate had traveled considerably from the time he left Bosnia. "Having handled the business of these former Patarins in Bosnia as I have already written your Holiness, I was in Hungary a few days . . . . In the realm of Ban Kulin of Bosnia there is only one bishopric and the bishop has recently died. If possible, some Latin (i.e., a man knowing theology and proper practices, if not even a Latin liturgist) should be installed there (as bishop), and also three or four new (bishoprics) should be created there, which would immeasurably increase the effectiveness of the church, which realm (takes) at least ten days or more (to cross).''(45)

Casamaris used the term Patarin, presumably about the priors unless he had had dealings with other religious groups in Bosnia about which we know nothing. His statements that the Bosnian bishop was dead, accompanied by the fact that the magister of the monastic order had been absent from the proceedings at Bilino polje, suggests that the head of the monastic order may also have been Bishop of Bosnia (or at least that this had been the case with the particular individual who had recently died prior to Bilino polje). Since there seem to have been few priests in Bosnia (and thus a limited amount of church administration), and since there would also have been few people to select a bishop from, it seems likely that the Bosnians might have frequently, or even regularly, combined the two positions in the hands of one man.

VI: Religious Affairs from 1203 to the 1220's

With the exception of a Hungarian confirmation (in 1207) of the Archbishop of Split's rights to jurisdiction over Bosnia,(46) (a claim ignored by the Bosnian clerics, whom Resti depicts as continuing to deal with Dubrovnik), the only source for the period 1203-1220 is the eighteenth century chronicle compiled by Resti. Since Resti's data about the late twelfth century is similar to that of Nicolo di Ragnina and Orbini, it is likely that his material on this early period came from documents in the archives of Dubrovnik. However, the reliability of the lost documents upon which he based his account of relations between Bosnia and Dubrovnik early in the thirteenth century is open to question.

Resti describes a visitation in 1206 of Bosnia by the Archbishop of Dubrovnik, Leonardo, who visted all the dioceses of his archidiocese on papal commission.(47) If true, this statement could not be a more clear contradiction of the Hungarian confirmation of the rights of the Archbishop of Split of the following year, for Resti not only shows the Archbishop of Dubrovnik acting with authority in Bosnia but also shows him doing it with papal approval and recognition. We are told that Leonardo was received with great honor in Bosnia and that its prince presented him with many gifts. If accurate, this statement shows that the Bosnian ban still was on good terms with the Catholic Church. In 1209 Leonardo consecrated a man named Dragohna(48) as Bishop of Bosnia, showing that Dubrovnik was continuing to exercise its rights over Bosnia. By 1211 Dragohna was having a great deal of difficulty with his people, who were exclusively of the Patarin sect, but in time he had considerable success against them.(49)

We cannot be sure what sort of heretics or schismatics Resti thought he was describing by the term. It is difficult to believe that, only five years after the warm reception which the Bosnians gave Leonardo, nearly all of them would have become "Patarins." Thus we may suggest that Resti, or his unknown source, at least is unreliable insofar as he exaggerates the number of "Patarins" in Bosnia.

VII: Charges of Heresy in the 1220's

In the 1220's the papacy and the Hungarians again showed interest in the Slavic lands apparently because of the activities of the Kačić family, who were successful and troublesome pirates centered in the Dalmatian coastal town of Omiš. King Endre of Hungary was writing them in about 1220, calling them pirates and Patarins.(50) Perhaps they were heretics; perhaps the Patarin label was attached to them for showing little respect for church property in their pursuit of wealth. In 1221 a papal legate, Acontius, was sent to Dalmatia to deal with the Kačići. He was also to handle the heretics who were sheltered in Bosnia, openly propounding their errors to the detriment of the Lord's flock.(51)

In 1222 Pope Honorius III called on Dubrovnik to elect an archbishop who would aid Acontius against the heretics and people of Omiš.(52) Apparently by this time the pope had given up the attempt to subject Bosnia to Split and again recognized Dubrovnik's suzerainty over the church in Bosnia. If Resti's information is correct, then papal recognition of this fact dates back to at least 1206.

The Chronicle of Thomas the Archdeacon of Split describes Acontius' activity against the pirates and heretics on the coast. The legate called on Dalmatia and Croatia to help him and raised a navy and horsemen to go against Omiš.(53) Having done this, Acontius turned to the problem of heretics in Bosnia. Thomas had no interest in Bosnian affairs and thus he tells us no more than that Acontius labored a long time on behalf of the Catholic faith in Bosnia against the heretics.(54) Since Thomas' lack of interest makes him a poor source for Bosnian affairs and since Thomas tells us Acontius died in 1222 (probably this date is a year or so too early), we have good reason to believe Acontius labored only a year in Bosnia.(55)

In 1225 Pope Honorius asked the Hungarian Archbishop of Kalocsa, Ugrin, to take action against the heretics of Bosnia, Soy (Sol, the region of modern Tuzla), and Wosora (Usora).(56) The pope, we see, believed that the heresy was penetrating the region to the northeast of the Bosnian state up toward the Sava, in the region between the Bosna and Drina rivers. This region seems to have been nominally under the Bosnian ban, and later in the century we find members of the ban's family (i.e., descendents of Kulin) governing it, though we do not know who ruled over the area as local lords in the 1220's and how much control the ban actually had in this region. In any case, the Hungarians did nothing in 1225; King Endre was then in no position to interfere in Bosnia since he was having trouble with his own nobles.

In 1227 Pope Honorius III assigned the city of Požega (north of the Sava in Slavonia) to the Archbishop of Kalocsa for him to defend against heretics.(57) This signifies either further movement of the heretics, the spread of heretical ideas among the populace of this area beyond the Sava, or both. A few days after the letter about Požega, Honorius wrote John the "Prince of Constantinople" (John Angelus Lord of Srem) to fulfill his promise to the Archbishop of Kalocsa to take action against the heretics, for which he had already accepted 200 marks.(58) The pope was trying to organize a crusade but was having great difficulties in getting it underway.

Thus in the 1220's Pope Honorius was constantly referring to "heresy" in Bosnia and the need to take action against it. Whether he was referring to dualists, to heretics of some other doctrine, or simply to our unreformed Catholic monks and their followers is impossible to determine. And probably we should not expect the term "heretic" to have been used with precision or even consistently for the same group in all these letters. In fact, Honorius probably did not really understand the situation in Bosnia and may not have realized that the "heretics" and ignorant monks were members of separate groups. He also probably feared that heresy was more widespread than it actually was and may well have believed that it had infected many people who were actually quite untouched by it. And such confusion would play into the hands of any party, such as Hungary, which sought an excuse to intervene in Bosnia.

Gregory IX succeeded Honorius in July 1227, and, in accord with the usual custom, reissued a variety of charters, among which was one confirming the Archbishop of Dubrovnik's suzerainty over Bosnia and Hum.(59) Hum, ruled by members of the Serbian royal house, was predominantly Orthodox under a fairly well organized bishopric, with its seat in Ston. Catholic believers, however, were to be found in the coastal regions of Hum.

In 1228 the Franciscans were given assignments in Dalmatia and Slavonia.(60) They do not seem to have been sent to Bosnia, and it is not known how many actually went to Slavonia or how long they may have remained there; it is possible that none reached Slavonia at all in these years. It is often dangerous to assume that assignments by a pope were actually carried out. However, the attempts to organize a crusade, as well as the assigning of Franciscans to preach in this area, both reflect a renewed papal interest in the Slavic lands.

In 1223, as noted in Chapter II, a papal legate named Conrad spoke of a dualist anti-pope from somewhere in the Balkans (within the boundaries of Bulgaria, Croatia, and Dalmatia next to Hungary) who played a role in reorganizing the French dualist church.(61) I shall not discuss this document here since there are almost no details to discuss. We cannot be sure the man was from Bosnia, or even that Conrad knew exactly from where he did come. We do not know whether he was really an anti-pope or whether Conrad was simply projecting the Catholic organizational structure upon the dualist church. There is evidence that there were dualists in Bosnia; however, we know almost nothing about how they were organized. We do not even know whether the dualists in the Bosnian banate were a sizable movement by 1223. I am sceptical that a heresiarch — so important that French dualists would bow to his wishes — could have come from Bosnia as early as 1223, and do not believe this one vague report evidence to demonstrate that this man was Bosnian. Moreover, I am sceptical about the level of knowledge and understanding of Balkan matters possessed by Western clerics who had not themselves been to the Balkans or who did not have close contact with those clerics who had.

VIII: Papal Action and Hungarian Crusades in the 1230's

In 1232 the papacy began actively to reorganize the Catholic Church of Bosnia; its initiative was followed by Hungarian military action against the heretics. Our sources now multiply.

On June 5, 1232, Pope Gregory IX wrote to the Archbishops of both Kalocsa and Zagreb, asking them to investigate the Bosnian bishop, because it had come to the pope's attention that the Bosnian bishop knew nothing of letters, was a public defender of heretics, and had received his episcopal office through a certain open heretic by the crime of simony. No one in his church celebrated the divine offices or administered the sacraments. It was even said that he was a foreigner (or hostile) to the ecclesiastical services, was completely ignorant of the baptismal formula; and this was not astonishing because, as is asserted, he lived with his own brother, an open heresiarch, in a certain village and, instead of leading him back from iniquity to the way of truth — as he ought to have done — he supported and defended him in his error.(62)

In May 1233 Gregory assigned a legate, Jacob of Preneste, to remove the bishop, stating that after an investigation the bishop had been found unworthy to hold office. He had told the investigators that he had sinned through ignorance. The pope did not question the truth of this, but found him unfit and ordered him removed. The pope concluded that probably the Bosnian territory was too large for one man to administer (particularly since it was infected with heretical errors) and reiterated Casamaris' recommendation of 1203 that two to four new bishoprics be carved out of the Bosnian territory.(63) The bishop's plea of ignorance was accepted. His case clearly illustrates the low level of religious understanding in Bosnia. Such a man probably should not be called a heretic — i.e., an adherent of a heretical movement with defined doctrinal basis — because he apparently considered himself loyal to Rome. Once again we are faced with the same situation that we observed with the monks at Bilino polje, who, we know, had also failed to carry out a whole series of basic Catholic practices, and yet were apparently not heretics or in any way hostile to Rome. These situations were a natural result of the fact that there was no education in Bosnia.

The removal of the ignorant ineffectual bishop was the first in a series of events which were to cause Bosnia a tormented decade, although the immediate aftermath seemed to augur well. In October 1233 Pope Gregory placed Ban Matej Ninoslav of Bosnia and all his lands under the protection of the Holy See. Ninoslav reportedly renounced heretical errors, accepted the Catholic faith, and turned to persecuting heretics.(64) What Ninoslav's heretical errors had been is not stated. The pope was so satisfied with Ninoslav's new Catholicism that on the same day he wrote the Dominicans to release, provided that he be fully converted to the faith, the son of "Ubanus" called Priezda, Ninoslav's relative, who had been held as a hostage for Ninoslav's good faith in accepting Catholicism.(65) Perhaps we can link Ninoslav's acceptance of Catholicism with the mission of the legate Jacob of Preneste to remove the bishop.

On the same day the pope also wrote Coloman, Duke of Croatia and brother of the King of Hungary, that, because Ninoslav had now been placed under papal protection, Coloman was to leave the problem of heretics in Bosnia to Ninoslav.(66) This probably indicates that Coloman and the Archbishop of Kalocsa had been seeking an excuse to reassert Hungarian authority over Bosnia.

The next important event was the appointment late in 1233 or early in 1234 of the Dominican, Johannes von Wildeshausen, as Bishop of Bosnia, the first foreigner to receive the post. Johannes was a noted scholar of languages — although as far as we know he knew no Slavic language — who had studied in Paris and Bologna, and prior to his arrival in Hungary sometime in 1232 he had served in Bremen. We know that he did not want the Bosnian post and that in 1235 (after violence had begun) he requested, though in vain, to be relieved of his unpleasant duties.(67) We have no evidence that Johannes ever resided in or even visited Bosnia.

Pope Gregory's attitude toward Bosnia changed greatly in the course of the year that followed these October 1233 letters, which were friendly toward Ninoslav. In October 1234 the papal letters were demanding military action against the Bosnian heretics. What caused this change? Had the pope sent agents to Bosnia who had reported that religious deviation was prevalent? Or had Coloman, seeking an excuse to invade Bosnia, sent such a report? Or had Ninoslav done nothing about the heresy? Whatever the cause, Gregory wrote six letters in three days (between October 14 and 17, 1234) to Coloman, to the Bosnian bishop, and to would-be crusaders about the new crusades. The pope said it was necessary to convert to the faith heretics in various parts of "Slavonia."(68) Bosnia was singled out by name ("land of Bosnia") only in the letter to the Bosnian bishop. Presumably this referred to the Bosnian diocese. It is usually assumed that by "Slavonia" Gregory meant Bosnia (and environs). Yet "Slavonia" could also have referred to Slavdom (i.e., the South Slavic lands) in general or even to Slavonia, north of the Sava, in particular. Since the Bosnian bishop was informed about the crusading plans, it is obvious that some of the crusaders' activities would be directed against heretics in his diocese which included not only the Bosnian banate but also all the territory up to the Sava River. In these letters Gregory promised indulgences to those who would fight the heretics, and, in accepting Coloman as a crusader, put him and his crusaders under the protection of the Holy See.(69) The crusade seems to have been directed against the Bosnians in general with no defined group being singled out as the enemy. Papal letters throughout use only the general term "heretics," rendering it impossible for us to ascertain whether the papacy was alarmed solely by the heretics (probably dualists) or whether it was also concerned with the errors and ignorance of the Bosnian Catholic monks. Only one source, Suibert's "Commentary," implies that the crusade may have been, at least nominally, sent against dualists.(70)

The crusade presumably began in 1235 and continued through 1236. We do not know where the armies went or how far they penetrated during these two years. In August 1236 the pope told his crusaders to leave Sibislav, Knez of Usora, and his mother alone, since they were good Catholics in the midst of a nobility in the Bosnian diocese infected with heresy, lilies among thorns.(71) The two were placed under papal protection; Sibislav was called the son of a former Ban Stjepan of Bosnia, and is generally considered to be grandson of Ban Kulin,(72) which would make him either the brother or nephew of Ninoslav.

It is generally assumed that in these years the crusaders ravaged Bosnia proper. However, we cannot assume this. It would have taken time to occupy and convert the populace of the territory of Slavonia, Usora and Sol, in which we have been told lived heretics, and which would have had to be pacified before the crusaders could proceed to Bosnia itself. It is reasonable to assume that Ninoslav and other Bosnian lords would have resisted, which, when we consider the mountainous terrain of Sol and beyond into Bosnia, would have insured that the crusaders' progress was slow.

In May 1237 Dubrovnik prohibited further trade with Bosnia until an embassy it had sent there returned.(73) Since Bosnia usually had friendly relations with Dubrovnik, it is probable that some difficulties within Bosnia (possibly connected with the crusades) may have hampered the embassy. Warfare continued through 1237 and 1238. Johannes von Wildeshausen in these years finally managed to extricate himself from the post of bishop since on April 26, 1238 Pope Gregory IX could write to Bishop Theodoric of the Cumans about the selection of the Dominican Ponsa as the new bishop. This letter, which announced that after great labors the Bosnian land had been brought back to the light of Catholic purity, shows that by then the crusade was considered a success. The pope also wanted Hum (Cholim) to be subjected to Ponsa. Whether this reflected crusader gains into Hum or future plans for the crusades is unknown.(74) In addition, it is apparent that part of the central region of Bosnia had fallen, because in the same letter the pope mentioned raising money for the establishment of a future Bosnian cathedral chapter.(75) In other words the pope wanted to create in Bosnia a new seat for the Bosnian bishop and his staff, with sufficient income to support it.

Later documents give further details about where the cathedral was to be established: a letter to Ponsa in December 1239 calls him the Bishop of the Chapter of St. Peter, and a charter of Bela IV of 1244 states that the See was located at Brdo (Burdo) above Vrhbosna (modern Sarajevo), ("supra Urhbozna Burdo . . .")(76)

Because the pope presumably desired a safe site for the future cathedral and its clergy, his choice of Vrhbosna suggests that he considered that town safe; thus probably by April 1238 Coloman had actually occupied the place. That Vrhbosna must have been lost to Ninoslav is also deduced from the fact that plans for the cathedral chapter continued after December 1238 when the pope declared that Ninoslav had relapsed into heresy.(77) If Vrhbosna lay in the lands of a relapsed heretic, presumably papal plans for the new See would have been postponed. Besides, if Ninoslav had been involved in the plans for the new See, we would expect the bishop's seat to be established in the Visoko-Zenica region, the heart of Ninoslav's state. That the church was to be erected at Vrhbosna, then, shows that Coloman had conquered this region from Ninoslav and also that Coloman's crusaders most probably had not been able to occupy the Visoko-Zenica center of Ninoslav's banate. Thus Ninoslav had managed to retain his independence despite some territorial losses.

Finally, the involvement of the Dominicans who were closely associated with Coloman shows both that the project was Coloman's and that the planned new bishopric was to be a Dominican venture. Thus, the new bishopric under the Dominican Ponsa, staffed by Dominican clerics, supported by the Hungarians, and located inside Bosnia was intended to assume whatever authority the Bosnian monastic organization had had and to reintegrate Bosnia with the international Catholic Church. It is not surprising that the pope, in an area of shaky faith, would have wanted to establish the bishop's seat in an area under Coloman and to have the church under the direction of the Dominicans. We can assume that Dominicans, and very likely Ponsa himself, arrived in Hungarian-occupied Bosnia during 1238.

One strange fact remains to be explained: in the papal letter of December 1238 Gregory says that Ninoslav had, at some previous time, deposited money for the cathedral with the Dominicans.(78) But why should Ninoslav have contributed for a cathedral to be established by his enemies? We have noted that, in 1233 when Ninoslav accepted Catholicism, one of his relatives had briefly been held by the Dominicans as a hostage. It seems likely that Ninoslav had then had to post a sum of money with the Dominicans to guarantee his adherence to Catholicism, to be forfeited if he relapsed; perhaps, then, it was this earnest-money, posted by Ninoslav with the Dominicans, that they later decided should be used for the cathedral. It is, however, even more likely that, at the time Ninoslav had accepted Catholicism in 1233 and prior to the crusades launched against him, he had given money for a future cathedral to be built in his lands.

In Bela's charter of 1244, which we shall discuss shortly, we find that Ninoslav had both granted and recognized the church's rights to collect tithes and to possess various lands within his banate.(79) Unfortunately, we cannot demonstrate whether these privileges toward the church had been granted by Ninoslav prior to the launching of the crusades or whether he had been forced to grant them by the Hungarian armies. Since the 1244 charter purports to be a reissue of a 1239 charter of Coloman's, and since we know of no treaty or submission to Coloman by Ninoslav in 1238-39, I think it most probable that Ninoslav had recognized the church's rights to these lands back in 1233, at the time the pope's letter had noted his acceptance of Catholicism. Yet his acceptance of the Roman faith, and possibly even his recognition of the church's rights to income and lands, had not protected his banate from the Hungarian armies. Naturally he would have become anti-Catholic (in loyalty though not necessarily in doctrine) and at least have severed relations with the Bosnian church hierarchy, which by then was composed of foreigners and no longer residing in Bosnia. In fact, Ninoslav's alleged relapse, noted by Pope Gregory IX in December 1238,(80) may not signify more than a refusal by the ban to recognize the Dominican Ponsa as bishop.

The conditions of war had also cut off the Bosnian clergy from its official hierarchy. And since this hierarchy supported the invader, it would not be at all strange if both Bosnian ruler and clergy should have begun to seek ways to make their church independent of the now foreign Bosnian bishop, the Dominicans, and the Hungarians. The simplest step would have been to choose a bishop from their own ranks, as had been the local custom until the 1230's.

Whatever the pope originally intended the crusade to be, it became a war of conquest by the Hungarians against Bosnia. Such, of course, is the natural result of any crusade which meets resistance, for the crusader sees clearly that the only way to establish what he is fighting for is to conquer and administer the recalcitrant region himself. Many believe that the conquest of Bosnia had been Hungary's goal from the beginning, and some even claim that "heresy" in Bosnia was a Hungarian invention to excite the pope into blessing and supporting its war of conquest.(81) Such a view, although perfectly plausible and temptingly in accord with what we know, cannot be demonstrated from the sources.

In 1238, in the midst of the fighting, a strange document appeared, in which Pope Gregory confirmed Dubrovnik's rights over Hum, Bosnia, and Trebinje.(82) So far as we know, Dubrovnik had had nothing to do with the crusade, had not even been asked to participate, and, in fact, had maintained trade and friendly relations with Bosnia throughout. In 1234 and 1240 Ninoslav confirmed treaties of friendship and commercial privileges between Bosnia and Dubrovnik.(83) The religious aspects of the war seem to have been entirely under the control of the Archbishop of Kalocsa and the Dominicans. This confirmation ran counter to Gregory's own policies; in 1238 he was discussing with Hungarian bishops the election of Ponsa as the new Bishop of Bosnia. The Archbishop of Dubrovnik, who as suzerain should have confirmed and consecrated the new bishop, was neither consulted nor had a part in Ponsa's consecration.

Thus, Dubrovnik's overlordship was empty. Hungary was in the process of, and in fact by 1238 had succeeded in, usurping control of the Bosnian bishopric — a bishopric which would shortly become an empty title itself. Why would the pope go out of his way to issue a meaningless charter? We can only assume, if the charter is authentic, that he was answering some complaint that no longer survives from the Archbishop of Dubrovnik about the usurpation of his rights by the Hungarians. In reply, we conjecture, the pope simply reissued the old charter since, after all, no official change of suzerainty had yet been authorized.

In December 1238 the pope called upon both Ponsa and Duke Coloman to persecute the Bosnian heretics.(84) These letters were perhaps motivated by Ninoslav's alleged relapse. Presumably during 1239 papal wishes were carried out, because the pope wrote Coloman in December to praise him for his zealous persecution of heresy;(85) but we have no information about Coloman's activities. The Bosnian state cannot have been entirely overrun, for in March 1240 we find Ninoslav issuing a charter of peace and friendship to Dubrovnik.(86) We do not know whether Dubrovnik's leaders were motivated by commercial interests or by frustration at the loss of ecclesiastical authority in Bosnia. However, it is clear that Dubrovnik (whose Catholicism was beyond question) did not regard the Hungarian invasion of Bosnia as justified, did not consider Ninoslav to be a heretic, and did not believe that Bosnia was a land of heretics.(87)

In December 1239 Gregory called on the Prior of the Dominicans in Hungary to send several preachers to "the land of Bosnia" against the heretics.(88) The letter as usual says nothing about where they should go in Bosnia (does "land of Bosnia" refer to banate or diocese?) or about the nature of the heresy. Presumably these new preachers were sent to work in Hungarian-occupied Bosnia (Vrhbosna and environs). A few weeks later Gregory wrote some Dominican brothers of Bosnia to send Ninoslav's deposit to the Bishop of Bosnia.(89)

Though the Dominicans provided the religious impetus for the crusade, little is known of their role in Bosnia, and their influence was short-lived. In the 1230's they supplied "Bosnia" with two bishops; the first almost certainly never set foot inside Bosnia. The second, Ponsa, presumably entered Hungarian-occupied Vrhbosna in 1238 in the train of Coloman's armies. The influence of Ponsa and his entourage could not have been extensive, however, because they could have remained only until the Tatar invasion of 1241. Thus the Dominicans would have had influence only in part of Bosnia and this for a maximum of three years. There is no evidence that they ever set foot inside the territory that Ninoslav retained, and considering the existing state of war, it seems highly unlikely that they did.

We have a list of Dominican monasteries from 1303, which includes none in Bosnia.(90) Thus it is clear that the Dominicans never established any convent or permanent mission in Bosnia. Their failure was due to the fact that the Hungarians were never able to establish lasting control over the banate. Most likely the Dominicans began in the 1230's to establish themselves in Slavonia and the region around the Sava, and presumably they had some success in converting heretics or teaching the unenlightened of that region.

In 1241 the Tatars invaded Hungary. Coloman and Ugrin, the Archbishop of Kalocsa, perished in battle, and Dominican sources add that thirty-two Dominicans were drowned by the Tatars in a river.

Coloman's crusade had failed to destroy Ninoslav and his state. And the territory it had conquered was lost as the Tatar invasion led to the withdrawal of the Hungarians from the Vrhbosna region.(91) The crusade clearly increased Bosnian hatred for the Hungarians, a sentiment which would prove to be an influential factor in determining Bosnian politics up to and beyond the Turkish conquest. And by making religion a factor in this international dispute, it no doubt increased the rupture between the Bosnians and official Catholicism, and was an important factor in mustering support in Bosnia behind one or more anti-Catholic local movements. The events of the following decade would solidify and make this schism permanent.

In 1244 Bela IV, King of Hungary, confirmed a grant made in 1238 or 1239 (but no longer extant) by his brother Coloman to the Bosnian bishopric.(92) The document lists considerable territory inside Bosnia proper which nominally belonged to the bishop. But apart from Vrhbosna, held from 1238 to 1241, and other territories which might have been occupied by the Hungarian crusaders, we can suppose that, from the crusade's beginning, the bishop had neither administered these lands nor derived any income from them. The document's greatest significance lies in the fact that it lists a considerable number of churches in the central part of Bosnia at a time when many scholars think that Bosnia was without churches, because of the presence of heretics.

The charter tells us nothing directly about heretics. But the presence of so many churches suggests that heretics had not bothered these Catholic churches in various parts of Bosnia. In addition, the text of the charter states that Ninoslav, his brothers, and barons recognized the bishop's right to these possessions. It also refers to possessions in Usora, given to the church at some unspecified time by Ninoslav. Above, we have speculated that this recognition was given prior to the launching of the crusades.

Bela's re-issue of Coloman's charter shows that the Hungarians in 1244 hoped — in vain, as it turned out — to re-occupy the parts of Bosnia they had briefly held and lost and to extend their domination into new regions.

IX: Bosnian Catholic Church Subjected to Kalocsa

During the period 1246-52, Pope Innocent IV made official what had in fact occurred at least a decade earlier: the subjection of the Bosnian bishopric to the Archbishop of Kalocsa. By this act was ended the nominal suzerainty of Dubrovnik that had hitherto been recognized and often confirmed. The Hungarians had pressed for this move, as we shall see from the text of Innocent's letter of July 1246, since it would simplify their task; in theory as long as Bosnia was under Dubrovnik, the Hungarians had to obtain papal permission for their activities there. Placing the Bishopric of Bosnia under the Archbishop of Kalocsa would give them complete freedom of action.

The document of July 1246 describes the chaotic and evil state of the Bosnian Catholic Church. It had lapsed into heresy with a heretical bishop who had had to be removed for not performing church services; the Archbishop of Dubrovnik, it is noted, had installed the man and had tolerated the situation. To exterminate heresy and to improve matters it had been necessary to entrust the Bosnian Catholic Church to the care of the Archbishop of Kalocsa. The Kalocsa church, after much labor and expense, and despite many perils, had succeeded in carrying out this enormous task. King Bela had therefore requested that the Bosnian Catholic Church be transferred from the Archdiocese of Dubrovnik to that of Kalocsa. After much thought, the pope called upon the abbot of the Pannonhalme monastery and certain other clerics to investigate the matter so that the pope would have the information necessary to decide the question.(93) The fact that Innocent called upon this Pannonian abbot to head the inquiry leads us to believe that the pope intended to accept the Hungarian request.

In August 1246 the pope wrote the Archbishop of Kalocsa, clearly hoping for further crusading activity since he bestowed upon the archbishop the power to let true believers divide up the heretics' property among themselves.(94) In January 1247, the pope called on Bela to take action against the Bosnian heretics.(95) The pope, thus, seems to have hoped for zealous action by the Hungarians while their case was pending.

In the summer of 1247 Innocent wrote to a Hungarian Cistercian abbot: The Bosnian Catholic Church had fallen into heresy. The Archbishop of Kalocsa had labored hard to exterminate heresy, with a great shedding of blood, and at great expense. A large part of Bosnia was conquered and many heretics were captured and led away. But the fortifications that the Church controlled were not strong enough to defend this territory from attack or to maintain the purity of faith there. Since there was little hope that the land would (naturally) come around to Catholicism, and since Kalocsa had gone to such great lengths in leading crusades, the Kalocsa Archbishop, the Bosnian bishop (Ponsa) and King Bela had requested that the Bosnian Church be subjected to Kalocsa. If, the pope stated, that was the way things were, let the Bosnian Church be subjected to Kalocsa.(96)

However, subjection does not seem to have occurred immediately, since on May 23, 1250 the suffragan bishops of the Dubrovnik archdiocese were summoned to Rome to settle a disagreement about filling the Antibari (modern Bar) bishopric. Among those called was a Bishop of Bosnia.(97) Since a letter of Ponsa, dated 1252, cites the papal summons,(98) clearly Ponsa was the Bosnian bishop called. Thus, in 1250, the pope still could summon Ponsa as a suffragan bishop of Dubrovnik.

Other events may have postponed the papal decision. We have noted that in the autumn of 1246 and the spring of 1247, the pope was calling for another crusade. Whether any movement of troops actually took place we do not know, but in March 1248 the pope made an about-face and called off the crusade. The new papal view was expressed in a letter to Benedict, Archbishop of Kalocsa, in which he said that he did not want the armies to march against Ninoslav because Ninoslav had appealed to the pope and claimed that he was really a good Catholic but had been forced by circumstances to favor the heretics and to accept their aid against the foreign enemies of his country.(99) Ninoslav may well have been trying to play for time but the strength of his argument certainly was clear to the pope. For even if Ninoslav was a good Catholic, how could the pope have expected him to persecute heretics when he needed their help to defend himself and his realm from the invading Hungarians?

So the pope stopped the new crusade and sent legates to investigate the situation. He dispatched a Split Franciscan and the Bishop of Senj, (100) two coastal clerics, both of whom probably spoke the language, a considerable advantage in understanding Bosnian conditions. By not sending a Dominican and by not ordering the Archbishop of Kalocsa to send an agent, the pope was seeking a neutral non-Hungarian opinion before yielding to Hungarian pressure for further military intervention. We may suggest that the pope postponed a decision on the subjection of the Bosnian See until these legates had made a report; quite possibly they were expected to gather facts to help the pope decide which archbishop should have jurisdiction over Bosnia, in addition to investigating Ninoslav's Catholicism.

What Ninoslav's religious opinions and loyalties at this time were we cannot say. Doubtless, he was hostile to Hungary. His "relapse" in 1238 possibly had been no more than a refusal to recognize Ponsa as bishop, which the pope might well have interpreted as a refusal to accept papal supremacy, since the pope had supported Ponsa. However, had the pope been willing to relieve Bosnia of any and all ties with Kalocsa (and the Dominicans) and to allow the Church of Bosnia to be subjected to Dubrovnik — *de facto* as well as *de jure* — Ninoslav might well have become reconciled with Catholicism and Rome. But this decision was not taken. Unfortunately, the sources give us no reasons as to why. We may speculate that Hungarian pressure was too great. Nothing is known about the legates' mission of 1248 or whether they judged Ninoslav Catholic or relapsed. Nor can we tell whether further Hungarian military action occurred in or near Bosnia in 1248 or 1249. Ninoslav was still in power in 1249 when he signed a treaty of friendship with Dubrovnik.(101) Thereafter he disappears from the sources entirely; perhaps he died about this time. We do not know the name of his successor.(102)

Thus, in 1249: Ninoslav still maintained friendly relations with Dubrovnik; the papal inquiry into the ban's Catholicism and the general situation in Bosnia was either in progress or completed. The Hungarian request for Bosnia's subjection to Kalocsa was still pending; and Dubrovnik still maintained a tenuous hold on its *de jure* rights as suzerain archbishop, as is shown by the fact that in May 1250 Ponsa was summoned by the pope as one of several suffragan bishops under Dubrovnik.

Next on October 11, 1251, the Ragusan archbishop wrote a letter listing the precedents for his archdiocese's rights of suzerainty over Bar, Ulcinj, and Bosnia. The letter lists various popes who had confirmed Dubrovnik's rights over these three dioceses, as well as various Dubrovnik acts demonstrating the suzerain rights of the Dubrovnik archbishop. Among these other acts was the consecration of a Bosnian bishop named Bratislav.(103) Unfortunately, no other document refers to Bratislav so there is no way to determine when he had been consecrated. Kalocsa is not mentioned in the letter; thus, we cannot be sure whether the pope by then had decided for Kalocsa which had elicited this Dubrovnik response, or whether Dubrovnik, believing that the pope was on the verge of so deciding, had written to try to dissuade him.

On May 8, 1252, we find Ponsa, still Bishop of Bosnia, residing in Djakovo in Slavonia.(104) This is the town where the Hungarian-sponsored, and the only papal-sanctioned, Bishop of Bosnia was to reside until the fall of the Bosnian Kingdom. By the establishment of the See in Djakovo, Hungary recognized that its puppet bishop could not find support in Bosnia. Thus, the Hungarians had to give him a court in Slavonia to go with his empty title until such time as they might reoccupy Bosnia. This bishop in Djakovo, except for Peregrin Saxon during his brief tenure as bishop (1349-55) and Peter, his successor, never set foot in Bosnia or had anything to do with Bosnian affairs; he was a Hungarian court personage, whose main activities concerned affairs in Slavonia.

Thus we may conclude that the pope, though apparently ready to subject the Bosnian See to Kalocsa in 1247, and though encouraging Kalocsa's activities in Bosnia and recognizing the Kalocsa-sponsored Dominican bishop of Bosnia, Ponsa, only reached his final decision after May 1250. It is likely that the decision had been made prior to October 1251, and that the Dubrovnik archbishop's letter asserting his See's suzerain rights was a response to the papal decision.

X: Establishment of the Bosnian Church

From the beginning of the Hungarian crusades in 1235 — if not from the time of the first foreign bishop's appointment in 1234 — and throughout the Dubrovnik-Kalocsa quarrel for jurisdiction over Bosnia, the Bosnians must have been themselves administering religious matters and churches within Bosnia. There is no evidence in these years that the Archbishop of Kalocsa had any authority within the banate, nor is there evidence that Dubrovnik, in defiance of papal policy, was supervising religious matters in Bosnia. Since we have no knowledge of any religious organization inside Bosnia other than the monastic order, we may suspect that the monks, in addition to their usual activities, had taken on the burdens of ministering to the peasantry. Thus the magister and priors, in addition to their monastic administrative duties, would have assumed the role of bishop and staff for Bosnia. If, as seems likely, the magister of the order prior to 1203 had on occasions, or even regularly, been the Bishop of Bosnia, then his assumption of these tasks after 1234 would have been natural. Since in the fourteenth century we shall meet a Bosnian Church headed by its own bishop or djed, it seems plausible to date the foundation of this autonomous church to the period after 1234.

That the sources do not mention such a church until 1320 is hardly an argument against its establishment between 1234 and 1252, since the seventy-year period 1250-1320 is one almost without sources. Presumably, the initial act in the establishment of this church would have been the assumption by a local Bosnian of the bishop's title in opposition to the Hungarian-sponsored bishop. The likely time for this act to have occurred would have been after the launching of the Hungarian crusades. It may well have occurred in or before 1238; possibly it was such an act that the pope referred to then when he stated that Ninoslav had relapsed. If the anti-bishop had not been established in the 1230's, a natural time for his establishment would have been ca. 1250-52, when it became apparent that the Bosnian Catholic Church would be subjected permanently to the Archbishop of Kalocsa and the Hungarians. Interested Bosnians must have realized that their bishop would henceforth serve Hungarian interests. There would be no further elections of local clerics — with Ragusan consent — to the post, as had been the practice until 1233. Such a realization would have further stimulated the Bosnians to form some sort of autonomous "national" church. The precedents of Serbia and Bulgaria were there. Most of the Bosnian nobles would have felt the need for such a church and would have supported it, no matter what its doctrine might be.

The Bosnian clergy and monks would have favored such a move, because they were already separated from their hierarchy of foreigners who had attacked them, who lived abroad, and who had little interest in or sympathy for Bosnian conditions.

The Bosnian Church, which appears — as we shall see — in the sources from the 1320's to the fall of the state, resembled a monastic order in many ways. Then we shall find this church headed by a single *djed,* below whom were a number of *gosti,* many of whom headed Bosnian Church monastic houses. This arrangement resembles the monastic organization seen at Bilino polje in 1203 when we found a single magister (who most probably was also Bishop of Bosnia) below whom were a number of priors, heading monasteries. Orbini refers to the djed and stroinik (an administrative position in the church) as being connected with monasteries.(105) Pius II, writing in the middle of the fifteenth century, who believed the Bosnian Church to be dualist, referred to Manichee monasteries.(106) Below the djed and the gost was the starac. The term starac (elder) is an Orthodox monastic term. In addition, in 1203 the monks called themselves Christians (as a term for those ordained), and the ordained members of the Bosnian Church were called by the Slavic equivalent, krstjani. Finally, there is no evidence of any other clerical organization in Bosnia from which the church could have been formed.

Thus I conclude that the Bosnian Church as an autonomous church was created between 1234 and 1252; and for its hierarchy it utilized the existing administrative organization of the monastic order. Thus the Bosnian Church would, in fact, be the same institution that we met at Bilino polje in 1203. The main difference between the "Christians" of 1203 and those of ca. 1250 lay in their relations to the Catholic world. In 1203, they, at least verbally, recognized the pope and promised to adapt themselves to the ways of international Catholicism. Now in these violent years 1234-52, in defiance of the pope and the papally sponsored foreign bishop appointed to govern them, the Bosnians severed relations with foreign Catholicism and established their own autonomous church.

And, with the Bosnian decision to go it alone, the whole controversy over the subjection of the Bosnian See disappeared. The "Bosnian bishop" in Djakovo clearly was under Kalocsa since Slavonia was. But since this bishop had no authority in Bosnia, Dubrovnik had no further reason to protest.

Professor Ćirković believes, as I do, that the Bosnian Church was formed in the mid-thirteenth century and sees it as developing out of the existing local Catholic organization (i.e., the monastic order). He also believes that there was a dualist heresy in Bosnia that was growing during the thirteenth century, against which the Hungarian crusades were directed. During the first half of the century, although the monks may have been somewhat influenced by dualist ideas, they had managed to remain distinct from the heretics until the wars. The wars threw all Bosnians together to repel the invader, and also, by making the populace anti-Catholic, attracted many toward the heretical movement. Those so attracted included the now bishopless Catholic organization. The two groups, Ćirković believes, merged about the middle of the thirteenth century. This merger resulted in an administration based on the former Catholic Church organization of monks, with a theology taken from the dualist heretics.(107)

It is clear that there is great similarity between my construction of events and Professor Ćirković's, and I acknowledge my debt to his presentation for many of my own views. I accept with no hesitation his contention that the local Bosnian Catholic organization severed its links (however weak) with Catholicism and declared its independence at or shortly after the middle of the thirteenth century.

However, I see no reason why the Catholic monks had to accept dualism as Ćirković believes. There is certainly no evidence in our sources from the fourteenth and fifteenth centuries to conclude that the Bosnian Church was dualist, though it may well have had certain practices similar to, and possible even acquired from, dualists. There could easily have been a dualist current which continued to exist in Bosnia, but which remained as it had prior to ca. 1250, distinct from the monks' organization even after that organization became the Bosnian Church. There is no reason why two different movements, both deviant from a Catholic point of view, could not have coexisted in Bosnia. The Bosnians have always tolerated the coexistence of different faiths whenever outsiders did not force them to violence. Thus, the Bosnian Church could have peacefully coexisted with and even have had relations with "Manichees" or Orthodox, just as Catholics, Orthodox, and Moslems coexisted in relative peace in Bosnia after the Turkish conquest, so long as they were not stirred to conflict by the Turks, or later by the European powers, or finally by Croatian or Serbian chauvinistic nationalism.

It seems likely, then, that the Bosnians, in utilizing the monastic order to establish their church, would have tried to preserve the beliefs and practices which that order had had up to that time.(108) Thus, at first we would simply have had a schism, and owing to the ignorance of the monks we would also have found a variety of deviations in belief and practice. As a result of Hungary's attempts to assert its authority over Bosnia, we would expect the church of these monks to exhibit all the traits that anthropologists would associate with a "nativistic movement." This type of movement is defined as a reaction against domination from a foreign culture by means of an organized attempt to revive or perpetuate certain aspects of the native culture in the face of pressure to change. Generally, such a movement will be religious in nature. It is pointed out that it usually focuses on a few selected elements of the native culture that have emotional and symbolic value. These tendencies are often strongest in the classes who occupy favored positions and who feel these positions are threatened.(109)

This definition applies perfectly to Bosnia in the middle of the thirteenth century. Its political leadership was threatened with loss of position and authority by Hungary. Its religious leadership (the local monkish hierarchy) was threatened as well by what would happen if Hungary and the Hungarian church organization achieved dominance over the church in Bosnia. The Hungarian Catholic invasion would have made many Bosnians hostile to both the Catholic Church and Hungary and thus ready to turn to a local organization, and the choice of local religious organizations seems to have been limited to the monastic order.(110) There would have been a strong desire among the monks, and those who were influenced by them, to maintain what was theirs, what was Bosnian; this would include all their deviations in practices. In fact, if there was pressure by Hungary to change particular deviant practices, this alone would have been strong impetus to make those particular deviations into important practices that had to be kept, and into symbols of their church. If these deviations were in matters considered important by Rome, this alone would have been enough to make Rome regard the Bosnian Church as heretical as well as schismatic.

However, it should be stressed that there is no evidence that this "nativistic" emotion swept Bosnia. As was stressed in the beginning the emotions are usually strongest in the classes with the most to lose. Thus, we can suspect that nativism would have been found among the monks and among some of the nobility (in other words, the leaders of the society) and among those townsmen and the limited number of peasants living near enough centers to be influenced by the spirit. The majority of the population would probably have been relatively slightly affected.

Thus, I postulate that the Bosnian Church was organizationally and doctrinally a continuation of the monkish order of Bosnia. In the way I have described, it based its "theology" on its own brand of Catholicism, a combination of uneducated Catholicism and what originally were chance deviations, now through nativistic pride, inflated to assume the role of basic practices. As a result of making deviations from Catholicism into prescribed practice, it could well have earned the label "heretical" from Rome. The existence of heretics in the realm gave Rome additional justification to call this church heretical. However, because it was basically Catholic in doctrine, Dubrovnik was able to continue to regard the Bosnian Church as schismatic and not heretical. Thus, Dubrovnik maintained good relations with Bosnians for the next two centuries and even had close relations with certain clerics of the Bosnian Church. In fact, Dubrovnik rarely called Bosnia heretical and usually referred to Bosnia as schismatic. It is only after 1430, and then only in wartime, that Dubrovnik bandied the label heretic about. In the same way we can find the Orthodox on occasions tolerating Bosnian Churchmen. The Bosnian Church bishop, Djed Miroslav, was allowed to witness a charter in 1305,(111) and Metropolitan David of Mileševo was willing to appear alongside Gost Radin of the Bosnian Church as a witness to Herceg Stefan's will in 1466.(112) All the above examples would have been inconceivable, on the part of either the Orthodox or Dubrovnik if the Bosnian Church had in fact been neo-Manichee.

The seventy years following the establishment of the official bishop in Djakovo, and the last mention of Ninoslav in the sources is a dark period for which we have very few documents. When we again have sources in the 1320's, we shall find the Bosnian Church firmly established under a bishop bearing the Slavic title of djed.

XI: References to Heretics from the 1280's to 1305

In the period from 1250 to 1280 the only mention of heretics appears in a letter from Ottocar, King of Bohemia, to the pope in 1260. Ottocar reported that Bela's armies included Greeks, Bulgarians, Serbs, and Bosnian heretics.(113) By this time, after the papal missions and crusades, combined with the general lack of education in Bosnia, it would not be surprising if the term heretic had become a natural label to be attached to the Bosnians by foreigners, like "damn Yankee." We cannot say if there was any real heresy among these Bosnian soldiers or whether this was just a pejorative label that came naturally to Ottocar.

Not until 1280 is a crusade mentioned again; on this occasion it is clearly a papal idea. Two letters refer to the matter. In the first, Jelisaveta, mother of the Hungarian King Ladislas and Duchess of Mačva, Bosnia, Požega and "Wolkou" (The Vukovska župa in Slavonia), promised the pope to persecute heretics in her territory; she said nothing specific about the heretics.(114) Whether she took any action between then and the end of her reign in 1284 is not known. At the same time, King Ladislas, at the request of the pope, called for the persecution of heretics in "our kingdom and the diocese of Bosnia and adjacent lands under our jurisdiction." He mentioned heretics of "whatever errors or sect" and of "various heretical sects."(115) His statement clearly shows that he believed there was more than one heretical movement.

In 1282, Stefan Dragutin, King of Serbia, was forced to abdicate his throne to his brother. Having the good fortune to be brother-in-law of the King of Hungary, Dragutin shortly thereafter was given the Banates of Mačva, Usora, and most probably Sol (Tuzla).(116) In 1284, Dragutin's daughter, Jelisaveta, married Priezda's son Stjepan Kotroman. This marriage would have tremendous implications, because the son born of it, Stjepan Kotromanić, would re-establish a strong Bosnian state, and because by the marriage, Kotromanić's nephew, Tvrtko, would justify his claim to the Serbian royal title a century later. Archbishop Danilo, who wrote a life of Dragutin, states that Dragutin found many heretics from the Bosnian land, and he converted many and baptized them.(117)

Though Orthodox, Dragutin had good relations with the papacy and in 1291 turned to Rome for aid against heretics. Pope Nicholas IV called upon the Franciscans of Slavonia to send two capable brothers, knowing the language, to Dragutin's lands with full authority to act against heretics and those defending them.(118) We know nothing of their mission. In 1298, Boniface VIII in almost the same words ordered the Franciscans of Slavonia to send two brothers to investigate — for the purpose of destroying — heresy in Serbia, Raška, Dalmatia, Croatia, Bosnia, Istria, and the province of Slavonia.(119) That no mention was made of a previous mission suggests that for some reason the 1291 order had not been carried out. We have no information as to whether the 1298 papal order was acted upon. The papal letters of assignment from 1291 and 1298, however, were to become the basis of Franciscan claims, when some thirty years later the Franciscans and Dominicans came to quarrel over the right to conduct the mission in the South Slavic lands.

In 1303, Pope Boniface VIII turned to the former militant ally of the papacy, the Archbishop of Kalocsa, and called on him to take action against heretics in Bosnia and other provinces of his archbishopric. The pope allowed him to call upon the secular arm to wage war against heretics and Patarins.(120) We can suggest that the pope was making an intentional distinction between the Patarins and the heretics.

During the period 1290-1305 the Bribirski princes of the Šubić family began to try to annex Bosnia.(121) In the first decade of the fourteenth century, their expansion was checked by Stjepan Kotroman, son-in-law of Stefan Dragutin. Heretics appear again in a 1304 obituary of Mladen Šubić, called "Ban of Bosnia" but whose territory was chiefly in the Donji kraji region. His obituary simply reports that he was killed by unfaithful heretics.(122) The most likely explanation for his death is that he was killed in battle against some Bosnians (who tend to be described as heretics by those fighting against them). We may speculate that Mladen had died in a battle against Stjepan Kotroman, against whom he had been waging war in 1302.(123)

XII: Djed Miroslav of the Bosnian Church

One of the oddest and most difficult pieces of information that students of the heresy must try to explain appears in a charter, issued between 1305 and 1307, to the monastery of St. Mary near Boka Kotorska by the Serbian King Stefan Uroš (Milutin). Witnesses to this charter included: Marin, Bishop of Bar (a Catholic), Ioan, Bishop of Hum (Orthodox), Mikail, Bishop of Zeta (Orthodox), Djed Miroslav (of the Bosnian Church as we shall see), župan . . . (a secular official), and Dumina, Bishop of Kotor.(124) One's first reaction to a document including a djed in such company is that it must be a forgery. However, this simple way out does not seem probable. For, as we shall see, there is an independent source that mentions Miroslav at this time. This second source was discovered in Russia in the nineteenth century and it is highly unlikely that a forger of a charter would have known about it.(125) Thus I conclude the charter is authentic.

How, then, are we to interpret the word djed? Does the term refer to the Bishop of the local Bosnian Church who (besides this charter) is first referred to by that title in a charter in the 1320's? The term "djed" was found in eleventh century Croatia, where a djed, the abbot of a royal monastery who was often found at court, witnessed several charters.(126) The Croatian djed was clearly a Catholic. In 1211 in Bulgaria Tsar Boril held a synod against the Bogomils. Among those anathematized at the synod was a Djed of Sardica.(127) This Bulgarian djed was clearly a Bogomil, but we do not find the term used in any other document about Bulgarian Bogomils. More recently in Macedonia and in parts of Serbia the term has been used both for an Orthodox bishop (particularly in the form "deda vladika") and for an older monk.(128)

Since a Bogomil djed, as a member of a sect both opposed in principle to churches and anathematized by the Serbian Church, would have been a most unlikely personage to witness a charter to a Serbian monastery, and since the other Orthodox bishops in the document are referred to as "jepiskup" (and besides, the term djed meaning an Orthodox bishop is not found for another three hundred years), we can conclude that the only possible meaning for the term at the time the charter was issued is the Bishop of the Bosnian Church. Thus it seems highly likely that Miroslav was the Djed of the Bosnian Church, and this theory is confirmed by a second totally independent source, which also refers to

Miroslav.

This source is a list of previous Bosnian Church leaders written at the beginning of a Bosnian Gospel, copied in 1393 for an important nobleman and lay member of the Bosnian Church named Batalo. The relevant list reads: (presented as a column) 1. G(ospodi)n' Rastud'e, 2. Radoe, 3. Radovan', 4. Radovan', 5. Hlapoe, 6. Dragoš', 7. Povr'žen, 8. Radoslav', 9. Radoslav', 10. Miroslav', 11. Boleslav', 12. Ratko.(129) Elsewhere I have shown that the list reads down from Djed Rastudije, the top name on the list, who lived in 1393, through the eleven names listed below his which represent his predecessors as djed.(130) The eighth and ninth names on the list are both Radoslav, and in ca. 1322 we have a Bosnian charter witnessed by Djed Radoslav at the hiža (monastic house) of Gost Radoslav.(131) Thus, we know that a Radoslav was djed in ca. 1322 and since gost was the title below djed it is likely that Djed Radoslav was succeeded by Gost Radoslav which would identify the two Radoslavs on this list. The tenth name on the list is Miroslav. Thus, the predecessor of Djed Radoslav (who was djed in 1322) was named Miroslav. Thus, there was a Miroslav listed as head of the Bosnian Church during the same period that a Djed Miroslav witnessed the Kotor charter. The Miroslav on the charter is almost certainly this Bosnian Church bishop. The Kotor charter, then, is the earliest source to mention a Bosnian djed.

It does seem strange, however, to find a Bosnian Church djed witnessing a charter for an Orthodox monastery in the company of a Catholic and several Orthodox bishops. That both Catholics and Orthodox witnessed the charter shows that the presence of a man of a third faith is not excluded. But it does suggest that the man of the third faith should hold more or less orthodox views, and that the distinction between him and the other clerics would be more political and organizational than theological. Thus, I think Miroslav's presence here as a witness suggests that the church he headed must have held relatively orthodox theological views. It is improbable that he could have headed a dualist or Bogomil Church.

This charter is not the only document in which we find a member of the Bosnian Church acting in the capacity of witness in the company of Orthodox clerics. In 1466 Herceg Stefan's last will and testament was witnessed by David, the Orthodox Metropolitan of Mileševo, as well as by the Bosnian Church leader Gost Radin.(132)

We cannot say positively what Miroslav was doing in Kotor,(133) or in the suite of Stefan Uroš (Milutin). However, we can speculate that the Šubići had occupied or at least invaded the part of Bosnia where Miroslav resided and caused him to flee. Since the Bosnian Church he headed was in secession from the Catholic organization, it is probable that Miroslav would not have been treated kindly by the Catholic Šubići, and it is not surprising that he fled. If his theology was not heretical, and as we have argued above, his presence as a witness on the charter suggests that it was not heretical, then there is no reason for Miroslav not to seek refuge in Serbian territory. For after all, the Serbian Church, though more prestigious in its autonomy, was in basically the same position vis-à-vis Rome, as was the Bosnian Church.

This charter, then, is an important source to show that the Bosnian Church in 1305 was not significantly heretical. Though involved in a quarrel with Rome, it was able to enjoy peaceful and even friendly relations with the Serbian Church. This is exactly what we would expect if the church was simply a local autonomous church built upon the organization of the local originally Catholic monkish order. If some of its deviations had been raised to the level of doctrine as a result of nativistic pride, either these deviations were not sufficient to bother the Serbs or else the Serbs were unaware of the deviations at the time.

Before departing from the subject of the djed of the Bosnian Church and the notation inside the Gospel, we might point out that two names appeared after (hence, preceded in time) Miroslav's. We know nothing else about the two men (Boleslav and Ratko) who are referred to. But we can conclude that if the author in 1393 accurately knew the history of his church and if he knew the names of all the leaders since the church's becoming independent, which are both big ifs, then either Ratko was the first djed of the church or else Ratko was responsible for a significant change (be it doctrinal or organizational) in an already existing institution. If we speculate that the first three bishops had terms of average length this would date Ratko's accession or the establishment of the independent church to roughly 1270-80 (of course, a period about which we know pathetically little). Only if one, two, or even all three of our first bishops had had very long terms could we conclude that Ratko was installed as the first djed back in the 1250's, which is the most likely time for the independent church to have been established. Thus, the 1393 document only roughly confirms our dating of the foundation of the Bosnian Church.

FOOTNOTES TO CHAPTER III

1. Dj. Basler, "Kasnoantičko doba," in *Kulturna istorija Bosne i Hercegovine* (A. Benac, *et al.*, editors), Sarajevo, 1966, pp. 315, 340.
2. *Ibid.*, pp. 331, 338-339.
3. "Neretva" is the name of a River. The word was also used to designate the territory along its course and specifically to denote the towns of Drijeva and Konjic.

4. N.Z. Bjelovučić, *Povijest poluotoka Rata (Pelješca),* Split, 1921, pp. 26-27. This work complicates matters further than necessary by assuming that Bogomils were active in the region of Ston at the time. No source indicates this.

5. On Miroslav's church of St. Peter and St. Paul on the Lim River, see R. Ivanović, "Srednjovekovni baštinski posedi humskog eparhiskog vlastelinstva," *Istorijski časopis,* IX-X, 1959, pp. 79-80; also R. Ljubinković, "Humsko eparhisko vlastelinstvo i crkva Svetoga Petra u Bijelom polju," *Starinar,* n.s. IX-X, 1958-59, pp. 97-123. In the 1250's the Bishop of Hum was forced to vacate Ston; he moved to this church on the Lim which was made his episcopal seat.

6. *CD,* II, pp. 206-07, 226.

7. Resti, p. 63.

8. On the Gnostics, see: H. Jonas, *The Gnostic Religion,* (2nd ed.), Boston, 1958; R. Grant, *Gnosticism and Early Christianity,* (revised ed.), New York, 1966.

9. The dualism of the Paulicians has recently been questioned by N. Garsoian, *The Paulician Heresy,* The Hague, 1967. However, Prof. Garsoian does admit that a dualist wing of the movement appeared in Constantinople in the ninth century. Thus this leaves the whole question of Paulician influence on Bogomil origins up in the air. See my review of Garsoian in *Speculum,* XLIV, no. 2, April 1969, pp. 284-88.

10. On the Bogomils in Bulgaria the two basic sources are the tract of Cosmas the Priest against them (written in the tenth century) and Tsar Boril's Council's Anathemas against them from 1211. The texts, both edited by M. Popruzhenko, appeared in *B'lgarski Starini,* Cosmas in Vol. XII, 1936, and the Sinodik of Tsar Boril in Vol. VIII, 1928. Both have been translated into French with detailed and interesting commentaries on the sources and the dualist movement by H. C. Puech and A. Vaillant, *Le Traité contre les Bogomiles de Cosmas le Prêtre,* Paris, 1945 (Travaux publiés par l'Institut d'études Slaves, XXI); See also the monographs, D. Angelov, *Bogomilstvoto v B'lgarija,* (2nd ed.), Sofija, 1961; D. Obolensky, *The Bogomils,* Cambridge (England), 1948.

11. On the Cathars in the West, see A. Borst, *Die Katharer,* Stuttgart, 1953 (Schriften der Monumenta Germaniae Historica, Deutsches Institut für Erforschung des Mittelalters, XII); On the Cathar churches in Italy see the texts with good (though biased) commentary by A. Dondaine, "La Hiérarchie Cathare en Italie," *Archivum Fratrum Praedicatorum,* XIX, 1949, pp. 280-312, XX, 1950, pp. 234-324.

12. On this Council, see the text published with interesting commentary by A. Dondaine, "Les Actes du Concile Albigeois de Saint-Félix de Caraman," *Miscellanea Giovanni Mercati,* Vol. V, The Vatican, 1946, pp. 324-355.

13. On the location of Dragovica, see I. Dujčev, "Dragvista-Dragovita," *Revue des Études Byzantines,* XXII, 1964, pp. 215-221.

14. "Žitije svetog Simeuna" by Stjepan Prvovenčani, see Chapter VI of this life. Ed. L. Mirković, *Spisi sv. Save i Stevana Prvovenčanoga,* Beograd, 1939, pp. 181-182. Simeun was the name taken by Nemanja when he became a monk at the end of his life. His biography was written by his son. All that the biography says about the heretics' beliefs is that they did not teach that Christ was the Son of God.

15. It is worth noting that no source provides evidence of direct ties during our period between Bosnia and Bulgaria, of either a religious or secular nature.

(I do not include the Tractatus of Anselm of Alexandria — to be discussed below — since it states that Constantinople played the role of middleman between Bulgaria and Slavonia (Bosnia).) If the Bosnian Church was really Bogomil or dualist, the absence of ties between it and fellow South Slav dualists in Bulgaria — particularly when we consider that Bulgarian dualists maintained ties with fellow dualists both in Constantinople and Italy — would be, to say the least, extremely odd.

16. Thomas Archdeacon of Split, *Historia Salonitana,* (ed. F. Rački, MSHSM, 26, Scriptores, 3, Zagreb, 1894). The story of the two brothers quoted below is from Chap. 23, p. 80. For Zadar as a hotbed of heresy see p. 83. For more details on the brothers and the exalted position which they have unjustifiably been given in certain historical works, see my article "Aristodios and Rastudije — a Re-examination of the Question" in *GID,* XVI, Sarajevo, 1967, pp. 223-229.

17. *CD,* II, pp. 296-97.

18. A. Dondaine, "La Hiérarchie Cathare en Italie," *Archivum Fratrum Praedicatorum,* XIX, 1949 (Text of "De Heresi Catharorum in Lombardia" given pp. 306-312. For the beliefs of Caloiannes' church, see pp. 310-312).

19. *Ibid., AFP,* XX, 1950. Text of "Tractatus de hereticis," pp. 308-24. Passage discussed in our text, p. 308.

20. J. Šidak, " 'Ecclesia Sclavoniae' i misija Dominikanaca u Bosni," *Zbornik radova Sveučilište u Zagrebu* (Filozofski fakultet), III, 1955, pp. 37-38; and D. Mandić, *Bogomilska crkva . . .,* pp. 439-30. Both authors cite relevant passages from the "Commentary . . . ." The source is discussed further below in notes 70 and 85.

21. *CD,* II, pp. 168-69.

22. Miklosich, pp. 1-2.

23. Muhašinovići borders on a village called Biskupići about which there is a local tradition that a bishop lived there in the time of Kulin Ban. P. Andjelić, "Revizija čitanja Kulinove ploče," *GZMS* (arch) n.s. XV-XVI, 1960-61, p. 287. The name of that village does suggest that the village had had some relationship with a bishop, though possibly it had been merely part of church lands providing income for the bishopric. It would be interesting to know if the tradition connecting the bishop with the days of Ban Kulin existed before the plaque was discovered. However, since this region was near the center of Kulin's state, and since he built a church here, and since it is likely that the bishop would have resided near the center of the state (when relations were friendly between church and state) at a place having a church, it is perfectly possible that the bishop really did reside here. No source ever specifies where the Bishops of Bosnia resided before 1233.

24. P. Andjelić, "Revizija . . .," p. 290. "Siju crkv' ban' Kulin' zida egd . . . jeni Kucev'sko Zagorie i nade na nu grom('i) v Podgorie Sljepičist' i postavi svoi obraz' nad' (or "za") pragom' B(og) dae banu + Kulinu zdravie i banici Vojslavi." The meaning of the phrase following the lacuna is not clear. Andjelić's interpretation is most reasonable. He thinks that Kulin built this church when he conquered (or bought) Kučev'sko Zagorie, *ibid.,* p. 306.

25. At first glance one might think that the intent of these later carvers was hostile since not only do we find writing across the face of the plaque but we notice that the center of each cross has been roughly chopped at and dug out. However, Prof. Andjelić pointed out to me that popular belief attaches medicinal

properties to powder taken from these ancient monuments (particularly the *stećci*) and suggests that the crosses were not hollowed out to obliterate the sign of the cross but since the cross was a holy sign, peasants believed that powder from that part of the stone would be most potent. This is quite possible, in which case we can attribute the later writing to the same initial carving illness that is so universal today at important historic or natural sites.

26. Andjelić, "Revizija . . .," p. 299.

27. Andjelić, "Revizija . . .," p. 305. Whether Kulin really had a son named Stjepan who ruled the banate is not certain. On that see below note 72.

28. Š. Bešlagić, Z. Kajmaković, F. Ibrahimpašić, "Ćirilski natpis iz doba Kulina Bana," *Naše starine*, X, 1965, p. 204. ("V' d'ni B(a)na velik(a)go Kulin(a) bješe Gra(d) ješa Sudi(ja) veli u neg(a) i s'zida (crkvu) Svetago Jurija. I se leži u n(e)go i žena eg(o)va rče (= reče) položi(t)e me u nego. Ase zida Draže o(h?)mučanin . (To) mu vol'i B(ož)e. N(an') (A)z' pish' pro (dan') (po);'.")

29. Orbini, p. 350.

30. "Annali di Ragusa del magnifico Ms. Nicolo di Ragnina" in *Annales Ragusini,* edited by S. Nodilo (MSHSM, Vol. XIV, Scriptores I, Zagreb, 1883), p. 219.

31. Orbini, p. 350; Resti, p. 63.

32. Orbini, p. 350.

33. *CD,* II, pp. 251-52.

34. *CD,* II, pp. 333-334.

35. S. Ćirković, "Jedan prilog o Banu Kulinu," *Istorijski časopis,* IX-X, 1959, pp. 71-77. Stefan Nemanja had left three sons: Stefan who held Serbia, Vukan who was assigned Zeta, and Sava the future archbishop and saint. Vukan had his sights set on Serbia and was supported in this enterprise by King Imre of Hungary. Both Imre and Vukan were hostile to Miroslav of Hum as well. After Miroslav's death, Imre appointed his brother Endre Vojvoda of Dalmatia, Croatia and Hum. How much of Hum he actually controlled is not known. Presumably Endre's jurisdiction was limited to Hum west of the Neretva. Miroslav's widow, Kulin's sister, left Hum and returned to her brother's court in Bosnia at this time. Various scholars have assumed that she had been forced to flee; however, we really do not know. In 1202 Vukan aided by Imre invaded Serbia; Imre made himself King of Serbia and gave Vukan the title of Grand Župan. Presumably Kulin's attack noted in the text was connected with this war. We assume Kulin attacked territory held by Vukan. Shortly thereafter, Stefan was able to recover his throne and later Sava was able to bring about peace between his two brothers.

36. *CD,* II, p. 351; P.T. Haluščynskyj, *Acta Innocentii III,* p. 209.

37. *CD,* III, pp. 14-15; Haluščynskyj, *Acta Innocentii III,* pp. 224-25.

38. Unless one wants to argue that Kulin, instead of sending some of the suspected heretics, sent some innocent substitutes.

39. Text published in many places: D. Mandić, *Bogomilska crkva,* pp. 435-36; *CD,* III, pp. 24-25; Theiner *MSM,* I, p. 20; Haluščynskyj, *Acta Innocentii III,* pp. 235-37. The chief variation between the texts is the rendering of the priors' names. Hence for those names with variants, I have given in my text the variants in parenthesis after the given name.

40. The size of Bosnia in time, a typical rural way of measuring distance, was provided by Casamaris in a letter to the pope about his mission. *CD,* III, p. 25.

41. See Chapter I, note 22, for a Russian envoy who visited neighboring

Montenegro in 1760 and noted the absence of crosses, icons and books in churches there. We have no evidence that dualism penetrated Montenegro; thus we find that such furnishings may well be absent in uneducated regions without dualism being the cause.

42. See below note 70 for this Chapter.

43. This was a postscript to the abjuration. Mandić, *op. cit.*, p. 436; *CD*, III, p. 25.

44. *CD*, III, pp. 36-37.

45. *CD*, III, p. 36.

46. *CD*, III, p. 70.

47. Resti, p. 74.

48. Resti, p. 75.

49. Resti, pp. 75-76.

50. *CD*, III, pp. 187-88.

51. *CD*, III, pp. 196, 198-99. A Tautu, *Acta Honorii III . . .*, p. 111. Against the Kačići, *CD*, III, pp. 205-06.

52. *CD*, III, pp. 209-210.

53. Thomas Archidiaconus, *Historia Salonitana*, ed. F. Rački, (MSHSM, 26) Scriptores, 3, Zagreb, 1894, p. 96.

54. *Ibid.*, p. 98.

55. *Ibid.*, p. 103. The year 1222 which Thomas gives for Acontius' death may not be accurate since we have a letter from Honorius III to Acontius dated July 1223 (*CD*, III, p. 229) which shows that unless the legate had died without the pope learning of the fact, Acontius did not die in 1222. A papal letter of May 1225 refers to the legate as deceased. (*CD*, III, pp. 242-43). Since we have no evidence of Acontius working in the Slavic lands prior to 1221, since much of his activity in Slavic lands took place in Dalmatia, and since Thomas' chronicle is a poor source for Bosnian affairs, it seems probable that the statement in his chronicle about Acontius working a long time in Bosnia is exaggerated.

56. *CD*, III, pp. 242-43; Tautu, *Acta Honorii III . . .*, p. 183.

57. *CD*, III, p. 264.

58. *CD*, III, pp. 264-65. John Angelus was called "Prince of Constantinople" because he was a member of the family which ruled the Byzantine empire from 1185 until the Latin conquest of 1204. After that debacle he came to Hungary, whence his mother had come, and eventually was assigned the districts of Srem, Beograd, Baracs and Mako which he governed until 1254.

59. *CD*, III, p. 274.

60. *CD*, III, p. 286.

61. See Chapter II.

62. *CD*, III, p. 362. Tautu, *Acta Honorii III et Gregorii IX*, pp. 233-34.

63. *CD*, III, p. 379, Tautu, *op. cit.*, pp. 268-69.

64. *CD*, III, p. 388; in a second letter, also p. 388, the pope mentions Ninoslav's acceptance of Catholicism. (Also in Tautu, *op. cit.*, pp. 271-72.)

65. *CD*, III, p. 389. Tautu, *op. cit.*, p. 273. Ubanus surely was a distortion of the title ban. Priezda was therefore presumably a ban of some part of Bosnia under Ninoslav's suzerainty. Since in the period 1255-87, we shall find Priezda holding land in Usora and around Zemljanik on the Vrbas as well as in Slavonia, we can assume that in 1233 he already ruled land in that region.

66. *CD*, III, pp. 388-89.

67. On Johannes von Wildeshausen, see A. Rother, "Johannes Teutonicus (von Wildeshausen)" *Romische Quartalschrift für Christliche Alterthumskunde und für Kirchengeschichte,* IX, 1895, pp. 139-170. And N. Pfeiffer, *Die Ungarische Dominikanerordensprovinz von ihrer Grundung 1221 bis zur Tatarenverwustung 1241-1242,* Zurich, 1913, pp. 62-70.

68. *CD,* III, p. 415.

69. *CD,* III, pp. 416-17, 419.

70. As noted earlier Suibert stated that conversions in Bosnia had been effected by Dominicans sent "to Bosnia and Dalmatia in which the Church of Sclavonia is to be found." Since this title was used by the Dominicans connected with the inquisition to refer to a dualist church, it is probable that the Dominican Suibert used the term with this meaning. Thus here we have further evidence that there were dualists in Bosnia (and in Dalmatia, parts of the same current) and that the crusade, at least in part — and quite possibly nominally — was directed against them. Regardless of the fact that many, and probably most, of the Bosnians affected by the crusade were not dualists, Suibert felt that the dualists were significant enough — or it was useful to depict them as such, possibly to justify the bloodshed of the crusade, and possibly to support the accusations the Hungarians had most probably been making to the papacy — to describe them as the *raison d'être* of the crusade. However, since this is the only source to imply the crusade was directed against dualists (rather than simply heretics of unspecified beliefs) it would be a mistake for us to conclude that the crusade was directed solely against dualists or that all, or even most, of the Bosnian "heretics" were dualists. Suibert (Peter Patak), "Commentary on the Founding of the Hungarian Dominican Province" (written in 1259). Passages discussing Bosnia given by J. Šidak, " 'Ecclesia Sclavoniae' i misija Dominikanaca u Bosni," *Zbornik radova Sveučilište u Zagrebu* (Filozofski fakultet), III, 1955, pp. 37-38.

71. *CD,* IV, pp. 15-16; the mother under protection, p. 17.

72. We know nothing whatsoever about Sibislav's father Stjepan. The pope's letter merely states that Sibislav was the son of a former Ban Stjepan of Bosnia. The title suggests that Stjepan had, at some point, ruled the Bosnian banate. However, later we shall find people called Bans of Bosnia who ruled only over small areas in the north (see J. Fine, "Was the Bosnian Banate Subjected to Hungary in the Second Half of the Thirteenth Century?", *East European Quarterly,* III, June 1969, pp. 175-76). We also know that there were at times more than one ban within greater Bosnia at the same time (e.g., Ninoslav and Priezda). Thus it is possible that Stjepan never actually ruled the banate. Various historians (e.g., Perojević, in *Poviest BH,* Sarajevo, 1942, pp. 216-18) believe that Stjepan was Kulin's son and successor. We know that Kulin had a son and have already mentioned that son's trip to the Hungarian court in 1203. However, we do not know the son's name. Klaić (*Poviest Bosne . . .,* pp. 67-68) believes that Stjepan succeeded Kulin and then ruled the banate until ca. 1232 when Ninoslav and heretical nobles ousted him from the banate, installed Ninoslav as ban and left Stjepan to rule only Usora. We have no evidence to refute this theory; however, we also have no evidence to demonstrate it. It is just as possible, if Stjepan had really ever ruled the banate, that he died a natural death and was succeeded by Ninoslav, his brother or eldest son, while son Sibislav was assigned Usora.

73. *CD,* IV, p. 27.

74. The document states that Hum, having been cleansed of heresy, should be given to Ponsa (see note 75). It is not evident whether the purification has already been carried out or whether it is merely planned and that after the plans have been realized Ponsa should be given Hum. We also do not know what part of Hum is meant. M. Vego, *Povijest Humske zemlje,* Vol. I, Samobor, 1937, pp. 90-91, suggests that the document refers only to that part of Hum between Cetine and the west bank of the Neretva River. Vego believes that the Hungarians had successfully subjected this territory by 1238.

75. *CD,* IV, p. 57; Tautu, *op. cit.,* pp. 316-17.

76. December 1239 reference, *CD,* IV, p. 94; 1244 reference, *CD,* IV, p. 239.

77. *CD,* IV, p. 66; Tautu, *op. cit.,* p. 330.

78. *CD,* IV, p. 66; Tautu, *op. cit.,* p. 330.

79. *CD,* IV, p. 239.

80. *CD,* IV, p. 66; Tautu, *op. cit.,* p. 330.

81. See for example, fra L. P.(etrović), *Kršćani bosanske crkve,* in *Dobri pastir,* Sarajevo, 1953, and published separately.

82. *CD,* IV, p. 54.

83. *CD,* III, p. 427; *CD,* IV, pp. 107-08, 126; also Miklosich, pp. 24-25, 28-30.

84. *CD,* IV, pp. 65, 67-68.

85. *CD,* IV, pp. 93-94. Suibert's "Commentary on the Founding of the Hungarian Dominican Province" praises Coloman highly for converting many heretics and adds that many who refused to convert were burned by Coloman's men. Šidak notes that many later Catholic scholars have skipped this passage. See Šidak, " 'Ecclesia Sclavoniae' . . .," pp. 37-38.

86. *CD,* IV, pp. 107-08; Miklosich, pp. 28-29. In this charter, which opens with a cross, Ninoslav swears by, among other things, the honorable life-giving cross which suggests that if he had been heretical in some way, his heresy did not include rejection of the cross. In this connection we can note that Ninoslav's personal seal also had a cross on it (see P. Andjelić, *Srednjovjekovni pečati iz Bosne i Hercegovine,* Akademija nauka i umjetnosti BiH, Djela, knj. 38, odj. Društv. nauk, 23, Sarajevo, 1970, p. 10).

87. Dubrovnik was not alone among Dalmatian cities, in maintaining good relations with Ninoslav. Hungary's former friend Split also did not seem to find anything heretical about the Bosnian ban, and in 1243 the citizens of Split even briefly elected Ninoslav "knez" of their city. Thomas the Archdeacon, who reports this event, and who usually is the first to suggest a taint of heresy about people, never suggests that Ninoslav was heretical. See, Thomas, p. 195.

88. *CD,* IV, pp. 94-95.

89. *CD,* IV, p. 95. Reference to the money in the same terms as in 1238 (see above, note 78).

90. Document published in appendix of N. Pfeiffer, *Die Ungarische Dominikanerordensprovinz,* Zurich, 1913, p. 150. This 1303 notation simply lists Dominican convents: seven in Pannonia, six in Slavonia (including Chamensis, Požega, Veronica, Zagreb, Bihač, and Jadrensis — the last name surely refers to Zadar, which points to the geographical ignorance of the writer), and eight in Dalmatia (not including Zadar there). The only location mentioned here that we would associate with Bosnia is Bihač, and that city was never part of the medieval state. In the Middle Ages the area around Bihač was Catholic;

we have data about many Catholic churches in the region and no references to heretics there (see R. Lopašić, *Bihać i bihaćka krajina,* Zagreb, 1890).

91. The Tatar invasion gave Ninoslav the chance to take the offensive on behalf of Split against Hungary's ally Trogir. And in 1243-44 we find Ninoslav allied with Split (in fact he was briefly Prince of Split), Andrej of Hum and various other nobles against Trogir, King Bela and Bela's vassals the Šubići and the Nelipci. (see Klaić, *Poviest Bosne . . .,* pp. 77-78).

92. *CD,* IV, pp. 236-40.

93. *CD,* IV, p. 297.

94. *CD,* IV, p. 299.

95. *CD,* IV, pp. 310-11.

96. *CD,* IV, pp. 322-23.

97. *CD,* IV, p. 420.

98. *CD,* IV, p. 494.

99. *CD,* IV, pp. 341-42.

100. CD, IV, p. 342.

101. *CD,* IV, pp. 386-87. Unfortunately there are no references to ecclesiastical affairs in this charter. The charter is actually issued by a Bosnian ruler who is called Matej Stjepan though it bears the seal of Ban Ninoslav. Most scholars believe that Matej Stjepan is Ninoslav. This problem, which unfortunately cannot be conclusively resolved, is discussed in some detail in my article, "Was the Bosnian Banate Subjected to Hungary in the Second Half of the Thirteenth Century?" *East European Quarterly,* III, June 1969, pp. 168-70. See also, G. Čremošnik, "Bosanske i humske povelje srednjega vijeka," *GZMS,* n.s. III, 1948, pp. 124-29 and V. Mošin's review of Čremošnik's article in *Historijski zbornik,* II, 1949, pp. 316-17. Čremošnik believes that Ninoslav had had to share power with a relative named Stjepan; Mošin believes that either Matej Ninoslav changed his name to Matej Stjepan or else was succeeded by a new ruler having that name.

102. Assuming that Matej Stjepan (the name given in the 1249 charter, see note 101) refers to Ninoslav and not to a successor. It seems that Ninoslav did have more than one son. According to Thomas the Archdeacon of Split, when Ninoslav departed from Split in 1243 he left in control a relative named Rizarda and "one of his own sons" (see Thomas, p. 195); also both the 1240 and 1249 charters to Dubrovnik stated that the terms were binding on his children and grandchildren; of course, this may have been a formal way of stating the terms were binding on his successors rather than reference to individuals who actually existed. Ninoslav also had brothers. Bela's 1244 charter (*CD,* IV, 239) refers to property which Ninoslav with his brothers and barons had granted the church. The 1249 charter also refers to Matej Stjepan's brothers. It is logical to assume that either a son or a brother succeeded Ninoslav; however, whether one of them actually did is unknown.

103. *CD,* IV, p. 460.

104. *CD,* IV, p. 494.

105. Orbini, p. 354.

106. *Europa Pii Pontificis Maximi nostrorum temporum varias continens historias,* Basel, 1501, p. xxiii.

107. S. Ćirković, "Die Bosnische Kirche," *Accademia Nazionale dei Lincei, Problemi Attuali di Scienza e di Cultura,* Rome, 1964, pp. 552-555; see also his *Istorija srednjovekovne bosanske države,* Beograd, 1964, pp. 58-69.

108. In fact it is very hard to conceive of a religious order — even a loosely organized one like that of our monks — remaining under its own hierarchy, voluntarily changing in a major way its religious beliefs and practices.

109. R. Linton, "Nativistic Movements" in W. Lessa and E. Vogt, *Reader in Comparative Religion,* 2nd ed., New York, 1965, pp. 499-506, originally published in *American Anthropologist,* XLV, 1943, pp. 230-40.

110. Similar things have happened elsewhere, e.g., in the Philippines: ". . . even under Spanish rule there was a great deal of religious ferment which produced native movements of a so-called 'heretical' nature, especially in rural areas. On various occasions these movements had been headed by native-born members of a Catholic religious order, who had become openly antagonistic to the Spanish hierarchy." (V. Lanternari, *The Religions of the Oppressed,* pp. 221-22.) The phenomenon of a rural Catholic order breaking away from official Catholicism to go it alone seems also to be an accurate description of what we find in medieval Bosnia.

Hostility to a form of Christianity because of its association with a foreign power (seen against Spain in the Philippines) is regularly found in Africa. Lanternari, *op. cit.,* p. 60, remarks, ". . . one of the most influential factors in the growth of these (native) churches and cults has been the nature of relations between the white man and the natives, and those personal experiences which gradually caused the indigenous peoples to look upon the missions as facets of European rule. The native church is thus to be regarded as the limit to which the independent movements will go in accepting the full doctrine of Christianity. It reveals a middle course between traditional beliefs and the new theology." In this vein the Bosnian Church can be seen both as a reaction to the threat of Hungarian dominance and as a compromise between traditional values and Catholic doctrine.

111. Miklosich, p. 69.

112. Lj. Stojanović, *Stare srpske povelje i pisma,* I, pt. 2, pp. 79-83, and Pucić, II, p. 126.

113. M. Ristić, *Bosna od smrta bana Matije Ninoslava do vlade sremskoga kralja Stevana Dragutina, 1250-1284,* Beograd, 1910, p. 105.

114. *CD,* VI, pp. 357-358.

115. *CD,* VI, pp. 378-79. Twice this letter stresses the lack of uniformity among the heretics.

116. Usora and Sol, in fact, are the only regions which previously had been controlled by Ninoslav, be it directly or as overlord (see reference to privileges he granted the Catholic Church in Usora in Bela's 1244 charter, *CD,* IV, p. 239), found in the second half of the thirteenth century in the hands of bans appointed by the King of Hungary.

117. Arhiepiskop Danilo, *Život Kraljeva i Arhiepiskopa Srpskih* (Srp. knj. zadruga), Beograd, 1935, p. 34.

118. Pope Nicholas IV wrote three letters pertaining to sending this mission on 23 March 1291, see F. Delorme and A. Tautu (eds.), *Acta Romanorum pontificum ab Innocentio V ad Benedictum XI (1276-1304),* numbers 101, 102, 105, pp. 171-174, 176-80.

119. Delorme and Tautu, *op. cit.,* pp. 202-03; *CD,* VII, p. 302.

120. Delorme and Tautu, *op. cit.,* p. 238; *CD,* VIII, pp. 47-48.

121. On this principality, see V. Klaić, *Bribirski knezovi od plemena Šubića do godine 1347,* Zagreb, 1897.

122. S. Zlatović, "Bribirski nekrolog XIV-XV v," *Starine*, XXI, p. 84.
123. M. Dinić, "Odnos izmedju Kralja Milutina i Dragutina," *ZRVI*, vol. 3, 1955, p. 61.
124. Miklosich, pp. 67-69.
125. The main signs of a forged document are absent; the majority of South Slav forged charters were the work of families in the late sixteenth, seventeenth or early eighteenth centuries who sought to provide proof of their families' importance in medieval times. Since not one witness of the Kotor charter gave his last name, this charter could not have been used to support this sort of fraud. The only party who might have stood to gain by forging such a charter would have been the monastery itself. I do not know enough about the monastery to pass judgment on the matter. An argument that might indicate that the charter was not particularly important for the Kotor church, however, is the fact that the charter was discovered in the middle of the nineteenth century, not in Kotor, but in Vienna.
126. Fra L. P(etrović), *Kršćani bosanske crkve*, pp. 142-43.
127. T. Florinskij, "K voprosu o Bogomilah" in *Sbornik statej po Slavjanovedeniju sostavlennyj i izdannyj učenikami v čest V. I. Lamanskago*, St. Petersburg, 1883, pp. 37-39. Florinskij gives the original text and we see the Bulgarian was called "Dedec (Ded) Sredec" (presumably Sardica, modern Sofia).
128. V. Glušac, "Problem Bogomilstva," *GID*, V, 1953, p. 107.
129. For the text see Dj. Radojičić, "Odlomak bogomilskog jevandjelja bosanskog tepačije Batala iz 1393 godine" in *Izvestija na Instituta za Istorija* (B'lgarska Akademija na naukite), XIV-XV, 1964, p. 502.
130. See J. Fine, "Aristodios and Rastudije," *GID*, XVI, 1967, pp. 225-228. Most of the argumentation is also presented below in Chapter V.
131. L. Thalloczy, "Istraživanja o postanku bosanske banovine sa naročitim obzirom na povelje körmendskog arhiva," *GZMS*, XVIII, 1906, pp. 404-405.
132. Pucić, II, p. 126, and Stojanović *Stare srpske povelje . . .*, I, pt. 2, pp. 79-83. The Metropolitan David also participated in depositing some of Herceg Stefan's valuables in Dubrovnik in 1465 in the company of several krstjani, see Miklosich, p. 497.
133. There is no evidence that Miroslav's church had any following in Kotor at the time.

CHAPTER IV

BOSNIA FROM CA. 1315 TO 1391

I: Ban Stjepan Kotromanić: Early Political Successes and Relations with the Different Faiths

Ban Stjepan Kotroman died sometime between 1305 and 1315, and his death paved the way for local disorders which forced his widow and son to flee to Dubrovnik. During the second decade of the century, the son, Stjepan Kotromanić, returned from exile and established himself as ban.(1) His holdings were clearly in the central part of the state; he issued charters in the 1320's from places located in the Visoko-Zenica area. We know neither how he recovered his position nor how extensive was the banate he controlled in these first years.

Kotromanić was most probably Orthodox when he became ban. His mother, Jelisaveta, daughter of Stefan Dragutin, was Orthodox, and presumably raised her son in that faith. This supposition is supported by Orbini (writing in 1601) and Peter Masarechi (writing in 1624), both of whom probably drew on Pietro Livio of Verona (ca. 1530), who report that Stjepan Kotromanić was of the Greek rite.(2) In 1318 the new ban wanted to marry the daughter of Menhard of Ortenburg, a Catholic. Because the two were cousins of some sort, special dispensation had to be obtained from the pope. This dispensation was obtained through the offices of Mladen II Šubić.(3) Mladen's intervention on behalf of Stjepan shows that relations were patched up between the Šubić family and Kotroman's successor.(4) With Šubić recognition of Kotromanić's authority in central Bosnia, Kotromanić's state would be free to develop without interference from his family's long-time most powerful rival. Moreover, since the pope approved the marriage, obviously the groom was not heretical. Indeed, Kotromanić later would cooperate in the establishment and work of the Franciscan mission. Since we have no evidence that the ban accepted Catholicism at the time of his marriage, and since the contemporary Bobali documents, utilized by Orbini (or Pietro Livio before him), attribute the ban's acceptance of Catholicism to Bobali's influence after the establishment of the Franciscan mission in 1340,(5) it seems probable that the ban remained Orthodox until the 1340's.

However, we shall also see that the ban was cordial toward the Bosnian Church. He had dealings with its hierarchs and allowed them to be active in the center of his state, which reflects his favor (or at least tolerance). If this church was really a Slavic liturgy schismatic off-shoot of Catholicism, there is no reason why he should not have been kindly disposed toward it. Since in Bosnia there do not appear to have been Orthodox believers, the ban, if he was to receive sacraments and participate in the activities associated with a church, would either have had to have his own Orthodox chaplain (which is quite possible) or else to have made use of the offices of the Bosnian Church. It is quite possible that he followed the latter course; it would have been slight change from the Orthodox Church, and would not have involved him in heresy.

The fact that the ban had cordial relations with the Bosnian Church, when weighed with the other data we have about this church, gives us reason to believe that this church did not differ much from orthodox forms of Christianity, and certainly was not dualist. We have argued above that the Bosnian Church (called Patarin in Dubrovnik) developed out of a Slavic rite Catholic Church in Bosnia. This view is supported by the report of an Irish Franciscan Symeon Semeonis, who visited Dubrovnik in 1322. He reports that Slav Barbarians, ''Paterini'' and other schismatic traders came to Dubrovnik. The language of the Slavs is quite similar to that of the Bohemians; though in rite they depart considerably for the Bohemians employ the rite of the Latins while that of many Slavs is Greek.(6) If the Patarins were heretical in doctrine we would expect local Franciscans to have told Symeon of this, yet he classified them with other Slavic schismatics. Thus he made no distinction between the Bosnians and the other Orthodox Slavs (presumably Serbs, Montenegrins, and Bulgarians) whom he saw. Since he made a general statement about the rite of ''many Slavs'' (i.e., presumably distinguishing these Slavs from Latin rite Slavs), he evidently thought these South Slav schismatics mentioned, among whom he includes the Patarins, all practiced the same rite. And this is exactly what we would expect if the Bosnian Church followed a Slavic liturgy version of the Catholic rite.

Further evidence of the schismatic (non heretical) character of the Bosnian Church comes from the account of a Catholic collector of tithes from the years 1317-20. The cleric's task had been to collect tithes from the towns and Diocese of Bosnia.

He reports that in his trip to the cities and Diocese of Bosnia he found them ''near to schismatics and almost destroyed.''(7)We do not know if by Bosnian diocese, is meant the former diocese including the central state and the region north to the Sava, or only that territory north of

Bosnia proper where the bishop in Djakovo actually had some authority. Since we have no evidence of Orthodox believers having influence in either (so that their propinquity might be relevant to the state of deterioration within the diocese), we can suggest that by "schismatic" he meant the Bosnian Church. That he did not use the term "heretic" shows that he did not believe the Bosnian Church was heretical. Neither did he note the existence of any large scale heretical movement in the diocese; he attributed the destruction of Catholicism in the diocese solely to schismatics.

In 1318, in Pope John XXII's letter to Mladen Šubić, approving Kotromanić's marriage, the pope also called on Mladen to exterminate the long-standing heresy in the Bosnian land.(8) By "Bosnian land" did the pope mean territory to the west that Mladen might still have held, or was Mladen expected to put pressure on Kotromanić to discipline his subjects? Quite possibly the pope did not even know who controlled what region in Bosnia. In 1319 Pope John again wrote Mladen to take action against these heretics. The pope stressed the decline of Catholicism, which was said to be owing to long neglect and to the heretics, and pointed to abandoned churches and the absence of clerical orders. He said that the cross was not revered and that the sacraments of baptism and communion were in many places ignored by the people.(9)

Here the pope was clearly speaking about the condition of the Catholic Church in Bosnia, and the absence of clerics refers to Catholic priests. By 1319, Bosnia had had no contact with official Catholicism for seventy years; thus this absence is not strange. And without priests, churches would naturally have been abandoned and sacraments not administered. If by chance these churches had been taken over by Bosnian Churchmen the pope would still have considered the churches abandoned. And any sacraments that might have been administered by these "heretical" priests would not have been considered valid by the pope, who would naturally say the sacraments were being ignored. Thus on these matters, the letter is simply describing the decline of Catholicism; it provides no information on the number of Bosnian Church priests or on the attitude of that church toward churches and sacraments. In addition, it is likely that after the disappearance of Catholic priests, in many areas there would have been no priests of other faiths to replace them. If there were no priests, then the people could not partake of sacraments, and if many years passed during which the people never saw a priest they might well have forgotten about these rites. Even the significance of the cross might well have been forgotten by popular opinion.(10) Such a situation could easily have occurred without any conscious rejection of the sacraments by the populace. Although lack of priests and church education can well explain how this situation came about, we cannot discount the possibility that dualist heretics in the area, with their hostility to the cross and sacraments, might have had a certain influence on various Bosnians, including some who remained members of other non-dualist bodies.

The 1320's was a period of rivalry and warfare among the Croatian nobles. The Hungarian king played an active role in these affairs as overlord of most of these nobles. In addition, he had to struggle against a rival for the Hungarian throne. In all this chaos, Ban Stjepan remained loyal to the king, and the two seem to have had cordial relations through the 1320's. Although papal letters to the King of Hungary about Bosnian affairs suggest that some sort of Hungarian overlordship over Bosnia existed, we have no evidence that it was anything more than nominal. There is no evidence of any actual interference by Hungary in the affairs of Kotromanić's banate. In 1324 Kotromanić added the lands of Sol and Usora to his title of Ban of Bosnia.(11) These lands previously had been under his grandfather, Stefan Dragutin, but we do not know how Kotromanić obtained them or how great was his actual authority over these regions. Between 1322 and 1325, he obtained suzerainty over the Donji kraji, and the magnates there, the Stjepanići, paid homage to him.(12) This acquisition reflects the decline of the Šubići. For the rest of the century, except for the early years of Tvrtko's reign, the Donji kraji region was to be considered part of Bosnia, though in reality it was a more or less autonomous region under the powerful Stjepanić-Hrvatinić family who paid nominal homage to the ban, and might better be considered allies than vassals.

With the decline of the Šubići, the balance of power among the Croatian nobles was upset; old alliances collapsed and new ones were made. At this time the Nelipac family rose to predominance, and in 1324 an alliance including Djuraj Šubić and Stjepan Kotromanić was formed against them. In the warfare that followed the Bosnian ban reaped handsome profits in enormous gains to his west, acquiring territory all the way to the coast, including Završje and the Krajina, with the towns of Hlivno, Imota, Duvno, Glamoč, and Makarska. This territory included many Catholics, two bishoprics (Makarska and Duvno, the latter being reestablished at about this time) which were subject to the Archbishop of Split, and a relatively well-functioning parochial *(župa)* organization. Most of this territory was to remain Bosnian until the Turkish conquest.(13)

Owing to the decentralized nature of the Bosnian state, the incorporation of this Catholic territory had little effect on the development of Catholicism in the territory of the old banate; with poor communications and little migration, the Catholics, left undisturbed to practice their religion, remained in their towns and villages to the west and had little or no influence on religious affairs in the banate. The Bosnians also did not seem to seriously interfere with the Catholic Church in these annexed regions. The only sign of trouble is seen in a papal letter of 1344 asking the ban to let Valentinus, the Bishop of Makarska, return to Makarska from which he had fled to Omiš, and once again to collect tithes.(14) We do not know why the bishop had had to leave; various scholars have attributed his departure to persecution by Bosnian heretics, which, though not impossible, seems no more probable than any number of possible political or financial disputes between the church and its new overlord. There is no evidence that the Bosnian Church made any attempts to acquire adherents in this newly annexed region. No source speaks of the spread of heresy here. In the beginning of the fifteenth century we shall find in the nobleman Pavle Klešić, who held territories around Glamoč, our only example of a Bosnian Church adherent in this region. Probably it is not surprising that Pavle's son was to be a Catholic.

As a result of the death of the Serbian King Milutin in 1321 and the power struggle in his lands that followed, the Bosnian ban was able to annex most of Hum. The most recent research dates this acquisition to 1326.(15) Relations between Bosnia and Serbia evidently were tense in the years that followed. The leading noble family in Hum prior to the annexation had been the Branivojevići, and those of that family who survived the Bosnian take-over fled to the Serbian court, where they presumably played a role in anti-Bosnian agitation. Their departure permitted the Draživojevići, who by the 1330's had become vassals of the Bosnian ban, to become first family of Hum. From this family came the powerful Kaznac (Treasurer) Sanko and his sons Beljak and Radič Sanković who were to play such a central role in Bosnian and Hum affairs later in the century. We know of no actual warfare between Bosnia and Serbia after the annexation, but we can be sure that a search for allies, as well as local feuds between allies of the two more powerful realms, followed. In 1331 Dubrovnik offered to mediate, and, though we have no evidence that a treaty was signed, it seems probable that one was; for we see no further sign of hostility between the two states for twenty years. In addition, the rulèr of the aggrieved party, the new Serb King, Dušan, was more interested in expanding to the south into Macedonia, and certainly would not have wanted war on two fronts. It is also evident that Dušan had limited interest in these western lands; in 1333 he sold Ston with the Pelješac Peninsula to Dubrovnik for cash and annual tribute.(16) After negotiations, Bosnia, too, recognized

Dubrovnik's rights to administer these lands.

The magnates of Hum and of the territory east of the Neretva (which together comprise what we now call Hercegovina) retained semi-autonomy until the fall of the Bosnian state. Which families had greatest influence and how much and what parts of Hercegovina they held varied with circumstances, but at least one or two families always existed there with sizable lands and with virtual independence, whose loyalty could not be taken for granted by the ban. At times these nobles would be loyal to Bosnia, at others to the ruler of Serbia, and sometimes one of them might even try to go it alone, with the lesser local lords aligning themselves with or against him.

In the following century, men like Sandalj Hranić and Herceg Stefan were able to set up an independent state in what is now Hercegovina; their power and influence could seriously rival the ruler of Bosnia. Most — if not all — of these families holding power in Hercegovina during the fourteenth century were Orthodox; and in fact most of the nobility as well as the bulk of the population were to remain Orthodox until the Turkish conquest. It seems likely that the Orthodox in Hercegovina remained under the jurisdiction of the Bishop of Hum, located at the Church of St. Peter and St. Paul on the Lim River, until the seat was transferred to Mileševo (in the same region) in the second half of the fourteenth century. Only in the fifteenth century shall we find signs of the Bosnian Church in Hercegovina; then we shall meet Bosnian Church clerics at the court of one magnate family there and shall find references to a few of their monastic houses in Hercegovina.

Thus from the above — and from what we shall discover later — we see that each faith was active only in certain parts of Bosnia and Hercegovina. The Bosnian Church was present in the central banate and had some following (but never enough to be called a strong movement) in the regions to the north (Sol, Usora, Donji kraji). In the fifteenth century on a very small scale it would obtain followers in a limited number of new regions. The Catholic Church, absent in the banate from ca. 1250 until the arrival of the Franciscans in the 1340's, retained at least a nominal hold over many people in the northern regions; and in the territory to the west of the Vrbas River, straight south through Završje and the Krajina, including Livno, Duvno, and Glamoč, it had the loyalty of almost everyone, just as the Orthodox Church had the loyalty of almost everyone in Hercegovina (excluding some Catholics near and along the coast). There is no evidence of any large-scale migrations in this period. Thus this general picture will remain valid into the next century. In the early fourteenth century, there is no sign that any faith tried to acquire followers in other regions. And this picture will be modified only by the arrival of the Franciscans in the 1340's, who would begin proselytizing and whose activities would gradually begin to increase the number of Catholics in the central part of the state.

Thus in the 1320's Bosnia was ruled by an energetic ban, who was threatened by no outsiders and in fact had taken the offense himself against his neighbors. The previously hostile King of Hungary, now threatened by a rival for his throne, by factious nobles, and by civil war, clearly posed no threat to Bosnia. Besides, in these quarrels the ban sided with the Hungarian monarch. As a result of the chaos and civil wars in Croatia, the main rival of the Kotromanić family, the Šubići, had so declined in power that they could no longer interfere in the affairs of the banate. Shortly thereafter, (between 1322-25) the chief vassal of the Šubići in the Bosnian region, the Stjepanići-Hrvatinići, shifted alliances and came over to Kotromanić's camp. In 1321 the death of the King of Serbia, which initiated a struggle for power there, allowed the ban to expand south into Hum, which had previously been more or less Serbian, Thus the ban, with no outside enemies on his borders, was able to solidify his position in central Bosnia and to assert his overlordship over the nobles whose lands lay on the borders of the Bosnian state.

This situation explains why we begin suddenly to find a fairly large (for Bosnia) number of charters, from the 1320's and 1330's, issued by the ban after a long hiatus. These charters re-affirmed the nobles' rights to lands they had previously possessed, but now asserted the ban's right of suzerainty.

In one of these charters (dated 1322-25) we find our first reference to the term "Bosnian Church"; its hierarchy guaranteed and witnessed the terms of the grant. The church's role of judicial guarantor has been noted in most studies of the Bosnian Church. Yet in extant charters the Bosnian Church rarely played this role. The bulk of medieval Bosnian charters were guaranteed not by churchmen but by nobles or court officials. Of the six charters published by Thalloczy from about 1322 to 1331(17) only two had hierarchs of the Bosnian Church among the guarantors. We shall find no more of such charters from the fourteenth century and only two or three examples from the fifteenth century.

The first charter begins, "In the name of the Father and Son and Holy Ghost, I, Saint Gregory (Gr'gur) and called Ban Stjepan, Lord of Bosnia . . ." This formula is then followed by the text about the lands of Vukoslav Hrvatinić, which were confirmed in the treaty, after which the charter states that this act was carried out before the Great Djed ("De(do)m' velikim' ") Radoslav and before the Great Gost ("Gostem' velikim' ") Radoslav and before elders ("Starcem' ")

Radomir, Žun'bor, and Vl'č'kom and before all the Bosnian Church ("Pred v'som cr'k'vom i pred Bosnom.") The charter concludes with the statement that this deed was enacted and written in Moštre at the monastery of Gost Radoslav ("u gosti velikoga hiži u Radoslali (sic)").(18)

Besides being the earliest document to use the expression "crkva bosanska," this charter gives evidence about the hierarchy of this church: it mentions and names the head of the church, the *djed,* a leader of the second rank of hierarchs, the *gost,* and several of the third rank, the *starac.* The existence of these ranks in this hierarchical order is confirmed by later documents. The gost headed a monastic establishment, which supports the theory that the Bosnian Church was a continuation of the Bosnian monastic order present in 1203 at Bilino polje, where each prior had represented the members of the convent he headed. The monastery (hiža) of Moštre, where the charter was signed, was roughly six kilometers from Visoko, and thus very near the center of the state. This fact suggests that the church enjoyed the favor of the ban. The church thus sometimes guaranteed charters, though not always. Church guarantees were not felt to be necessary; in a society fairly indifferent to religion this is not surprising. The ban was clearly friendly enough with the church (and its leadership) to visit its monastery and to allow its hierarchy to play a role in the affairs of state.

The invocation to the Trinity at the beginning shows that the Bosnian Church's doctrine did not oppose the concept of the Trinity, though how its clerics, or anyone else in medieval Bosnia, interpreted and understood the Trinity, we do not know. However, it is likely that few Bosnians meditated on the matter. No doubt, the medieval Bosnian simply felt that it was proper to begin a charter with an invocation to the Trinity, and that the charter would not be a real charter without such an invocation. This feeling was probably shared by the Catholic Ban Kulin, the Orthodox Sankovići and Djed Radoslav of the Bosnian Church. However, the acceptance of the Trinity in orthodox terms does distinguish the Bosnian Churchmen from the Bulgarian Bogomils who, if we can believe hostile polemics directed against them, affirmed the existence of two distinct trinities or blasphemed against the Trinity.(19)

The meaning of the initial phrase of the ban, "I am Saint Gregory (Gr'gur) and am called Ban Stjepan" is disputed in the scholarly literature. I think it simply shows that the ban identifies himself with his *slava* saint, who happens to be Saint Gregory.(20)

The second charter from the 1320's referring to the Bosnian Church also begins with an invocation to the Trinity and has the same pairing of the ban and Saint Gregory. It states that nothing shall be carried out against the grantee or his heirs without asking the Bosnian Church.(21) A formula, frequently found in medieval Bosnian charters, follows to the effect, that, if the charter should be violated, the violater is to be cursed by a series of powerful forces commonly invoked: The Father, the Son, the Holy Ghost, the Twelve Apostles, the Four Evangelists, Judas Iskariot, the Pure Blood of God, and all heavenly creation.(22) Such a curse is not found in the first charter and thus is not absolutely necessary to legitimize a charter. However, it was a common practice which helped strengthen the charter. Probably it was also intended to put the fear of God into the heart of those making the contract, but this seems to have been an idle hope considering how frequently these charters were violated. The formulaic way of listing these higher powers suggests an invocation or spell and so reflects the medieval Bosnians' beliefs in magic. The fact that charters had to be written in a certain form is also connected with magic. The form of a charter had to be correct or else the magic would not work and the charter's contents would not be fulfilled.

The language of the charters may be compared to the language and form of the medieval spells written on metal strips and placed in fields, which call on the same powers in the same sort of language to protect the field from the devil and inclement weather.(23) Thus the charter intended these powers to punish the violator and also to act magically upon the parties of the contract to prevent them from violating it. Because such formulas were not thought out at the time but were repeated mechanically in the way believed to be effective, we can hardly hope to extract from them information about the doctrines of the Bosnian Church.

Now that we have discussed the charters, including one issued at a hiža at Moštre, it might be well to turn to the second inscription carved on the plaque from Kulin's church at near-by Muhašinovići.(24)

This second inscription is rudely scratched across the face of the plaque. It consists of a list of names, most of which are indecipherable; the only decipherable sentences read: "Radohna Kr'stjann" writes, Desjen' Rat'n'cevit' writes in the days of Ban Stjepan, God give him health and many years." Professor Andjelić who deciphered the above, attributes it to the reign of Stjepan Kotromanić.(25) In addition to these names, at some time subsequent to the original work on the plaque, presumably at the same time as the second inscription, someone carved a small and rude figure of a man, wearing what seems to be a crown, with a raised left hand. He reminds us of the men with raised hands that appear on the *stećci* in an Orthodox cemetery of the Miloradović family at Radimlja near Stolac in Hercegovina, dating from the second half of the fifteenth century or even the beginning of the sixteenth.(26) We do not know the significance of the figure at either place.

The second inscription at Muhašinovići is further evidence of the activity of members of the Bosnian Church in the Visoko area at the time of Kotromanić. It is possible that Krstjanin Radohna was a member of Gost Radoslav's hiža at Moštre. We also must consider the possibility that Radohna's name on the slab signifies that the Bosnian Church, after its schism with Rome, had taken over Kulin's church for its purposes. Possibly Krstjanin Radohna was the priest for that church.

Scholars have often argued that the Bosnian Church did not have church buildings, because they have considered it a dualist church which would have rejected material churches and because they aver that there are almost no ruins of medieval churches in Bosnia. If as I believe, however, the weight of the evidence suggests the Bosnian Church was not dualist, the argument that it rejected churches collapses. In fact, we would expect it to have had churches. Church buildings existed at the time of the church schism in the thirteenth century. And, because up to the schism Bosnian Church members most probably had been Catholic monks who already were using the existing churches, one would expect that, having elected their own local bishop and gone into schism, they would have continued to use the churches they had been using up to that time.

We know the Bosnian Church had monastic houses *(hiže)*. It is likely that churches were connected with each monastery. This view is confirmed by the fact that at almost every village or town where we know a hiža existed, there is now either a tradition of or a location named after a church.(27) In addition, in his will of 1466, the Bosnian Church leader Gost Radin left money to build a *hram* ("temple"). Those who do not want to consider the possibility of his leaving money to build a church have suggested by "hram" he meant an elaborate sepulchre. But since a 1472 document refers to money being designated by his will for "a sepulchre and a chapel," it is evident that Radin intended the "hram" to be a burial chapel.(28)

The frequently advanced argument that the absence of church ruins in Bosnia shows that the Bosnian Church did not have churches is equally invalid. In Fact, churches did exist all over Bosnia and Hercegovina in the Middle Ages. In all parts of Bosnia we find traditions of churches, fields or other localities called "church" (*crkvica, crkvina,* etc.), medieval cemeteries lying beside ruined foundations or in localities called "church", or ruined foundations of buildings whose shape suggests they had once been churches. When we investigate a locality called "church" very often foundations of an old building can be found. Of course, every local tradition cannot be accepted.(29) But I suggest that many, if not most, of these churches in Bosnia could well have been Bosnian Church churches. In Hercegovina, on the other hand, whereas a few ruins might have been "Bosnian" churches, the majority were almost certainly Orthodox. It is necessary that scholars make a full study of these churches in the context of regional history, migration patterns, folklore, and popular traditions. It is to be hoped that such a study will give us a firmer base to speak about the denominations of at least some of these medieval churches.

The problem now is not to make categorical statements, for the present state of our knowledge about these ruins and traditions does not enable us to do so. From the data it is only clear that churches did exist; very rarely can we say who built a particular church or when it was built.

Further excavations of the ruins and more attention to the traditions are required. And in calling for further investigation, I hope that in the future scholars will be willing to face the possibility that many of these churches may have belonged to the Bosnian Church. If scholars will approach the problem with minds open to that possibility, and not assume that a church (if medieval) had to be Catholic or Orthodox, then it is possible that eventually we will arrive at a better understanding of the positions held by the different confessions in medieval Bosnia.

II: Continued Papal Concern about Heresy in Bosnia, 1325-1338

In the midst of the ban's political successes, in May 1325 Pope John XXII sent fra Fabian, a Franciscan, as inquisitor of heretics to the "province of Slavonia,"(30) perhaps Slavonia (since the pope said "province" and not "provinces"), perhaps the South Slav lands in general. There is no evidence that Fabian ever reached Bosnia, although in June the pope wrote both Ban Stjepan and the King of Hungary to remind them to take action against the heretics. The tone of the papal letter to the ban is friendly and there is no reason to believe that the pope believed Kotromanić to be anything other than an orthodox Christian. The pope informed the Hungarian king, Charles Robert, that many heretics from many and various regions were coming to the Bosnian state and were flocking together in the neighborhood of Dalmatia ("ad principatum Bosnensem in confinio Dalmatiae . . . confluxit"). The same statement appears in the letter to the ban with the phrase about

Dalmatia omitted.(31) Thus we can suggest that the regions near Dalmatia did not belong to Kotromanić but lay under the jurisdiction of the King of Hungary.

The pope's letters suggest a mass migration of heretics into Bosnia. Where were they coming from? Would heretics from Italy or remnants of those from southern France have come in any numbers to Bosnia? It seems improbable on the face of it, and if many foreigners had come, why is there no sign of their presence in other sources? However, it is likely that something alarmed the pope, prompting him to write the letters. Perhaps some heretics (probably a relatively small number) did flee from Italy toward the Balkans, and the pope had reason to believe they might be headed for Bosnia or Dalmatia. Whether they actually arrived, and if they did whether they remained there for any length of time, is unknown; we shall hear nothing more about foreign heretics in Bosnia unless, by chance, it should be they who were receiving and teaching doctrine to visiting northern Italian heretics later in the century. The trips of these Italians to Bosnia, mentioned in testimony before the inquisition, will be discussed at the end of this chapter.

From 1327 into the early 1330's Dominicans and Franciscans quarrelled over which had inquisitorial rights in Bosnia. Both based their claims on their past achievements in the Slavonic lands — the Dominicans on their work in connection with the Hungarian crusades and their more long-term activities in Slavonia under the aegis of the Hungarian crown; the Franciscans on the super-human work of the two brothers who set off in the 1290's to convert the whole Balkans, thereafter disappearing from the sources, and also on the assignment to Fra Fabian of "Slavonia." Both sides misrepresented matters to the pope and took advantage of his ignorance of Balkan geography. After wavering, he finally awarded to the Franciscans the coveted task of converting schismatics and heretics in the Slavic lands. The Dominicans were ordered to keep out. This paved the way for the creation of the Franciscan mission to Bosnia and for the establishment of the Bosnian vicariat in the years following 1339. In 1327 the pope wrote the King of Hungary, praising him for giving the Franciscans a friendly reception. Thus, at the Hungarian court the active support of the Dominicans and relative hostility toward the Franciscans characteristic of the thirteenth century had now disappeared.

References to heretics continue to be found in papal letters through the remainder of the 1320's and the 1330's although we learn little about the heretics themselves. In 1327 the pope wrote the Ban of Slavonia urging him to persecute heretics.(32) and after the Archbishop of Zadar jailed an abbot and a monk on charges of heresy, the pope called for an investigation.(33) But no mention of Bosnia appears in these two papal letters. And since the Bosnian Church did not exist in Slavonia and Zadar, the pope cannot be referring to it. Thus his letters confirm the presence of at least one other heterodox religious movement on the borders of Bosnia. Despite occasional references to heresy in Bosnia in these decades, papal letters to Kotromanić were always friendly. For some reason this papal attitude changed in 1337. We may suspect that the papacy had continually hoped that the ban himself would resolve religious matters in Bosnia until finally, in 1337, Pope Benedict XII ran out of patience and called upon the Princes of Krbava, Senj, Bribir (the Šubići), Knin (the powerful Nelipac, who at the expense of the Šubići had acquired considerable authority in the region to Bosnia's northwest), and other nobles from Croatia to give armed support to the Franciscans, who, he said, were unable to progress with their work in Bosnia because the ban and some other nobles were defending and favoring the heretics which thereby hindered the Franciscans' inquisitorial work.(34)

We do not know to what extent the papal letter was a response to intrigues against Bosnia by the Nelipac family and by some of the other Croatian nobles. It is clear, however, that correspondence had taken place between the pope and Nelipac and that the Count of Knin had expressed his willingness to march against Kotromanić. How the Franciscans were hindered is unknown. We have no evidence that any friars had yet appeared in Bosnia. The only Franciscan whom we have heard anything about in this period was Fra Fabian who had been sent to "Slavonia" in 1325.

However, it seems clear that the quick action of Ban Stjepan forestalled the crusaders from launching any attack. In 1338 Bosnian troops crossed the territory of Trogir en route for Klis, despite the Bishop of Trogir's objections to the passage of "heretics."(35) His objections had little effect since it seems that Bosnia was on good terms with the town fathers; in fact, in 1339 Ban Stjepan granted Trogir's merchants free entrance and exit for purposes of trade with his banate.(36) That the Bosnian army was operating here suggests that the crusaders had not yet launched an attack, and it is likely that Ban Stjepan had sent troops thither in order to force the crusaders to take the defensive and thereby prevent them from organizing an attack on Bosnia. The King of Hungary was also instrumental in preventing crusader action, for he declared that anyone who attacked Bosnia, whose ban was his friend, was showing himself to be unfaithful to the King of Hungary, who was suzerain of most of the would-be crusaders.(37) Thus we can see the improvement in Bosnia's position in the fourteenth century over what it had been in the period of the thirteenth century crusades.

Interestingly enough, at this time there was some sort of scandal in Trogir. In the spring of 1338 the pope summoned Lampredius, the Bishop of Trogir, to Rome to answer charges, most of which concerned moral matters and matters of church administration. One charge stated that the bishop favored the Bosnian heretics.(38) The issue seems to have been complicated and continued on until 1342, and soon led to a dispute between the bishop and the Trogir Town council. It seems most probable that the charge of favoring Bosnian heretics is connected with the passage of Bosnian troops and that the pope was angry at the bishop for allowing the Bosnians to pass. Yet, as we stated earlier the bishop is elsewhere said to have opposed their passage. It seems more plausible that the bishop in fact did oppose the Bosnians and the existence of the charter of 1339 from Kotromanić giving the merchants of Trogir rights of free trade in Bosnia shows who the ban's allies in Trogir in fact were. The word for word repetition of the 1338 charge against Lampredius in 1342,(39) suggests scribal copying rather than a confirmation of the accuracy of the accusation.

As a result of the ban's timely action and the support of the King of Hungary, it seems most probable that Bosnia was never invaded at all. In 1340 the ban and the King of Hungary planned a joint expedition against some of the Croatian nobles. Plainly, a war was not going to realize the pope's aims; he would have to try another tactic. The new tactic was to be the Franciscan mission.

III: The Establishment of the Franciscan Bosnian Vicariat

It is apparent that the idea for the mission already existed in June 1339, when Benedict XII had ordered Franciscans of Hungarian and Slavic nationality to be sent for general training.(40) In the course of 1339, the General of the Franciscan Order, Gerard Odinis, passed through Bosnia and Slavonia en route to Hungary.(41) Gerard was well received by the ban; three papal letters written 28 February 1340, refer to this reception.(42) Because these letters refer to promises made by the ban, we may assume that these promises had been made to Gerard when he stopped in Bosnia during his trip in 1339.

The letter addressed to the ban addressed him as son and extended the papal blessing, which shows that the pope had abandoned plans for military intervention and was satisfied that the ban was neither a heretic nor a protector of heresy. The pope stated that the ban had promised to cooperate in the destruction of heresy, the restoration of ruined churches, and the revival of Catholic services in them.(43)

The letter to Gerard states that the ban had gone out to meet the Franciscan General on the road, had received him with honor, had listened to the Catholic teaching of truth and salvation with pleasure, and expressed his desire to exterminate from the principality of Bosnia all heretics. To do this, however, he needed the help of the Apostolic See and the King of Hungary because of the schismatics who lived in neighboring lands, whom the said heretics would call upon for help, if they learned that they were going to be driven from the land(of Bosnia).(44)

The schismatics referred to presumably are the Orthodox Serbs under the powerful Dušan. Dušan, however, was no friend of heresy, and his law code, when written in 1349, would have articles against heresy. We could expect him to come to the aid only of heretics whose beliefs did not differ greatly from his own. Thus this may be taken as evidence that the Bosnian Church, if it might seek aid from Dušan, was a Slavic liturgy church not differing greatly from the Serbian.

Kotromanić may also have been anticipating a possible attack from Dušan for other reasons, and feared that if a persecution were to be launched against heretics and/or Bosnian Churchmen, the persecuted, in the event of a Bosnian-Serbian war, might side with Dušan. Such a war did take place in 1350-51 when Dušan invaded Bosnia; Kotromanić successfully withstood the attack. We do not know whether any Bosnians supported the Serbian invaders.

What were the ban's motives in allowing the Franciscan mission? Orbini states that the ban was of the Greek rite (i.e., Orthodox) and therefore was not obedient to the pope. He granted the Catholics license to preach freely against the heretics and to introduce the Roman faith because he felt it was better to have Catholics in his lands than heretics who were against both the Greeks and the Catholics.(45) Since the ban seems to have had cordial relations with the Bosnian Church, it is not clear whether by heretics here is meant the Bosnian Church or another heretical current. If he had another heretical current in mind, possibly dualists, this may be evidence that this movement was gaining an increased following which the ban believed to be dangerous to the state. It is not likely that he would have considered the Bosnian Church a danger, unless he was reasoning that its presence, since it annoyed foreign powers, increased the likelihood of foreign invasion. Most probably, however, the ban simply wanted to end plans for crusades against his realm. If allowing the Franciscans to come to Bosnia to preach would end such plans, then why not agree to this compromise?

One other factor that may have had a part in Kotromanić's allowing the Franciscans to come and preach in Bosnia was his own attitude toward the Catholic faith. It seems the ban in 1340 was still Orthodox, since we have no evidence that he had accepted Catholicism prior to that date. However, in 1339 he was clearly friendly toward Catholicism, since he received the Franciscan general with such warmth. In addition the pope says that he listened to the Catholic teaching of truth and salvation with pleasure. This cannot be taken to mean that he accepted Catholicism and was baptized at their meeting in 1339, but it does show him as sympathetic toward and interested in Catholic teaching. The earliest Franciscan chronicle ("Chronica Generalium Ordinis Minorum," written in 1374) states that Gerard converted the ban to Catholicism.(46) Since Gerard had dealings with the South Slav world only until 1342 (when he accepted another post), if he did convert the ban, he would have to have done so between 1339 and 1342. Orbini confirms the approximate date for the ban's conversion but attributes it to the influence of a Ragusan canon Domagna di Volzo Bobali, who was a close associate of the ban and who greatly aided the Franciscan mission.(47) Orbini's account is based on charters and privileges found in the Bobali family archive in Dubrovnik. Thus if these no-longer extant documents are authentic, and not later fabrications to glorify an ancestor, Orbini's story, being based upon contemporary documents would be more reliable. In any case, the ban in 1340, not yet a Catholic, seems to have warmly received the Franciscans. At about the same time Bobali arrived and assisted the Franciscans. In the course of the 1340's, and most probably before 1347 (owing to the contents of a letter, we shall discuss presently, written by the ban in that year), the ban, under the influence of the Franciscans and/or Bobali, accepted Catholicism.

The 1374 chronicle also reveals that Gerard summoned brothers from many lands to preach against heretics. The vicariat was founded, many churches were built, and many were converted.(48) Gerard was General of the Franciscan Order until 1342, at which time he was appointed "Patriarch of Antioch" (an office which enabled him to reside in Sicily). All Franciscan accounts agree that Peregrin Saxon was installed as the first vicar. His installation almost certainly occurred between Gerard's return to Bosnia in 1340 and his acceptance of the Antioch post in 1342. The first time we find Peregrin as vicar in contemporary documents is 1344, and it is clear from the context that he had not just been installed.(49) Thus it is safe to date the founding of the Bosnian vicariat and the installation of Peregrin Saxon as the first vicar to the two year period 1340-42.

The vicariat covered a region much vaster than the Banate of Bosnia. In the course of the fourteenth century, the area in which its members could be found at work was expanded. The vicariat started with little more than the good will of the ban, almost no money, and very few Franciscans. The claims of Franciscan annals that many friars came to the vicariat are quite exaggerated, as shown by a letter of Pope Gregory XI in 1372 which states that the Franciscans were not permitted to have over sixty brothers in the vicariat.(50) In 1378 they were permitted to receive thirty more brothers, though in 1380 the vicar noted that they had taken only twenty. Six more were added in 1395.(51) We do not know whether these new recruits were filling vacancies in the quota of sixty or if that quota had been increased. If Šidak is correct in assuming that the quota had been increased, then these two new enlistments, not taking into account probable deaths and departures, brings the number of friars in the vicariat to eighty-six. Of course, we can suppose that others came, whose arrivals were not noted in the limited number of surviving sources. However, even so, it is clear that the Franciscan mission was a small-scale operation and intended to be such by the pope who set strict limits on its numbers.

In addition, we know that the Franciscans operated from monasteries, and that it was not permitted them to have more than twelve brothers to a monastery.(52) Although letters of papal permission to build Franciscan monasteries exist from the fourteenth century for a variety of places near Bosnia, no such letters have survived for Bosnia itself. Thus we do not know exactly when any of the Bosnian monasteries were built. All we have is a list from 1385 of the thirty-five monasteries that then existed in the vicariat. From this list, we know that by 1385 only four monasteries existed in Bosnia proper. They were located at: Curia bani (Kraljeva Sutjeska), Saint Nicholas (probably referring to the monastery of that name at Visoko), Lascrova (Lašva), and Plumbum (Olovo).(53) If there had been the maximum number of friars in each monastery Bosnia proper would have had forty-eight Franciscans in 1385. Yet since there were a relatively large number of other monasteries elsewhere in the vicariat which had to be staffed, it is unlikely that forty-eight Franciscans (out of a total of roughly a hundred) could have been spared for Bosnia. To have staffed each monastery evenly, from an assumed hundred friars, would have meant roughly three to a monastery. This would have given Bosnia roughly twelve brothers in 1385; and since Bosnia might have been considered a critical area, we might be justified in increasing that number to about twenty for Bosnia. It immediately becomes apparent that such a small number of friars would have had limited success in converting a large area with poor communications like Bosnia.

In addition, the establishment of monasteries in a mining town like Olovo, or a trading center like Visoko, with significant foreign colonies of merchants and miners, would have meant that the Franciscans in these places would have had to spend considerable time serving the needs of these foreign Catholics, which would have reduced the time they would have had to devote to teaching the heathen. Of course, by choosing these towns with foreign colonies, the Franciscans secured for themselves a strong base of support. They clearly would not have been driven out of such places; thus they gained secure centers from which they could spread their teaching into the surrounding area.

Presumably then, this is the way in which the Franciscans operated. And since considerable numbers of peasants in these areas would have been without priests of any confession, they were unlikely to have been convinced believers in any other religion. Thus we can expect that the Franciscans made steady but slow progress. By 1385, if the four monasteries had been erected nearer to 1340 than 1385, it is probable that the majority of villages near these monasteries would have been won over at least nominally to Catholicism. In the relatively small area of the Visoko-Lašva-Sutjeska triangle, roughly the center of the state, three of the four monasteries had been erected. Thus the Franciscans by concentrating their limited numbers in this key central core could utilize their small corps to maximum effectiveness. Here they surely made especially good progress. And as it was the center of the state, it was, of course, the key area.

It was also convenient for the ban, for hither would come foreign legates, who seeing Catholicism here might be more easily persuaded that the issue of heresy was dead and Bosnia was now a Catholic land. However, as this region was the center of the state, the Bosnian Church also had its people here. Whether the hiža remained at Moštre we do not know, but in 1404 we shall find the djed issuing a letter from Janjići, which lay not far north of Lašva. However, we may suppose that with the support of the rulers, (Kotromanić, a Catholic from about 1340, and his nephew and successor Tvrtko, a Catholic all his life, who ruled to 1391), the three Franciscan monasteries, located in relatively close proximity to one another, by the end of Tvrtko's reign probably had succeeded in winning over the majority of people in this central triangle to Catholicism. In areas further afield, other than Olovo (with monastery and significant foreign Catholic colony) it is to be doubted that the populace even heard about Catholicism.

Another factor which must have contributed to the Catholicizing of the center was the development of mines, begun during the reign of Kotromanić, which increased the influx of foreigners (particularly Dalmatians but also some Saxon miners) to the area in the second half of the fourteenth century. Taking advantage of the increase in commerce resulting from the mining, these Dalmatian merchants enlarged their old colonies and established new ones in other towns near mining centers. These new colonies brought Catholic influences into new towns and also provided secure centers for future Franciscan monasteries; it was not long after 1385 that the Franciscans would erect monasteries in three such towns — Kreševo, Deževica and Fojnica.

Quite naturally a shortage of money would hinder any religious organization that tried to function in Bosnia. Thus, almost immediately after the foundation of the Bosnian vicariat the Franciscans, lacking any real financial support and prohibited from seeking alms within Bosnia proper, began to claim the right to tithes raised in the "land and diocese of Bosnia." The bishop in Djakovo immediately wrote the pope to complain that he had long had the right to tithes paid by certain subjects of Ban Stjepan (we do not learn which subjects and in what regions of Bosnia); now the Franciscans had begun to assert their rights to tithes from those people they converted to Catholicism, falsely claiming that the ban had agreed to this. The bishop also claimed that the Franciscans were usurping his authority. The Pope decided the case in favor of the bishop in Djakovo.(54)

Since we do not know where the tithes being disputed were collected, we cannot use the above material to demonstrate that the bishop's organization had actually been having some role up to then within the banate. In fact, since the report of the tithe collector of 1317 complained that the Bosnian diocese was almost destroyed, we may suspect that the bishop was claiming tithes which he had title to, yet which prior to the 1340's owing to the chaotic situation in Bosnia had not actually been collected. Thus if the bishop was arguing on *de jure* rights, we cannot use this to demonstrate that the bishop had had some authority within Bosnia prior to 1340. And it is possible that the people of Ban Stjepan referred to were in such border regions as Sol, Usora, and the Donji kraji, which were under his suzerainty but not his direct rule and which throughout, despite the existence of heresy, always had Catholic churches functioning.

It is almost certainly true that the Franciscans were usurping the authority of the bishop. Since as far as we can tell the bishop had not set foot in Bosnia for over a century, now that the Franciscans had appeared in Bosnia, it is hardly surprising that they had achieved some influence and authority there. In fact, Peregrin Saxon, the Franciscan friend of the ban, who would be elected bishop in 1349, would become the first bishop to have been in Bosnia since the 1230's. The pope's decision on behalf of the bishop was presumably made on the basis of *de jure* rights as opposed to the *de facto* situation. This surely was a serious blow to the work of the Franciscans, the only Catholics who were active in Bosnia itself. The dispute between Franciscans and bishop was to be the first stage of a long term quarrel that was to continue through the next century, paralleling quarrels of the Bosnians against the Hungarians, the Dominicans and Djakovo a century earlier. The only difference between the two situations lay in the fact that the Franciscans were trusted by the pope and had his blessing, though not always his active support.

With their small numbers and limited finances, the Franciscans, had they not had the support of the state would have failed completely. Kotromanić's support of their mission and his distress at the mission's lack of success is shown by a letter the ban sent in 1347 to Venice to be transmitted to the pope. The ban praised the work of Peregrin on behalf of Catholicism, and complained that there was an insufficient number of Franciscans for the considerable task that they had undertaken. He requested that more friars be sent to the vicariat and stressed the necessity of the newcomers knowing Slavic, or at least having the aptitude to learn it. This implies that some of those already there did not know it; except for one Hungarian, all of the known fourteenth-century Franciscans in Bosnia were Italians.(55) Foreigners, not knowing Slavic, could not have had much success with the Bosnians; thus the number of effective Franciscans in Bosnia may have been even smaller than we suggested before. The ban suggested that the vicar be allowed to summon more Franciscans from abroad as well as friars from other orders, that monasteries be established near-by where neophytes could learn Latin and theology, and that the vicar be allowed to summon to Bosnia secular priests who could administer the sacraments to people who although excommunicated might be able to be brought back into communion, or that the Franciscans might be granted by the Holy See the privilege to administer the sacraments.(56) The need for men eligible to administer the sacraments was continually a pressing problem owing to the shortage of priests. Finally in the last quarter of the century the pope had to allow the Franciscans to administer the sacraments themselves. The excommunicates, whom the ban hoped might be restored to communion, presumably referred to both heretics and Bosnian Church members.

The ban's interest in the religious affairs of his realm, as well as his good will toward Catholicism, is also illustrated by a second request he made to Venice in 1348, to ask the pope to appoint Peregrin as the new Bishop of Bosnia.(57) We know that up to this time the Bosnian bishop had resided outside of Bosnia and had had nothing to do with Bosnian affairs. With the establishment of the Franciscan mission in Bosnia, a quarrel had immediately broken out between the absentee bishop and the friars over the income from tithes. The obvious solution for the good of Catholicism would be to unite the foreign bishopric with the mission, and this was what the ban aimed to do. In addition we know that he knew and trusted Peregrin, having worked with him for almost a decade. The pope wisely accepted this suggestion.

Stjepan Kotromanić died late in 1353. Orbini incorrectly dates his death in 1357, and claims that he was buried in the Franciscan monastery of St. Nicholas at "Milescevo,"(58) an unknown monastery not to be confused with the famous Orthodox monastery of Mileševo. On the 1385 list of Franciscan monasteries for the vicariat, three of the four erected in Bosnia proper are easily identifiable. The fourth is called only St. Nicholas. A monastery of St. Nicholas at Visoko is referred to in a Ragusan document from 1367.(59) Since Visoko was both the commercial center of Bosnia as well as the ban's first city, and since the ban supported the Franciscans, we would expect a monastery to have been erected here in the 1340's. Thus we can be fairly certain that the St. Nicholas on the 1385 list refers to the noted Visoko monastery, and since this town was more or less Kotromanić's capital, a monastery here would have been a likely final resting place for him.

IV: Tvrtko Establishes Himself in Power: Relations With the Different Faiths

Stjepan Kotromanić seems to have left no sons; he was succeeded by his nephew Tvrtko, the son of Vladislav Kotromanić and Jelena Šubić.(60) Tvrtko was certainly brought up as a Catholic. His mother was Catholic, and just shortly after his birth his uncle, the ban, became one. Tvrtko gave proof of his Catholicism shortly after his succession when he issued in February 1355 from Djakovo a letter confirming trade privileges to Dubrovnik, in which he called Bishop Peregrin Saxon "our spiritual father."(61) Prior to 1349 when on the recommendation of Ban Stjepan, the pope had made Peregrin Bishop of Bosnia, such a visit by a ban to the episcopal headquarters in Slavonia would have been inconceivable. But since the pope had made the leader of the Bosnian Franciscans, since ca. 1340-42 the *de facto* head of Catholicism in Bosnia, *de jure* head as well by appointing him Bishop of Bosnia, he made it possible for Bosnian bans actually to deal with the bishop in

Djakovo for the first time. However, the union of the two offices was temporary. After Peregrin's death in 1355 they were separated again as the pope replaced Peregrin with Peter, a Djakovo cleric.(62) Relations between the state and the bishop deteriorated, and once again we find a titular and inactive bishop outside the borders of Bosnia and an active Franciscan mission inside, each at odds with the other. Tvrtko was to have some dealings with Peregrin's successor, Peter, but after Peter's death, the Bishop of Bosnia in Djakovo again had absolutely no role or influence within the Bosnian state.

Kotromanić's state had been built on alliances and by his own military strength, as a result of which he obtained the loyalty of independent nobles in the vicinity of his holdings. He then had used this increased power to annex other lands, like Hum. However, he had never created much of a state apparatus, and had been satisfied to let his vassals administer their own territories and simply render specific duties to the state.

On Stjepan's death in 1353, his successor, Tvrtko, was a boy of about fifteen; thus, very few of Stjepan's vassals felt obliged to serve Tvrtko. Tvrtko found himself holding only the central part of Bosnia, his personal lands, with the task of acquiring the loyalty of the nobles whose territories lay all about him and who were waiting to see what new balance of forces would develop. Thus the great "state" of Kotromanić had been an artificial and temporary creation, that split into its separate units and now would have to be re-assembled again. The nobles considered themselves independent lords of their own lands and had little loyalty for abstractions such as a Bosnian state.

Tvrtko's task was made more difficult by the King of Hungary, Louis. The king was bidding for the loyalty of many of Kotromanić's former vassals in the north, particularly of the now disunited Hrvatinić family who ruled in the Donji kraji. In addition, just prior to Kotromanić's death, Louis had married the ban's daughter, Jelisaveta. Now he demanded that Tvrtko give him Hum as the girl's dowry. Tvrtko, not having won support from enough nobles, was forced to agree. In 1357 he went to Hungary, accepted Louis' overlordship, surrendered to him the western part of Hum, and in turn was confirmed as ruler of the Banate of Bosnia and Usora. This confirmation was given on the condition that Tvrtko take action against the heretics and that his brother Vuk live at the Hungarian court.

Ominous thunder from Rome about the need to take action against heretics, which now probably would find Hungary agreeable, increased Tvrtko's danger. Peregrin, who might have defended Bosnia, was dead and in his place sat a Hungarian cleric. Once again, papal and

Hungarian letter refer to heretics, Bosnian heretics, and "Patarins." In 1358 Louis mentioned "heretics and Patarins in our Kingdom of Bosnia."(63) It is probable that Louis was distinguishing heretics from the Patarins (the Bosnian Church).

In 1358 Tvrtko seems to have reverted to the methods usual for the weaker side in a power struggle and to have tried to organize some sort of a plot. Peter, the Bosnian bishop in Djakovo, apparently intercepted a letter from Tvrtko to a lector in the Djakovo church named John, who confessed at the hearing that he hated his bishop and had been working with Tvrtko, the rival of King Louis, the open "favorer" of heretics, and the secret enemy of the Bosnian bishop. Tvrtko had spoken various words against Christianity, as well as against the honor of the king and of the bishop.(64) Vague as this is, it is all we have. And, of course, even this is the testimony of one man before a hostile court.

We now have a gap in our sources for the vital years that preceded a Bosnian-Hungarian war which broke out in 1363. The specific cause for the war is not known, nor are Tvrtko's machinations that preceded it among the nobles to his north.(65) In 1360, a brief warning that something may have been about to happen was contained in Pope Innocent VI's letter to the Bishop of Bosnia, Peter, which told him that he was free to call on the secular arm against heretics and to make use of all church punishments against them.(66)

Then, in 1363, Hungarian armies struck the territory to the north of Bosnia in two waves. The first wave attacked the region of the Donji kraji, whose lords (the Hrvatinići) were divided among themselves, some for Tvrtko and some for Hungary. Both rulers had been actively trying to win over these nobles for years. Now the crucial test came. Loyalties already promised were not firm commitments, and we find Vlatko Vukoslavić, loyal to Tvrtko up to this point, surrendering the important fortress of Ključ to Louis. However, Tvrtko's side emerged victorious as Vlkac Hrvatinić ably and successfully defended the fortress of Sokograd in the Plivska župa, and the Hungarian army was forced to turn back to Hungary. Vlkac was given the whole *župa* by Tvrtko as a reward a couple of years later. Presumably that had been his price. The second Hungarian wave, a month later than the first, hit Usora; once again the Bosnian defenders were successful, this time by a key defense at Srebrnik fortress in Usora (not to be confused with Srebrnica, and its fortress Srebrnik, on the Drina).(67)

Somehow between 1358 and 1363 Tvrtko had become powerful enough to resist the Hungarian attack. Unfortunately the sources are silent on the subject of how he had managed to achieve this strength. Tvrtko was fortunate that the Hungarian attacks had come when they did, for in February 1366 a revolt of local nobles forced him to abandon his throne and flee to the court of the King of Hungary, where the king welcomed his recent adversary. In Bosnia, the nobles placed Tvrtko's brother, Vuk, on the throne. Was Vuk an initiator of the plot or a figure-head for the conspirators? We do not know; in any event, once on the throne Vuk took up his new role with enthusiasm. But having quickly recognized the suzerainty of the Hungarian king again and acquired some aid of unspecified sort (troops?) from the Hungarian monarch, Tvrtko was back in Bosnia by the end of March and had regained some, but not all, of Bosnia. He also had the support of the lords of the Donji kraji. A variety of nobles took part in this affair, shifting sides as things seemed best to them. By the end of 1367, Tvrtko had regained his banate and Vuk was in exile. Now Vuk began to seek outside help, and, knowing that there was no better issue to capitalize on than "the heresy," he wrote to Pope Urban V.

Urban then wrote the King of Hungary in January 1369, reporting that Vuk had stated that the majority of bans up to the present had been schismatics or heretics, and had received and defended heretics, while he (Vuk) was a good Catholic, who hated heretics and only longed to persecute them. The senior ban, his brother (Tvrtko), however, favored and defended heretics, who came to his realm from various places. The pope asked the King of Hungary to help restore Vuk to power.(68) The King of Hungary, however, did not seem to be interested, and nothing more came of it. Eventually, by 1374, Vuk and Tvrtko were reconciled and are found endorsing charters as senior and junior bans.

In the 1360's the pope continued to write letters to the Hungarian king and high clerics on the coast about the need for action against heretics and schismatics. Finally in 1369, Urban V wrote the Archbishops of Dubrovnik and Split. He complained that in parts of Bosnia not far from their dioceses were found many heretics who freely came into their dioceses and cities for purposes of trade, and once there, spread heresy. The pope ordered the archbishops neither to allow such heretics to enter their dioceses nor to allow trade with them. He also said that merchants from their cities should not be allowed to go to Bosnia, lest they become infected.(69) It is very doubtful that this letter had any practical effect; it certainly could not have had any on Dubrovnik whose economy was based on trade, much of which went overland to or through Bosnia.

At this time, with peace re-established at home, and his relations good with the lords to the north, Tvrtko began to involve himself in the feuds and rivalries among the leading families to the south, in what is now Hercegovina. The neighboring states of Serbia and Zeta were also involved. Religion played no part in these quarrels. As far as we can tell from the sources, these nobles were Orthodox. Nikola Altomanović, the most powerful nobleman in this region at the start of the struggle, whose holdings were totally lost as a result of the failure of his aggressive and self-seeking policy toward his neighbors, was almost certainly Orthodox.(70) He was also friendly toward the Franciscans and allowed them to build two monasteries in his territories around Rudnik.(71) The powerful family of Sanko Miltenović (and his sons Beljak and Radič Sanković), who after the fall of Altomanović became the leading family here were also Orthodox. Sanko had been Kaznac (treasurer) for Tvrtko, but never had lost sight of what he thought were his own interests. Twice he had joined Tvrtko's opponents in crucial conflicts and twice he had managed to restore himself to favor afterwards. His estates lay in the region of Nevesinje from which he and his sons expanded their holdings west to the coast and north toward Konjic.(72) In the Konjic area at Biskup near Glavatičevo, the family erected an Orthodox church which served as the burial chapel and cemetery for most of the family.(73) Sanko's sister Radača married a member of the powerful Čihorić family, who ruled Popovo polje, and later in life became an Orthodox nun under the name of Polihrana. Her gravestone with inscription can be seen at the Orthodox cemetery at Veličani.(74) Thus, we see that the leading families of what is now Hercegovina in the fourteenth century were Orthodox. The sources do not mention a single noble in Hercegovina being connected with heresy or with the Bosnian Church.

Tvrtko was able to play a dominant role in these quarrels and as a result was able to expand his state to the east, where he annexed land to the Lim region and on to the Drina.(75) In this territory that he acquired were several famous Orthodox churches: Mileševo (where St. Sava's relics were), Dobrun (built in 1343 or 1353 by a protovestijar at Dušan's court),(76) the Banja Monastery and the Church of St. Peter and St. Paul on the Lim. Thus he acquired lands with a loyal Orthodox population, one or two bishops, and a relatively large number of priests and monks. In this way a large number of Orthodox were incorporated into Tvrtko's state. However, we have no reason to believe that this led to any significant Orthodox influence in Bosnia itself. Nor is there evidence that there were any transfers or migrations of people from this newly annexed region to the other parts of the state.

The extinction of the Nemanjić family in Serbia, together with Tvrtko's Serbian conquests and the fact that he was the great grandson of Stefan Dragutin (through Dragutin's daughter's marriage to Kotroman), provided a basis for Tvrtko's claims to the Serbian kingship.

In 1377 Tvrtko was crowned King of Raška and Bosnia at Mileševo by an Orthodox metropolitan. Thus Tvrtko had active and friendly relations with the Orthodox as well as with Catholics. His religious policies seem to have been based on political expediency.(77) In 1374 Tvrtko married an Orthodox girl, Dorothea, daughter of the Bulgar Knez of Vidin, in a Catholic ceremony performed by his old enemy, Peter, the Bishop of Bosnia, at Djakovo. Tvrtko awarded Peter a large land grant in the area of Djakovo. His brother Vuk also attended the wedding, and, restored to grace, placed his signature on the grant to the bishop in Djakovo. Tvrtko clearly was one who bore no grudges.(78)

V: Papal Letters about Bosnia in the 1370's

In 1372 and 1373 Pope Gregory XI issued a series of letters about the Franciscan mission. He gave permission to build several monasteries in the vicariat (two in Altomanović's territory around Rudnik, one in Glaž, and other new ones in unspecified locations in Bosnia and Raška.) He allowed the vicariat to have a maximum of sixty friars, and taking into consideration the shortage of priests, he gave the Franciscans the right to administer all ecclesiastical sacraments.(79) In a second letter he complained that in parts of Bosnia (presumably referring to the vicariat) all the inhabitants, excluding those who had been converted to Catholicism, were schismatics or heretics. In many places there were no parochial churches and no priests to administer to the spiritual needs of the populace; such conditions existed even in regions, such as "Glacs" (probably referring to Glaž, lying on the Ukrina river beyond the northern borders of Tvrtko's banate) where the ruling nobles were Catholic. In such places were to be found mixed populations of Catholics, schismatics and heretics; there were neither fixed parishes nor parish priests. If a man sought a priest he often had to travel two or three days to find one. Hence the inhabitants of such places lived without doctrine and Christian rites.(80)

Most important is Gregory's long letter to Vicar Bartholomaeus, answering twenty-three questions that the vicar had put to the pope about problems the Franciscans had had in the course of their work in the vicariat.(81) One point about these twenty-three answers, frequently overlooked by scholars, is that they concern the whole vicariat rather than just the Bosnian banate.

The letter gives us a picture of the great ignorance found in the area. It mentions the ignorance among the "Greeks" (i.e., Orthodox) who were not aware of the existence of a schism between the Eastern and

Western churches and who did not know even the basic matters of faith.(82) The pope is told that there were some semi-priests there — not real priests but rustics and schismatics not canonically ordained but instituted by the custom of the land(83) — and he is asked what functions these rustics could be permitted to carry out. The shortage of priests was so great that the Franciscans could use their help. Whether these rustic priests were to be found in Bosnia or elsewhere in the vicariat is not stated. If these priests were found in Bosnia, were they Orthodox, or Bosnian-Church? Or might they not have simply been ignorant peasant-priests, nominally Christian, whose position went from father to son, who may originally have been Catholic generations earlier? It is noteworthy that they were not spoken of as heretical. We can regret that we are given no details about the manner in which they were instituted "according to the custom of the land." However, it does show that a local practice for instituting priests that deviated from church canons had been developed.

The letter also refers to areas with no priests and no parishes with fixed boundaries.(84) This suggests that in these areas the Catholic organization had completely broken down, which is hardly surprising, considering the century between ca. 1240 and 1340 without any Catholic administration in the land. Without Catholic priests and parishes, the population would either have had contact with schismatic or heretical priests or none at all.

Reference is made to Greeks and schismatics having lost all Catholic rites and learning.(85) If "Greeks" refers to Orthodox, does schismatic then refer to a second group other than the Orthodox? Could it refer to followers of the Bosnian Church? Elsewhere in the document schismatics are also mentioned, sometimes these schismatics may be Orthodox — for the vicariat, of course, did include Serbian and Bulgarian territory — but must they always be? We do not know if the Franciscans considered the Bosnian Church as schismatic or heretical, but if they did consider it schismatic, then quite possibly at times this term refers to Bosnian Church followers.

A variety of questions were asked about re-baptism, administration of sacraments, eligibility for rites, etc. It is in this letter that we learn that the Bosnians did not consider marriage a sacrament but took women on the condition that they promise to be good wives to their husbands, with the intention of dismissing them when they chose. We are told that scarcely one man in a hundred lived with his original wife.(86) The pope told the Franciscans to withhold baptism until church marriages were accepted. Complaint was made about the keeping of concubines. In this case it was foreign Catholic merchants (spoken of as Latins and

Germans) who were criticized for keeping them. These merchants presumably lived in the mining and commercial towns of Bosnia and Serbia.

Only three questions in the letter make any reference to heresy; and two of these show only that some distinction was made between those people considered schismatics and those considered heretics. Yet we are not told whether those considered to be heretics were members of a specific sect or whether they had simply deviated into various errors through ignorance. The first of the three questions asked whether Franciscans should serve the mass before those who had aided heretics; the second questioned the wisdom of allowing laymen to dispute about matters of faith in the presence of heretics.(87) In both cases we have no way of knowing whether these heretics lived in Bosnia itself; the number of references in other sources to heretics in Bosnia, however, makes it quite possible that they did live in the banate.

The third reference to heresy is the only one which could be construed in a dualist sense. Question twenty-one asks, what should the Franciscans do about those converted to Catholicism who continued to "adore heretics" and to openly observe heretical rites while keeping their Catholicism secretly? Should the friars give them secret penance because they would not want to perform a public penance, or should they perhaps remain in (i.e., be allowed to publicly observe) their heresy, though they have been recently converted.(88) This shows that there was some religious sect somewhere in the vicariat which the Franciscans considered heretical. This sect evidently had its own service which differed from the Catholic. It also, at least in some places, had considerable following with a leadership hostile to the Franciscans which caused converts to Catholicism to fear to admit publicly their conversion and caused them to continue to publicly perform the heretical rites.

The adoring of heretics, mentioned in connection with these heretics, sounds very much like the Cathar practice by which the believers adored the *perfecti.* If so, it shows that dualist heretics — or peasants influenced then or at some earlier time by dualists — did live in the vicariat. However, since this is the only suggestion of dualism in the entire document, the sect may not have been important in the vicariat as a whole. In addition, since we have had so few references to specific practices that might be dualist in any document about Bosnia up to now, it seems dangerous to insist that dualism is the only explanation for this practice. It is always possible that there existed some local custom of paying extreme reverence to priests, which had a different and non-dualist origin. Extreme reverence shown to a man of God by an ignorant populace does not strike me as being so unique a phenomenon that it could be explained in only one way. Besides, how these people adored heretics is not described; thus we cannot be sure that their adoration was carried out in the same way that the Cathar adoration was.

We noted in our discussion of traditions in Chapter II the extreme reverence shown the elders around Krupa until late in the last century. Such a rite as the tradition describes might well have seemed heretical to Franciscans. As we recall in Krupa, family elders (rather than clerics) received the adulation of the different households and extended their blessings to these homes. A French traveller at the beginning of the nineteenth century also notes the unlimited veneration, confidence and submission to priests by Catholic Bosnian peasants. He states that when these peasants meet a priest they kiss his hands, and fall on their knees asking for his blessing.(89) In addition at the Council of Basel in 1435 the Terbipolensis (?) bishop stated that when he had visited Bosnia he was received with such humble veneration that it was all that he could do to prevent the Bosnians from kissing his feet.(90) To do everything short of kissing a bishop's feet might accurately be described as "adoring" him. It is possible that what the bishop described was a local practice of honoring clerics; and that what was shown him was the same practice that the Franciscans had objected to in 1373. Thus while it is quite possible that the "adoration" does reflect the existence of a dualist practice in Bosnia, it is possible that it may not. Finally, it should be pointed out that the document never suggests that the heretics "adored" had any connection with the Bosnian Church.

Kniewald believes that Gregory XI's letter supports the theory that there were dualists in Bosnia.(91) What, however, is striking to me about the letter is that it does just the opposite. With the one possible exception of the mention of adoring heretics, there is nothing in the document to suggest that there were dualists (or heretics of any specific heresy) in Bosnia. Had there been a significant heretical movement, we would expect some of the Franciscans' questions to have been concerned with how to deal with specific practices and beliefs connected with it. However, this is not the nature of the question. Instead of depicting a heretical land, the twenty-three questions mainly reveal the extreme ignorance about religious matters found in the whole area of the Bosnian vicariat. That all sorts of variant beliefs and practices with no intention of deviating and with no dogmatic base could grow up in such an environment is obvious. And when these deviations become associated with a man's Christianity, who can say if he is a heretic or not, and who can say whether there would have been agreement among the Western European clerics who visited Bosnia as to where to draw the line? Each time we see a source like the 1373 letter, we must wonder whether most of the ''heretics'' in Bosnia that so disturbed the popes were really part of any movement.

In the years that followed, Gregory continued to be troubled by the situation in the vicariat. In July 1374 the pope gave permission to the Franciscans who had been converting heretics and schismatics from error and schism to Catholicism in Timini (Timisoara?), Wilkrike (the Vukovska *župa* in Slavonia?), Vrbas and Sava to build churches in those localities.(92) If Timini is Timisoara and Wilkrike a *župa* in Slavonia, then only the last two places are in the vicinity of Bosnia; and while the Vrbas River does extend into Tvrtko's banate, most of its course as well as the entire course of the Sava River lie well to its north. If Vrbas refers to the *župa* of that name, rather than the river, most of its territory also lay to the north of Tvrtko's state. Thus the letter has little direct reference to Bosnia proper and shows that much of the papal concern for the vicariat was directed at regions within it that lay beyond the borders of the Bosnian banate. What sort of heretics were believed to be living in these northern regions is unknown; we also do not know whether by schismatics he meant Orthodox in the vicinity of Timisoara or whether by them he also included people from Vrbas who might have been adherents of the Bosnian Church.

In 1376 Pope Gregory approved a donation by Johannes Horvat, Ban of Mačva, to the Franciscans to help them in their work to convert schismatics and Patarins who lived in those parts.(93) Since most of Mačva lies to the Serbian side of the northern reaches of the Drina, it is evident that the schismatics are Orthodox. However, if the pope was accurate when he mentioned Patarins, and if he used the term according to Slavic usage — which would be likely in a reply to the Ban of Mačva — then this constitutes evidence that the Bosnian Church had some following in this area. We may recall that Danilo's biography of Stefan Dragutin (who ruled Mačva from 1282 to 1316) stated that Dragutin converted many heretics from the Bosnian land, some of whom might have been adherents of the Bosnian Church. In Chapter V we shall discuss one tomb inscription referring to a Bosnian Church cleric from Bogutovo, a village near Bijeljina in the north. Thus we are able to gather a little evidence to suggest that the Bosnian Church may well have had some adherents in regions, which were not directly part of Tvrtko's banate, north of the banate and on both sides of the Drina.

Of particular interest is Gregory's letter to the bishops and Franciscans of the western Balkans from 8 August 1377. The pope had been notified that when the Franciscan vicar and brothers in the presence of the Ban of Bosnia (i.e., Tvrtko) were celebrating mass and the divine offices, as well as administering the sacraments, heretics and Patarins, whom the ban protected, sometimes appeared, and the ban received them and allowed them to participate. Because of this, the brothers withdrew from the celebration, refusing to say the mass before such company. By so doing, they hoped to discourage the ban from sponsoring and receiving heretics. However, since the ban did not dare be deprived of divine rites, the secular priests sent thither by the pope or from elsewhere (i.e., clearly Catholic secular priests) arrived and celebrated the mass for him. Thus the ban was able to hear the mass, ignore the brothers, and perservere in his stubborness. The pope wanted to end this evil and thus threatened with excommunication any priest who celebrated the mass before the ban or any other protector of heretics or schismatics after the brothers have withdrawn from serving it.(94)

This letter shows that Tvrtko attended Catholic Church services and felt it important to hear the mass; thus he clearly considered himself a Catholic. At the same time it also shows his tolerance of the heretics and Patarins and provides evidence that he maintained cordial relations with them. Thus the ban saw nothing terrible about their "errors." It also indicates that the heretics themselves found nothing distasteful about attending Catholic Church services, which suggests that their beliefs and practices did not differ greatly from the Catholic and certainly that they did not find anything in Catholic practice so damning and sinful as to endanger their own salvation as the inquisition tracts against "Bosnian Patarins" would have us believe. We cannot draw any conclusions about the behavior of the secular Catholic clerics who were willing to perform the mass when the Franciscans withdrew; perhaps they saw nothing damning about the Bosnian heretics and simply considered them ignorant yokels, or perhaps, as foreigners they were ignorant of the Slavic tongue and thus did not realize that they were serving mass before heretics. However, in the latter case, we would expect the Franciscans to have clarified matters for them. The statement of their presence also shows that, despite occasional claims to the contrary, there were by now at least a few Catholic secular priests in Bosnia; perhaps they were all connected with the court.

VI: Tvrtko's Involvement in the Hungarian Civil War

From 1377 until his death in 1391, Tvrtko bore the title King of Raška and Bosnia. He took advantage of a civil war between Sigismund of Luxemburg and Ladislas of Naples for the Hungarian throne to oppose Sigismund and to expand his influence along the Dalmatian coast, where he became overlord of most of the major towns except Zadar. In 1390 the Archbishop of Split came to Sutjeska to receive from his new overlord Tvrtko a promise to respect the property of the Church of Split. Among the witnesses to this charter was Pavle Klěsić,(95) who, a decade later, would appear in the sources as a supporter of the Bosnian Church. His presence does not seem to have troubled the Archbishop. Tvrtko also spread his influence into the Croatian lands to the northwest. All the lords of the Donji kraji were under his suzerainty and the branch of the Hrvatinić family which came to dominate in this region was that which had supported Tvrtko in the 1363 war with Hungary, namely that of Vlkac Hrvatinić, whose son Hrvoje Vukčić in 1380 received the title of Great Vojvoda of Bosnia. Hrvoje had good relations with Tvrtko and in the 1380's amassed tremendous influence not only in the Donji kraji, but also in Bosnia itself, where after Tvrtko's death he was to become the most powerful figure in the realm and would assume the role of king-maker. Hrvoje also achieved great influence in the Dalmatian lands and in Croatia; and after Tvrtko's death, Hrvoje was to become Ladislas' viceroy for the Slavic lands and Herceg of Split.

Tvrtko sent a detachment to Kosovo (1389) and after the sultan was slain, he proclaimed a great victory and sent embassies to Italy to spread word of his success. In Florence he was proclaimed a savior of Christendom.

During these last years, we hear nothing about heresy in Bosnia, except for the references to infidels, schismatics, and heretics in the angry letters of Tvrtko's enemy, Sigismund. However, just after Tvrtko's death, but still in relation to these events, we encounter the term "Manichee" referring to certain of Sigismund's enemies. It occurs in a letter of Pope Boniface IX of 18 December 1391, written in response to a letter by Sigismund; thus the pope's terms may repeat those of Sigismund's preceding letter, and must be taken in the context of the war Sigismund was waging against Ladislas, Tvrtko (recently deceased) and various Croatian nobles including Hrvoje. The pope's letter in speaking of Sigismund's many enemies mentions: Turks, and besides them, Manichees, heretics, and other Christians in the before mentioned lands (i.e., the Kingdom of Hungary, Dalmatia, and other lands under Sigismund's subjection) . . . and also "schismatics in parts of Bosnia."(96) It is clear that Manichees, if they existed, were not the only sort of heretic in the region. It is not stated where in this large region Manichees supposedly existed nor if they were in Bosnia proper. In fact, what is stressed for Bosnia proper is the presence of schismatics, whether Orthodox from Hum and the east, or supporters of the Bosnian Church, or both, is not stated.

How are we to interpret the term Manichee? There is no reason from the text to associate it more with Bosnia than with Hungary, Dalmatia, or Croatia. It is possible that the dualists, whom we obtain brief glimpses of from time to time, had joined Sigismund's enemies in the war against him. However, it is also possible that the pope had introduced the term. The inquisition in Turin had in 1387 held inquiries about some dualist heretics from Chieri, and in the course of the investigation one convicted heretic, Jacob Bech, had claimed that various Italian dualists had gone to Bosnia to study doctrine. Thus the pope could easily have used the term Manichee in a letter that contained a complaint about Bosnia.

VII: Italian Documents About Dualists in Bosnia

In the chapter on sources, I have argued that the two late fourteenth century tracts from Italy (probably written by Italian Franciscans) about Bosnian heretics may well be based on the beliefs of Italian dualists who had some sort of link with Bosnia (Slavonia, Dalmatia). And since these Italian heretics also had contacts with Western dualists from whom they most probably acquired beliefs and practices, it would be dangerous to assume that anything in those records (in the nature of specific beliefs and practices) need be pertinent to Bosnia. Thus I am ignoring the contents of the two tracts and the doctrinal material in Jacob's testimony.(97) These documents, however, can be used as evidence that there were almost certainly some dualists in Bosnia. But we must stress that there is nothing in the two tracts or the records of the Turin trial to identify the Bosnian heretics they mention with the Bosnian Church. In addition, the data we have examined so far — and which we shall continue to examine in the next two chapters— depict the Bosnian Church as a non-dualist body. Thus I conclude these documents relate to a second movement totally separate from the Bosnian Church. And since local and Ragusan sources do not mention Bosnian dualists,I think we are safe in concluding that theirs was a small movement which attracted attention in Italy because it had links with the Italian heretics the inquisition was investigating in the 1380's.

If Jacob can be believed (and after all one article of the beliefs he claims to have held was that it was no sin to lie before the inquisition) then several of the Italian heretics had come to Bosnia to learn doctrine. The names and approximate dates he gives are as follows: Rabellator de Balbis of Chieri (ca. 1347), Iohannes Narro and Granonus Bencius (ca. 1360), Petrus Patritii (ca. 1377), Bernardus Rascherius (ca. 1380), Jacobinus Patritii (brother of Peter ca. 1382).(98) No Slavic source hints at anything about this. As we have stated before,it seems hard to believe that these Italian urbanites had anything to learn about doctrine from a Bosnian. That they may have come to visit an Italian heresiarch in exile (in Bosnia or more likely Dalmatia, part of which was Bosnian and where Italian was spoken) is quite possible.(99) Jacob Bech testified that about 1377 he had set out for, but had not reached owing to bad weather, Boxena (i.e., Bosnia), whose ruler was called the Albanus of Boxena and is subject to the King of Rassene (Raška).(100) Albanus means "Ban," and it is clear that he had set out after Tvrtko's coronation in 1377 as King of Raška. That Jacob has taken Tvrtko's two titles and made them into two different rulers merely shows his ignorance of the true state of things in Bosnia. Since he never reached Bosnia, this ignorance probably is not at all surprising. The traveling that Jacob mentioned was not entirely one-way. Jacob attributed his own conversion at some time around 1378 to the influence of two Italians (whom he names) and one man from "Sclavonia."(101)

Since Jacob attributed his conversion to a man from "Slavonia" and then tried to go to "Bosnia" to learn doctrine, it seems possible that by the 1380's some Westerners were using the word "Bosnia" as a synonym for "Slavonia" to describe the general region inhabited by the more western of the South Slavs. We have already noted that Anselm of Alexandria in his tract of ca. 1270 against heretics had used the two names as synonyms. And now particularly after the creation of the Bosnian vicariat in the 1340's, which included all the territory that was covered by the old term "Slavonia," it would not be at all strange for foreigners to use the word "Bosnia" for the whole vicariat or any part of it. And Italians, who heard of a Ban of Bosnia, could well have believed that this ban ruled over all the territory of the vicariat.

VIII: Tvrtko, Catholicism and the Bosnian Church

There was never any question about Tvrtko's Catholicism. The Franciscan mission progressed during his reign; Tvrtko built at least one church, that of St. Gregory at Trstivnica (from which he issued a charter in 1378).(102) His *protovestijar* ("chief chamberlain") was a Ragusan priest named Ratko.(103) On occasions he even had dealings with the bishop in Djakovo and gave that bishop lands for his church.

No Slavic source connects Tvrtko with the Bosnian Church. None of his charters (excluding the suspect 1370 Rajković charter)(104) was witnessed by its clerics. Even Pope Gregory's letter of August 1377, which accuses Tvrtko of protecting heretics, clearly depicts the ban as a

Catholic, attending Catholic services and fearful of the consequences when the Franciscans refused to perform mass before him. Thus Tvrtko is shown to be both a practicing Catholic and a tolerant ruler, unwilling to dictate on the religious beliefs of his subjects and seeking to maintain cordial relations with members of all faiths. There is no evidence that Tvrtko took any action against the Bosnian Church; and the Batalo hierarchy list — to be discussed in Chapter V — will show that the djeds succeeded one another in unbroken succession throughout his reign. In addition, since the Bosnian Church will appear in so many sources in the beginning of the fifteenth century, clearly in a position of some influence, it is probable that it had maintained its position in society under its own hierarchy throughout Tvrtko's reign.

None of the Bosnian nobles appears in any source from Tvrtko's reign with Bosnian Church connections. However, as we shall see in Chapter V, certain nobles prominent in the last decade of Tvrtko's reign will be found in the beginning of the fifteenth century to have such a connection. Presumably, some of them were already members of the Bosnian Church in Tvrtko's lifetime.

FOOTNOTES TO CHAPTER IV

1. After Mladen Šubić was killed by "Bosnian heretics" (see Chapter III), his older brother Pavle assumed the title of "Lord of all Bosnia"; then sometime before 1308 Pavle assigned Bosnia to his son Mladen II who began to be called Ban of Bosnia. Pavle died in 1312 and Mladen became the head of the Šubić family. (For the history of the Šubići, see V. Klaić, *Bribirski knezovi od plemena Šubića do godine 1347*, Zagreb, 1897.) Klaić believes that Kotroman had been forced to accept the position of vassal to the Šubići (*ibid.*, p. 97). Unfortunately the sources give no information about Kotroman in the first decade of the fourteenth century. We do not know whether the Šubići really assumed control of all Bosnia as their titles suggest; if they did not, then Kotroman may well have retained a part of Bosnia as his own state. If the Šubići did subdue all of Bosnia, then we must assume that either Kotroman did become their vassal, or else he was driven out of Bosnia across the Drina into Stefan Dragutin's Mačva lands. We do not know when Kotroman died. After his death Orbini reports (on the basis of an unknown source) that the leading barons of the realm rose up against Kotroman's son and heir Stjepan Kotromanić, forcing him and his mother to flee to Dubrovnik. (That the two were in exile in Dubrovnik is confirmed by a document found by V. Ćorović in the Dubrovnik archive, see his *Historija Bosne*, p. 611.) According to Orbini (p. 351) Kotromanić was later able to return to Bosnia as a result of the consent of all the barons. Why the barons had opposed him, and why they later agreed to his return and assumption of power, is unknown.

2. Orbini, p. 353 and K. Draganović, "Izvješće apostolskog vizitatora Petra Masarechija o prilikama katoličkih naroda u Bugarskoj, Srbiji, Srijemu, Slavoniji i Bosni g. 1623, 1624," *Starine,* 39, 1938, p. 44.

3. *CD,* VIII, p. 508.

4. Most scholars believe that Kotromanić was allowed to return because he consented to accept vassalage under the Šubići; if, as Klaić believes, Kotromanić fought along side the Hungarian king and the Šubići against Milutin of Serbia in 1319, we have confirmation of this view. The 1318 letter, then, should be taken as an overlord's assistance to his vassal. Ćirković (*Istorija . . .*, pp. 84-88) agrees and believes that Kotromanić remained a vassal under Šubić protection from 1318 to 1322. Thus he does not believe that Kotromanić's return marked a decline in Šubić authority during these years. In 1322 civil war broke out between Pavle II Šubić and Mladen II; Kotromanić sided with Pavle who defeated Mladen. Mladen was then taken prisoner by the King of Hungary. This civil war initiated a rapid decline for the Šubići, who lost whatever dominance they had had over Bosnia. Most scholars agree that Kotromanić's rise to independence and a position of real authority followed the events of 1322.

5. Orbini, pp. 353-54. See discussion on Orbini's sources in Chapter II. Orbini describes Bobali's role in converting the ban, and since he (or Pietro Livio) mentions seeing privileges granted to Bobali by the ban in the Bobali archive in Dubrovnik, Orbini's account and dating of the ban's conversion probably should be accepted.

6. *Itineraria Symonis Simeonis et Willelmi de Worcestre,* (Ed. by J. Nasmith), Cambridge, 1778, p. 13.

7. *CD,* VIII, p. 473; see also, *Monumenta Vaticana, Historiam Regni Hungariae illustr.* ser. I, Vol. I (Rationes collectorum Pontificorum in Hungaria (1281-1375)), Budapest, 1887, p. 37.

8. *CD,* VIII, p. 508.

9. *CD,* VIII, pp. 535-36.

10. For similar neglect in eighteenth-century non-heretical Montenegro, see Chapter I, note 22.

11. *Monumenta Ragusina;* Libri reformationum, I, MSHSM, 10, Zagreb, 1879, p. 115.

12. L. Thalloczy, "Istraživanja o postanku bosanske banovine sa naročitim obzirom na povelje körmendskog arhiva," *GZMS,* XVIII, 1906, pp. 404-05.

13. "To remain Bosnian," of course, means that the local lords accepted the Bosnian ban as their overlord. Presumably through most of the following period they continued to manage their own affairs as they had previously done. Also, in the course of the following century on occasions certain parts of this annexed territory were to be lost to Bosnia for brief periods of time: e.g., Završje and the Krajina in 1357 when Tvrtko had to yield them to the King of Hungary as part of his cousin's dowry; both regions were recovered by Tvrtko after thirty years.

14. *CD,* XI, pp. 160-62.

15. V. Trpković, "Kad je Stjepan II Kotromanić prvi put prodro u Hum," *Istoriski glasnik,* 1-2, 1960, pp. 151-54. See also his, "Branivojevići," *Istoriski glasnik,* 3-4, 1960, pp. 55-84. Hum is the western part of what is now Hercegovina. (See V. Trpković "Humska Zemlja," *Zbornik Filozofskog fakulteta* (Beograd), VIII, 1964, pp. 225-259). The eastern parts of Her-

cegovina including the regions of the Drina and Lim Rivers as well as Gacko and the land bordering on Zeta remained under the Serbian state which still reached all the way to the sea in southern Dalmatia. These lands were to be annexed by Bosnia only under Tvrtko.

16. The population of Ston and environs was evidently chiefly Orthodox. In February 1344 Pope Clement VI describes the difficulties of Marinus, the Catholic Bishop of Ston, who had been prevented from preaching and administering the sacraments by the local populace who were Rascian schismatics (scismate Rassiencium, i.e., Serbian Orthodox); no mention is made of heretics. Bishop Marinus was forced to leave Ston. (*CD,* XI, pp. 118-19) Dubrovnik shortly thereafter sent in the Franciscans who, after much labor, won the populace over to Catholicism.

17. L. Thalloczy, ''Istraživanja . . .,'' *GZMS,* XVIII, 1906. Texts of these first six charters are given, pp. 403-08.

18. *Ibid.,* pp. 404-05. Mandić, *Bogomilska crkva . . .,* pp. 214-216, 219 believes that it is significant that the gost referred to in this charter is called ''veliki gost'' (great gost), and suggests that ''great gost'' was a rank higher than gost. However this seems indefensible since our charter also says ''veliki djed'' and thus we would have to have a ''great djed'' as a higher rank than the djed. The term ''veliki'' (great) used here simply is intended to exalt the bishops being referred to, and to show them added respect, in the same way that it is customary to refer to the great ban or great king, etc. To say ''great ban'' does not increase the ban's actual status. When a medieval Bosnian wanted to differentiate between the degree of status between two bans (e.g., under Tvrtko I (1353-91), when his younger brother Vuk also received the title ban), the terms ''senior'' and ''junior'' were used. However, though the terms senior and junior appeared in the secular world, we never find them in connection with Bosnian Church offices.

19. On Bogomil views about the Trinity see, A. Vaillant, ''Un Apocryphe pseudo-Bogomile: La Vision d'Isaie,'' *Revue des Études Slaves,* XLII (1-4), 1963, p. 119, and particularly H-C Puech's commentary in H-C Puech and A. Vaillant, *Le Traité contre les Bogomiles de Cosmas le prêtre* (Travaux publiés par l'Institut d'Etudes Slaves, XXI), Paris, 1945, pp. 178-81.

20. Solovjev tries to connect this phrase with Manichee concepts of reincarnation. (''Saint Grégoire, Patron de Bosnie,'' *Byzantion,* XIX, 1949, pp. 275-78.) This view cannot be accepted because, regardless of what doctrine the Bosnian Church might have held, Ban Stjepan was no Manichee. The wording appears in three of these 1322-25 charters, one of which was not witnessed by the Bosnian Church. Then in 1331 we find the wording ''I am Ban Stjepan and am called the slave of St. Gregory.'' It seems likely that the earlier phrase was intended to convey this meaning as well. Probably St. Gregory was Ban Stjepan's Patron Saint, and hence his *slava* saint, although Professor Milenko Filipović argues that at this time the *slava* as we know it did not exist in Bosnia. (''Studije o slavi, službi ili krsnom imenu,'' *Zbornik za društvene nauke* (Matica srpska) Vol. XXXVIII, Novi Sad, 1964, pp. 61-65). He notes that the medieval documents do not use the term *slava* but *''krsno ime''* (a term that later clearly was synonymous with *slava)* and argues that *krsno ime* also referred to the church name one received at baptism. He suggests that this baptismal name is what was meant by *krsno ime* in medieval Bosnia. To support his view he points out that a 1391 charter of the Sanković brothers reveals the two

brothers swearing by their *"kr'stnemi imeni"* of Saint George and the Archangel Michael. (For this charter, see Miklosich, p. 219) Two brothers, Filipović rightly points out, would never under regular circumstances, have had two different *slava*(s). The Sanković brothers bore the popular names Beljak and Radič and both were Orthodox. Clearly at baptism they had received church-approved names. Thus Filipović's interpretation about them is surely correct.

However, Ban Stjepan already had a church-approved name of Stjepan and did not need to add Gregory, as Ban Ninoslav may well have needed to add Matej. In addition, the term used in the Sanković charter is "kr'stno ime" which certainly is derived from the term for baptism. This does not mean that it must be equivalent to the term "krsno ime" found in other medieval documents, for example Gost Radin's will from 1466. And in fact I do not think the two terms are synonyms. I think *krsno ime* does refer to the patron saint of one's *slava,* as it does a few centuries later. That Saint Gregory was Ban Stjepan's *krsno ime (slava)* is confirmed by the fact that his nephew and sucessor Tvrtko in a charter of 1366 also called himself the slave of Saint Gregory (*AB,* pp. 34-35). Since it is less likely that both men would have received the same baptismal name, it suggests that St. Gregory was the ruling family's *slava.* This would also explain how St. Gregory came to be the patron saint of Bosnia in 1461. (*AB,* p. 244). Thus we conclude that at least some Bosnians celebrated a family saint, the *slava,* and also some, like the Sankovići, honored the saints, whose names they received at baptism. In either case, be it a baptismal link or a *slava* link with a saint, Stjepan's expression about Saint Gregory probably refers to a saint in the role of a protector.

21. Thalloczy, "Istraživanja . . .," p. 406.

22. *Ibid.,* p. 406.

23. For discussion of these spells, see above Chapter I.

24. The plaque and its first inscription is discussed above in Chapter III.

25. P. Andjelić, "Revizija čitanja Kulinove ploče," *GZMS,* 15-16, 1960-61. Inscription given p. 299; see p. 305 for his dating. A photograph of the plaque which shows the rude form of the later writing and also the little figure of a man can be found in S. Ćirković, *Istorija srednjovekovne bosanske države,* facing p. 49.

26. A. Benac, *Radimlja,* Sarajevo, 1950.

27. We have information about very few specific hižas. These are listed below, with reference to where each is discussed in this study. I also give what information we have about churches in the same villages.

a) Moštre, 1322-25 pp. 174-76. There is a local tradition that prior to the arrival of the Turks a church existed at the location Svibama at Moštre. (M. Filipović, *Visočka nahija* (SKA, SEZ vol. XLIII, Naselja i poreklo stanovništva, vol. XXV) Beograd, 1928, p. 207.

b) Janjići 1404 p. 227. The local inhabitants claim that an old church had existed there at the site of a medieval grave yard (*Bosanska vila,* IX, 1894, p. 104).

c) Ljubskovo, 1412-16 pp. 256-57. We do not know exactly where Ljubskovo was so we cannot trace old traditions. However, there were various churches in the *župa* of Osat in which Ljubskovo lay.

d) Bradina, 1418 p. 257. Here there exists a local tradition that a "Greek' church (i.e., Orthodox) used to stand near the medieval cemetery. Prof. Andjelić could find no signs of ruins (P. Andjelić, "Srednjevjekovna kultna mjesta u okolini Konjica," *GZMS*, n.s. XII (arh), 1957, p. 191).

e) Between Goražde and Borač, in the 1440's pp. 258-60. Unfortunately where this hiža was located is not known; thus we cannot try to connect it with any local traditions or ruins. In this general area though several old churches did exist.

f) Zgunje pp. 261-62. We do not know that a hiža existed here; however, since we have a mid-fifteenth century cemetery with two stones referring to Bosnian Church clerics, it is quite likely that there was a hiža here. We do not know of a medieval church in Zgunje.

g) On the Neretva below Biograd, 1449 pp. 311, 371. Andjelić has suggested that this hiža lay on the Neretva a couple of kilometers above Konjic. In the vicinity of Konjic there are numerous traditions as well as several medieval church ruins.

h) Seonica, 1466, p. 366. Reference was made to a Gost Radin Seoničan. We do not know whether Seoničan refers to his birthplace or to a possible hiža under him. We have no other evidence to suggest that a hiža existed in Seonica. Both Catholics and Moslems of present day Seonica have an active tradition that the present mosque there (in Seonička donjoj mahali) sits on the site of an earlier church; there is also a story that the former church's bells are buried under a corner of the mosque. Since the mosque is built near a medieval cemetery, the tradition could well be true (P. Andjelić, "Srednjevjekovna kultna mjesta . . ." *GZMS*, n.s. XII (arh), 1957, p. 190).

i) Uskoplje, 1466, p. 369. A church of Saint John is referred to here in King Bela's charter of 1244 (*CD*, IV, pp. 239-40). This church was Catholic at the time it was referred to; but this does not mean, however, that it could not have been a Bosnian church in the fifteenth century.

j) Bijela, 1467, p. 372. At Bijela near Konjic there are numerous church ruins and traditions of churches: 1) On the hill "Djeva" there is a locality called "crkvina" and a medieval cemetery; cemeteries often are contemporary with churches (P. Andjelić, "Srednjevjekovni gradovi u Neretvi," *GZMS*, n.s. XIII (arh), 1958, p. 189). 2) In Gornja Bijela at a medieval cemetery there are some ruins which popular tradition claims were once a "Greek" (i.e., Orthodox) church. (*Ibid.*, p. 189) 3) On a rise at the location Pjevčeva glavica in Gornja Bijela at the present Orthodox cemetery (where there are also some medieval *stećci*) people believe there was once a church. (*Ibid.*, p. 189) 4) M. Vego describes some church ruins in Gornja Bijela which he dates thirteenth or fourteenth century. It is not clear whether he is describing one of the above mentioned or a fourth church here (M. Vego, review of S. Ćirković, *Istorija . . .*, in *GZMS*, n.s. XX (arh.), 1965, p. 301).

28. M. Dinić, *Iz dubrovačkog arhiva*, III, p. 216, no. 82.

29. And when investigation of a tradition fails to unearth a foundation, it does not necessarily mean that the tradition was incorrect. We know that wooden churches were not uncommon in Bosnia. From my reading and conversations with scholars who have worked in the field, I have come to the conclusion that a general tradition (e.g., there was a church in that place) is fairly reliable, whereas the specific parts of a specific tradition must be treated with scepticism. Thus "there was a church here built by Kulin," cannot be relied upon. Very probably there had been a church, but there is no reason to believe it had really been built by Kulin. Kulin as a result of limited schooling has entered the popular idiom to signify "long ago."

30. *CD*, IX, p. 234.

31. *CD,* IX, pp. 241-44.
32. *CD,* IX, p. 324.
33. *CD,* IX, pp. 493-94.
34. *CD,* X, pp. 326-27. Though called upon to participate, it is unlikely that the Šubići would have been interested in this crusade since they seem to have long since patched up their relations with Kotromanić. In fact Ban Stjepan's brother Vladislav at this time (1338) married a Šubić girl. In addition, to have helped this crusade would have only served to advance the interests of the Šubići's most powerful rival, the Nelipac family.
35. Lucius, *Memorie de Trau,* p. 234 (A seventeenth century history based on original documents some of which have by now disappeared). I have cited the passage from, *Poviest Bosne i Hercegovine,* Sarajevo, 1942, p. 264.
36. *CD,* X, pp. 494-95.
37. Ćirković, *Istorija . . .,* p. 110.
38. *CD,* X, pp. 383-86.
39. *CD,* X, pp. 665-70.
40. *CD,* X, pp. 469-70.
41. Fermendžin, "Chronicon . . .," *Starine,* 22, 1890, pp. 6-7.
42. *CD,* X, pp. 525-28.
43. *CD,* X, p. 525.
44. *CD,* X, p. 526.
45. Orbini, p. 353. Orbini erroneously dates the mission 1349 instead of 1339 on p. 352. Possibly Orbini's statement about the ban's motives is drawn from something he found in the Bobali archive. See discussion on Orbini's sources above in Chapter II.
46. Chronica XXIV Generalium Ordinis Minorum. *Analecta Franciscana,* III, 1897, p. 529. The same text given in the Bosnian compilation, E. Fermendžin, "Chronicon . . .," *Starine,* XXII, 1890, p. 7.
47. Orbini, p. 353. For text of the Bobali story, and a discussion of this tradition, see above in Chapter II.
48. *Analecta Franciscana,* III, 1897, p. 529; Fermendžin, *Starine,* XXII, 1890, p. 7.
49. *CD,* XI, p. 138.
50. Theiner, *MH,* II, pp. 117-118.
51. G. Čremošnik, "Ostaci arhiva bosanske franjevačke vikarije," *Radovi,* (NDBH), III, Sarajevo, 1955, p. 19, cites two notations written on the 1372 papal bull cited above in note 50 which set the figure of 60 as the maximum number of friars allowed to work in the vicariat. "Ego fr(ater) Bartholomeus, vicarius predicte vicarie Bosne, ex virtute presentis bulle XX fratres recepi ante annum istum s(cilicet) MCCCLXXX . . ." Below this is written "Et usqu nonagesimum Vm(1395) credo quod tm (Tantum G.Č.)" *Ibid.,* p. 22 cites a 1378 bull which permits 30 Franciscan volunteers to go to work in Bosnia "triginta fratres vestri ordinis sponte venire volentes ex suis mediatis vel immediatis superioribus licentia . . . precipimus recipiendi et dictam vicariam ducendi harum serie concedimus facultatem . . . ." J. Šidak, "O autentičnosti i značenju jedne isprave bosanskog 'djeda' (1427)," *Slovo,* 15-16, 1965, p. 293, believes this means that in 1378 the quota had been raised from 60 to 90, and that the 1380 notation shows that of the desired increase of 30, only 20 were received, giving them a total of 80. While this is a reasonable assumption, it seems equally plausible that the quota had not ever been increased, but the

number of actual Franciscans working in the vicariat had fallen well below sixty, so the pope called for thirty volunteers to meet the old quota, of which by 1380 they had obtained twenty which would give us a total of about fifty Franciscans working in the vicariat.

52. It is possible that some of these Bosnian Franciscans, despite regulations, lived outside of monasteries. In the 1430's Jacob de Marchia, endeavoring to enforce discipline among the Franciscans, expelled them from Jajce where they had been living in private homes (see below, Chapter V). Since we have no evidence of Franciscans living outside of monasteries prior to that, we do not know whether this breach in discipline had begun shortly before Jacob's appearance in Bosnia or whether it dated back much further.

53. "De conformitate vitae Beati Francisci ad vitam Domini Iesu," written by fra Bartholomaeus of Pisa, vicar of Bosnia, published in *Analecta Franciscana,* IV, 1906, pp. 555-56. In addition to the four monasteries noted for Bosnia proper, Bartholomaeus lists several other monasteries near its borders (in territories some of which were held by the Bosnian ruler) at Imota, Glamoč, Glaž, Greben (Krupa on the Vrbas), Modriča, Vrbica, Bijeljina, Srebrnica, and Mačva. To keep the Bosnian bishop company a monastery also existed at the town of his seat in Djakovo. Since the Franciscans had not been in Bosnia prior to 1340, except possibly on brief visits by the two in the 1290's and Fabian in the 1320's, it is clear that these four Bosnian monasteries were all erected after 1340.

54. *CD,* XI, pp. 137-139. A misprint in the book has the letter dated 1343 instead of 1344.

55. J. Šidak, "Franjevačka 'Dubia' iz g. 1372/3 kao izvor za povijest Bosne," *Istorijski časopis,* V, 1954-55, p. 215.

56. *AB,* pp. 28-29.

57. Ljubić, *Listine,* III, p. 107.

58. Orbini, p. 354.

59. M. Vego, *Naselja bosanske srednjevjekovne države,* Sarajevo, 1957, p. 126. He cites an unpublished document from the Dubrovnik archive (Minus Consilium, 25, VI, 1367).

60. Kotromanić had had a son but we do not know whether he survived Kotromanić; even though Kotromanić's brother Vladislav was alive at the time of Kotromanić's death it was Tvrtko rather than Vladislav who succeeded. (See M. Dinic, *Državni sabor srednjevekovne Bosne,* Beorgrad, 1955, p. 24.) Why the succession went the way it did is unknown.

61. *CD,* XII, p. 270.

62. *CD,* XII, pp. 332-33.

63. *CD,* XII, p. 670.

64. L. Thalloczy, *Studien zur Geschichte Bosniens,* pp. 333-34.

65. The King of Hungary was later, in 1370, to state that he had invaded Bosnia to destroy the many heretics and Patarins there (*CD,* XII, pp. 670-71). Probably nearer the truth than Louis' rationalization is the statement of one of Louis' secretaries that Louis had sent troops to Bosnia to punish certain rebels (i.e., Tvrtko and any Bosnian nobles who were trying to assert their independence). (Cited by Perojević, *Poviest BH . . .,* p. 300.)

66. *CD,* XIII, pp. 18-19.

67. On this war, in addition to the general histories, see, F. Šišić, "Studije iz bosanske historije," *GZMS,* XV, 1903, pp. 319-25.

68. Theiner, *MH,* II, p. 91; A. Tautu (ed.), *Acta Urbani V . . .,* pp. 299-300.

69. *CD,* XIV, pp. 219-20; A. Tautu (ed.), *Acta Urbani V . . .,* pp. 293-294. Perojević believes this information is all tendentious and based upon the letters the pope was then receiving from Vuk who was hoping to excite the pope into intervening on his behalf (*Poviest BH,* p. 306).

70. On Nikola Altomanović, see M. Dinic, *O Nikoli Altomanoviću,* Beograd (SKA), 1932.

71. *AB,* p. 38.

72. On the family of Sanko, see J. Mijušković, "Humska vlasteoska porodica Sankovići," *Istorijski časopis,* XI, 1960, Beograd, 1961, pp. 17-54.

73. M. Vego, "Nadgrobni spomenici porodice Sankovića u selu Biskupu kod Konjica," *GZMS,* X, 1955, pp. 157-65, and XII, 1957, pp. 127-39.

74. M. Vego, *Zbornik srednjevjekovnih natpisa B i H,* Vol. II, Sarajevo, 1964, pp. 48-49.

75. Into the 1360's, as noted above, part of what is now Hercegovina was Serbian; after Dušan's death most of this Serbian land, the territory from the upper Drina, running southwest along the border of Zeta, including Gacko, on through Trebinje to the Adriatic south of Dubrovnik was held by the Serbian nobleman Voislav Voinović, who was called by the Ragusans the Count of Hum (Comes Chelmi). Voislav died in 1363 and was succeded by his widow, Goislava, who seems to have maintained cordial relations with Tvrtko, allowing him to pass through her lands to Dubrovnik in 1367. Voislav also left a brother Altoman who died in 1367 leaving his lands which were in the vicinity of Rudnik to his son Nikola Altomanović. Nikola soon declared war on his aunt, made himself master of territory all the way to the sea and became involved in a war with Dubrovnik. Tremendously ambitious, Nikola had made himself entirely independent in his own territory and prior to Tsar Uroš' death he was already minting his own coins at Rudnik. In 1371 King Vukašin, Knez Lazar and the Balšići of Zeta, with Tvrtko's blessing, had formed an alliance against Nikola. Before the allies could go into action, though, the Turkish danger led Vukašin to march against the Ottomans; the disaster on the Marica, in which Vukašin was killed, followed. Two months later Tsar Uroš was dead. The Serbian lands were up for grabs between Nikola (who had obtained more land after the Marica battle), Lazar, the Balšići and Tvrtko. Briefly Nikola allied himself with Djuraj Balšić; Tvrtko and Lazar also formed an alliance. According to Orbini, Nikola then tried to poison Lazar; as a result Lazar and Tvrtko joined together in a campaign against him. Nikola fled to his town of Užice in Serbia, where in 1373 he was captured and blinded; he then disappears from the pages of history. His lands were divided between Tvrtko and Lazar, with Tvrtko receiving the western portion including the upper Drina and Lim regions (with Mileševo and Prijepolje) as well as Gacko, and Lazar obtaining the eastern territory. The Balšići briefly obtained Trebinje and Konavli, but this territory was soon acquired by Tvrtko. On the above see, M. Dinić, *O Nikoli Altomanoviću,* Beograd, 1932.

76. For this new interpretation, which I am convinced is correct, of the date and circumstances of the building of this church, see the excellent article by Z. Kajmaković, "Živopis u Dobrunu," *Starinar,* 13-14 n.s. (1962-63), pp. 251-60.

77. N. Radojčić, *Obred krunisanja bosanskoga kralja Tvrtka I,* Beograd,

1948, argues that for one to be crowned in an Orthodox church it is necessary that one be Orthodox — for the king-to-be must promise to defend Orthodoxy. Thus he concludes that Tvrtko, to have been crowned at Mileševo, had to be Orthodox. Since Tvrtko's past and future associations and activities do not make him appear Orthodox, we cannot accept Radojčić's categorical assertion. In reply, we may suggest the following possibilities: a) The particular metropolitan who crowned Tvrtko did not know this rule, did not care about it, or felt that the advantages to be gained by crowning him were more important than the letter of the law; b) Tvrtko, having become ruler of Mileševo and its environs, possibly forced the metropolitan to crown him by threatening the Orthodox Church there or the metropolitan's own person; c) possibly Tvrtko promised to (or even underwent the rites to) convert to Orthodoxy but never fulfilled his promise. In any case, we suspect that Tvrtko, despite being a Catholic, must have at least taken an oath to defend the Orthodox Church, its possessions and its adherents in his lands; most probably such an oath would have been sufficient to satisfy the metropolitan.

78. A. Tautu (ed.), *Acta Gregorii XI*. . ., pp. 336-338.

79. A. Tautu, *op. cit.*, pp. 64-69, *AB*, pp. 38-40.

80. A. Tautu, *op. cit.*, pp. 106-07; *CD*, XIV, p. 494.

81. The complete text of this long papal letter is found in D. Kniewald, "Vjerodostojnost . . .," *Rad*, 270, 1949, pp. 156-63; also in Tautu, *op. cit.*, pp. 71-77.

82. *Ibid.*, p. 157.

83. *Ibid.*, p. 157.

84. *Ibid.*, p. 159.

85. *Ibid.*, p. 157.

86. *Ibid.*, p. 158.

87. *Ibid.*, pp. 160-61

88. *Ibid.*, p. 162.

89. M. A. Chaumette-des-Fossés, *Voyage en Bosnie dans les années 1807 et 1808*, Paris, 1816, p. 71.

90. Monumenta conciliorum Generalium XV, Concilium Basileense, II, Vienna, 1873, Joannis de Segovia, *Historia gestorum Generalis Synodi Basiliensis* (ed. E. Birk) lib. ix, Chap. 5, p. 750. This source is also discussed in Chapters II and V.

91. Kniewald, "Vjerodostojnost . . .," *Rad*, 270, 1949, pp. 144-56.

92. A. Tautu, *op. cit.*, pp. 214-15.

93. *Ibid.*, pp. 384-85.

94. *Ibid.*, p. 459.

95. *AB*, p. 49.

96. Monumenta Vaticana Hungariae I ser. Vol. 3, Budapest, 1888; *Bullae Bonifacii IX (1389-96)*, pp. 178-79.

97. On the texts of these Italian sources, and why I believe we cannot use their contents to demonstrate specific beliefs existing in Bosnia, see the discussion in Chapter II.

98. Amati, "Processus . . .," *Archivio Storico Italiano*, 1865, ser. III, Vol. II, pt. 1, p. 53.

99. It is possibly the arrival of such a heresiarch and some of his followers that the pope was referring to in June 1325 in the puzzling letters that referred to many heretics from many and various regions coming to Bosnia (see note 31 for

this chapter). I am certain the pope exaggerated the numbers involved but it is possible that he was then concerned about the flight of some leading Italian heretics to (or near to) Bosnia. Possibly, if they really came, these heretics and their successors remained in exile in the Balkans for the next 40 years and maintained contact with their followers back in Italy.

100. Amati, p. 53.

101. Amati, pp. 50-51.

102. Miklosich, pp. 186-90.

103. We find references to "Presbyterus Ratcho" as Tvrtko's *protovestijar* in 1381. V. Ćorović, *Kralj Tvrtko I Kotromanić,* Beograd, 1925, pp. 102-03, gives the text of a document in which Ratko collects the *mogoriš* tribute for Tvrtko. (November 1381) Two Ragusan council records also refer to Ratko (July and September 1381), see M. Dinić (ed.), *Odluke veća Dubrovačke republike,* I, Beograd, 1951, pp. 149, 161. According to Ćorović, *op. cit.,* p. 42, Ratko began as a court chaplain for Tvrtko (found as such in 1375) and became *protovestijar* in 1378. Could he have been one of the secular priests who served mass for Tvrtko in 1377 after the Franciscans withdrew in protest? Ratko was later to become Bishop of Trebinje. Thus the Catholic hierarchical authorities, by making him a bishop, showed that they found nothing wicked in his long association with Tvrtko. Besides Ratko, we find a Ragusan named Mihail writing Latin charters for Tvrtko in 1390; he was later to be chosen as Bishop of Knin (Ćorović, *op. cit.,* p. 92). In addition a priest named Marin, who had served Dubrovnik as an envoy to the Bosnian court in 1380, had rendered sufficiently valuable services to Tvrtko to be awarded land in the village of Uskoplje in Konavli by him. Marin later in 1395 became Archbishop of Bar (on Marin see, M. Dinić, "Dubrovčani kao feudalci u Srbiji i Bosni," *Istorijski časopis,* IX-X, 1959, p. 140).

104. On this charter, see the discussion in Appendix A of Chapter II.

CHAPTER V

BOSNIA FROM 1391 TO 1443

I: Character of the Bosnian State, 1391-1443

Tvrtko died in 1391 and was succeeded by an obscure figure, somehow related to him, named Stefan Dabiša.(1) Professors M. Dinić and S. Ćirković argue that this time the Bosnian state not only did not fall apart but in fact remained united since by then a feeling for a Bosnian entity had come into being. This unity, they argue, is illustrated by the institution of a state council *(sabor)* of nobles,(2) which had its greatest influence in the last decade of the fourteenth and in the first two decades of the fifteenth century. Ćirković stresses the council's role in choosing kings and in deciding on important matters of internal and foreign policy, and he concludes, "But above all, the sabor did not lose its character as a unified body which represented the Bosnian state as a whole." He then goes on to describe the crown of Bosnia as a symbol of that unity and states that, although the nobles might make war against the king, they remained true to the crown and to the entity it represented.(3)

I have previously expressed my doubts about the acceptability of such a view of Bosnian reality.(4) I believe that the *sabor* reflects not unity but the weakness of the central state; thus the king, too weak to assert his authority, had no other choice but to summon the leading nobles to a meeting to obtain their agreement on important questions, and by this means guarantee that a particular course of action was acceptable to them and would not cause a revolt. The king also could not enforce a decision unless he had aid in enforcing it from at least one or two of these powerful nobles. The *sabor* was not a regular institution, and there is no evidence that it had regular membership. However, when a major question needing resolution did arise (and many such questions arose during these thirty years) the king summoned those men who had the most influence and power in the state to discuss it.

Thus it is not strange that at most of the meetings the same figures were in attendance. Dinić speaks of the nobles acquiring the "right" to elect and dethrone the king.(5) However, this was not a matter of "rights" in a juridical sense, but of the *de facto* balance of power. If two or three of the most powerful nobles agreed to cooperate in removing a king, they had the power to overthrow him. If ever there was a state where "rights" and "contracts" counted for little and raw power and actual strength counted for everything, it was medieval Bosnia, and particularly Bosnia in the period after Tvrtko's death. The nobles, when they felt it was in their interests to do so, could and did work together. Yet each was out for himself, and anything he did that might seem to have been in the interests of the Bosnian state was purely coincidental. Thus, on Tvrtko's death the state did not go to pieces because certain powerful nobles felt it was in their interests to cooperate. Thus the "state" remained but its unity and strength were illusory, for it was not the king's (or any other central apparatus') authority that held the separate counties together. And if instead of cooperating, several of the nobles had decided to secede or withdraw their loyalty from the king, the king could not have prevented them from doing so.

As Professor A. Babić emphasizes, regionalism remained strong in Bosnia throughout the Middle Ages, aiding the nobles' separatist aspirations and preventing the development of a feeling for state unity or a tradition of "Bosnia." The ban was the ruler of the *župa* of Bosnia (the central region), and at times through strength he was able to assert his authority over other regions. But each of these other regions retained its own traditions, which always remained more important than concepts about a unified state. And each time these regions (Donji kraji, Hum, Sol, Usora) and their rulers had a chance to break away from the central state they did so. The mountainous geography aided localism and local traditions. And the fact that there were three religious confessions instead of a single national religion, which could have served the interests of cohesion among the separate entities, contributed to separatism.(6)

For this period after Tvrtko's death in 1391 we have a much larger number of sources, particularly from Dubrovnik but also from Hungary, which enable us to trace the careers of these self-seeking nobles and their relations and connections with the Bosnian Church.(7) We find little about the Bosnian Church's role as a religious body, however, since this subject was not important to the foreign authors of these sources.

II: Three Religious Sources: The Serbian Anathemas, the Batalo Gospel, the Gospel of Hval Krstjanin

From this period we have three important religious sources: the compilation of Serbian Orthodox Church anathemas, which seem to have been composed in the last quarter of the fourteenth century, and two entries found in Bosnian Gospel manuscripts, one written in 1393 and the other in 1404.(8)

The compilation of anathemas against heretics and heresies condemned at church councils over the preceding centuries survive in three slightly different manuscript copies of the sinodik to be read in the Orthodox Church for the Week of Orthodoxy. Having translated the texts into Slavic from the Greek, the Serbs then reworked the material to some extent and added to the original texts. Dating the texts is a complicated matter since names to be anathematized were often added in the course of an otherwise unimportant recopying, so that a thirteenth century text may appear in its original form in a fourteenth or fifteenth-century copy, differing only in that it has added a handful of names to the original. Here I concern myself only with dating the sections referring to Bosnian heretics, which, as the latest part of the sinodiks, enable us to date the copying. From the names mentioned, we may conclude that all three copies should be dated between 1370-1400. The best editions of the texts have been published by Mošin.(9) He has also written a study of the texts, as has Solovjev.(10) My conclusions differ somewhat from theirs.

In the Troicki manuscript, we have four articles that damn the Babuny, who falsely call themselves Christians, scorn the true faith, take the words from the holy books and alter them to their false faith, are apostates from the church of right belief, and scorn saints, the cross and icons, refusing to bow before them. They are anathematized as is anyone who accepts their teaching, reveres their teachers or gives shelter to any known babun. The term Babun here, following the usual Serbian usage, means Bogomil. Then follows a section of Eternal Memories to certain holy figures, after which comes: "Cursed be Rastudije bosanski, and Radomir, and Tolko, and Tvrdko and Tvrdoš and all calling themselves krstjani and krstjanice who do not worship the holy icons and the honored cross."(11)

The Zagreb text says: "Beliz'men'c, Rastudije, Dražilo, Tvrdoš, Dobrko, Radosim, Rastina gost, pop Drug, Tvr'dko, pop Braten, Hoteš, and all Bosnian and Hl'mscii (Hum) heretics be cursed." The following article lists more names: "Voihna, Prijezda, Nelep'c, Stog'šii, Boroje, Lepčin and all believing in them and witnessing to them, be cursed." The document then brings in material from the Synod of 843, presents the anathema against the Babuny and then has some more names "Vl'čko, Grubeša's father and his mother Radoslava, and Neradova, the grandmother of Gradislav be cursed. Ban

Stefan, Protomara his "snaha" (sister-in-law or daughter-in-law) be cursed. Prvoslav, Bregoč, Stogšin, Voihna, Rastudije be cursed. Vlahinja Voljavica be cursed."

The third manuscript, the Dečanski text published in 1864 but subsequently lost has a shorter list than Zagreb and a few variants. The variants are as follows: Rastima Gost (instead of Rastina), pop Tvrdoš (instead of just Tvrdoš), Stogniš (for Stogšii), and Stogniš again later (for Zagreb's Stogšin), and finally three new names: Vlah Dobrovojević and the brothers Vlah and Beloš Op'n'ković.

In all cases, the articles about the Bosnians are separate from the articles about the Babuny; thus, other than the common rejection of icons and crosses, there is nothing to indicate that the two groups were considered the same heresy by the Serbs. In addition, nothing in the text indicates that the Babuny had any connection with Bosnia. Both groups, however, were damned as enemies of the Serbian Church. The Bosnians called themselves krstjani and did not worship icons or the cross. The mention of krstjani (the name used by ordained members of the Bosnian Church, as well as by the monks at Bilino polje in 1203), and the mention of Rastudije (whom we shall meet shortly in a Bosnian Church document) both show that the Bosnians damned here were those connected with the Bosnian Church. Because the Serbian Church anathematized them and called them heretics, it is clear that the Bosnian Church did differ somewhat from the Orthodox Church.

The only charges leveled against the heretics were of an iconoclastic nature — refusal to worship icons and the cross. These objects had been noted as being absent in Bosnian churches back in 1203, and Pope John XXII had stated in 1319 that the cross was not respected. Thus it is possible that in the thirteenth and fourteenth centuries at least certain Bosnian Churchmen had come to reject them. It is quite possible that during most of this period even the majority of Bosnian Churchmen had this attitude about religious pictures and the cross, and the Serbian anathemas may well demonstrate this.

However, these anathemas, composed probably in the last quarter of the fourteenth century, conflict with two almost contemporary sources, we shall discuss shortly, both from 1404. The Gospel manuscript copied by Krstjanin Hval is beautifully illustrated, showing the Bosnian Church did not reject religious pictures. However, it is possible that the Bosnian Church did not *worship* icons. As for the cross; we have a letter written by Djed Radomer, quite possibly the Radomir anathematized in the Troicki manuscript, which begins with the sign of the cross. Thus we must suggest that either there was no generally accepted opinion about crosses within the church or else that the church's views on the cross changed. Since we have references to rejection of the cross in 1203, 1319, and now late in the fourteenth century, and since we shall find crosses used in 1404 and in a couple of other instances after that date, we must conclude that if the Bosnian Church had had a generally accepted tenet against crosses, this tenet seems to have been dropped about 1400.

The anathemas have nothing doctrinal against the Bosnians and do not speak of the Bosnians in the same terms as they do the Babuny (Bogomils). The Babuny are called false Christians who alter the faith and change holy texts. Because the Bosnians are not accused of such things, it is probable that they only deviated from the Orthodox in certain matters of practice. Perhaps political differences between Bosnia and Serbia made it expedient to damn the Bosnians as heretics.(12)

The most important name on the anathemas since he appears first on the Troicki manuscript, and though second on the other two he is named twice in both of them, is Rastudije bosanski. Since we have other information about him in the 1390's, he becomes the key figure that leads to our dating the manuscripts to this period. And since the other names anathematized, except for possibly Radomir whom we shall speak of later (and the names discussed in note 12), are not otherwise known,(13) let us now turn to our second source and learn more about Rastudije. Our second source is the entries written in the extant pages from a Gospel manuscript which now is found in the Saltykov-Ščedrin library in Leningrad.(14) All that has survived of the Gospels themselves is the end of John (Chap. XXI, verses 21-25). On the verso of that page is an inscription from an Orthodox monastery (of Skr'batno) written in 1703 which shows that the Gospel was later utilized by an Orthodox church. On the second surviving page appears the following: (the second column of twelve names, we shall show, is a list of Bosnian Church djeds):

V' ime o(t')ca i s(i)na i s(ve)tago d(ou)h'	i s(ve)tago d(ou)h'
1. Eremis'	1. G(ospodi)n' Rastud'e
2. Azarie	2. Radoe
3. Kouk'leč'	3. Radovan'
4. Ivan'	4. Radovan'
5. Godin'	5. Hlapoe
6. Tišemir'	6. Dragoš'
7. Didodrag'	7. Povr'žen'
8. Boučin'	8. Radoslav'
9. Krač'	9. Radoslav'
10. Bratič'	10. Miroslav'
11. Boudislav'	11. Boleslav'

12. Dragoš' 12. Ratko
13. Dragič'
14. Loubin'
15. Dražeta
16. Tomiš'

Se pišou predrečenie
redove ki sou se
narekli u red'
crkve prie g(ospodi)na
našego Rastou-
die. Se pišou v'
ime o(t')ca i s(i)na

Translation: In the name of the Father, the Son, and Holy Ghost — followed by a list of sixteen names. I write these before mentioned "redove" (in the plural, meaning rows, series, lists, orders) who are named in the "red" (order) of the church before our lord Rastudije. I write this in the name of the Father, the Son (then the writer moves over to the top of column two), and Holy Ghost — followed by a list of twelve names.

The next page states: In the name of the Father and Son and Holy Ghost. This book was made for Tepčija Batalo by his dijak Stan'ko Kromirijenin, having worked in silver and gold, to present to Starac Radin, and the book was written in the days of King Dabiša from the birth of the Son of God in 1393, the second year after the death of King Tvrtko. This book was made for Tepčija Batalo, who was much renowned, who held Toričan and Lašva and with him was Gospoja Resa, Vojvoda Vukac's daughter. And one of his brother-in-laws was Vojvoda of Bosnia, and the second was Knez of Bosnia and the third was Ban of Croatia. And then Tepčija Batalo held Sana, and wine was brought to him from Kremen to Toričan. And he was very good to good people, and much praised by good krstjani and therefore God be bountiful toward him forever, amen.(15)

A full discussion can be found in my article on Rastudije.(16) Rastudije alone is called "our lord" and since the expression is not often found for deceased figures other than Jesus Christ, it is probable that Ratudije was alive in 1393 when the list was compiled. At the bottom of the page, but more or less under column one, it states that the before-mentioned "redove" (plural; series, rows, orders, lists) lived before

Rastudije. The use of the plural shows that both columns of names were lists of predecessors of Rastudije. Since it would make little sense to make two lists of ancient church figures without bringing the list up to the present (1393) and including the present leader, and since Rastudije is the latest figure on either list, Rastudije must have been alive, and as we shall see djed, in 1393. Since the two names of djeds that we already know from the fourteenth century (Radoslav from ca. 1325 and Miroslav from 1305) appear on the second list it is clear that the second list is the list of djeds of the church. Since Miroslav, the earlier djed, appears below Radoslav it is clear that the list should be read from the top down starting with Rastudije, the present djed in 1393, back through numbers eight and nine Radoslav (and we recall the charter of ca. 1325 which Djed Radoslav witnessed at the hiža of Gost Radoslav. Hence we can conclude that Gost Radoslav succeeded his namesake as djed, thus eight is the Gost Radoslav of 1325 and nine, the Djed Radoslav of that date) to number ten Miroslav. The seven names between Radoslav number nine and Rastudije are a reasonable number to cover the span of nearly seventy years, and the time between Miroslav, in office ca. 1305-07, and Radoslav who was in office in 1325 is also perfectly reasonable. Thus column two is a list of djeds from 1393 when Rastudije was djed which reaches back into the thirteenth century, with Ratko either the earliest known name, or the first djed, or the founder of a new order. Assuming our writer was accurate about the list of names, we are able to learn the names of all the fourteenth century djeds and see that the hierarchy existed through Tvrtko's reign even though we heard nothing about it then. It also suggests that the anathema articles were drawn up ca. 1390. They clearly could not have been written before Rastudije came into prominence and we can assume that the reason Rastudije received particular emphasis in these anathemas was owing to the fact that he was the leader of the Bosnian Church at the time the anathema articles were composed. Thus Solovjev's theory that Rastudije was the same individual as the goldsmith Aristodios (mentioned in the beginning of the thirteenth century by Thomas Archdeacon of Split) cannot be accepted.(17)

On the other list (sixteen names) three of the names are the same as three of the seven priors at Bilino polje in 1203 (numbers 13-15, Dragič, Loubin and Dražeta). And we must assume the list refers to them for the coincidence would be too great to try to find another explanation. This confirms our theory that the Bosnian Church was a continuation of the Catholic monastic order. The other names on the list are unknown unless Boudislav is meant to refer to the Bladosius, also mentioned at Bilino polje. This first list does not provide the names of Ratko's predecessors, for if it did we would expect Eremis or Tomiš (depending which way the other list would read) to appear as number thirteen below Ratko. We also would expect to see Rastudije's name appearing on the upper left instead of heading column two. In addition, if these three priors of the Bosnian monkish organization had become bishops (presumably of a church that was still Catholic since the local Catholic hierarchy existed until 1233) we would also expect to see among the sixteen names the names of the local Catholic bishops whom Resti, Orbini, and others mention. However, none of these names are there. Thus we can conclude that the sixteen names were not earlier bishops, but most likely were simply figures of importance in the previous history of the Bosnian Church who had not achieved bishop's rank, and about whom data or tradition existed in 1393.

In the entries to the Batalo Gospel manuscript, we again see invocations to the Trinity. Except for Ivan, all the names mentioned are popular names. This follows the tradition of the early thirteenth-century Bosnian Catholic bishops and is not surprising. Batalo was clearly a supporter of the Bosnian Church; he made a Gospel for one member and the inscription states that he was good to good people and praised by the krstjani. Presumably, he gave donations to these Bosnian Church monks.

We see that the church used at least the Gospel of John (elsewhere we shall see that they used the whole New Testament), that they had nothing against decorating the Gospels luxuriously with silver and gold, and that they dated by the birth of Christ. We cannot say whether there was a theological reason to date this way or whether they simply employed the dating system used by the state (which we would expect the dijak who copied the Gospel to do) which was borrowed from Dalmatia. We shall return to this text when we discuss Batalo's position in the state. Here, it is sufficient to point out that this source shows that the church was supported by one very powerful nobleman.

Our third source is an entry in a beautifully illustrated Gospel (manuscript now found in Bologna), known as the Gospel of Hval Krstjanin. This is a complete New Testament, which shows that the whole book was "canonical" for the Bosnian Church. In addition, because it is illustrated and has a complimentary picture of John the Baptist, it shows the Bosnian Church was not opposed to religious art and was not opposed to John the Baptist. It also has a picture of Christ on the Cross, showing Hval did not reject the cross. It contains the Ten Commandments, the Psalms, and some old Testament songs. Thus the Bosnian Church did not reject the Old Testament outright. This Gospel in the margin of the text has "začete" (beginning) and "kon'c" (end) at the places where in Orthodox Church services the readings begin and end during the course of a year.(18) These marginal notations show that the Gospel was most probably subsequently used by an Orthodox church, unless we want to argue that the Bosnian Church had an identical service with identical readings throughout the year to the Orthodox Church. In either case, it shows that there were no significant differences between the Bosnian Church and Orthodox Church New Testaments.

Only the entry written on a leaf of this Gospel manuscript gives us any reason to associate it with the Bosnian Church: "Uram (a Hungarian title meaning 'my lord')(19) Hrvoje, Great renowned Vojvoda of Bosnia. In the name of the Father and Son and Holy Ghost. This book was written by the hand of Hval Kr'stienin in honor for the glorious lord Hrvoje, Herceg of Split and Knez of the Donji kraji and many other lands." He asks the reader's forgiveness for mistakes he may have made in copying. "The writing was completed in the year of Christ's birth 1404 in the days of the Bishop and instructor and 's'vr'šytela' (accomplished one or bringer to perfection) of the Bosnian Church, the lord Djed Radomer."(20)

This highly important entry not only links this orthodox-appearing Gospel manuscript with the Bosnian Church, but it also shows that the powerful nobleman Hrvoje Vukčić was a patron of that church. Later we shall find him called a Patarin, giving us evidence that "Patarin" meant a member of the Bosnian Church. The entry shows that members of the church dated by Christ's birth. It also proves that Djed does mean Bishop of the Bosnian Church and gives us the name of the man who was Djed in 1404, Radomer. We may speculate that this Radomer is the same as Radomir, referred to right after Rastudije in the Troicki manuscript anathema.

III: Political Events 1391-1421 and Role of Bosnian Church

Now we can turn to the chaotic political events. Upon Tvrtko's death, his relative Dabiša was accepted as king. The circumstances surrounding this choice are unknown. Klaić, noting Bosnia's failure to establish a law of succession by the eldest son, argues that in theory the old Slavic *starešina* system, (i.e., succession for the eldest male in the family, often a brother or cousin of the deceased) was retained.(21) However, in fact, such a system left matters vague and the succession in dispute. This permitted the great nobles, through their support of the weakest candidate or of the candidate promising them the most, to acquire in these years more autonomy and greater power for themselves.

Thus the 1390's were marked by succession crises, weak rulers, and ever increasing power to the great nobles.

We do not know to what faith Dabiša belonged; we know only that a charter he issued 6 March 1392 refers to "our honorable chaplain pop Milac'."(22) This suggests that Dabiša kept this cleric regularly at court, and most probably Dabiša and Milac' were of the same faith. Though "pop" usually refers to an Orthodox priest, I believe that in this case the term denotes a Catholic priest. At this period we shall also meet a Catholic priest called pop Butko who copied a Gospel for Hrvoje Vukčić.(23)

Dabiša presumably had as his direct holdings the central regions around Visoko, Zenica and Sutjeska. Great power also belonged to a handful of noble families.

Hrvoje Vukčić was lord of the Donji kraji. His lands extended from this extensive region toward Dalmatia. As deputy for Ladislas of Naples, Sigismund's rival for the throne of Hungary, Hrvoje was able in Ladislas' name to extend his own power into Dalmatia, and to become lord of many of the Dalmatian towns. In addition, his holdings along the Vrbas River would extend down into Bosnia at least as far as Jajce.(24)

Tepčija Batalo, as we learn from the Gospel manuscript entry, held Lašva and Toričan (mid-way between Lašva and Travnik). Since his mausoleum with a brief inscription was found at Turbe on the other side of Travnik, we can assume that his lands extended at least that far.(25) Thus Batalo controlled this great stretch of relatively fertile land along the Lašva River. As the Gospel entry tells us, he married Resa, the daughter of Vojvoda Vukac, and hence the sister of Hrvoje Vukčić; thus Batalo was allied with the most powerful nobleman of all.(26) He probably acquired the *župa* of Sana(mentioned in the inscription) from Hrvoje's family as Resa's dowry.

Pavle Klešić's lands lay to the west of central Bosnia around Duvno and Glamoč. We may suspect that, to maintain himself, he would have had to become a protogé of Hrvoje. To his east, of course, lay central Bosnia the holding of the king. And then to the king's east lay the lands of Raden Jablanić and his son Pavle Radenović (from whom would be descended the Pavlovići).(27) They were the most powerful nobles in eastern Bosnia: they held the region between the Bosna and Drina Rivers with their center in the *župa* of Borač with the fortresses of Borač and Prača. Their lands extended as far as Olovo, Srebrnica, Dobrun and Ustikolina. Possibly they already at this time held Vrhbosna.

South of them in what is now Hercegovina (in the region of the upper Drina, the Tara and Piva Rivers and in Hum) were the holdings of the Kosača family, then led by Vojvoda Vlatko Vuković; his nephew

Sandalj Hranić succeeded him in 1392. In 1396 Sandalj married Hrvoje's niece Jelena. After Hrvoje's death in 1416, the Kosače became the most powerful family in the land(28)

In the years that follow sources show all these families having ties with the Bosnian Church.

In the 1390's, the chief political issue was the widespread civil war in Hungary between Ladislas of Naples and Sigismund of Luxemburg, fought all over Croatia and Dalmatia as well as in Bosnia.(29) After Tvrtko's death, Hrvoje became Ladislas' deputy in that region, until 1393, when Sigismund forced Hrvoje to submit to him. The following year Sigismund got King Dabiša to recognize him not only as overload, but also as heir to the Bosnian throne. Whether Sigismund had threatened Dabiša with invasion, or whether Dabiša, insecure among his own nobles, sought an outside prop to his throne is unknown. Certainly, Dabiša's decision to leave his throne to Sigismund would not have been popular with the Bosnian nobility, who disliked all forms of foreign interference.

In September 1395, King Dabiša died; Sigismund, probably hoping to claim his inheritance, appeared in near-by Srem with troops. The Bosnian nobles, unwilling to let him obtain the Bosnian throne, immediately rallied, convoked a *sabor,* and elected Dabiša's widow Jelena as queen. Sigismund, presumably lacking an army strong enough to face the united forces of the Bosnian nobility, returned to Buda. But now with Sigismund posing a threat to Bosnia, many of the Bosnian nobility, including Hrvoje, returned to Ladislas' camp. In 1396 Sigismund was crushingly defeated by the Turks at Nicopolis. Thus for several years his military position was weakened and he was unable to meddle in Bosnian affairs.

Queen Jelena, meanwhile, chosen to thwart Sigismund and, lacking any real power base, was simply a puppet for the strong nobles, with whom she consulted before she issued charters. She ruled until 1398 when she was ousted under uncertain circumstances and was replaced by Ostoja, a member of the ruling family, not mentioned in the sources prior to his accession. All the powerful nobles remained around Ostoja, so we may assume that none of them had strongly opposed his election; though why some or all of them had wanted it, we cannot say. Once in power, Ostoja declared himself for Ladislas.

In these years we still do not find that Bosnian Church with any sort of political role. The nobles, whom we shall later find connected with it, were enjoying good relations with their Catholic neighbors. Hrvoje Vukčić's Bosnian Church ties did not seem to trouble Ladislas of Naples; nor did they trouble Sigismund whenever he and Hrvoje were at peace. In 1393 Venice granted Venetian citizenship to Hrvoje and his brother Ban Vuk, both of whom swore on a Gospel.(30) In 1399 Dubrovnik granted citizenship to Hrvoje. In 1396 Venice granted Venetian citizenship to Sandalj, whose ambassador on the occasion was Theodore Archdeacon of Scutari.(31) These two Catholic coastal cities thus overlooked the ties between these men and the Bosnian Church, and the Catholic cleric was willing to act as ambassador for Sandalj, who is elsewhere found giving a village in his part of Konavli to a local (probably Orthodox) priest called pop Ratko. Thus Sandalj had cordial relations with both Catholics and Orthodox.

In 1400 both Hrvoje and Ostoja complained to Dubrovnik about the sale of "human flesh" (that is, slaves) in Dubrovnik. Since in theory, slaves sold in Dubrovnik could not be Catholics, it is probable that many of the Bosnians captured as slaves and sold in Dubrovnik were adherents of the Bosnian Church (or at least passed off as such). Frequently, the records of a slave-sale transaction specifically stated the slave sold was a "Patarin."(32) This usage of the term Patarin in regard to slaves is one of the few cases where Dubrovnik used the term Patarin for followers of the Bosnian Church who were not ordained clerics.

In 1401 Ostoja sent envoys to Dubrovnik to announce the birth of a son and to invite Dubrovnik for the baptism. Dubrovnik sent an embassy with expensive gifts.(33) Resti gives no further details. Unfortunately we do not know his source. No record of this embassy can now be found in the Dubrovnik archive.(34) As we shall see later, the sum total of material about Ostoja's relations with the different confessions leads one to believe he adhered to the Bosnian Church. Thus this brief notice — provided Resti had a reliable source — suggests that this Church had a rite of baptism which Dubrovnik, by sending representatives, found in no way heretical or offensive. Moreover, if it had been a peculiar ceremony, very different from the Catholic, we would have expected the Ragusan records to use some term other than baptism to describe it. However, it is not impossible that Ostoja, despite his cordial relations with the Bosnian Church, gave his son a Catholic baptism.

In 1401 Hrvoje and Ostoja moved to the offensive on behalf of Ladislas against Sigismund, and began to march into Dalmatia to demand that towns there accept Ladislas' overlordship instead of Sigismund's. In 1401 Zadar accepted Hrvoje and Hrvoje signed a treaty to respect Zadar's rights. Hrvoje swore by the cross and the relics of a saint.(35)

Thus this man who would be called a "Patarin" had no objection to swearing by these two items, which both Catholic sources and the

Serbian sinodik would have us believe his church did not accept. Split refused to accept Hrvoje and declared its loyalty to Sigismund, and attacked Hrvoje's town of Omiš. Dubrovnik, though also loyal to Sigismund, had had good relations with Hrvoje and to maintain these sent grain shipments to the besieged in Omiš. Hrvoje's brother-in-law, Ivaniš Nelipčić, (Hrvoje had married his sister Jelena) then moved in on behalf of Hrvoje and took Klis near Split. We might point out that Ivaniš Nelipčić, a powerful Croatian nobleman, was clearly Catholic (as was his sister,Hrvoje's wife)(36) but had given his sister to Hrvoje and aided Hrvoje's Dalmatian adventures. Split gave up its resistance. By the summer of 1403 Hrvoje was in power in Split and clearly had good relations with the Archbishop of Split.

A secretary of Ladislas announced to the city council of Florence in July 1402, "The Vojvoda of Bosnia Chervoya (Hrvoje), Great Prince etc. rules Dalmatia and all parts of Slavonia and is loyal to Lord Ladislas. The man is a Patarin but the order is given that the Lord Cardinal (Angelus) confirm him with chrism (the consecrated oil) and bring him back to the light of the true faith. And he seems to be agreeing to it, in the hope that the lord king may make him marquis in the regions of Bosnia as his friend and neighbor and my father and friend the Lord Archbishop of Split tells me."(37)

Since an order was given to have Hrvoje confirmed with oil, it is quite possible that he had, at some previous time, been baptized. This possibility is not at all unlikely when we recall that Hrvoje came from the Donji kraji, the territory between Catholic Croatia, Catholic Dalmatia and Bosnia. In addition, his mother may well have been Catholic — we do not know who his mother was but the Hrvatinići frequently married into Catholic Croatian families — in which case we would expect her to have baptized her son. However, we cannot be sure Hrvoje was baptized from this; such fine distinctions often are not relevant to Bosnia. In Hercegovina in the first centuries of Turkish rule we find both Catholics and Orthodox re-baptizing converts from the other confession, which of course is against canons. It is clear that Hrvoje did not make the plunge to Catholicism since in 1413, as we shall see, he wrote the Queen of Hungary expressing his desire to become a Catholic.(38)

In a letter of 1402, King Sigismund also referred to Hrvoje as a Patarin, when he spoke of his fortress of Knin being endangered by the Patarin Hrvoje and his cohorts.(39) The usage of "Patarin" for a non-ordained individual was unusual. We shall, however, from time to time come across the term used for prominent nobles connected with the Bosnian Church. We have also noted it being used in connection with slaves sold in Dubrovnik.

Hrvoje diplomatically did maintain good relations with both religions. He was a friend of the Catholic Archbishop of Split and was presented with a proper Catholic Gosepl manuscript, which has survived and which has an inscription showing it to have been made for Hrvoje according to Roman law by a pop Butko.(40) At the same time in 1404, he was also presented with the Gospel manuscript from the Bosnian Churchman Hval Krstjanin. As Thalloczy points out, for political reasons, Hrvoje needed to be on good terms with both faiths. He had to have a pro-Catholic policy in Split, Dalmatia, and parts of his Donji kraji, and a pro-Patarin policy in at least some parts of Bosnia. Hrvoje was wealthy enough to be able to support both faiths and presumably he awarded gifts to each.(41) That he was successful in this is shown by the two Gospels made for him. His wife was Catholic and in 1413 she was to give a considerable sum of money for the sarcophagus of Saint Domnius. However, that Hrvoje was able to carry out a policy of having good relations with the two churches shows that the churches lived on terms with one another.

The papacy, meanwhile, seems to have believed that the Franciscan mission was going well in Bosnia, for in 1402 Boniface IX issued a letter saying that no Franciscan could leave the vicariat without the permission of the vicar. The pope went on to state that the Franciscans had converted 500,000 infidels and those who had fallen away from Catholicism.(42) This figure is unbelievable, but in various places in the vicariat near their monasteries the Franciscans were no doubt having success in baptizing the local population. It is notable that the pope did not use the term heretic but called them infidels and men who had strayed away from the Catholic faith (inter infideles et a fide catholica deviantes). This description is quite in keeping with my view of the Bosnian Church and suggests that the Franciscans were working to bring members of that church into the Roman fold. That the pope did not use the term heretic may be owing to more accurate information about Bosnian conditions obtained from the Franciscans.

Things went from bad to worse for Sigismund in 1402-03; he lost most of Dalmatia and Slavonia to Hrvoje and Nelipčić; Bosnia and the powerful nobility in the other Slavic regions were supporting his rival Ladislas; in 1401 Pope Boniface IX, with whom Sigismund had been on good terms, declared for Ladislas; and finally early in 1403 Ladislas declared his intention of coming to Zadar to take over the lead of the campaign for the throne of Hungary. In July of 1403 Ladislas of Naples came in person to Zadar, where Hrvoje awaited him. With Sigismund's fortunes at a low point, many of Ladislas' supporters including Hrvoje urged Ladislas to make a real bid for the throne and march against Sigismund. However, Ladislas did not have the courage; he extracted oaths of allegiance from his supporters, again named Hrvoje his deputy, gave him the title of Herceg of Split, and leaving him to rule Dalmatia, returned in October to Italy. Ladislas' behavior disappointed his followers, and gave Sigismund a chance to rally support again. Sigismund now granted amnesties to many Hungarian nobles, who now probably concluded that Ladislas was never going to march on Hungary, and therefore they had best make peace with Sigismund, who after all did hold *de facto* power there. Thus Sigismund's fortunes began to improve. Ladislas' timidity, however, also gave Hrvoje a free hand in his own regions. In addition, as Ćirković points out, by making Hrvoje deputy for Dalmatia and Slavonia, Ladislas put Hrvoje above the Bosnian king, whose vassal Hrvoje supposedly was.(43)

Meanwhile, relations between Bosnia and Dubrovnik rapidly deteriorated. Dubrovnik, which had maintained correct relations with Ostoja and Hrvoje, while recognizing Sigismund as its overlord, had quarrelled with Bosnia over the issue of customs. Various Hum nobles had been arbitrarily raising customs duties and tolls, which of course primarily hurt the Ragusan merchants. In addition, the dispute over the sale of slaves had arisen, and there was the constant problem of bandits plundering the caravans which passed through Bosnia and Hum. In 1403, Pavle Radišić, apparently a relative of the royal family, fled from Bosnia to Dubrovnik. Probably Ostoja suspected him of being involved in a plot for the throne. Angry that Dubrovnik had granted him asylum, Ostoja wrote Dubrovnik and demanded that the town expel Radišić, give up its recognition of Sigismund as overlord, and accept Ostoja in his place. If it did not, then the town had fifteen days to get its merchants out of Bosnia, after which hostilities would begin. Caught completely by surprise with unprepared defenses, Dubrovnik sent its diplomats to Hrvoje, Pavle Radenović, and Sandalj. Presumably a diplomat was also sent to Ostoja. One question undoubtedly asked of Ostoja now, since it would be emphatically asked in negotiations the following year, was: how could Ostoja object to Dubrovnik's acceptance of Radišić, when Radišić's brother, also an enemy of Ostoja, lived safely in Bosnia eating the king's bread in a house of the Patarins.(44) Thus if the Patarins were allowed to grant asylum within Bosnia, how could Ostoja object so strenuously to Dubrovnik doing the same?

Thus once again we hear of a Patarin house (i.e., monastery), and we learn that it, like a Western church, could give asylum to the pursued. And at least under Ostoja, who, we shall see, seems to have been closer to the Bosnian Church than any other ruler before or after him, this claim and right of the Patarins was respected. We have no other examples of fugitives seeking shelter with the Patarins (though we have one case of a fugitive seeking shelter with the Franciscans, also under Ostoja); thus we cannot say that all rulers respected this right. But it is clear that Ostoja did.

Hrvoje seems to have tried to intervene on behalf of Dubrovnik, with which he was on cordial terms despite the fact that the town recognized Sigismund. But Hrvoje's mediation had no effect on Ostoja, and in July Ostoja's armies were occupying territory belonging to Dubrovnik, ravaging property and sending refugees flocking to the town.

In September 1403, Ostoja sent as envoy to Dubrovnik, perhaps at the suggestion of certain of his nobles, Vlatko Patarin,(45) the first Patarin to appear in the sources in a diplomatic role. Since almost every work that discusses the Bosnian Church places great emphasis on Patarins as diplomats, Section XII of this chapter shall be devoted to the significance of Patarins in diplomacy. Neither the king nor the great nobles had professional diplomats; thus when it was necessary to send an envoy, the king — or nobleman — sought out a capable man whom he trusted. Once a man proved himself as a diplomat, he might frequently be used in this function, but he never received any official title.(46) The kings were to use Patarins as envoys only twice, Vlatko in 1403 and two unnamed Patarins in 1430. The nobles would begin to use Patarins only in 1419.

Dubrovnik never seems to have minded negotiating with Patarins. On every occasion the town council voted to give them gifts. In the case of Vlatko in September 1403 it gave him oil and cloth.(47) It has been argued that the gift of oil shows that Vlatko and the Patarins fasted and did not eat food cooked with lard. Though it will be apparent later that ordained Patarins did keep fasts, we do not know whether they constantly kept fasts or whether they fasted only at specific times a year. That the Ragusan gift of oil must be connected to their custom of fasting cannot be proved. Oil is a useful gift to anyone, regardless of whether he fasts or not.

In June 1403, the powerful noble Pavle Klešić, who held lands around Glamoč and Duvno, appeared in Dubrovnik seeking asylum; he was accompanied by some Patarins of "the Bosnian sect."

The Ragusan council voted to give Klešić a home, and approved gifts and food for the Patarins who had accompanied him.(48) On the same day the council also voted unanimously to send gifts to the Djed of the Bosnian Patarins.(49) The only apparent reason for sending gifts to the djed was good will; he was a religious leader with influence over Ostoja, who was about to attack Dubrovnik. In November 1403 more Patarins came to Dubrovnik, accompanied by Klešić's wife.(50) The town council voted to let them buy bread and other necessary items to the value of twenty-five ducats.(51)

In January 1404 Ostoja wrote Dubrovnik that he was returning to Klešić Glamoč and his property in Duvno and Bosnia that had been confiscated. This property would be placed under the protection of the djed, his stroiniks, and his church. The church would also guarantee the safety of Klešić's own person. The stroiniks were those Bosnian Church clerics who composed its governing council. Interestingly, Ostoja's letter was dated the Sunday of the Hallowing of the Waters.(52) Had the Bosnian Church been dualist such a holiday would almost certainly not have been recognized because dualists rejected all material objects used in religious ceremonies; they specifically rejected water.

The djed drafted a letter two days later. He began his letter with the sign of the cross, showing that the Bosnian Church did not reject that symbol. He wrote Dubrovnik that he was sending thither his stroiniks and krstjani to bring Klešić home; the djed had intervened with the king to get Klešić's property returned because the king had wrongly taken it. We do not know whether Klešić was innocent of whatever Ostoja accused him, or whether Ostoja simply did not have the right to confiscate the land. The djed thanked Dubrovnik for receiving and aiding Klešić. He called himself "Lord Bishop of the Bosnian Church" and signed the letter "Lord Bishop." Nowhere does he call himself djed, although Ostoja in his letter had so referred to the bishop. The letter was dated, Janjići 8 January 1404 of the year of Christ's birth.(53) This shows that the djed (probably following regular Bosnian usage) dated by the birth of Christ. Since he sent the letter from Janjići (a small place midway between Lašva and Zenica) he presumably either lived there or else was visiting a Patarin hiža there.

A Ragusan document lists the stroinici and krstjani sent by the djed and Ostoja: Starac Mišljen, Starac Belko and five krstjani (Stojan, Ratko, Radosav, Radak and Dobrašin).(54) Whether they were really sent jointly by the two leaders, as the document states, or whether they all came from the djed, but with Ostoja's agreement, as seems more likely, is not known. Nothing is known about any of the seven Bosnian clerics unless Starac Mišljen should be the same as the Gost Misljen buried at Puhovac near Zenica.(55)

The letter and the presence of Patarins in his and his wife's entourage shows that Klešić was a member of the Bosnian Church. It also shows that the Bosnian Church's guarantee of charters about property, at least under certain kings, should not be considered a dead letter. We do not know if the Church had previously been a guarantor of Klešić's property since no earlier charter about his lands is extant.

We can assume that Klešić had previously quarrelled with the king — possibly he had been involved in the Radišić affair. Ostoja (apparently illegally) seized Klešić's lands and Klešić fled. The djed stated he had intervened with the king to restore the property. Thus, the djed seems to have had considerable influence on Ostoja. However, it is always possible that behind the djed stood Hrvoje — a supporter of the Patarins and almost certainly Klešić's overlord — in which case there would have been more than simply church pressure on Ostoja to pardon Klešić.

However, both the djed's letter on behalf of Klešić and the fact that Radišić's brother was able to live undisturbed in a Patarin house in Bosnia show that, at least under Ostoja, the Bosnian Church did have strong influence on the king and his actions. I suspect that Ostoja was a member of that church. This is suggested not only by the fact that he listened to and respected the Patarins' position on the two above mentioned matters, but also because he used the Patarin Vlatko as a diplomat. Thus we can argue that under Ostoja the Bosnian Church achieved a position of political influence that it had never had before and would never have again.

Meanwhile the relationships between Dubrovnik, Ostoja, Hrvoje and Hungary were changing. Nothing seems to have been settled by Vlatko's visit. Dubrovnik turned again to Hrvoje and suggested that he attempt to overthrow Ostoja and put Pavle Radišić on the throne. Although the Ragusan embassy that proposed this was en route to Hrvoje's enemy Sigismund, it was warmly received by Hrvoje, who expressed his wish for peace but rejected the plan. The embassy was allowed to continue on to Sigismund. Even though Hrvoje rejected the plan to overthrow Ostoja, a quarrel had been building up between him and the king. We can attribute its causes to: Hrvoje's dislike of Ostoja's war against Dubrovnik, Hrvoje's almost certain anger at Ostoja's actions against Klešić, and finally Hrvoje's own ambitions and frustrations at the presence of a king he could not dominate.

Ostoja was certainly disturbed that his relations with this powerful nobleman were deteriorating and very likely he was bothered by the Ragusan embassies to Hrvoje. Probably the embassies between Dubrovnik and Sigismund also worried him. Sigismund was making a recovery in Hungary and soon might be able to meddle in Bosnia again. Various nobles who had deserted Sigismund were now returning to his side; Ostoja may even have suspected that Hrvoje had this in mind and that Dubrovnik might be paving the way for this. And if Hrvoje and Sigismund should make peace, together they could easily have removed Ostoja. But for whatever reasons, late in 1403 Ostoja began to make advances to Sigismund, and recognized him as his overlord. Thus Ostoja, who had gone to war with Dubrovnik over that town's recognition of Sigismund as overlord, had six months later, still warring with Dubrovnik, come himself to recognize Sigismund.(56) Ostoja's acceptance of Sigismund, of course, caused distrust at home, and this dissatisfaction may well have been a factor in causing Ostoja to listen to the djed and to allow Klešić to return home. However, if this was so, he acted tardily. Already in January 1404 when Ostoja was permitting Klešić's return, Dubrovnik and Hrvoje were negotiating to overthrow Ostoja and put Radišić on the Bosnian throne. Then in March the picture changed: Hrvoje made peace with Ostoja, Ostoja made peace with Dubrovnik, and it seemed that even Hrvoje and Sigismund were on the road to peace.

But this was not to last. In April a *sabor* of nobles convened, ousted Ostoja, and elected Tvrtko II Tvrtković as king.(57) (If, as his patronymic suggests and as is generally believed, he was the son of Tvrtko I, then he probably was illegitimate.) Ostoja, whose relations with Sigismund at that moment were good, fled to the Hungarian court. Sigismund sent an army and within a matter of months the Hungarians had recovered Bobovac and installed Ostoja there as an anti-king, while Tvrtko II remained the king recognized by the Bosnians.

In March Dubrovnik had sent an embassy to the king to confirm a treaty of peace. In the instructions to its ambassador Dubrovnik stated that the treaty should be sworn to by the queen and barons and that the Patarins should promise according to custom to uphold it.(58) In April, with the king overthrown, Dubrovnik had to have its ambassadors reconfirm the peace treaty; the ambassador was told to have the treaty confirmed in the presence of the Patarins.(59) Evidently, Dubrovnik believed that the Patarins' church had real influence in the state and that their presence would help to enforce the treaty. Dubrovnik was practical and would not have asked for religious sanction (particularly by a foreign religion) unless it really believed that this religious sanction would have had some actual meaning. In a letter to the ambassador in May, Dubrovnik again stated that the treaty should be sworn to by oath by the barons and the *sabor,* but with the promise of the Patarins, according to their custom.(60) Finally, in September, the same statement was made about the barons, with Sandalj being singled out showing the great influence he must have gained with the new king; as for the Patarins, the ambassador was asked to get them to vertify it in whatever way they want.(61)

The repeated emphasis on the Patarins in these Ragusan letters to its ambassadors does reflect the political influence they held in the Bosnian state in 1404. In addition, on each occasion when oaths were called for, Dubrovnik wanted oaths extracted from the nobility but was satisfied with promises from the Patarins. At times the envoys were told to let the Patarins verify the treaty "in whatever way they want," or "according to their custom" which suggests that they had some custom (other than an oath) of giving their word to a document. Thus, it is evident, as Professor Dinić has convincingly demonstrated, that Patarins did not swear oaths.(62) Unlike the ordained clerics, however, nobles who supported or belonged to the Bosnian Church did take oaths. Dinić believes this is strong evidence that the Patarins were dualists, since the dualists rejected oaths. I disagree, because so much other evidence suggests that the Patarins were not dualist. And refusal to swear is not a practice limited only to dualists. But, when we seek specific points of practice and belief of the Bosnian Church the refusal of the churchmen to take oaths is clearly one of them.

When he was ousted Ostoja was accused of violating the rights of the nobles, and we may suspect that this accusation referred to his seizure of Klešić's lands (and possibly the lands of some others about whom we have no information). This clearly was an excuse to get rid of a headstrong king who wanted to rule rather than be ruled. We suspect that Ostoja's newly established relations with Sigismund may have also been a motivating factor. Ostoja's removal and flight to Sigismund had the effect of drawing Sigismund into a war against the new king and the Bosnian nobility. The appearance of Sigismund's troops in Bosnia pushed Hrvoje, and with him Tvrtko II, closer to Ladislas. Still loyal to Sigismund, Dubrovnik was helpless to prevent this turn of events. In the years that followed as Sigismund sent troops into Bosnia, nominally on behalf of Ostoja, we can view the struggle as a phase in the war between Ladislas and Sigismund. Ladislas too clearly saw things this way for in 1405 and 1406 he issued a series of charters to Bosnian noblemen awarding them lands that were at the time in the hands of Sigismund's supporters. In 1405 Sigismund spoke of "schismatic Bosnian rebels"(63) and of Hrvoje and "the Bosnian schismatics,"(64) and "perfidious Bosnian Patarins," etc. He does not use the term "heretic," which he had used in the 1390's about Bosnia. Possibly in the 1390's there had been real heretics opposing him or possibly in the 1390's Sigismund, having no Bosnian allies to offend, had had no reason not to exaggerate in damning the Bosnians he hated. However, in 1404-05 he was fighting on behalf of Ostoja, a supporter of the Bosnian Church, who recognized Sigismund as his overlord. So now Sigismund calls the Bosnian Church merely schismatic.

Sandalj too had close relations with the Bosnian Church. In 1405

Dubrovnik quarrelled with Sandalj about the ownership of some villages that lay near their common border. Dubrovnik proposed that the djed should mediate the problem. It called the djed, "Lord and spiritual father of your (Sandalj's) church of Bosnia."(65) Thus Dubrovnik, which had constant contact with Sandalj and hence was in a position to know, spoke of the Bosnian Church as Sandalj's church, thus believing him to be a member, and believed that the djed would have influence upon him. Sandalj is the first noble from what is now Hercegovina, encountered in the sources, who was connected with the Bosnian Church. That Catholic Dubrovnik would suggest that the leader of the Bosnian Church mediate a quarrel between itself and a member of that church shows that Dubrovnik believed it could expect a fair judgment from the djed. This suggests that the djed's church was not particularly hostile to Catholicism, and that Catholic Dubrovnik was not averse to the djed's religion.

Cordial relations between Catholicism and the Bosnian Church are also shown by the fact, noted previously, that Catholic Venice bestowed its citizenship upon Sandalj, and on that occasion a Catholic priest had served as Sandalj's envoy. Venice also had no hesitation in referring to Sandalj as a Christian.(66) In addition, Sandalj's wife — Hrvoje's niece — Jelena was a Catholic; she gave an expensive reliquary to the Catholic church of St. Mary in Zadar.(67)

In May 1405 Dubrovnik sent a treaty to be signed by the King of Bosnia. Again, Dubrovnik showed its faith in the djed, by suggesting that the document be left with him for safekeeping until the king signed it.(68) In these years, then, the djed was a powerful figure who seems to have earned the respect not only of his own countrymen but also of his Catholic neighbor, Dubrovnik. This influence lasted beyond Ostoja's overthrow and continued under Tvrtko II, whose religious allegiance at this time we cannot identify. Since the Bosnian Church retained its influence under him, Tvrtko II certainly was not hostile to that church.

In evaluating the political role of the Bosnian Church at this time, we must keep in mind the possibility that some of its influence may have been the result of the strength of the personality of the particular djed who held office. He was clearly a strong man since he did not fear to oppose the king in political matters and managed to force the king to change his mind. Unfortunately we have no evidence of how long this particular djed was in office. However, since the church's most active role occurred in the years 1403-1405, we may assume that the office did not change hands during these years. If this is true, then the djed in this period would have been the Radomer referred to in the 1404 Hval Gospel inscription.(69) Unfortunately, we know nothing about his family and its connections.

Intermittent warfare between Sigismund and Bosnia continued from 1404 to 1408. It is in the context of this war that we must treat a request of 6 June 1407 by the people of Eger to the pope to appoint them a new bishop; their bishop, Thomas, having been insulting to Sigismund and involved in destructive activities against even the church of Eger, four years ago had deserted the city and church of Eger to go to the region of the unfaithful Bosnian schismatics.(70) Evidently Thomas had sided with Ladislas of Naples against Sigismund in their civil war (ca. 1403) and then, for political reasons, had sought refuge with Ladislas' allies somewhere in Bosnia, quite likely with Hrvoje. That a Catholic bishop sought refuge in Bosnia, rather than some place else like Naples, shows that he did not believe Bosnia was a land populated chiefly by outrageous heretics; the document itself calls the Bosnians schismatics.

In September 1408, Sigismund won what he proclaimed was a great victory over Tvrtko II and the Bosnians at a place called Dobor in Usora. Many tales were told about the aftermaths of the battle; one has Sigismund's men pitching one hundred and seventy Bosnian nobles over the town's battlements to their deaths; another claims that Tvrtko II was captured. But because Tvrtko was asking Dubrovnik for his tribute in February 1409, we can assume that he had not been captured.

The papacy meanwhile was becoming interested in crusades at this time and in 1407 Gregory XII issued an all inclusive call to fight the Turks, unfaithful Arians (!), Manichees and other unbelievers in the kingdom and lands of Constantinople, Romania (i.e., Byzantium), Hungary, Bosnia, Dalmatia, Croatia, Rama, Serbia, Galicia, among the Cumans, and Bulgaria.(71) Since a great deal more territory than just Bosnia was included in this call to arms we do not know where the pope thought the Manichees were to be found. That he mentions Arians, however, shows that we should not take his labels seriously.

In January 1409 Sigismund announced in Trogir that Hrvoje had submitted to him. Hrvoje's action presumably was motivated both by Sigismund's military successes and by the fact that Ladislas was planning to sell Dalmatia to Venice, which would have removed the legal basis for Hrvoje's activities there. Now Sigismund could hope to gain Dalmatia through Hrvoje, and Hrvoje could have the Hungarian king's blessing to continue his activities in Dalmatia. The actual conclusion of peace may well have occurred early in December 1408 since Hrvoje may have been in Buda for the festivities connected with the foundation of the Dragon Order. This order was founded by Sigismund to defend the Christian faith from pagans and heretics, and the Hungarian royal house from domestic and foreign enemies. There were to be two orders of knights, the first was limited to Hungarians, but the second included such illustrious foreigners as Despot Stefan Lazarević of Serbia, and a variety of Slavonian and Croatian bans and nobles.(72) Lucius says that Hrvoje and Sandalj both came to Buda (presumably in December 1408) and that Sigismund made Hrvoje a member of the order and also godfather for his recently born daughter Elizabeth.(73) Thereafter Hrvoje remained loyal to Sigismund, until Sigismund turned against him in 1412.

It is interesting to note, if we can believe Lucius, that Sigismund allowed the "Patarin" Hrvoje to be a member of the knightly order to defend Catholicism from heretics and pagans. Hrvoje, as we shall see from a letter he wrote in 1413, had clearly not become a Catholic as Ladislas' secretary and the Archbishop of Split had hoped in 1403. His standing godfather for the Catholic princess is also noteworthy.

After the reference to Tvrtko II in February 1409 he disappears from the sources for several years. By the end of 1409 Ostoja is master of Bosnia once again. We do not know how this change was brought about.(74) In June 1409 Sandalj sent Archdeacon Theodore (n.b. a Catholic cleric) to Zadar;(75) thus it seems likely that Sandalj was aligning himself with Ladislas. Ostoja was clearly opposed to Hrvoje since in December 1408 he issued a charter to Djuraj Radivojević awarding him large stretches of land, possessed at the time by Hrvoje.(76) With Ostoja opposed to Hrvoje, we find Sigismund withdrawing support from Ostoja; and in 1410 Sigismund and Hrvoje marched against Ostoja and seized Visoko. Ostoja fled; Sigismund then forced the two leading Bosnian nobles Sandalj and Pavle Radenović to recognize him. That winter Ostoja had to go to Djakovo to ask Sigismund's forgiveness. Sigismund recognized him as king, but Ostoja had to give up Sol and Usora which were to be assigned to a deputy to be chosen by Sigismund.

While the struggles between Ladislas and Sigismund, and between Ostoja and Tvrtko II, were going on both Sandalj and Pavle Radenović had been steadily increasing their holdings. By 1410 each was in a position to rival the king.(77) A struggle between these two nobles was foreseeable. However, Sandalj first had his sights fixed on eliminating Hrvoje's (and also Sigismund's) influence in the state. So, in 1411 Sandalj divorced Hrvoje's niece and shortly thereafter married Jelena the widow of Djuraj Stracimirović Balšić of Zeta, who happened also to be the daughter of Knez Lazar who perished at Kosovo, and the sister of the Serbian Despot Stefan Lazarević. His new marriage alliance reflected both the deteriorating relations with Hrvoje and the better relations he had been building up with the Serbs.(78)

Hrvoje meanwhile built a castle at Jajce which he occupied in 1411-12. On the hillside, on whose summit his fortress was perched, Hrvoje discovered early Christian catacombs. He placed his crest inside these as well as on the castle entrance. In the walls of the catacombs he also hollowed out niches of a size and shape to hold coffins. Thus it seems that Hrvoje intended to convert the catacombs into a mausoleum for himself and his family.(79) This plan was never realized, since at the time of his death he resided elsewhere. Various theories have been advanced about mysterious rituals (heretical or connected with the Dragon Order initiations) which might have occurred in the catacombs. However, there is no evidence for any such thing.

In 1412 peace seemed to reign in Bosnia. In the spring King Sigismund held more festivities. Ostoja, Hrvoje, Sandalj and Pavle Radenović all attended as invited guests. A Polish chronicler praised the prowess of the Bosnians in the knightly games.(80) This may well be a clue to the various tournament scenes on the *stećci.* We not only see that the Bosnians did hold tournaments and practice the knightly arts, but that they excelled at them.

This is the last time we shall see Hrvoje in the king's favor. Shortly thereafter Sigismund called him a renegade and a rebel, accused him of dealing with the Turks, and withdrew his titles and the power of authority in his realm.(81) Nowhere in Sigismund's tirade against Hrvoje is there reference to heresy or Patarinism. To our surprise Sigismund's action had effect, and one by one the coastal towns ousted Hrvoje's officials. Thus it seems that Hrvoje had been unpopular in Dalmatia.

Hrvoje pleaded that he was slandered and wrote Queen Barbara of Hungary asking her to intercede for him with her husband. He reminded her "for the love of St. John" that he had stood as godfather for her daughter and that he was a member of the Dragon Order. As a member he was entitled to be judged by his peers. He had not been judged by them. He was willing to stand trial before the order and if they found him guilty then let them have his head. He closed his plea by stating that she should intervene for him and not let him in his old age die in pagan rite (in paganismo (!) ritu) and in unbelief for he had with difficulty awaited the time to go over from the pagan rite to the Roman Catholic faith.(82) This plea had no effect.

Once again Hrvoje was playing with religion to serve his own ends. We recall that in 1403 when he had hoped for the title of Marquis from Ladislas he had expressed the desire to become a Catholic. From this letter we see that he had not accepted Catholicism then. The use of the term "pagan" is strange. Professor R. L. Wolff has suggested that Lucius, when he copied the original letter in the seventeenth century, misread the text and wrote "paganismo" for "patarino."(83) This theory is plausible for the following reasons: 1) elsewhere we have found Hrvoje referred to as a Patarin; 2) the two words when handwritten are very similar; 3) Whereas there were domestic peasant ceremonies which we might call pagan now, it is very doubtful that Hrvoje or anyone else then would have referred to the *slava* and other rites as pagan. Thus, we have no evidence of any rites practiced then in Bosnia or environs which a practioner would be likely to call pagan. 4) However, the Patarins clearly did have a rite. 5) Finally, since the word "rite" is used, some specific practices are referred to, thus it cannot simply mean that Hrvoje was unbaptized or like a heathen ignored the church. Against this view there are only two points: 1) Since the term "paganismo" appears twice in the text, Lucius would have had to make the same error twice; 2) The term "pagan" was used once earlier in this same milieu. When Sigismund formed his Dragon Order, it was expressly established to defend Christianity against pagans and heretics. Thus it is always possible that in Hungarian court circles in the beginning of the fifteenth century the term "pagan" had some specific meaning; it is not excluded that it might have been a pejorative term for members of the Bosnian Church. And if we consider peasant religious attitudes in Bosnia, such a label may have been a fitting one. However, we must conclude that we do not know which of the two terms was used in Hrvoje's plea.

It is generally assumed that Sandalj had machinated Hrvoje's fall from Sigismund's graces, and that Hrvoje was innocent of the charge of negotiating with the Turks. However, once in disgrace, deserted by his former supporters, and in danger of being attacked by Sigismund, Hrvoje had no choice but to do what he was accused of and turn to the Turks. Thus Sigismund's accusation became a self-fulfilling prophecy greatly to his own detriment. Hrvoje procured Ottoman mercenaries and joined by Tvrtko II — who reappeared just as mysteriously as he had disappeared five years earlier — indulged in raiding Bosnia and Croatia. In 1415 Hrvoje's Turkish leader Isak beg — who governed the Ottoman province centered in Skopje (1414-1439) and who was to play an active role in leading and supplying mercenaries in Bosnia for most of this period — met Sigismund's forces in a full-scale battle near Doboj. The Turkish victory was complete, and many Hungarian prisoners were taken, including some high nobles. The Hungarian Chronicle of Janos Thuroczi describes Hrvoje's magnanimity as a victor as well as his sterling sense of humor. Among the Hungarian prisoners was a nobleman named Pal Chupor who had previously amused the Hungarian court by mocking Hrvoje's stocky stature and hoarse voice; he would imitate a bull whenever Hrvoje would enter the room. Hrvoje now had Pal sewn into an ox-hide and then said to him, "if you bellowed like a bull when in human shape, now you have the shape of an ox with which to bellow." He then had him thrown into a river in the ox-hide."(84)

The Turkish victory marked the end of Sigismund's influence in Bosnia. In triumph, Hrvoje forgot about Tvrtko II and had the Turks confirm Ostoja as king in the name of the Sultan. Hrvoje's victory, the result of the first large-scale Turkish force in Bosnia, marked the beginning of the Turkish role in Bosnia politics. In the years that followed, this role was limited to that of mercenaries who fought for the Bosnian noblemen who hired them. However, within twenty years Isak beg would begin to have plans for Bosnia that would benefit the Sultan. Hrvoje has been blamed for being the first to summon the Turks into Bosnia. However, had he not done so, Sandalj or the Pavlovići would have; and if no one had invited them, the Turks would eventually have come on their own. And can we really find Isak beg more foreign to Bosnia's interests than Sigismund?

In August 1415 Ostoja and Sandalj convoked a conference at Sutjeska, and invited Pavle Radenović, his son Peter and other nobles. Shortly after their arrival the hosts suggested a ride in the country; Radenović and his son accepted. Not far from Sutjeska, the king, Sandalj and some of their retainers turned upon the pair. Radenović was killed, allegedly trying to escape; Peter was taken prisoner. Sandalj later told the Ragusan ambassador that this action had been necessary since the pair were traitors.

Vlatko Tumarlić, the Patarin who in 1403 had been Ostoja's ambassador to Dubrovnik, had by this time entered Pavle Radenović's service and had accompanied his lord on the fateful trip to Sutjeska. Whether Vlatko had come as chaplain or councillor is not known. Immediately after the murder he had fled to Gundulić, the Ragusan ambassador to the court, who gave him shelter in his quarters at the Franciscan monastery in Sutjeska.(85) That Gundulić resided there shows that the Franciscan monasteries, like those of the Patarins, took in travelers. The Franciscans apparently did not object to the Patarin taking refuge with them. But we cannot be sure from this that Ostoja respected the Franciscan right to shelter fugitives since we do not know that Ostoja had any intention of arresting Vlatko. Ostoja seems to have had cordial relations with the Franciscans, for in 1418 he twice used a

Brother Stephen as a diplomat.(86) (This would be the first mention in the sources of a Franciscan used as a diplomat.) The day after the murder, Vlatko took Pavle Radenović's body and returned to Vrhbosna.

Pavle Radenović's capable son Radoslav Pavlović was still at large.(87) To avenge his father and to prevent the division of his lands among the murderers he immediately procured a Turkish contingent and, joined by his brother Peter — who either escaped or was released from prison — proceeded to ravage Sandalj's lands. In this whole affair, Hrvoje had played no part, and in 1416 he died. With his death the great influence of the Hrvatinići in Bosnian affairs ended. He was succeeded by his weak son, Balša; but real authority in the Donji kraji went to his nephew, Djuraj Vojsalić. Much of Hrvoje's lands ended up in Ostoja's hands, however, for as soon as Hrvoje died, the king divorced his wife and married Hrvoje's widow, who brought with her much territory including Jajce.(88) We may presume that Ostoja disposed of this wife in the Bosnian manner which the Franciscans had objected to in 1373. Since no source mentions Ostoja having any difficulties from Rome or the Franciscans about this divorce, it is most probable that Ostoja was still a member of the Bosnian Church.

Having procured these lands that had belonged to Hrvoje, Ostoja now in 1417 joined the Pavlovići and Turks against Sandalj. Their combined forces pushed Sandalj back toward the sea. Ostoja captured the key customs town of Drijeva, where he banned entirely the sale of slaves. As noted earlier, the majority of these slaves seem to have been sold as "Patarins." This act of Ostoja might possibly be seen as a defense of his co-religionists and further confirmation that he was still a member of that church. Then, suddenly in September 1418, Ostoja died and was succeeded by his son, Stefan Ostojić.

In 1420 Sandalj procured Turks of his own and attacked the Pavlovići and the new king. It was obvious that neither Sandalj nor Radoslav was going to gain from his continual warfare and plundering, so late that spring a conference was convened, attended by the Pavlovići, Sandalj and the leader of Sandalj's Turks, Isak beg. In the course of their discussions Isak beg murdered Peter Pavlović because — as Sandalj later explained it — Peter was faithless to the sultan.(89) As a result of the murder, and of course Sandalj's continued alliance with Isak beg, Sandalj was able to recover most of the territory he had lost. In July 1420 Stefan Ostojić was ousted from power, and Tvrtko II, always ready, became king again. He established himself in Visoko, surrounded by the same nobles who had surrounded his predecessors. Stefan Ostojić fled to the coast, and shortly thereafter he disappeared permanently from the sources.

The disappearance of Stefan Ostojić left only three figures of influence and authority in Bosnia: Sandalj, Radoslav, and Tvrtko II. In October 1420 Radoslav made peace with Sandalj and accepted Tvrtko II as king. In 1421 Radoslav married Sandalj's niece, Theodora (the sister of Stefan Vukčić, who was to be Sandalj's successor).

IV: Patarins in Secular Service and Relations Between Secular Leaders and the Various Faiths

Meanwhile we begin to find Patarins for the first time performing secular services for certain members of the nobility.

On January 29, 1419, two envoys from Sandalj, Dmitar Krstjanin and Pribisav Pohvalić, obtained cash for Sandalj in Dubrovnik by pawning some of his valuables.(90) Pribisav was regularly used by Sandalj as a diplomat at this time. Dmitar, who accompanied him, is the first Patarin envoy, noted in the sources, to be sent by a nobleman.(91) Radoslav Pavlović began to use Patarins as diplomats the following year, in November 1420.(92)

Radoslav also used a Patarin in an administrative capacity. On June 21, 1422, Dubrovnik wrote to two of Radoslav's agents, Župan Radosav Glavić and Radovac Krstjanin, to suggest that people moving from the territory of one overlord in Konavli to that of the other henceforth pay taxes to their new overlord.(93) Neither of Radoslav's agents is known from other sources, but it is evident that they had been sent by Radoslav to administer his part of Konavli. As Radovac was probably an administrator sent thither by Radoslav, his presence does not indicate the existence of popular support for the Bosnian Church in the region of Konavli.

In October 1422, Radoslav sent to Dubrovnik Brother Stephen, the same Franciscan whom Ostoja had used as an ambassador in 1418.(94) Thus we find Radoslav, despite his Patarin connections, willing to use a Catholic cleric as his envoy, and the Catholic not averse to serving him. We have noted earlier that Sandalj also on occasions used Catholic clerics as ambassadors.

Meanwhile, at some point during 1422, fighting had again broken out between Radoslav and Sandalj; on January 16, 1423, Dubrovnik told an envoy being sent to Sandalj that the town approved efforts to bring about peace between the two noblemen and that Radoslav, for this purpose, had sent a Patarin and others to Sandalj.(95) A Ragusan envoy to Sandalj's court returned on the 25th.(96) He evidently brought the news noted in a Ragusan letter of the 27th that "conte Volchac and Miasa Gost" had been sent by Radoslav to negotiate with the Vojvoda (i.e., Sandalj).(97) Presumably, Miaša Gost was the unnamed Patarin referred to on the 16th. Once again Radoslav was using as envoys a Patarin and a layman. Miaša Gost is otherwise unknown; he is the highest ranking Patarin encountered up to now as an ambassador. We may suggest that he headed a hiža near Radoslav's capital of Borač. Miaša is the third different Patarin Radoslav had used in the period 1420-23; we note that Radoslav also constantly used different secular envoys. Thus Radoslav apparently did not have a regular corps diplomatique.

On February 15, 1423, a charter drafted by Dubrovnik stated that to Dubrovnik had come from Radoslav "the honorable man Krstjanin lord (gospodin) Vlatko Tumr'ka (sic), Radin Krstjanin, Knez Budislav and Knez Vukašin" and from Sandalj "lord (gospodin) Starac Dmitar and Knez Radovan Vardić," and that these two embassies had concluded peace between Sandalj and Radoslav.(98) It is possible that the Bosnian Church as an institution had used its clerics to bring about this peace. It is interesting to note that in these and other letters written in Slavic, Dubrovnik calls the envoys by the Slavic word "krstjanin," using the term "Patarin" for Bosnian Churchmen only when writing in Italian or Latin. Dubrovnik's good will toward the Patarins is shown by the fact that it honored both Vlatko and Dmitar with the title lord, which, since it was not part of their title, Dubrovnik was under no obligation to use. All the Patarin envoys had been used before; Dmitar had, since 1419, been promoted to the rank of starac. Both embassies contained laymen and Patarins.

The next day, Starac Dmitar, having concluded his main business, is found still in Dubrovnik acting on behalf of Sandalj's Orthodox wife and her financial affairs. With him were two secular figures, Radohna Pripčić and Brajan Dijak (i.e., scribe). Dmitar was given a charter stating that his mistress could collect the money depositied in Dubrovnik at any time and in any way she chose.(99) The same trio of envoys on the same day also obtained cash for Sandalj by pawning an icon.(100) This will not be the only example we have of icons in the possession of families connected with the Bosnian Church.(101)

Thus, in 1419 Sandalj and in 1420 Radoslav began to make use of Patarins as diplomats. Between 1419 and 1423 these two nobles sent seven embassies (six to Dubrovnik) in which Patarins participated. Throughout this period entirely secular embassies were also sent. And strangely enough, after the frequent use of Patarins during these four years, we shall have to wait seven years until 1430 before we meet another Patarin ambassador. Both these nobles continued to send embassies to Dubrovnik during these years. We have no reason to believe that the Patarin envoys fell from favor, since in the 1430's we shall meet both Radin and Dmitar again.(102) However, since the bulk of our sources are Ragusan, it is highly possible that between 1423 and 1430 Sandalj and Radoslav did use Patarin envoys within Bosnia, whose missions are not mentioned in the Ragusan documents.

Yet despite his Patarin connections, Sandalj had good relations with Catholic cities. In 1424, he once again, his brother and nephew Stefan Vukčić were granted the status of nobles of Venice.(103) In 1426 Dubrovnik made formal plans to receive Sandalj. If he should visit on the Day of Our Lady he would be invited for mass with the rector, both for matins and vespers. If he should visit on the Day of St. Blasius (patron Saint of Dubrovnik), he would be asked to participate at vespers and to choose one of the nobles who would guard the relics of the saint during the ceremony. He would also march in the procession to the church with the rector and would carry a special taper.(104) We do not know if Sandalj actually did participate. Resti reports that he did, but it is likely that Resti simply assumed that the plans were carried out.(105) However, it is worth stressing that Dubrovnik — presumably with the consent of its archbishop — was eager to invite this member of the Bosnian Church to participate in Catholic Church services. Presumably, Sandalj was no dualist; for had he been, Dubrovnik, wishing to honor him, would not have associated his visit with relics and churches, both of which any dualist would have rejected.

Now, it could be argued with considerable validity that Sandalj here and the Bosnian nobles whom we have met from time to time particpating in Catholic rites were basically indifferent to these religious matters; thus if they had anything to gain by attending a Catholic mass or kissing a cross they would do so, whether they were dualists or not. And I would agree with that argument. However, one must then ask, "if Sandalj were a dualist opposed to churches and services in them, why would Dubrovnik choose this particular way to honor him?" Previously in 1401, for example, we found Hrvoje swearing by a cross and relics to respect the rights of Zadar.(106) Hrvoje, almost certainly, would have sworn by anything to get what he wanted. But why would the town of Zadar, which of course wanted its rights respected, have extracted an oath in this manner from a dualist? For if Hrvoje were a dualist then he would have had no respect for either cross or relics and therefore his oath would have been a meaningless act giving no guarantee at all for the town's rights.

Thus, in all these cases, greater emphasis should be placed on the behavior of the Catholic towns or clerics — who called for Bosnians to participate in these orthodox acts — than on the actions of the indifferent Bosnian nobles. And since in many cases the Christian ceremonies the nobles were included in were honorary and the Catholics were not obliged to invite the Bosnians to avoid insulting them, we must conclude that the invitations were not abhorent to the Catholics and oaths sworn by the Bosnians were not felt to be meaningless. Thus we may conclude that the participating Bosnians were not dualists.

In Ragusan archival documents from 1426, 1428 and 1429 Jireček found references to a customs house located at a church in Glasinac (at the crossroads for routes to Olovo, Borač and to Vrhbosna).(107) Jireček claims that the church lay in Sandalj's territory. If so, Sandalj had presumably usurped a large piece of Radoslav's land — a usurpation not mentioned in any source known to me. It seems probable that it still lay on Radoslav's lands, but by some agreement Sandalj was allowed a certain share of the customs receipts. But in any case, one or both of these nobles, despite Patarin ties, not only allowed the church to exist, but also had good enough relations with it to ask the church to play a part in the administration by serving as customs agent or at least to let the noble's customs agents work in the church. We do not know what confession this church represented. We know there was a Franciscan monastery near-by at Olovo, but Glasinac also was near enough the Drina for the church to have been Orthodox. We cannot exclude the possibility that the church was Bosnian; however, since in no Ragusan document is a Patarin religious building called a church, I think it unlikely that this was a Patarin church.

Sandalj also had good relations with the Orthodox. His second wife as we have noted was Orthodox, and possibly the icon he pawned in Dubrovnik in 1423 was hers.(108) Sandalj also had a dijak (scribe), Pribisav, serving him who was the son of a priest (hence Orthodox) named pop Milaja.(109) Shortly, we shall find Sandalj involved in building at least one church that was almost certainly Orthodox. Radoslav, too, had cordial relations with the Orthodox. In 1424 the Serbian despot visited him for celebrations connected with the birth of a son to Radoslav.(110)

Tvrtko II ruled through the 1420's without any serious challenge. In 1423 he was given the title of nobleman of Venice.(111) In 1425 he became involved in a war with Serbia over the rich mining town of Srebrnica, which he failed to capture. It is about this town at approximately this time that we have the one possible reference to Bogomils in the vicinity of Bosnia. Despot Stefan Lazarević of Serbia convoked an Orthodox Church Council here, and one manuscript (of uncertain date) of Konstantin Filozof's biography of the despot, written in 1431, having described the council adds that everyone in the town was Bogomil.(112) Perhaps there were some Bogomils in Srebrnica, but there was also a Franciscan monastery and a thriving Ragusan colony.(113) And would the despot have convened an Orthodox Church Council in a town whose whole populace was heretical?

In 1426 Tvrtko II patched up his relations with Hungary. This caused a Turkish attack that swept through Bosnia devastating many villages. At the end of the raids, the Turks withdrew again beyond the Bosnian borders. Tvrtko, however, was forced to recognize Ottoman suzerainty and agreed to pay the sultan annual tribute.

Tvrtko seems to have been friendly toward both Catholics and Patarins. If he had not been Catholic previously, Tvrtko had become one prior to 1428, when he decided to marry Dorothy, daughter of the former Ban Ivan Gorjanski (Johannes Garai), a loyal servitor of Sigismund. The Catholic Church expressed doubts about the marriage because the religious allegiances of Tvrtko were in doubt. Tvrtko wrote the pope admitting that he ruled a land of schismatics and unbelievers. Although his enemies accused him of being an unbeliever, he asserted, he was a good Catholic. He asked the pope to investigate the matter and clear his reputation so that the marriage might take place. Because the marriage did take place in 1428, we can assume that Tvrtko had proved his Catholicism to the satisfaction of the investigators.(114)

However, it is a fact that the djed had influence on Tvrtko and even was often at court. In 1428 Dubrovnik instructed its ambassadors to greet the djed and give him a letter if they found him at court. It was hoped that the djed would intervene on Dubrovnik's behalf. The envoy was asked to show the treaty to the djed and other courtiers, who were known to be friendly toward Dubrovnik and who would work in the town's behalf.(115) Thus it is clear that Dubrovnik regarded the djed as both an influential personage at Tvrtko's court and also as its friend. It is worth noting that in no Ragusan letter written prior to 1430 were members of the Bosnian Church spoken of in hostile terms. The Bosnians were never called heretics, and nothing was said that might even have suggested that the Bosnian Church could be considered heretical.

Dubrovnik, however, could — and did — express hostility toward followers of the Bosnian Church for political reasons. This is seen for the first time in late 1429, when friction developed between the town and Radoslav Pavlović. Radoslav thought Dubrovnik owed him money, and when the town refused to pay he seized a Ragusan caravan. The next year (1430) Radoslav attacked Konavli, part of which had been turned over to Dubrovnik by Sandalj in 1419 and part by Radoslav in 1427. Dubrovnik was not prepared for a war and immediately sent diplomats to Hungary, Sandalj, Tvrtko, and the sultan. Because Radoslav had by now become a vassal of the sultan, it was hoped that he would intervene. The sultan, however, preferred to remain neutral; he neither intervened with Radoslav, nor did he supply troops when Radoslav requested them. Dubrovnik also sent boxes of sweets to the djed and gosti, presumably in hopes that they would exert some influence on Radoslav.(116) Sandalj immediately seized the chance to grab some of Radoslav's lands and attacked him. Conversations followed in which a league against Radoslav including Sandalj, Dubrovnik and Tvrtko was projected.

In June two unnamed Patarin ambassadors accompanied by someone named Juanis (Ivaniš) went between the king and Radoslav.(117) The same month also saw an embassy of unnamed Patarins including a starac sent by Radoslav to Sandalj.(118) Our sources tell us nothing about their purpose or their result. After these June embassies, no more is heard of the proposed league, and Truhelka, quite reasonably, speculates that the djed had opposed a war between the two leading supporters of his church, particularly if one (Sandalj) were to align himself with Catholics against a fellow Bosnian Churchman, and had sent envoys who had successfully convinced Sandalj not to join such a league.(119) This is the second time that the Bosnian Church may have had a role in making peace between these two nobles; we recall that in 1423 when war broke out between Radoslav and Sandalj, Gost Miaša had passed between Radoslav and Sandalj, and then a month later, when the two men made peace, Patarins were prominent among negotiators on both sides. Maybe Miaša, a high-ranking churchman who appears as a diplomat only this one time in 1423, had been sent, not by Radoslav but, by the djed, from Radoslav to Sandalj to bring about peace.

It is interesting to note the frequent use of Patarins as diplomats in the negotiations between warring parties. A major reason for this possibly was their desire as churchmen to end wars and killing. Also, their religious habit and status would have allowed them to pass unharmed through warring lands.

In Dubrovnik's letters of complaint about Radoslav, the town stated that the Catholic faith in these regions did not have a greater enemy than the Patarin Radoslav. In these letters he was called a Patarin, a perfidious Patarin, and even a heretical Patarin.(120) Excluding the references to Patarin slaves, this is the first time Dubrovnik has spoken of non-ordained people being Patarins. (Of course, Sigismund and Ladislas's envoy had earlier referred to Hrvoje as a Patarin). Dubrovnik would subsequently call other nobles who supported the Bosnian

Church by this name. However, it is worth noting that this label was used by Dubrovnik to describe a layman only when Dubrovnik was at war with the layman. This is the first time that the term was used in a pejorative sense by Dubrovnik, and it should be noted that at the same time the town was calling Radoslav a perfidious Patarin heretic, it was referring to ordained Patarins in friendly terms and sending gifts to the djed.

Shortly thereafter, Radoslav made peace with Tvrtko and sent his son Ivaniš to the king's court. In 1432 peace was negotiated between Radoslav and Dubrovnik. A key role in the peacemaking was played by Krstjanin Radin. This man, still in the early phase of the diplomatic career which was later to bring him wealth and renown (as Herceg Stefan's Gost Radin), always seems to have had good will toward Dubrovnik. Years later, in the service of the herceg, he frequently was able to maintain peace with Dubrovnik and consistently would play a major part in ending hostilities. Radin was one of four envoys — the other three were laymen — and Dubrovnik always listed his name first.(121)

One custom in the region of Bosnia, Zeta, and Serbia was to seal peace between two former enemies with the act of *kumstvo* (godfathership). Thus on this occasion Dubrovnik stood *kum* for Radoslav's son Ivaniš. Since the two were of different faiths, it could not have been a regular godfathership, but must have been the *šišano kumstvo* (or haircutting *kumstvo*). This is simply the act of cutting for the first time the god-child's hair. Luccari states that Radoslav, following the custom of his nation, derived from the Greeks (most probably meaning Orthodox), offered the hair of his daughter (!) to the Queen of Heaven and invited to the ceremony Dubrovnik.(122) In a letter of 1434 Radoslav calls Dubrovnik, "Old and new friend and heart-felt *kum* (godfather),"(123) showing that this act of *kumstvo* had occurred; since in 1447 Ivaniš calls Dubrovnik *kum* the ceremony evidently involved him and not his sister.(124) Luccari clearly was referring to such an event but he refers to a daughter instead of a son and does not seem to have understood the purpose of the ceremony nor Dubrovnik's role in it.

V: Jacob de Marchia and the Franciscan Mission

In these years the Franciscan mission acquired new energy from the presence of Jacob de Marchia, who appeared in the Bosnian vicariat as a special visitor in 1432-33, and then returned in 1435 to become vicar, a post he held until 1439. Jacob was a most dynamic figure, who pursued two tasks throughout his stay in the vicariat with unflagging energy — the establishment of strict discipline for members of his order, and the elimination of all heresies from the vicariat.

The project to reform the way of life of the Franciscans led to a quarrel between Jacob and Tvrtko. Jacob found that among the Franciscans morals were dissolute and discipline was scandalously lax.(125) When he began to use his full power to discipline the brothers, Tvrtko immediately rose to their defense and tried to prevent Jacob from enforcing his reforms. In Jajce Jacob discovered that the Franciscans did not live in monasteries but in private homes and in villages; presumably, then, the Jajce monastery had not yet been built. Jacob apparently did not realize that the Franciscans might have been more effective in Catholicizing the area if they actually did live and work with the villagers. He also found that the Franciscans did not live on alms, but had incomes; perhaps they were working as craftsmen or farmers. Since the Franciscans had been forbidden to collect alms in Bosnia for fear that they would alienate new and potential converts, and since they were not receiving any support from the hostile bishop in Djakovo, or from the Franciscan order itself, we may wonder how Jacob expected them to live.

Tvrtko rose to the defense of the Franciscans and wrote Jacob. The letter published by Wadding (dated 1435 and redated by Čremošnik 1432) states that when Jacob asked the king for permission to come to Bosnia, Tvrtko had written him so that he could come freely and safely thither to preach the faith of God, just as his predecessors had. Tvrtko states that he wishes to hold and treat Jacob as a friend, but word has come to him that Jacob wants to expel from the purlieus and environs of the castle of Jajce certain clerics of our realm (i.e., Bosnian Franciscans). The King begs Jacob not to do this, stating that "piously and with humility the faith and word of God is being preached" by the friars. He states that if Jacob takes this action against them then he will "disfigure the honor of our lord Gracomus (? Čremošnik simply states Wadding erroneously copied the name. The vicar at the time was Johannes of Korčula.) and us." And in Jajce there are only a few whom Jacob would expel. And Tvrtko concludes that in other parts of the kingdom where the Franciscans are esteemed, you could not otherwise enter or preach the faith. "Thus because of this, dearest one, you ought not to begin to do such things nor (allow) them to happen through you."(126)

This letter shows Tvrtko to be a Catholic interested in the faith's propagation in Bosnia, who supported the Bosnian Franciscans because he felt they were effectively preaching Catholicism. We find that there were few Franciscans around Jajce. We also find that the Franciscans in other parts of the kingdom were respected by the populace, which suggests that they were making converts. When Tvrtko states that in these other parts of the kingdom "otherwise you could not enter or preach the faith," we may assume that by "otherwise" Tvrtko means in a way other than that which the Franciscans had been following around Jajce. Thus we may assume that in various other parts of Bosnia, Franciscans were also living outside of monasteries in communities amidst the populace. And it is evident that Tvrtko felt this is the only way they could effectively reach the people.

Despite Tvrtko's plea, Jacob expelled the friars from the private homes, and thus in effect from Jajce, where there was as yet no monastery, since they would have had to go to other places where there were monasteries (e.g., Visoko, Sutjeska). Tvrtko then restored them to the private homes from which they had been expelled. Jacob, as we shall see from Tvrtko's second letter, lost his temper and left Bosnia. Čremošnik states, without giving a reference, that Jacob went to Hungary.(127)

On April 1, 1433 Tvrtko again wrote Jacob. He greets him warmly but with a touch of reproach, ". . . to the most reverend father in Christ Lord Jacob . . . our dearest friend in the Holy Spirit, all salutations with friendly affection, not returning evil for evil." Having told Jacob, that as Jacob well knew, his arrival in Bosnia had not been displeasing to the king, Tvrtko states; "then, when you had passed some days in our kingdom, we having been informed through certain teachers of your faith, were led (!) to speak seriously to your worship lest those things be begun through you in our kingdom which so far have not arisen through your previous holy brothers. On account of this, perhaps, urged on more by our enemies, you in anger departed from our kingdom.

"Now, however, holy father, and we feel as displeased as possible, you so very wickedly decided to spread ill reports through Christianity (sic Christendom) by considering us not a Christian man but a pagan — which let (even such a thought) be far from (our mind)." Tvrtko invites Jacob to return. He states that he does not object to Jacob preaching the Roman faith in his kingdom. In fact he wants to help it and Jacob whom he wants to consider a valued . . .. (Fermendžin fills in the lacuna with vicar, but Jacob was not yet vicar.) He closes by stating that he intends to send some of his men to the pope to clear himself of such charges.(128)

Thus it seems that Tvrtko, acting on an appeal by some teachers of Jacob's faith (i.e., Franciscans), had intervened with Jacob, and as a result Jacob had left. Tvrtko suggests that the quarrel may have been exacerbated by his enemies. We do not know who these enemies were; Jacob then spread word about abroad that Tvrtko was not a Christian but a pagan. Thus once again we find the word pagan, though here it probably is no more than a pejorative term for a man, who, by his behavior, in Jacob's opinion, is not worthy of being called a Christian. Tvrtko, however, called this all slander, and intended to send a mission to justify himself to the pope. That Tvrtko was really a good Catholic is shown by the cordial relations he had with the pope at this time. In 1433 the pope released Tvrtko from Church fasts provided his confessor and doctor thought it best for his health.(129) That Tvrtko had written the pope about the matter shows that he took his Catholic practices seriously; the pope's granting Tvrtko's petition shows that he had no fault to find with the king's Catholicism.

In the Franciscan sources the quarrel took on legendary aspects: Wadding, linking Tvrtko's opposition to Jacob to the influence of heretics (whom this later editor calls Manichees), describes how Tvrtko, surrounded by Manichees, feared that Jacob's activities might cause him to lose his throne. Therefore he sent an evil magician, skilled in incantations to cause death and madness, against Jacob. Jacob, however, confounded this wizard and struck him dumb for life. The Franciscans were not satisfied to simply attack Tvrtko but they also depicted his wife as an evil woman who was a great enemy of Jacob. She once invited Jacob to Bobovac. On the way four robbers ambushed him to kill him. He said, "do as you want with me, for I am not afraid to die for the true faith. I know who sent you and why she sent you. If God wills it, fulfill the command of that evil woman." The four robbers immediately became rigid, unable to move their arms. Then making the sign of the cross, the saint freed them from this state. The four fell to the ground and begged for forgiveness, and he blessed them. Then he continued on to Bobovac where he found the queen much surprised to see him. She made various other attempts to kill him, and each time he was saved by a miracle, until finally she grew tired of the game and made peace with him.(130)

Since Tvrtko's wife was Dorothy Gorjanski, whom he was allowed to marry only after the pope satisfied himself about Tvrtko's Catholicism, it is clear that the queen was a good Catholic. We can hardly imagine her trying to murder Jacob. But these stories may signify that she had been the instrumental figure behind Tvrtko's opposition to Jacob's interference in the lives of the Bosnian Franciscans.

Jacob was an unflagging enemy of heresy, and the documents frequently mention his work against heretics in the Bosnian vicariat. However, the bulk of these references speak of his war on the Hussites, who did not exist in Bosnia proper but in Bohemia and Moldavia, both of which were part of the Bosnian vicariat. If our sources are accurate, there were no Hussites nearer to Bosnia than Srem. As for Bosnia proper, the general Franciscan Chronicle edited by Wadding, in describing Jacob's disciplining the Jajce Franciscans, states that in Jajce the friars had lived among people infected with "various heresies."(131) Probably Jacob had encountered there a variety of heterodox practices, varying from village to village, among ignorant peasants practicing Christianity as they saw fit.

In 1437 Jacob was working against heretics in Srem. A Srem cleric wrote a letter which mentioned Jacob's work among Bosnian heretics (Patarins? a real heresy?) and Hussites in Srem and other regions near the Danube.(132) In 1438 the cleric re-wrote the same text but omitted the phrase about Bosnian heretics, and mentioned only Hussites.(133) It is possible, of course, that in copying the older text he inadvertently omitted the expression "Bosnian heretics." However, it is also possible that Jacob had concluded that he had been in error the previous year when he had called some people of Srem "Bosnian heretics." Later, a Srem bishop, referring to Jacob's mission, spoke of the mixture of faiths between the Sava and the Danube; he named Bosnian, Rascian (Orthodox), and Catholic faiths, as well as heretics of a variety of heresies and particularly the Hussites.(134) It seems here that Jacob or the bishop was trying to distinguish the Bosnian Church (i.e., Bosnian faith) from the heresies. In addition to these three faiths which were distinguished from the heresies, he noted the existence of more than one heretical current besides that of Jacob's *bête noire,* the Hussites. Because he does name the Hussites but only them, we can suspect that the other heresies were not part of clear-cut movements, specific enough to be named. Possibly some of the Bosnians noted in these northern regions had fled thither to escape the civil war — to be discussed shortly — that seems to have raged in Bosnia between 1433 and 1437.

VI: Bosnia and the Church Council at Basel

In 1433 Dubrovnik was invited to summon representatives from the Serbian Despot George Branković, Sandalj, Radoslav, and Tvrtko to attend the Church Council at Basel. Dubrovnik wrote that owing to wars it could not get envoys through to them and concluded: "The clergy of the Kingdom of Bosnia are called Patarins by the Bosnians themselves although they may more truly be called without faith, order or rule. The first among them is called djed; the second, gost; the third, starac, and the fourth, stroinich. These four are the 'maiores' (leaders) in heresy and in the infidelity of the Bosnians themselves."(135)

This is the only source (unless Hrvoje in fact said "Patarin rite" instead of ''Pagan rite'') that says the Bosnians used the term Patarin for their ordained people; the Bosnians generally called them Christians (krstjani). Patarin was a label attached to Bosnian clerics by Dalmatian Catholics. Possibly, however, because the Dalmatians called the krstjani Patarins, the Bosnians in communicating with Dalmatian towns at times called their clerics by the Dalmatian term. The remark that the Patarins were without faith, order, or rule is obvious slander designed for Basel; however, it is certain that the Bosnian Church lacked order and rule comparable to that of the Catholic organization. The first three ranks of their hierarchy were accurately named, but the term ''stroinik'' is misunderstood. Ordained clerics below the rank of ''starac'' were simply called krstjani. We have already met Sandalj's envoy Dmitar, who was a krstjanin in 1419 and a starac in 1423. We have also met Krstjanin Radin, whom we shall soon meet as a starac. The term ''stroinik'' is generally considered to be a member of the djed's administrative council. Thus in 1404 the djed sent stroiniks and krstjani to Dubrovnik to bring Klešić home. In the 1450's we shall find Radin, then holding the rank of gost, being made a stroinik; and we shall find the Pavlovići sending two stroiniks — a gost and a starac — to Dubrovnik to negotiate. Thus clearly stroinik was not a rank below the other three ranks, but was an additional function held by people who were at the same time also a starac or a gost. The final sentence in the Ragusan letter is notable for, excluding war-time polemics against Radoslav Pavlović (and in the 1450's against Herceg Stefan), this is the first and only time Dubrovnik called Bosnian churchmen heretical. However, because the letter was designed for Basel, we may assume that the Ragusans spoke of the Patarins not as they regarded them, but as they thought the clerics at Basel would. Thus this letter is evidence that the papacy and other foreign Catholics did regard the Bosnian Church as heretical.

Bosnia was twice mentioned in the minutes of the Council of Basel: First in February 1435 when it was noted that Tvrtko had submitted to the Hungarian king and was converted from error;(136) second, also in 1435 when the Terbipolensis (?) bishop spoke of the reverence with which the Bosnians had received him; it was all he could do to prevent their kissing his feet. He claimed they were all Manichees and a mission should be sent to them.(137) It is evident that if the Bosnians really had been good Manichees, they would not have shown such esteem to a Catholic cleric. We can attribute his reception to the ignorant peasants' awe of the lordly man of God and to the South Slavs' hospitality toward the stranger. While there is evidence that dualists existed in Bosnia, our sources give no indication that all or even most Bosnians were

"Manichees."

The statement of February 1435 about Tvrtko submitting to the Hungarian king and being converted from error is puzzling, since we have Tvrtko's letter of 1428 stating he was a Catholic, which seems to have been confirmed by the pope's investigators who then allowed his marriage to Dorothy. Perhaps the errors Tvrtko was converted from were not errors akin to heresy, but instead concerned his dispute with Jacob de Marchia. Jacob after his quarrel with Tvrtko had gone to the Hungarian court and complained that Tvrtko was not a Christian. Since Tvrtko, as far as we can tell, had only been defending some laxly living Catholic clergy, we are given no reason to doubt his actual Catholicism. Perhaps the King of Hungary had pressured Tvrtko into admitting he had been at fault in his dispute with Jacob and thereby had reconciled Tvrtko with Jacob, and it was this reconciliation that was announced at Basel in February 1435. This hypothesis is made more plausible because in 1435 Jacob returned to Bosnia as vicar.

There were also political reasons that might have led Tvrtko to move closer to the King of Hungary. In these years from 1433, a son of Ostoja named Radivoj began to make a bid for the Bosnian throne. He had the support of the Turks, who proclaimed him king, and he soon received recognition and aid from Sandalj. Thus Tvrtko was deprived of what had been little more than nominal suzerainty over the vast Kosača holdings. Radivoj, not recognized in the banate, resided at Sandalj's court from 1433 and was still there in 1435 at the time of Sandalj's death. Adding to Tvrtko's difficulties were an increase in Turkish raids — some of which were nominally supporting Radivoj — and increased pressure along the Drina from Sandalj's allies the Serbs. Eventually the Serbs took Zvornik. With these developments it is not surprising that Tvrtko II moved closer to Hungary. And it is possible that it was this rapprochement that was referred to when it was announced at Basel that Tvrtko had submitted to Sigismund. By 1437, in any case, Tvrtko had again been forced to become a vassal of the sultan paying an increased annual tribute.

VII: Roman Catholic Gains in Bosnia

In these years after Jacob's return, Catholicism made rapid gains in Bosnia. In 1436 Tvrtko granted the Franciscans freedom to teach in Bosnia and took them under his protection. It is interesting to note that one of the witnesses to the charter to them was Vladislav Klešić, the son of Pavle Klesić who had been so closely linked with the Bosnian Church in 1404.(138) Presumably by 1436, Vladislav had accepted Catholicism. The same year Pope Eugene IV, noting the shortage of priests in Bosnia, allowed the Franciscans to administer the last rites, permitted the vicar to administer all sacraments, and gave him power to delegate this authority to other brothers who worked in regions without secular clergy.(139)

Not only had Klešić's son most probably accepted Catholicism, but so too had Hrvoje Vukčić's nephew and successor Djuraj Vojsalić, Lord of the Donji kraji and Tvrtko's leading supporter, whom the pope in 1437 allowed to build a church of Our Lady for the Franciscans toward whom he had particular affection.(140) Further evidence of his Catholicism is shown by a charter he issued in 1434 to the brothers Jurijević (the sons of Djuraj Radivojević). It was witnessed by the Bosnian vicar and other Franciscans: "With all the described (lands) we render them into the hands of Lord Vicar Žuvan and each vicar who will succeed him ("i vsakomu vikaru kon' vikara") and all Franciscan brothers of the Holy Catholic Church ("svete cr'kve katoličaske") of Roman faith of the Order of St. Francis."(141) The expression "vikar(u) kon' vikara" is unusual; the meaning, however, almost certainly is that the present vicar and his successors (those who come after him) stand behind the grant. This is similar to the formula "did kon' dida" from a Bosnian Church charter of 1446.(142)

Further progress of Catholicism and the Franciscans is demonstrated by four papal letters from 1434 and 1440 which speak of a bishopric (controlled by Franciscans) inside of Bosnia.(143) Unfortunately very little is known about this bishopric. We know nothing of its formation, its relationship to the bishop in Djakovo, how much territory it included or how long it lasted. We know only that it was subject to Dubrovnik and included Visoko. Srebrnica was added to the title by 1440.

VIII: The Kosača Family: Church Building and Relations with the Orthodox Church

In 1434 Sandalj sent an envoy (notably not a Patarin) to Dubrovnik to say that he wanted to build a church and hospital there. Dubrovnik was unwilling at first to allow this.(144) Truhelka thinks that this was a sign that Sandalj in his old age had begun to worry about his soul and wonder if the Bosnian Church really could secure his salvation, and had decided to get some insurance by building a church.(145) We may question Truhelka's view. The fact that Dubrovnik was not interested in the church suggests that it was to have been Orthodox. Dubrovnik always worked to prevent the building of Orthodox churches in its lands, and to convert Orthodox and schismatics to Catholicism. If the church was to have been Orthodox then it seems probable that Sandalj had wanted to build it for his Orthodox wife, Jelena, daughter of the Serbian Knez Lazar.

This view is confirmed by Resti, who has his chronology slightly twisted but otherwise presents a plausible account of what must be the same event. Resti says that the Serbian despot at the insistence of Jelena, wife of the late Sandalj (thus erroneously dating the event after Sandalj's death) requested the Republic to permit her to build outside the city a church in which Greek services would be carried out . . . the (Ragusan) Council, always averse to any rite other than the Catholic in its lands, at once gave a negative reply.(146) Eventually it seems that some sort of compromise was reached. However, we do not know if the church was ever built because five months later in March 1435 Sandalj Hranić died.

Churches were built in Sandalj's lands and it is possible that Jelena may actually have built one of them. Ruins of a church are found in a medieval cemetery between Sandalj's Ključ and Cernica. The cemetery clearly dates from Sandalj's time since one stone with an inscription marks the grave of Radohna Ratković who was killed beneath the fortress of Ključ for his lord Sandalj.(147) Since the church ruins are old and since cemeteries frequently were established beside churches, it is probable that the church was contemporary with the cemetery. The church was probably Orthodox, though it may have been Bosnian. Near the walls of Ključ is an old mosque which the local inhabitants say had originally been a church built by Sandalj's wife Jelena.(148) The tradition claims erroneously that she was Jelena Obilić (rather than Lazar's daughter), and that Sandalj had abducted her by force (about which we know nothing). The tradition reports that Sandalj was a man without faith (*bezvjerac*) while his nephew Herceg Stefan had been Orthodox.

That Sandalj died a member of the Bosnian Church is confirmed by Resti whose description of Sandalj deserves full quotation: "He was a prince with lively spirit, with great intelligence and with much delicacy (!), who was always able to penetrate the heart of the matter with great facility, whose memory would have been immortal, if his life had not been stained and his fame obscured with the error of schism and the Patarin rite in which he was born and in which he died."(149) Here too the Patarin rite is depicted as schismatic rather than heretical. Sandalj left a huge state to his nephew and successor, Stefan Vukčić Kosača who was to go down in history as Herceg Stefan, as a result of the title

Herceg obtained in 1448.(150) Stefan began his career holding the mouth of the Neretva, and all the land from there to the west bank of the Lim River. He inherited his uncle's main court of Samobor (near Goražde) as well as other fortresses at Kosman (near Foča), at Sokolgrad, at Blagaj, and at Ključ. His lands stretched north to Konjic and south as far as Nikšić.

Sandalj's widow Jelena, Knez Lazar's daughter, remained an active supporter of the Orthodox Church. Presumably she had been instrumental in Sandalj's request to build a church in Dubrovnik, and tradition also links her to the church in Ključ. After her husband's death she returned to Zeta.

In 1442-43 she left a will which was deposited in Dubrovnik. The will was witnessed among others by two clerics, Nikandar, a starac (elder — the term was also used in the Orthodox Church) from Jerusalem, her spiritual adviser, and by a monk named Jovan. Nikandar presumably lived at court and served as her chaplain and confessor. She left him, monk Jovan and a priest (pop) named Teodosius sums of money. She left to her great-nephew Vladislav various valuables, including the relics of a saint, and to her sister a gold icon. Money was set aside for charity and for her grave. To the churches of St. George and the Holy Virgin were left a silver bowl and some other items to do as they wanted with, and money was assigned to roof-over the church she had built for her grave. A few lines below she mentioned 200 ducats to cover St. George's church at Gorica. She also left 1000 ducats for a new church to be dedicated to the Virgin at her grave in Gorica (The church was built on the island Gorica (Beški) on Lake Skadar).(151) She assigned the task of seeing that the churches were completed and her will executed to her "grandchildren" Elena and Vladislav. Vladislav who was the son of Stefan Vukčić was actually in our terms her grandnephew. They were to provide for anything the churches lacked — books, furnishings. Near the close of this interesting testament, Jelena stated that someone named Voio had sent a message to her through Starac Radin in the Korčanska (sic. Goričanska) Church of the Holy Trinity that she should put down her wishes and make a will and pray to God in the name of the Holy Virgin that her wishes be carried out. And therefore the will was written by her spiritual advisor Nikandar . . . .(152)

Jelena was evidently a devout woman, involved in church building and surrounded by Orthodox clergymen. Yet she had been married to a member of the Bosnian Church. There is no evidence that any of these clerics caused friction by trying to force her to convert her in-laws. To complete her church-building projects, she called upon her husband's grandnephew Vladislav, about whom we have no evidence that he was

Orthodox. Vladislav's father was clearly a supporter of the Bosnian Church, and a later Ragusan source calls Vladislav's brother Vlatko a "Patarin" as well. This means Jelena trusted a man who probably was a follower of the Bosnian Church to see to it that these Orthodox churches were completed and properly furnished. In addition, she left to this grandnephew some treasured holy relics, which she could easily have left to one of the three mentioned Orthodox clerics. All this strongly suggests that the Bosnian Church was neither opposed to the cult of relics nor to church buildings.

Jelena's will also mentions a Starac Radin. The title starac can refer either to a Patarin or an Orthodox monk, but because the famous Patarin Radin (subsequently to be a gost) had by 1437 achieved the rank of starac and entered the service of Stefan Vukčić, it is possible that he was the Starac Radin in the will. If so, then Radin the Patarin gave advice to the Orthodox Jelena and she publicly admitted this fact in her will; clearly, she did not consider him a heretic — any more than Catholic Dubrovnik did. And this Radin himself had been in the church and passed on this Voio's words; thus Radin's speaking to Jelena inside a church (presumably an Orthodox one) would be evidence that Patarins were not hostile to church buildings,(153) and that Radin had cordial relations with Orthodox clerics. Jelena's career and her will show, despite Sandalj's and his successor Stefan Vukčić's ties with the Patarins, that there were strong Orthodox influences at their court, and that relations between the Orthodox and Patarins were cordial. Thus we see here what could be regarded as an Orthodox-Bosnian Church symbiosis, similar to the Catholic-Bosnian Church symbiosis frequently noted in relations between Bosnia and Dubrovnik. Thus, I think Jelena's will is further proof that the Bosnian Church was not Bogomil or, indeed, heretical. The presence of the two faiths at Stefan Vukčić's court also illustrates Bosnian tolerance to, or indifference about, the existence of different religious faiths.

IX: Political Events from the Late 1430's to 1443 and the Secular Role of Patarins

In the late 1430's Turkish activities continued at an increasing rate in Bosnia. The Turks were used by Stefan Vukčić in his wars with his neighbors and they were also active on their own behalf. In 1439 Serbia fell, and Ottoman raids into Bosnia became more frequent. Contrary to general belief, however, it seems that the Turks occupied no territory in Bosnia other than a few towns along the Drina. It had been generally believed that Sarajevo and the near-by fortress Hodidjed had been occupied in 1436 and held permanently. However, Dinić and Šabanović have shown that this town and fortress were not taken and held permanently until after 1448.(154) The Turkish raids seem, however, to have done considerable damage, and in 1437 the Franciscans complained that over the previous two years the Turks had destroyed sixteen churches and monasteries in the Bosnian vicariat.(155) We do not know how many, if any, of the sixteen were in the Bosnian kingdom.

Despite the obvious danger to the kingdom, the Bosnian nobles remained faithful to their own interests. In addition to the struggle between Tvrtko II and Radivoj, civil war broke out in 1435 between Radoslav and Stefan Vukčić after Radoslav had divorced his wife Theodora (Stefan's sister) and had remarried; thus we may assume that he had married her according to Bosnian custom. In 1439 the two made peace; to seal it Radoslav — his second wife having died — took Theodora back.(156)

These casual Bosnian marriages also caused complaint beyond the borders of Bosnia. After Tvrtko I's conquests, Bosnian overlordship extended along southern Dalmatia as far as Kotor (from ca. 1385) and various Bosnians had settled in that town in the final decade of the fourteenth and the first half of the fifteenth centuries. In October 1438 the Bishop of Kotor received a complaint from Stanislava, wife of one Miloslav of Dračevica (the *župa* stretching inland behind Hercegnovi), that Miloslav, with whom she lived in lawful marriage, had, for no fault of her own, driven her from his house and had taken a second wife according to Bosnian and Patarin custom.(157) We have discussed Bosnian attitudes towards marriage previously and have argued that they were derived from popular tradition rather than from any religious creed. However, we may assume that the Bosnian Church did not oppose popular attitudes toward marriage, and therefore, it is not at all odd for the Kotor bishop to attribute this custom to "Patarinism."

Radoslav died in 1441 and was succeeded by his son Ivaniš, who was able, after a brief quarrel with Stefan, to arrange a peace. Stefan's interests were diverted elsewhere as he pushed both into the region beyond Nikšić and into Zeta's coastal holdings. In 1437 King Sigismund died; his weak successor Albrecht died shortly thereafter setting off a civil war in Hungary. Hungary was thus unable to intervene in Bosnia which may have been a hardship for Tvrtko who was still faced with the anti-king Radivoj. When Tvrtko died in 1443 he was succeeded by Ostoja's son and Radivoj's brother, Stefan Tomaš.

After an interval of several years without mention of Patarin envoys, we again find references to several Patarin ambassadors in Dubrovnik from Radoslav and Stefan Vukčić in 1437 and in the years immediately following.(158) In April 1437 Stefan Vukčić sent Radin Starac to Dubrovnik.(159) This shows that Radin had left Radoslav and had come to serve Stefan: he would remain in Stefan's service until the latter's death in 1466. Radin had been by now promoted to the church rank of starac.

Early in the spring of 1438 ambassadors had been sent to Dubrovnik jointly by the Bosnian king, Radoslav, and the djed.(160) We do not know what they came to discuss, nor who were sent as envoys. Ćirković believes that these embassies were sent to intercede on behalf of Radoslav who was having difficulties in his war with Stefan Vukčić.(161)

In August 1439 Dubrovnik wrote Radoslav to send it one of his Patarins for discussions.(162) It is clear that on this occasion for some reason Dubrovnik did not want any envoy, but a Patarin. Had Dubrovnik found Patarins more easy to deal with? This is also evidence that Radoslav must have had several Patarins around his court who were on call. We do not know if Radoslav acceded to this request.

At the end of April and early May 1440, we find Radin and a layman again in Dubrovnik on behalf of Stefan. Radin is once referred to as "starac christiano."(163) which shows that the term krstjanin could be a general one for all who were ordained, and on occasions could be used for a man even after he received a higher title.

And finally in 1442, a Ragusan envoy sent to the royal court was instructed to get a treaty sealed before ordained members of either the Roman faith or of the Bosnian faith.(164) Thus as we move into the reign of Stefan Tomaš (1443-61) we find that, despite Catholic gains, the Bosnian Church was still active in Bosnia, and even at Tvrtko's court.

X: Patarin Hižas and Dubrovnik

We have noted Patarin hižas in Moštre in 1322-25 and have heard of one at an unknown location where Radišić's brother had sought refuge. We may suspect that the djed's letter of 1404 from Janjići was also issued from a hiža. In 1412 Ragusan sources mention yet another Patarin house. It was normal for Vlachs to be hired to lead merchant caravans and in this year they were hired to go to "Glubscovo" to the Patarins.(165) A second contract that year hired the Vlachs to take the merchants to the home of the Christians (domos Christianorum) in

Glubscovo.(166) In that contract Dubrovnik used the term "Christians" that Patarins used when they spoke of themselves. In 1414 the Vlachs were hired to take merchants to the place of Ruxin Patarin at Glubscovo.(167) Ruxin was presumably prior of the house. Another contract of 1414 simply referred to the Patarin place at Glubscovo.(168) In 1415 a contract instructed the Vlachs to lead the merchants to the house of Radoycho in Glubscovo.(169) We may assume that this man was a Patarin and had succeeded Ruxin as prior of the house. This document stated that the merchants were en route to Srebrnica. Thus we can assume that Glubscovo (probably correctly Ljubskovo) was located somewhere in the vicinity of Srebrnica, and most probably a full day's journey away from it.(170) References to Glubscovo and the Patarins continued in 1416.(171) Ljubskovo lay in the territory of the Dinjičić family; since in 1450 this family sent a Patarin to Dubrovnik as an ambassador, we can conclude that the Dinjičići, if not actually adherents of the Bosnian Church, at least maintained cordial relations with its clerics.(172)

In 1418 a contract was made for the Vlachs to take merchants going to Bosnia to Bradina to the house of Milorad Patarin.(173) Milorad presumably was the prior of the Bradina hiža. The Vlachs were told that if bad weather prevented them from going as far as Milorad's house, they should stop at the house of the king in Konjic. Bradina lay on the route between Konjic and Vrhbosna. This last document tells us of yet one more hiža, and shows that in 1418 the king held Konjic and also maintained caravanserais. The collection of contracts shows clearly that at least certain Patarin hižas served as hostels for merchants and travelers. And they also show that Dubrovnik did not object to its merchants stopping over at these monasteries of the Bosnian Church.

From the point of view of function — being residences and centers for the Bosnian monastic order — the hižas can be called monasteries. However, the term "monastery" may suggest to the reader large buildings (or complexes of buildings) housing many monks. Though this may have been the case, the terminology of contemporaries to describe these residences — domus, casa, hiža: words that all mean "house" — suggests that they were fairly small in size and in number of inhabitants. In confirmation of this hypothesis is the fact that archaeologists have not unearthed any large foundations in villages where known hižas were located.(174)

However, relations between the Patarin hižas and the Ragusan merchants were not always friendly. In November 1440 before a Ragusan court a merchant made complaint against one Radoe Rugiza, who he claimed had taken a horse with furnishings and other goods worth thirty ducats from him near Goražde on the Drina on account of (or for the sake of) that Patarin.(175) Since we shall meet Radoe again in a second court case, where we shall find him to be a landowner with his own peasants, it is likely that he was not an ordained Patarin. Thus we must conclude that either the Ragusan court was using the term loosely for a lay supporter of the Bosnian Church and hence was referring to Radoe as a Patarin, or else Radoe was working for a Patarin well enough known to the judge so that it was not necessary to refer to him by name. Since Radoe committed his crime near Goražde, we may choose the second alternative and suggest that the Patarin referred to is Gost Gojisav, whom we shall meet in the next case. However, since complaint was made against Radoe we can suggest that he and the Patarin were in cahoots.

In January 1441, two Ragusan merchants complained against two brothers and their father-in-law Bosigchus, who had promised to obtain for the merchants a safe-conduct so that they could pass through the territory of Gost Goysavus (Gojisav), who lived near Borač. The gost with a functionary of Vojvoda Stefan (Vukčić) (cum valioso vayvode Stiepani) gave a safe-conduct to Bosigchus who returned with it to the merchants and suggested that they set out, saying that he would lead the way and retain the document. When they entered the gost's territory Bosigchus disappeared and since they were unable to produce the safe-conduct, the gost had confiscated all their goods.(176) It is apparent that the merchants had fallen into a trap. Presumably the Patarins and Bosigchus' family split the profits.

The Patarin hiža lay below Borač which was still the capital of the Pavlovići.(177) Yet it is clear that the gost's hiža was not in Radoslav's lands but in those of Stefan Vukčić, since it was the latter's official who had validated the safe-conduct. In the previous case complaint had been made against Patarins near Goražde. Goražde lay in Stefan's territory on the road between Borač and Foča, not far beyond Borač. The Patarins in the 1440 complaint did not live in Goražde but near it. Both complaints then seem to have been leveled against the same hiža, which lay on Stefan's lands on the road between Goražde and Borač.

The gost's involvement in issuing a safe-conduct and his confiscation of goods lacking papers show that Stefan Vukčić utilized Patarins in matters of administration. Since the document had to be shown the gost, the hiža was apparently located immediately on the Pavlović-Kosača border, and served as a border post (and very likely also as a customs house). Thus when the merchants could not produce the safe-conduct, the gost as customs agent simply seized their goods, since they did not have permission to bring them into Kosača land. After all, the merchants could not expect to travel without a visa. If safe-conducts were needed to go from the territory of one noble to another, we have further evidence that Bosnia ought to be regarded as a land of separate autonomous units rather than as a unified state. The role of the gost in Stefan's administration also should be emphasized. We have no evidence that other hižas played this sort of role, but considering the scarcity of documentation we cannot rule out the possibility. And if the gost controlled a border post and was able physically to seize goods we may speculate that he had assigned to him (or had acquired himself) a certain number of armed retainers. This would have been almost a necessity for an administrative official on this tense border between two frequently warring parties. And if he had retainers, we may speculate that the gost played some role in keeping order in the region of Goražde. Thus this gost, regardless of his unscrupulous behavior, was evidently an important and influential figure in the Goražde region.

In February 1441 Boghissa Bogmilović, a merchant, complained to the Ragusan court about Radoe Rughiza (whom we met in the case of the lost horse in November 1440) and his villagers of the town of "Vinizcha" and about three other men, one of whom was from Goražde, and some other peasants. Added in a second hand to this list is: "and against Gos patarin and against his familia." These people were accused of seizing a large quantity of goods having considerable value.(178) Since the Patarin was called "Gos" and since again Goražde was mentioned, we may assume that gos meant gost and that once again our Gost Gojisav and his hiža near Goražde were involved in robbery. The most interesting item is the expression about the "familia" of the Patarin. Presumably it refers to his household which would include the other krstjani living in the hiža and any servants (or even armed retainers) the gost might have had. It might, however, refer to relatives of the gost. It has generally been assumed that the krstjani were celibate. This belief is supported by the fact that we know of no examples of any Patarin having children. In addition, since the Bosnian Church seems to have been a continuation of what was originally a Catholic monastic organization, it seems probable that the monks would be celibate. Thus since the evidence suggests, though not conclusively, that ordained members of the Bosnian Church were celibate, then family, if that is the meaning, would presumably have referred to relatives, (brothers, nephews, etc.) who lived in the area.

In December 1441 a Patarin named Vukosav and a layman were accused of theft.(179) In a case before the Ragusan court in January 1442 once again complaint was made against Patarins near Goražde seizing goods.(180) We should not generalize about Patarin behavior on the basis of this material since all the identifiable cases most probably refer to the same hiža. But it is clear that Gost Gojisav was a figure far more concerned with gathering the fruits of this world than storing away credits for some vague hereafter. Gost Radin, as we shall see later, also would manage to amass an enormous fortune from other activities carried out while he was a starac and a gost.

XI: Gravestone Inscriptions Giving Information About the Bosnian Church

We have eight undated gravestone inscriptions at seven sites which refer to ordained members of the Bosnian Church. Most scholars date all or most of them in the fifteenth century.

The first inscription is found at Košarići near Bogutovo village not far from Bijeljina. It states that a man whose name is illegible (Truhelka has been able to deduce only the first name Radoe), buried on his own land, has had a new stone placed over him by Gost Ratko (or Raško) and the son of the deceased.(181) This suggests that the dead man had been a follower of the Bosnian Church. Somehow something had happened to his original gravestone, and the man's son and the gost had placed a new stone over him. This shows that Bosnian Church clerics were, not surprisingly, concerned with the burial of members of their church. It also shows that the Bosnian Church allowed its adherents to be buried on their own land. The most interesting thing about the inscription is its location. Bijeljina is far to the north, near the junction of the Drina and Sava. Other sources have only vaguely suggested that the Bosnian Church was active in this region which was only at times, and then only nominally, under the control of the Bosnian ban.(182) This one stone is not sufficient evidence to show that a Bosnian Church monastery existed at (or near) Košarići, although we cannot rule out the possibility. Otherwise this region seems to have been a mixed Catholic and Orthodox area. There was a Franciscan monastery at Bijeljina by 1385 since it was noted on Bartholomaeus of Pisa's list of monasteries in the Bosnian vicariat.(183) Orthodox believers lived all the way along the Drina, and near Bogutovo village at Trnjačka, a grave inscription also from about the fifteenth century was written by "Goičin the priest" (pop),(184) whose name and title suggest he was Orthodox.

At Poljska between Lašva and Travnik, we again find a gost placing a new stone over a dead man, presumably a follower of the Bosnian Church. This time the gost wrote the inscription.(185) Presumably that means he wrote the text, and someone else did the actual carving.

A third inscription also concerns a gravestone. Found at Medjedji at settlement Djipi between Goražde and Višegrad, it mentions a certain Radovan and a Krstjanin Radašin pulling a stone (presumably from the quarry to the grave site).(186) Perhaps Krstjanin Radašin was a member of the hiža near Goražde where we found the greedy Gost Gojisav.

Two inscriptions are found on two different stones at the village of Zgunje on a hillside overlooking the Bosnian side of the Drina between Bajina Bašta and Perućac. Both stones have been removed from their original locations by farmers since they interfered with farming. The first stone is damaged and only one word "stroinik" now appears on it;(187) we have noted before that a stroinik was a member of the djed's governing council. The rest of the surviving face is clean and bears no signs of writing; the missing piece is not large, thus the complete inscription that originally stood there must have been brief. The second stone is a very attractive obelisk. It is the first of the stones we have noted that has designs upon it. On one face it has three rosettes. On a second face is depicted a solitary "T" staff. This shows that this symbol was not limited to churchmen of rank but could be used by any ordained member of the Bosnian Church, including *krstjani,* for the inscription "Here lies Ostoja Krstjanin at Zgunje" tells us the deceased was a krstjanin.(188) Ostoja is known from one other source. Okiç found in a Turkish defter of 1534 reference to the hereditary lands (baština) of "Kristians" Ostoja and Vladislav in the village of Lower (Doljnji) Zgunje, nahije Osat.(189) Bešlagić concludes that Ostoja was still alive at the time of this sixteenth century survey and thus dates the stone to the sixteenth century. The obelish shape also suggests that Ostoja's stone was influenced by Turkish styles and thus was late. Bešlagić dates the "stroinik" stone, however, to the fifteenth century because of its boxlike shape and the style of the carved letters. It seems probable to me, since there are only eight stones extant now (possibly there were once more), suggesting that this was originally a fairly small cemetery, that all the stones would be from roughly the same date. Since in the defters we frequently find lands named after former owners who have been dead some time, I do not think we can assert that Ostoja lived in the sixteenth century (though we cannot rule out the possibility). Since his obelisk does suggest Turkish influence I think we should date him no earlier than the middle of the fifteenth century. Since we do not find mention in documents of djeds or stroiniks after the Turkish conquest, it seems likely that the stroinik stone would be prior to 1460 (if not even earlier). Thus I would date the two stones 1440-1470. The inscriptions tell us little about the Bosnian Church except that it seems to have had members as far as the Drina. Since the other six stones have no markings at all, on the basis of two Bosnian Church inscriptions we may suggest that this had once been a Bosnian Church clerical cemetery. There may well have been a hiža at Zgunje. The community in Zgunje evidently was of some importance, since one of its members had been on the djed's administrative council. Zgunje does not lie very far from Ljubskovo where we noted a hiža in the second decade of the fifteenth century. Thus this general region may have been a Bosnian Church center. Whether these communities dated back to the fourteenth century, or whether they sprang up only during the fifteenth century under the protection of the Pavlovići and Dinjičići when the Catholic offensive began to assert itself in the center of the state is unknown.

At Puhovac near Zenica, not at all far from Bilino polje and Janjići, was found a stone without decoration bearing the following inscription: "Here lies good Gost Mišljen to whom it was arranged by order of Abraham his great hospitality. Good lord (i.e., Mišljen) when you go before Our Lord Jesus mention us your servants. G. M. writes."(190) This inscription shows two important things about the Bosnian Church. First, it mentions Abraham, showing that the Bosnian Church did not reject the Old Testament or its prophets. Secondly, we see that in the church's views regarding the hereafter, it is believed that the intervention of holy individuals (in this case, the gost) could affect and aid the salvation of other people. Thus the hypothesis that the Bosnian Church considered salvation dependent solely on correct baptism and thereafter strictly obeying certain prescriptions is incorrect.

Our seventh inscription marks a grave close to the old fortress Soko (or Sokolgrad) on the Piva River. Since Herceg Stefan is referred to in the inscription bearing the title "vojvoda" we can date it between 1435 (the date of Sandalj's death and Stefan's accession) and 1448 (the date Stefan took the title herceg). Unfortunately there are several lacunae in the inscription which reads: "God, unite Petko . . . my soul; and brothers and friends I beg you not to disturb my remains for I was as you are and you will be as I, dead. I placed this grave-marker during my life time and death cut me down at Sokol . . . Bless Vojvoda Stefan who honorably fed me and God guard his soul forever . . . I, Petko Kr'stienin."(191)

This inscription, then, marks the grave of a Bosnian Church cleric. The inscription shows that Stefan Vukčić had had close relations with members of the church, and had fed at least one cleric. We may assume that Petko had lived either at court or else at a near-by hiža whose members were supported by the charity of Stefan Vukčić. Other evidence suggests that there may well have been a hiža in the vicinity.(192)

Our final stone is probably the most interesting. It was found at Humsko near Foča, and now sits in the court yard of the Zemaljski muzej in Sarajevo. On the front face is depicted a man standing, bareheaded, dressed in a skirt which ends above the knees. He wears a rope-like belt. In his left hand he holds a rectangular object which is almost certainly a book; in his right, he holds a "T" staff. On each of the two sides is depicted a rope-like object running vertically, which probably represents his belt. This belt probably had some significance (possibly it designated his rank) that we cannot be sure about today. On the back is depicted a rosette inside two circles. All four sides are enclosed within a rope-like border. Inscriptions are found on three sides. "In your name Immaculate Trinity, the gravestone of Lord Gost Milutin born Cr'ničan," "He perished as God willed it." and "Zitije ("Life", a term used for a saint's biography): He lived honored by the lord of Bosnia, he received gifts from the great lord and nobles and from the Greek lord and was known to all."(193) The grave was excavated and around the skeleton were found fragments of a rich gown (in which the corpse had obviously been lain to rest) on which were depicted stars and a lion. Skarić dates the stone fourteenth or fifteenth century.

This stone reveals a man who had achieved the rank of gost. He presumably was from a family in "Crnice" (possibly Cernica near Ključ in Hercegovina). He clearly was a worldly figure; not only did he live honored by the nobility, but this aspect of his life, rather than some religious aspect, was what seemed to him (or to those who buried him) important to mention on his gravestone. He was wealthy and like it, noting gifts from various noblemen including a Greek lord; he also was buried in a rich robe, not what we would have expected for an ordained cleric.

We do not know who the "Greek Lord" was, for surviving sources give almost no data on any relations between Bosnia and the Greeks. Truhelka speculated that the reference most probably would have been to the Despot of Epirus. I would like to suggest that the Greek Lord was Thomas Despot of the Morea to whom Herceg Stefan sent an embassy (whose participants are not named) in 1457. At the same time at Stefan's house in Dubrovnik was staying Manuel Cantacuzenus, who was related to Anna Cantacuzena, the wife of Stefan's son Vladislav.(194) Since Cernica whence Milutin seems to have come belonged to the Kosače, it is not impossible that Milutin served Stefan Vukčić and participated in the 1457 embassy to the Morea.

Since it is after the list of his important connections, that it is stated that he was known to all, we may suspect that his fame was through his state-court service. This would also be a reason for his receiving so many gifts. Strange to say, his name is not mentioned in any extant source despite the fact that we have considerable material about Patarin diplomats in the Dubrovnik archives. As a worldly and wealthy figure we may place him in the company of two other worldly rich gosts — Radin and Gojisav.

The only religious reference in the inscription is to the Trinity which we have noted before was clearly accepted by the Bosnian Church. We may suggest that the man whose portrait is carved on the stone is dressed in the habit worn by krstjani in general, or possibly gosts in particular.

XII: Patarin Diplomats

As a preface to this discussion we must note that almost all the information about Patarin envoys comes from Ragusan records, which focus on the town's own foreign affairs and speak almost entirely about their missions to Dubrovnik. Occasionally, mention is made of a Patarin embassy within Bosnia, but these are scattered and give us no basis to generalize. Of the forty-two missions for which there is data, forty were to Dubrovnik.(195) It is probable that an equally high — if not higher — number of Patarin envoys traveled between different courts within Bosnia. However we just do not know.

Despite the fact that we know of forty-two embassies on which Patarins participated as envoys, we should not exaggerate their role in diplomacy. We have no evidence that any Patarin served as a diplomat prior to Vlatko in 1403. The Bosnian kings used Patarin diplomats only twice(196) — in 1403 and 1430 — and Truhelka argues convincingly that the 1430 mission may well have really been sent by the djed. When we turn to the great nobles — and only the leading magnates did conduct their own foreign policy — we find only two families using Patarins with any regularity — the Kosače (Sandalj and Stefan Vukčić) beginning in 1419 and the Pavlovići beginning in 1420. A third noble family the Dinjičići sent one Patarin mission in 1450. Hrvoje, though called a Patarin, was never known to use a Patarin as a diplomat.

The Kosače used Patarins in twenty-seven missions. For this figure I am not counting the possible mission of Dmitar Starac in 1430 since though we know he was due to be sent to Dubrovnik we have no information that he actually came. I am arbitrarily saying that Radin made three missions to Dubrovnik in 1453-54 to negotiate the peace treaty after Stefan's war against the town. Radin is found in Dubrovnik for large parts of those two years, and it is impossible to determine how many trips he actually made. The Kosače missions are spaced as follows: 2 in 1419, 1 in 1423, then a gap — possibly broken by Dmitar in 1430 — until 1432 when 1 Patarin came, 2 in 1437, 3 in 1438, 1 in 1440, 2 in 1441, 1 in 1443, 2 in 1445, then a gap until 2 in 1450 then on to 1453-54 when we estimate Radin made 3 trips, 1 in 1456, 1 in 1459, 1 in 1461, 1 in 1462, 2 in 1465, 1 in 1466. In all but one of these twenty-seven missions at least one lay ambassador participated.(197) Frequently though, the Patarin (especially Radin) was the first negotiator. Of the twenty-seven Patarin envoys, Radin was the ambassador in eighteen cases, which suggests that Radin was trusted by Stefan and may well have been used frequently for that reason rather than because he was a Patarin. And in fact Radin occupied at Stefan's court a position which could accurately be described as foreign secretary.

The Pavlovići sent Patarins on twelve different missions. On four occasions more than one Patarin participated. If Radoslav did accede to Dubrovnik's request in 1439 to send one of his Patarins to discuss matters, a thirteenth embassy took place. The twelve known Pavlović missions,are spaced as follows: 1 in 1420, 1 in 1421, 1 in 1422, 2 in 1423, then a gap until 1430 when one mission occurred, 1 in 1432, then a five-year gap until 1 more is sent in 1437, 2 in 1438, 1 in 1439, and then a long gap until 1 mission was sent in 1454. Of these twelve missions, Radašin participated in four, Vlatko in three and Radin in three — with Vlatko and Radin participating once on the same mission. In nine of the twelve missions lay envoys definitely participated; in one case it is evident that Radašin came alone, in two cases we do not know about lay participation.

In the 1403 and 1430 royal missions and the 1450 Dinjičić mission lay figures also went along.

When we turn to the thirty-nine missions sent by the two leading magnate families, we find that in thirty-seven cases the Patarin envoy is mentioned by name; we find that Radin served on 21 (3 for the Pavlovići and 18 for Stefan Vukčić); Radašin, 4; Vlatko, 3 (excluding his 1403 mission on behalf of King Ostoja); Dmitar, 2; the total is 29 since Radin and Vlatko once served on the same mission.This means out of 37 missions (where the envoys are mentioned by name) in 29 of them, one of these four men participated. Hence, we may suggest that it was their individual merit rather than the fact they were Patarins which led to their selection. Of course, on occasions, for example in war-time, their clerical status might have been advantageous, and hence a factor in their selection. Dubrovnik did evidently see a difference between Patarin and secular envoys. The town usually showed special deference to Patarins, on occasions giving them larger gifts than it gave to the secular envoys, and once in 1439 specifically asked Radoslav to send one of his Patarins. Unfortunately, we do not know why Dubrovnik particularly liked Patarins.

However, despite the frequency with which these two noble families used Patarins after 1419-20, we should not lose sight of the role of secular envoys. First we must stress that prior to 1419-20, all the embassies sent by these families had been entirely composed of laymen — lesser nobles or *dijaks* (scribes). Thus Patarins only served these families as envoys during the final four decades of the Bosnian state. In addition, of the thirty-nine missions on which Patarins participated sent by these families, lay envoys accompanied the Patarins on at least thirty-four occasions.(198) Unfortunately, for many of these occasions it is impossible to determine whether the Patarin or a layman was the chief negotiator. Finally it should be added that the presence of a cleric on an otherwise entirely secular embassy was common in the Middle Ages, both in Slavic and non-Slavic states.(199)

Professor Dinić has extracted from the Dubrovnik archives all mention of Patarin missions. Thus this gives us a picture of Patarins dominating diplomacy. However, during this period, 1419-66, we find various missions composed entirely of secular figures. Unfortunately the records of the Ragusan council meetings for the fifteenth century have not yet been published; and no scholar has had the interest — or time — to extract the references to entirely secular diplomatic missions to Dubrovnik. Thus we do not have a basis on which to make concrete comparisons between the number of missions with Patarins and those without. However, when we take into consideration that relations were constant between these two families and Dubrovnik both for commercial and political reasons, and when we note the long time intervals without mention of Patarin envoys — 1423 to 1430; 1432 to 1437; 1445 to 1450; and on occasions longer for a particular family — we must assume that the diplomacy during these intervals was handled entirely by secular envoys.

It cannot be argued that many Patarin missions occurred which sources do not mention. Ragusan council records are preserved complete for these years of the fifteenth century. And since it was one of the council's tasks to vote upon matters of foreign policy as well as to receive and give gifts to envoys, we may assume that few embassies arrived in Dubrovnik that were not noted in the council minutes. Professor Dinić has combed all these, as yet unpublished, minutes and records, as well as letters to Ragusan ambassadors and court cases on wills, and has published all references to Patarins. Thus we may conclude that there were not many more than forty Patarin missions to

Dubrovnik. In the intervals between Patarin missions we know that relations continued, for we have a variety of charters issued by Sandalj, Stefan, or the Pavlovići to Dubrovnik, which are signed only by secular figures. In addition we know that the Konavli tribute was to have been paid regularly. We have only two references to Patarins being involved in its collection. Evidently during the years when we see no Patarin envoys, lay embassies had been sent from the Pavlović and Kosača courts to Dubrovnik to collect it. Thus, we have good reason to believe that when the Ragusan records are published from the fifteenth century — and publication has now reached 1389 — we shall find that the number of entirely secular missions greatly exceeded the number of those participated in by Patarins. Since most frequently the records only mention receiving letters brought by envoys without mentioning the nature of the business being transacted we are in no position to judge whether Patarins tended to be sent to discuss particularly critical matters.

Unfortunately we do not have data to judge about Patarin missions between the different courts within Bosnia. Dubrovnik mentioned Miaša Gost and a secular envoy going from Radoslav to Sandalj in 1423, and the king sending two unnamed Patarins to Radoslav accompanied by lay envoys in 1430. We may suspect that these nobles used Patarin envoys to a greater extent for missions within Bosnia than they did for missions to the coast.

Thus, we conclude that although we should not underestimate the role of Patarin diplomats in the Kosača and Pavlović foreign service in the years after 1419-20, we also should not exaggerate their importance. If we exclude the towering figure of Radin, who, almost certainly because of his specific talents rather than his clerical status, became Stefan's "foreign secretary," it can certainly be said that even after 1419 lay envoys continued to conduct the bulk of diplomatic affairs with Dubrovnik. Of course, within Bosnia, who knows?

XIII: Social and Political Position of the Bosnian Church

Now that we have discussed the period to 1443, we can turn to the question of the social and political position of the Bosnian Church. Prior to 1340 — the advent of the Franciscans — we may safely assume that most of the nobles in Bosnia were members of the Bosnian Church. This supposition seems certain since in much of the banate no other church was present. It is also supported by passing references to "heretical nobles" in papal and Hungarian letters. Between 1340 and 1391 some of them presumably followed the example of their rulers — Stjepan Kotromanić and Tvrtko I — and became Catholics, while others probably retained their loyalty to the Bosnian Church.(200) During this whole period, however, we do not have reference to a specific nobleman being a member of the Bosnian Church, except for the small number of nobles, only one of whom — Vlah Dobrovojević — is known from other sources, who are damned in the Serbian sinodiks.

However, although many nobles were members of this church prior to 1391, its influence on them and its services to them were probably solely religious. We do not find Patarins at their courts; we do not find the nobles aiding the Bosnian Church or it serving them in any secular way. All that we can point to is that the Bosnian Church had been among the guarantors and witnesses of two charters in the 1320's. Thus, in the years prior to 1391, we may well wonder how important this church was to the nobility in non-religious matters, and whether one is really justified in speaking of an alliance between this church and the nobility, which many scholars feel was a major aspect of the Bosnian Church's history throughout its existence.(201)

However matters may have stood, it is only in the period directly after 1391 that we find seven major families having connections with the Bosnian Church — six in Bosnia and one in what is now Hercegovina. Thus, the common, possibly unjustified, and certainly unsupported statement frequently made by historians that the Bosnian nobles as a group were politically allied with the Bosnian Church is based on sources from the last seventy years of the state. We shall now turn to these seven families, taking those in Bosnia first:

a) Tepčija Batalo: We have no evidence that Patarins served him, but we know that he was interested in their church. He had a Gospel manuscript copied for them and presumably gave them financial or other support since the inscription in that Gospel stated that he was good to good people. We thus have evidence that this church was important to him for religious reasons; there is no evidence that it played any other part in his affairs. We know nothing about the religious loyalty of any of his children. All documents about Batalo are from the 1390's; presumably he died about 1400. Thus we cannot speak about ties between his family and the Patarins in the fifteenth century.

b) Hrvoje Vukčić: He was called a Patarin and clearly considered himself something other than a Catholic, since in 1413 he complained that he wanted to become a Catholic for he did not want to die in the "pagan" rite. Krstjanin Hval copied a Gospel manuscript for him. This is the only specific Bosnian churchman linked with Hrvoje. We might note that a Catholic, pop Butko, also copied a Gospel manuscript for him.

There is no evidence that Patarins were at his court, or served him as diplomats, or even had any monasteries on his lands. Thus we cannot say that the Bosnian Church had any role in his political activities or even that it played an important part in his life — religious or otherwise. His wife was Catholic; so was his nephew and successor Djuraj Vojsalić, as was Djuraj's son Peter. It also is most probable that Hrvoje's relatives of his own generation included Catholics. Hrvoje, in fact, is the only member of the family known to have had connections with the Bosnian Church, although his uncle Vukoslav Hrvatinić (whose charter was guaranteed by that church in the 1320's) may well also have been a member. Hrvoje's career leaves us with the impression that he was indifferent to religious questions.

c) Pavle Klešić: We first hear of him in the 1380's. We first find reference to his connections with the Bosnian Church in 1403-04, when the djed did him the great service of pacifying King Ostoja's wrath against him and restored to him his confiscated property. In this case, then, we do have evidence that the Bosnian Church supported a noble. However, the church may have supported him because his case was just, rather than because he was a nobleman. The king it supported him against also seems to have been a follower of the Bosnian Church. Klešić's ties with the Patarins did not lead him to persecute Catholics. A Catholic bishopric existed in his city of Duvno and we know of a variety of Catholic churches, including Franciscan monasteries mentioned by Bartholomaeus of Pisa at Duvno and Glamoč, on his lands. The Catholics eventually succeeded in winning his family, and we find Pavle's son Vladislav a Catholic in the 1430's. There is no evidence that any Klešić besides Pavle and his wife were followers of the Bosnian Church.

d) Pavle Radenović, his sons Radoslav and Peter Pavlović, and Radoslav's son Ivaniš Pavlović: This family had active ties with the Bosnian Church. When Pavle Radenović was murdered in 1415 a Patarin — Vlatko Tumarlić — took his body back to Vrhbosna. When Radoslav Pavlović in 1430 made war on Dubrovnik, the town called him a Patarin. The family used Patarins as diplomats frequently — on twelve known missions — between 1420 and 1454. And it is evident that the family had several Patarins on call — at court or at a hiža nearby — since Dubrovnik once asked Radoslav to send "one of his Patarins" to Dubrovnik for negotiations. A Patarin also was found in 1422 as one of two administrators for Radoslav's part of Konavli. It has also been speculated that the Bosnian Church played a role in bringing about peace between Radoslav and Sandalj Hranić in 1423 and 1430. Although we cannot prove the church itself played this role, on both occasions Patarins were active as envoys. However, it is possible that they were simply acting on orders from their secular masters. Thus this family was closely connected with the Bosnian Church and was served regularly by Patarin diplomats between 1420 and 1454. This, then, is the first (and only Bosnian) family found connected to the Bosnian Church early in the fifteenth century which would retain its loyalty to the Patarins well into the reign of Stefan Tomaš.(202)

e) Peter Dinjičić: Unlike the four families noted above, no Dinjičić was ever spoken of as a member of the church. However, the Dinjičići, if not members, clearly had close ties with its clergy. The Ljubskovo hiža, which we have discussed, was located on their lands. And in the next chapter we shall find Peter Dinjičić in 1450 sending a Patarin as an envoy to Dubrovnik.

f) The brothers Dragišić (Pavle, Marko, and Iuri, sons of Ivaniš). Though never called members, we may conclude that they belonged to the church since its hierarchy, as we shall see in the next chapter, witnessed a grant of Ključ on the Sana to them from Stefan Tomaš in 1446. Because the king was then Catholic, we can assume the djed was made guarantor at the request of the Dragišići. Since they lived in an area with relatively strong Catholic influence, we cannot conclude that their predecessors had been members of the Bosnian Church.

These, then, are the only families in Bosnia for whom we can find concrete ties with the Bosnian Church. And out of these six families, only one — the Radenović-Pavlovići — had ties extending for a period longer than one generation.

In what is now Hercegovina, the bulk of the nobility mentioned in sources remained Orthodox throughout the Middle Ages. Only one family had known ties with the Patarins, and that happened to be the most important family — the Kosače.

g) The Kosače: Sandalj Hranić, Herceg Stefan Vukčić, and probably at least some of the herceg's sons. Dubrovnik spoke of Sandalj as being a member of the Bosnian Church, and in 1405 it suggested that the djed mediate a border quarrel between it and Sandalj. Whether Sandalj accepted the suggestion or not is unknown, but here we see the djed being suggested for a political role. The djed and his stroiniks also, as we shall see, were involved in guaranteeing a treaty between Stefan Vukčić and his family in 1453 that put an end to his son Vladislav's revolt. Both Sandalj and his successor Stefan Vukčić frequently used Patarins as diplomats after 1419. We know of twenty-seven diplomatic missions on which Patarins served the Kosače between 1419 and 1466. Stefan Vukčić also used the Patarin hiža near Goražde on his border with the Pavlovići as a border-customs station. Thus the Patarins did serve this family politically and diplomatically. This family and the Pavlovići are the only families to whom the Patarins gave considerable service and to whom this service was extended over a period of time longer than one generation.

It is almost certain that some (or even many) less important families did have connections with the Bosnian Church not recorded in the sources. For example, we may note the three men at whose burials Bosnian Churchmen participated, as we learn from gravestone inscriptions. However, the services rendered by the Patarins at burials can be considered part of their religious functions. Yet, because the only regular secular function that we have noticed the Patarins fulfilling was diplomacy, and because diplomacy was limited to the king and two or three great families and was not carried out by these lesser nobles, we may conclude that the impact of the Bosnian Church in the secular activities of these lesser nobles was slight. Thus other than possibly giving advice on occasions, or using their literacy to help draft documents, or possibly providing asylum or mediating quarrels brought to them, the Patarins' role in the lives of the lesser nobility who adhered to their church most probably was limited to religious matters.

When we examine the above data we find that only seven families had known Patarin connections. Of these seven families, the church only rendered secular services to five — Klešić, Dinjičić, Dragišić, Kosača, Radenović-Pavlović. And of these five, only the last two were regularly served by Patarins over a long period of time, and are known to have kept up ties with the Bosnian Church for a period longer than one generation. Both Klešić's and Hrvoje's successors were Catholics. Obviously then the frequently repeated generalization that the Bosnian nobles supported the Bosnian Church is an exaggeration. Such ties can only be shown for a short period of time — all the data pertains to the final sixty years of the state and only two families maintained such ties for the full sixty years.

From these few cases we cannot support the common generalization that the Bosnian Church supported the nobility or decentralization. There is evidence that at least King Ostoja (and the king was the representative of centralization) was a supporter of the Bosnian Church also. He once used a Patarin diplomat, and it is apparent that various Patarins lived at or near his court. Several times the djed is referred to as being present at court, and from Ragusan documents it is apparent that he played an influential role there. Not only did Ostoja sometimes follow his advice, but also Tvrtko II, most probably a Catholic throughout but certainly one by 1428, kept the djed as a councillor at court. Their continued presence at the royal court suggests the Patarins were not anti-royal. No document ever states that the Patarins took a particular postition; thus we cannot say that their advice or influence was weighted in a direction favoring the nobles or the king. We also see that frequently their influence was beneficial to Catholic Dubrovnik, and that frequently they appeared to have worked to maintain or bring about peace. Thus if Radin was regarded by Dubrovnik as its friend, even though he served its enemies Radoslav Pavlović and Stefan Vukčić, it is evident that Radin was not simply working to advance the interests of the local nobility.

We have only two cases, other than direct service as a diplomat, border guard etc., where Patarins helped nobles. First they gave shelter to Pavle Radišić's brother in 1403. However, it was regular for churches in the Middle Ages to give shelter to fugitives. This does not prove that the church supported Radišić's political position. It does not even prove that he was a member of the Bosnian Church. Second, the church played a role in restoring Klešić to his property, but as we have stated before, we do not know what the djed's motives were. These two examples certainly are not sufficient evidence to support a generalization that the Bosnian Church supported decentralization and the nobility against the king.

However, the Bosnian Church did in one respect indirectly work against centralization. Professor Babić points out that in other lands, the Catholic or Orthodox Churches supported the interests of the central government against fragmentation. Such a policy facilitated both the church's effectiveness in its own territorial administration and also served to gain the church greater landed estates. In Bosnia, however, the presence of the Bosnian Church prevented either of these other churches from attaining such a position of power and influence which might well have furthered centralization, and in so doing it thereby served the interests of decentralization.(203)

It is also difficult to judge how important the Patarins were to the two families which they did serve regularly. Prior to 1419-20, lesser nobles and other secular figures had served as diplomats. And even after 1419-20 the majority of diplomatic missions were probably still carried out by secular figures. Presumably, these petty nobles and clerks could perfectly well have handled all diplomatic dealings in the fifteenth century as well. One advantage the Patarin diplomat might have had would be that as a cleric he could more easily pass through warring lands.

Certain Patarin diplomats who were used frequently like Vlatko Tumarlić, Starac Dmitar, Radašin and Radin were obviously important to their masters. However, this importance may have lain in their individual abilities rather than in the fact that they were Patarins. And with the exception of the influence Gost Radin seems to have had on Stefan Vukčić's policy in the final fifteen years or so of his reign, we have no evidence that Patarins were involved in the making of policy decisions. It is probable that some, if not the majority, of Patarins used as diplomats simply carried out assignments given to them by their secular lords.

In one case a hiža served as a border station. Presumably Gost Gojisav was useful in this role but if he had not had his monastery there, we can be sure that a secular vassal of Stefan could have managed the task quite well. We have noted that the church — be it Catholic or Orthodox — at Glasinac also served as a customs house. Thus this was not a unique or unusual service carried out by the Patarins. We have also noted that on occasions, the king, Sandalj and Radoslav had made use of Catholic clerics as diplomats. That the Patarins were willing to carry out these services shows that their church was not completely focussed on religious and other-worldly matters. Thus, it was typical of its age. In Western Europe Catholic clerics regularly performed the functions we find Patarins fulfilling in Bosnia. The literacy, of at least many, of the Patarins as well as the fact that most of them do not appear to have been members of great families, having their own interests to advance, made them logical choices to be used for diplomatic missions.(204)

I have argued that it cannot be proved that the Bosnian Church supported the aspirations of the nobility as a class. Presumably a Patarin working for a given noble served that noble as best he could. Presumably also, the Patarins at the royal court served the king just as loyally as those at Sandalj's or Radoslav's courts served them.

And just as there is no evidence that the Bosnian Church had any political principles or worked to support an idea like decentralization, there is no evidence that the nobility in the fourteenth or fifteenth centuries strove to maintain or support the Bosnian Church. We have frequently pointed to the indifference of most of these nobles about religious matters. In Stefan Tomaš's reign, when persecution of the Bosnian Church was initiated, we shall see little sign that this bothered the nobility or that the nobility tried to defend the Patarins. There is no reason to believe that the nobility would have behaved any differently in the first half of the fifteenth century, unless it was felt that the church organization that would replace the Patarin Church would be opposed to the interests of the nobility — e.g., a Hungarian-run, pro-Hungarian Catholic Church, or even a church that clearly supported the king against the nobility.

When we turn to the relations between the Bosnian Church and the ruler (i.e., the ban and later the king) we find that here too the church had a minor role. We have do documentation about the church playing any political role in the thirteenth century — other than provoking crusades or plans for crusades against Bosnia. When we start getting sources in the fourteenth century we find that the rulers were not members of the Bosnian Church. Stjepan Kotromanić seems to have been Orthodox until his conversion to Catholicism in the 1340's and Tvrtko I was a Catholic throughout his whole reign. And though both rulers seem to have had cordial relations with the Patarins (Kotromanić even had the Patarins witness two charters), there is no evidence that the Patarins were regularly at court or had any political influence. This situation continued until the accession of Ostoja, when, for the first time (from 1403), we find the Patarins exercising political influence and playing a political role. In 1403 the Patarin Vlatko Tumarlić served Ostoja as an envoy to Dubrovnik and the djed's intervention in politics in the period 1403-05 has been discussed. We have noted his role in patching up the quarrel between Ostoja and Pavle Klešić. We also have seen that from 1403 to 1405 (i.e., extending into the first years of the first reign of Tvrtko II), the djed was present at court and his advice seems to have been frequently followed. Dubrovnik clearly believed the djed to be influential at this time and sought that its peace treaty with Ostoja — and then the treaty's confirmation when Tvrtko II came to the throne — have Patarins among its guarantors. But this impressive position is found only for the two-year period 1403-05. The djed may have been present at court prior to 1403 also, but since the sources do not speak of him, his influence may not have been great. This is not strange since neither Stjepan Kotromanić nor Tvrtko I was a member of the Bosnian Church, whereas Ostoja seems to have been. After 1405 the djed's influence seems to have declined and we hear nothing more about him in state affairs until 1428. Then Dubrovnik instructed an ambassador that if he found the djed at court he should give him a letter, for it was hoped that the djed would intervene on the town's behalf. From this we may conclude that the djed may still have frequently been at court and even have been a relatively influential figure. But still his role had declined from a guarantor to one who might talk the king into some policy the town desired. The decline is not surprising since, by 1428 (if not long before), Tvrtko II was clearly a Catholic. In 1430 Tvrtko used some unnamed Patarin as diplomats; but their use may have been caused by the fact the issue at hand was a war between two members of that church — Sandalj and Radoslav. The djed (or members of his church) was present at least on occasions at Tvrtko's court as late as 1442, for a Dubrovnik envoy sent thither then was told to get a treaty sealed before ordained members of either the Roman or Bosnian faith.

After 1442 we cannot demonstrate Patarins were regularly at court. We have noted that they stood behind a royal charter for the Catholic King Stefan Tomaš in 1446. But since this may well have been at the request of the grantees — the Dragišići — this does not prove close association with the king.

Thus we may conclude that the Bosnian Church did have some influence in politics. However, this influence was not constant, and only in exceptional cases does it seem to have been decisive. Hence we should not exaggerate its importance. Except for guaranteeing two charters in the 1320's and one more in 1446, the political role of the church in the affairs of the central (royal) government was limited to the period 1403-1442; and it was really only between 1403 and 1405 that this influence seems to have been of great significance. And in the second reign of Ostoja, the only ruler who seems to have been a member, the church does not appear to have had a significant political role. In 1418 we even found Ostoja twice using a Franciscan as a diplomat. Thus one certainly should not consider the Bosnian Church a state church. And we may well conclude that its influence on the state — and on society as well — was relatively unimportant.

The rapidity with which the church disappeared after the Turkish conquest, and the speed with which the populace accepted other religions may be evidence that the Bosnian Church had not created strong bonds between itself and its followers. Thus the peasantry and even the nobility seem to have lived their lives according to traditional patterns little affected by the teachings of this church. The greatest failure of the Bosnian Church was that it never seems to have made a concerted effort to build ties with the populace. The reason such ties were not created may partly have been owing to lack of popular interest.

However, this situation was also a natural result of the organizational structure of the Bosnian Church itself. The evidence suggests that the Bosnian Church as an organization — its hierarchy and clergy — was simply a Catholic non-preaching order gone into schism with Rome. Non-preaching monastic orders lack the organizational structure necessary to assume the task of serving the needs of a populace (i.e., of becoming a real church). To have been successful, the Bosnian Church would have had to re-organize its hierarchy and organization to handle the new situation in which it found itself. The limited evidence we have does not suggest that the Bosnian Church ever tried to reform or "modernize" its organization; there is no evidence that it ever ordained secular priests or tried to establish a territorial organization. In fact, the evidence suggests that, throughout its history, the Bosnian Church and its clergy — except for the few clerics involved in secular service at certain courts — remained a monastic order with its djed presiding over a relatively small number of ordained clerics living in monasteries. Orbini, writing in 1601, confirms this picture when he defines the djed as an abbot and the stroinik (sic! gost) as a prior of a monastery.(205) There is no evidence that the gosts were responsible for overseeing religious needs for the territory surrounding the hižas, or that Bosnia had been divided up into territories supervised by particular gosts. The number of hižas seems to have been fairly small, and there is no evidence that hižas could be found throughout Bosnia. Thus many villagers probably had no contact with Bosnian Churchmen at all, and we can say that the church's influence on society was limited to villages located near its hižas, and even these villages may not have developed strong ties with the church. Thus we must conclude that even though the Bosnian Church tried to serve as a church, it failed to realistically face the situation in which it found itself and by failing to re-organize itself into a preaching order, it never was able to make itself into a real church or to establish firm ties between itself and the peasantry.

Except for two charters from the 1320's guaranteed by the Bosnian Church, all evidence of Patarin participation in secular affairs comes from the fifteenth century. Thus we have reason to believe that these functions at court were newly acquired in that century. It is probable that the acquisition of this political role must have strengthened the Bosnian Church's position in the state and to some extent have worked for its preservation in the face of Catholic pressure to destroy it. Though the Patarin church did not play a major role at the courts of the Pavlovići or the Kosače, still the king could not freely attack the Patarin Church, lest it provide an excuse for those nobles to revolt against him. The Bosnian Churchmen, though, in accepting this role and in order to keep this role (which to some extent was necessary for the church's survival), had to become more worldly to serve the interests of these noblemen. Of course, it was not difficult for men like Gosts Radin, Milutin or Gojisav to be worldly. Yet it is clear that the Patarins serving as diplomats had to serve the interests of their secular masters rather than their church, for otherwise these lords would have utilized Franciscans or secular diplomats. Thus the failure of the Bosnian Church to acquire actual popular support, now in the fifteenth century with a Catholic offensive underway, left the Patarins totally dependent on the favor of a limited number of powerful nobles (by 1443 only the Pavlovići, Stefan Vukčić, and possibly the Dinjičići) to survive.

As a postscript we may note the secular functions of the hižas themselves. We have one example of a hiža near Goražde serving as a border post. In addition at least certain hižas functioned as hosteleries, one in Ljubskovo mentioned as a hostel between 1412 and 1416 and another in Bradina in 1418. We may suspect that Gojisav near Goražde also found it profitable to take in guests. This function was not unique to the Bosnian Church. Catholic and Orthodox monasteries did the same. We know that in 1415 the Ragusan envoy to the Bosnian court, Gundulić, stayed at the Franciscan monastery at Sutjeska. And Ragusan records mention fifteenth-century caravans stopping at Orthodox monasteries in the region of the Lim — Mileševo, St. Peter and St. Paul's at Bijelo polje and the Banja monastery near Priboj.(206) Sixteenth-century travelers report spending the night at these same Orthodox monasteries. Thus we see that the Bosnian Church by allowing its monasteries to take in travelers simply followed this general custom, and presumably it was simply retaining a custom that it had had when it was still a Catholic organization; at Bilino polje in 1203 the priors had promised to bury travelers who happened to die at their monasteries in their church cemeteries.

FOOTNOTES TO CHAPTER V

1. Dabiša had been mentioned by name only in various diplomatic records. The earliest reference we have to him is in 1358 when a Ragusan embassy was sent to Tvrtko, his relative Dabiša and Sanko Miltenović. (*Monumenta Ragusina,* II, p. 208). In 1366 a Ragusan ambassador was instructed to consult with Dabiša and Sanko because the town had hope and trust in them. Orbini claims that Dabiša joined Vuk in his rebellion against Tvrtko in 1366; we do not know if Orbini was correct in this, bur subsequently Dabiša did seek refuge from Tvrtko in Dubrovnik. Orbini calls Dabiša Tvrtko's first cousin. Evidently, since he was mentioned back in 1358, he was an older man when he came to the throne.

2. M. Dinić, *Državni sabor srednjevekovne Bosne* (SAN), Beograd, 1955; and S. Ćirković, *Istorija srednjovekovne bosanske države, Beograd, 1964, esp. pp. 224-226.*

3. Ćirković, *op. cit.,* pp. 224-25. Ćirković in a separate article, "Sugubi venac," *Zbornik radova Filozofskog fakulteta,* Beograd, VIII, 1964, pp. 343-69, presents his views in detail. He argues that Tvrtko's coronation created a new concept of state which was completely different from the earlier one under the bans. He argues that prior to 1377 the state was considered the personal property of the individual ban, whose rights over the state were the same as those of a noble over his lands. No differentiation was made between private and public power. However, Ćirković insists that by the beginning of the fifteenth century the state was considered a unit in itself, represented by the king and nobles collectively through the institution of the *sabor;* the idea of the state, symbolized by the crown, had become independent of the individual ruler; the towns and lands belonged to the crown and were considered inalienable to be passed on to the next king.

4. J. Fine, review of Ćirković's *Istorija . . .*, in *Speculum*, XLI, 1966, p. 527, where I said, "I must express scepticism, however, about Ćirković's contention that these self-seeking nobles really had a deep belief in abstractions such as the unity of Bosnia or loyalty to her crown." I still think his view is too abstract to be relevant to Bosnia and feel that evidence is lacking to prove it. Nobles were present to witness charters prior to 1377, and there is no evidence that the "state" was considered indivisible after 1377. One cannot say that after Radič Sanković sold Konavli to Dubrovnik, Vlatko Vuković and Pavle Radenović marched against him to preserve the unity of Bosnia. The fact that the two noblemen divided the recovered territory suggests that their motives were to increase their own holdings. Then in the 1420's both the king and the leading nobles agreed to the sale of Konavli to Dubrovnik by Sandalj and Radoslav Pavlović. I also cannot accept Ćirković's contention that the existence of the crown aided in stabilizing the Bosnian state. The period from Tvrtko's death well into the 1430's was medieval Bosnia's most chaotic and anarchistic despite the presence of the crown. And what unity there was in the last years of Tvrtko II and under Stefan Tomaš can be attributed to the decline of the great nobility (excluding the Kosače who for all practical purposes had created for themselves an independent state in Hercegovina.)

5. Dinić, *op. cit.*, pp. 62-65.

6. A. Babić, "Struktura srednjovjekovne bosanske države," *Pregled,* XIII, No. 1, 1961, pp. 1-8. Professor Babić points out that the Pavlović lands were a unique territory in that they formed a region without specific traditions, whose only cohesion was owing to the fact they had been accumulated by Pavle Radenović.

7. On the sources, see Ćirković, *op. cit.*, pp. 166-69.

8. On the Gospel manuscripts in general, see above Chapter II.

9. V. Mošin, "Serbskaja redakcija Sinodika v Nedelju Pravoslavija," *Vizantijskij vremennik,* XVII, 1960, pp. 278-353; the lists of anathemas referring to Bosnia, see pp. 302, 345-46, 348.

10. V. Mošin, *ibid;* "Analiz tekstov," *Viz, vrem.* XVI, 1959, pp. 317-394, and his "Rukopis pljevaljskog sinodika pravoslavlja," *Slovo,* 6-8, 1957, pp. 154-76. A. Solovjev, "Svedocanštva . . .," *GID,* V, 1953, pp. 44-68.

11. Mošin, *Viz, vrem.*, XVII, 1960, p. 302.

12. The "Ban Stefan" probably refers to Kotromanić, who was Orthodox until he accepted Catholicism in ca. 1340. Since he was clearly not a heretic it is plausible to link his damnation to some political opposition to Serbia, presumably in 1326 when he annexed Hum or 1351 when Dušan went to war against him. The other possible "Ban Stefan" was his father Kotroman who also was certainly no heretic since he married Stefan Dragutin's daughter. The Tvrdko on the list is certainly not the ban-king, since the man is given no title and since Tvrtko received his coronation from the Serbian metropolitan at Mileševo. Possibly, though, Tvrtko's activities in the Drina and Lim region were the cause of Vlah (or Vlaj) Dobrovojević's anathema. This nobleman anathematized is known from several of Tvrtko's charters.

13. We assume that the Priezda mentioned does not refer to Kotroman's father Ban Priezda or Kotroman's brother of the same name. These two thirteenth-century figures were both clearly Catholic. If it is one of them who is anathematized, then presumably he was damned for political motives. Priezda is not an unusual name. A Priezda S'finar was a witness to Ninoslav's 1234

charter to Dubrovnik (Miklosich, p. 28) and the name is also found in various Serbian epics. Nelep'c in the anathema may refer to Nelipac or one of the Nelipčić family who were involved in Bosnian affairs early in the fifteenth century; however, these Croatian noblemen were clearly Catholic as well. Protomara, the sister-in-law or daughter-in-law (snaha) of Ban Stefan (presumably Kotroman or Kotromanić) is not known from other sources.

14. Most recent edition, Dj. Radojičić, "Odlomak bogomilskog jevandjelja bosanskog tepačije Batala iz 1393 godine," *B'lgarska akademija na naukite* (Izvestija na Instituta za istorija), 14-15, 1964, pp. 498-507. We can only regret the large number of typographical errors in the rendering of the text. Also there is much in his introduction that cannot be accepted, particularly his acceptance of D. Mandić's (*Bogomilska crkva . . .*, p. 209) view that the list shows Bosnian djeds back to the beginning of the eleventh century.

15. Radojičić, pp. 504-506.

16. J. Fine, Jr., "Aristodios and Rastudije," *GID*, XVI, Sarajevo, 1967, pp. 225-29.

17. A. Solovjev, "La Doctrine de l'Eglise de Bosnie," *Académie de Belgique*, Bulletin, Lettres, 5e ser., XXXIV, 1948, p. 518.

18. On this Gospel manuscript see, G. Daničić, "Hvalov rukopis," *Starine*, III, 1871, pp. 1-146, esp. pp. 2-7. Daničić presents a valuable discussion about the manuscript and then presents the actual text.

19. N. Radojčić, "O jednom naslovu velikoga vojvode bosanskoga Hrvoja Vukčića," *Istorijski časopis*, I (1-2), 1948, pp. 37-53.

20. Miklosich, p. 253.

21. V. Klaić, *Poviest Bosne . . .*, p. 196. It is interesting to note that in medieval Bosnian history we have only three cases (only one of which went smoothly) of succession from father to son; a) Stjepan Kotromanić who, before he could assume his banate, faced a revolt which forced him to seek refuge in Dubrovnik; b) Stefan Ostojić (1418-21), who was rapidly deposed; c) Stefan Tomašević in 1461 who, though spared a local rebellion, was overthrown and beheaded by the Turks two years after his accession.

22. Lj. Stojanović, *Stare srpske povelje i pisma*, I, pt. 1, p. 170.

23. See below, note 40.

24. On Hrvoje, see F. Šišić, *Vojvoda Hrvoje Vukčić Hrvatinić i njegovo doba 1350-1416*, Zagreb, 1902.

25. Ć. Truhelka, "Grobnica bosanskog tepčije Batala obretena kod Gornjeg Turbeta (kotor Travnik), *GZMS*, XXVII, 1915, pp. 365-73. The inscription simply says that here lies the powerful man Tepčija Batalo (the) Bosnian and Radomil the dijak (clerk, scribe) writes (the inscription). ("Ase leži uzmo(žni) muž t(ep)čija Batal(o) bosansk(i) a pisa Radomil dijak'.") The mausoleum is found under a mound which the local inhabitants refer to as "crkvina" (church). Whether at some time in the past a real church had stood here is not known. If there had been a church, the fact that Batalo was buried in it might lead us to believe that it had been a Bosnian church — further evidence that the Bosnian Church did have churches. However, it is more than likely that the label came from the mausoleum itself. On the locality's name see, M. Mandić, "Turbe kod Travnika," *GZMS*, XXXVI, 1924, p. 85.

26. See above note 15, Radojičić, "Odlomak . . .," pp. 504-06. In the Gospel inscription when he refers to his brothers-in-law, it is Hrvoje Lord of the Donji kraji who is Vojvoda of Bosnia, Hrvoje's brother Vuk Hrvatinić who is Ban of

Croatia, and presumably Hrvoje's second brother Knez Vojislav "Vojvodić" who is referred to as Knez of Bosnia.

27. On Pavle Radenović see, J. Radonić, "O knezu Pavlu Radenoviću, priložak istoriji Bosne krajem XIV počet, XV veka," *Letopis Matice srpske,* Novi Sad, Vol. 211, 1902, pp. 39-62, and Vol. 212, 1902, pp. 34-61.

28. On Sandalj see, J. Radonić, "Der Grossvoyvode von Bosnien Sandalj Hranić Kosača," *Archiv für Slavische Philologie,* 19, 1897, pp. 387-456.

29. The following historical narrative is, unless otherwise noted, drawn from Ćirković, *Istorija . . .* 'the collective work *Poviest . . .* published in Sarajevo in 1942, and the monographs noted above in notes 24, 27, 28, on Hrvoje, Pavle Radenović and Sandalj. Any matter that is disputed or controversial will receive a special footnote, otherwise the reader may assume that an event or fact mentioned is generally accepted in the historical literature. Matters related to religious questions will in all cases be documented.

30. Ljubić, *Listine,* IV, p. 309.

31. Ljubić, *Listine,* IV, p. 378. This archdeacon represented Sandalj again in 1398, Jorga, II, p. 72.

32. Documents on the sale of slaves, see Dinić, *Iz dubrovačkog arhiva,* III (henceforth in notes to Chapters V and VI cited simply as Dinić), pp. 5-180, and G. Čremošnik, *Istoriski spomenici dubrovačkog arhiva,* ser. III, sv. 1, SKA, Beograd. 1932. That Catholics could not be sold as slaves is shown by an interesting case from 1393 when two Bosnian women bought as Patarins were freed by a Ragusan court after they testified that they were baptized Christians (i.e., Catholics born of a Catholic mother). On this case, see Ć. Truhelka, "Još o testamentu gosta Radina i o patarenima," *GZMS,* XXV, 1913, pp. 380-81. See also, Dinić, pp. 63-64.

33. Resti, p. 188.

34. Written communication from S. Ćirković.

35. F. Šišić, "Nekoliko isprava iz početka XV st.," *Starine,* 39, 1938, p. 170.

36. Hrvoje's wife made a large donation for the elaborate sarcophagus for St. Domnius in Split.

37. F. Šišić, "Nekoliko isprava . . .," pp. 207-08.

38. See below, note 82 to this chapter.

39. Šišić, *op. cit.,* p. 192.

40. V. Djurić, "Minijature Hvalovog rukopisa," *Istoriski glasnik,* IX (1-2), 1957, p. 48. It is also worth noting that this Catholic Gospel has appended to it certain non-canonical rites connected with the South Slavs' domestic ceremonies. The Butko Gospel includes the rite for a hair-cutting god-fathership *(šišano kumstvo)* and the wine blessing rite to be performed on St. John's Day. For this, see L. P.(etrović), *Kršćani bosanske crkve,* p. 111.

41. L. Thalloczy, "Vojvoda Hrvoja i njegov grb," *GZMS,* V, 1892, p. 173.

42. F. Šišić, "Nekoliko isprava . . .," p. 181; and *AB,* p. 65.

43. Ćirković, *Istorija . . .,* p. 197.

44. Pucić, I, appendix from službena knjiga, V.

45. Dinić, p. 222, No. 3; Jorga, II, p. 98.

46. A. Babić, "Diplomatska služba u srednjevekovnoj Bosni," *Radovi,* (NDBH), XIII, Sarajevo, 1960, pp. 11-70, esp. pp. 33-34, 66.

47. Dinić, p. 222, No. 3.

48. Dinić, p. 221, No. 1

49. Dinić pp. 221-222, No. 2
50. Dinić, p. 223, No. 4.
51. Dinić, p. 223, No. 4.
52. Lj. Stojanović, *Stare srpske povelje i pisma,* I, pt. 1, pp. 433-34.
53. *Ibid.,* p. 434.
54. *Ibid.,* p. 435.
55. See below, note 190.
56. Ćirković, *Istorija . . .,* p. 202.
57. M. Dinić, *Državni sabor . . .,* pp. 26-28, discusses the overthrow of Ostoja. On April 22, 1404 Ostoja issued a commerical charter to Venice; witnesses included Pavle Radenović, Pavle Klešić and Radič Sanković. Sandalj and Hrvoje were absent. On April 25 Ostoja was deposed. Dinić suggests that those present to witness the charter on the 22nd still supported Ostoja, while the two powerful absentees opposed him. A Venetian source confirms half of this theory, stating that Ostoja was ousted by the authority of Hrvoje (Ljubić, *Listine,* VI, p. 134). Hrvoje, however, makes it appear a collective decision; he announced to Venice that the nobles had ousted Ostoja and had chosen Tvrtko II (Ljubić, *Listine,* V, p. 45). Dinić points out that this shows Tvrtko II obtained the throne, not by any right of inheritance, but, by his selection by the nobility; his membership in the royal family simply made him one of those eligible. I think it clear that Hrvoje opposed Ostoja and was instrumental in his overthrow; I only wonder whether we can be sure that the three nobles present at court to witness the charter on the 22nd really supported him. After his recent experiences one would hardly expect Klešić to have backed Ostoja.
58. Dinić, pp. 181-82, No. 2.
59. Dinić, p. 182, No. 3.
60. Dinić, pp. 182-83, No. 4.
61. Dinić, p. 183, No. 6.
62. M. Dinić, "Jedan prilog za istoriju patarena u Bosni," *Zbornik Filozofskog fakulteta Beogradskog univerziteta,* I, 1948, pp. 33-44.
63. Šišić, "Nekoliko isprava . . .," p. 256.
64. L. Thalloczy, *Codex Diplomaticus comitum de Blagaj,* Budapest, 1887, p. 220.
65. Dinić, pp. 183-84, No. 7.
66. Ljubić, *Listine,* X, p. 229.
67. *Poviest B i H . . .,* Sarajevo, 1942, p. 409.
68. Dinić, p. 184, No. 8.
69. Hval's Gospel manuscript and its dedication, discussed earlier in this chapter, see above note 18.
70. Thalloczy, *Studien zur Geschichte Bosniens . . .,* pp. 348-50.
71. Theiner, *MH,* II, p. 179.
72. Fejer, X, pt. 4, pp. 682-93.
73. Lucius, *Memorie de Trau,* p. 268, As noted earlier I have not been able to get hold of the original text. I have drawn Lucius' material here from F. Šišić, *Vojvoda Hrvoje Vukčić . . .,* pp. 205-06.
74. A Venetian source just states that the barons and especially Sandalj were responsible for Ostoja's return to the throne (Ljubić, *Listine,* VI, p. 134).
75. Jorga, II, p. 119.
76. *AB,* pp. 87-89.
77. By 1410 Sandalj held most of the land from the Drina and Lim Rivers to

the Neretva. As a result of his seizure of the Sanković lands he gained the territory around Nevesinje, and from there up almost to Konjic, as well as the land from there down to Popovo. Pavle Radenović held the territory from Dobrun to Vrhbosna and from Olovo down to Ustikolina. His capital was at Borač. He had also obtained parts of the Sanković lands, including the rich town of Trebinje and the territory from Trebinje up to Fatnica and Bileća.

78. Sandalj gave active support to the Serbian despot in the latter's struggle against the Turkish leader Musa son of Sultan Bajazit, which resulted in Musa's death in 1413. Thus the despot or Sandalj, rather than Marko Kraljević (+1394), should have been the hero in the epic describing the death of the three-hearted Musa Kesedžija. On Sandalj's participation in the campaign against Musa, see *Poviest BH,* Sarajevo, 1942, pp. 437-38.

79. Having visited these catacombs I accept this conclusion which corresponds to that of Ć. Truhelka, *Kraljevski grad Jajce povijest i znamenosti,* Sarajevo, 1904, esp. pp. 57-60.

80. J. Dlugosz, *Historiae Polonicae,* XI, in *Opera Omnia,* XIII, Cracow, 1877, p. 141.

81. Fejer, X, pt. 5, p. 404.

82. Fejer, X, pt. 5, pp. 385-86 (who cites Lucius, *Memorie de Trau,* p. 392 ff.).

83. Oral communication to me.

84. Johannes de Thurocz, *Chronica Hungarorum* (written 1488) edition, Monumenta Hungarica, I, Budapest, 1957, pp. 92-93.

85. Dinić, p. 186, No. 14a.

86. Jorga II, pp. 162-63; Ostoja's son King Stefan Ostojić made use of the same Brother Stephen once in 1420 (Jorga, II, p. 180).

87. On Radoslav Pavlović see, A. Ivić, "Radosav Pavlović veliki vojvoda bosanski," *Letopis Matice srpske* (Novi Sad), 245, 1907, pp. 1-32; 246, 1907, pp. 24-48.

88. This is the second wife Ostoja had so disposed of. In 1399 he had abandoned Queen Vitača. (Jorga, II, p. 79). Sometime thereafter he married a woman named Kujava, mentioned in a 1409 charter (Miklosich, pp. 272-73). Presumably it was Kujava who was ousted in 1416. In 1415 it was reported in Dubrovnik that the Bosnian queen feared for her future because she was a relative of "Conte Polo" (i.e., Pavle Radenović) (Jorga, II, p. 151). In 1416 we learn of Ostoja marrying Jelenica, Hrvoje's widow (Gelcich-Thalloczy, *Diplomatarium . . .,* p. 261). Thus Ostoja's divorcing Kujava was probably influenced both by his desire to acquire some of Hrvoje's lands, as well as by the fact that Kujava was related to the Radenović-Pavlović family who now were in disfavor.

89. Miklosich, p. 301.

90. Stojanović, *Stare srpske povelje i pisma,* I, 1, p. 363.

91. Only the most important embassies including Patarins (or those providing information about the Bosnian Church) will be discussed in the text. For a complete list of those missions I have found see note 195 of this chapter.

92. See below note 195.

93. Stojanović, I, 2, p. 3; Pucić, I, p. 169.

94. Jorga, II, pp. 209, 211.

95. Jorga, II, p. 211.

96. Jorga, II, p. 211.

97. Dinić, p. 189, No. 20.

98. Miklosich, pp. 319-21; Radoslav re-issued the same charter from Borač that April (Miklosich, pp. 323-25). The same four envoys are referred to again. Radin whom Radoslav had first used as an envoy in March 1422, here a young man at the beginning of his career, will later achieve fame as Herceg Stefan's Gost Radin.

99. Miklosich, pp. 321-22. Miklosich is responsible for causing considerable confusion with his publication of this document, because he publishes it twice. The first time (p. 251) he reads the date on the document one thousand ''četiri sta i tret'je leto mesac fervara 31 d'n' '' (February 17, 1403); and the second time as one thousand ''četiri sta dva deseti i tret'e leto mesac fervara 21 d'n' '' (February 16, 1423). Since the document refers to Sandalj's wife as Lazar's daughter; and since Sandalj married Lazar's daughter in 1411 (Ljubić, *Listine,* VII, pp. 123-24) we must choose the second reading. Miklosich's error was caused by overlooking the word ''twenty'' the first time he saw the document. *AB,* p. 67, citing Miklosich repeats the mistake. This error of Miklosich is the source for the frequently seen erroneous statement that Sandalj sent Dmitar to Dubrovnik in 1403. As we have seen above, Sandalj began to use Patarins as diplomats only in 1419.

100. Stojanović, p. 371.

101. In 1413 (Stojanović, I, 1, p. 357) envoys from Sandalj pawned for him in Dubrovnik an icon depicting the Virgin. Whether the icon really belonged to Sandalj rather than to Jelena, his Orthodox wife, is not stated. However, on other occasions she pawned items in her own name.

102. We shall meet Radin constantly from the 1430's until his death in 1467. Dmitar we hear about once again in 1430, when the Ragusan council hears from a layman envoy Vlatko Pokalić that they must wait until Dmitar ''Crestianin'' comes with a letter from Sandalj (Dinić, p. 190, No. 25). We do not know if Dmitar ever did come, since Dinić publishes no more material about it. However, this mention does show that Dmitar was still alive, in favor, and serving Sandalj in 1430.

103. Ljubić, *Listine,* VIII, pp. 255-56.

104. V. Ćorović, ''Sandalj Hranić u Dubrovniku 1426,'' *Bratstvo,* 17, 1923, pp. 102-07; Jorga, II, pp. 230-31.

105. Resti, pp. 229-30.

106. See above note 35 to this chapter.

107. K. Jireček, ''Glasinac u srednjem vijeku,'' *GZMS,* IV, 1892, p. 100.

108. Stojanović, p. 371.

109. Pucić, II, p. 91.

110. Jorga, II, p. 217.

111. Ljubić, *Listine,* VIII, p. 227.

112. V. Grigorovič, *O Serbii v eja otnošenijah k sosednim deržavam v XIV-XV stoletijah,* Kazan, 1859, p. 52. In making his ''chronicle'' of Serbian history out of various older texts, Grigorovič has Russified the original language so (now that the manuscript of Konstantin's life of the despot is lost) that we cannot judge the date the manuscript might have been written. In Chapter II we have shown that the remark about Bogomils in this text cannot be accepted without hesitation.

113. On Catholics in Srebrnica: The earliest reference we have is to a Presbyter Marcus ''Capellanus'' in Srebrnica in 1376. Presumably he was

chaplain for the Ragusan commercial colony that grew up there in connection with the mines. Shortly thereafter we begin to hear of the Franciscan monastery there; it is mentioned in Bartholomaeus of Pisa's list of monasteries from 1385. We have frequent references to this monastery church (St. Mary's) from 1387 on. In 1395 we find mention of a second church, dedicated to St. Nicholas. Immediately after the Serbs recovered Srebrnica in 1413 artisans were hired from Dubrovnik to build a church there; clearly this was to be an Orthodox church. Srebrnica was to become a major Catholic center, and it seems by 1440 a Catholic bishopric was to be centered here. On Srebrnica see M. Dinić, *Za istoriju rudarstva* . . ., I deo, pp. 54, 57, 73, 94-95. Dinić is convinced that the majority of the inhabitants of the town were Catholics. See also, M. Vego, *Naselje* . . ., p. 108.

114. Thalloczy, *Studien zur Geschichte Bosniens* . . ., pp. 142-44. n. b. Tvrtko calls his subjects schismatics and unbelievers rather than heretics. Thalloczy (p. 143) erroneously states that that marriage never took place. However, documents in the Dubrovnik archive present a full description of the participation of Ragusan envoys at the marriage festivities (see Jorga, II, pp. 242-43). Thus we may safely conclude that the marriage took place.

115. Dinić, p. 190, No. 23, 24.

116. Dinić, p. 191, No. 29 (also Jorga, II, p. 282).

117. Dinić, pp. 190-91, No. 26, 27.

118. Dinić, p. 191, No. 28.

119. Ć. Truhelka, "Konavoski rat 1430-33," *GZMS*, 29, 1917, p. 185.

120. Gelcich and Thalloczy, *Diplomatarium* . . ., pp. 337, 347-49, 351, 355, 358, 363, 372, 373. The texts of course concern the war and have nothing more about Radoslav's religion than the negative lables about it, e.g., "ab perfido heretico patarino Radissavo."

121. Dinić, p. 224, No. 9, Jorga, II, pp. 308-09; Stojanović, pp. 619, 628.

122. Luccari, p. 151.

123. V. Ćorović, "Iz prošlosti Bosne i Hercegovine," *Godišnjica Nikole Cupića*, XLVIII, 1939, p. 136.

124. Stojanović, I, pt. 2, p. 111.

125. Wadding, *Annales Minorum*, Vol. X, original pagination, p. 195.

126. On these events see G. Čremošnik, "Ostaci arhiva bosanske franjevačke vikarije," *Radovi*, NDBH, Sarajevo, III, 1955, pp. 32-38. The summarized letter is from p. 34. Čremošnik's convincing reasons for re-dating the letter, pp. 34-35.

127. *Ibid.*, p. 32.

128. *AB*, p. 139.

129. Jorga, II, p. 317.

130. Drawn from M. Batinić, *Djelovanje franjevaca u Bosni i Hercegovini*, Zagreb, 1881, pp. 88-89; and Wadding, *Annales Minorum*, Vol. X, original pagination, p. 233. Wadding misnames Tvrtko's queen. Further evidence of the Franciscan order's (as opposed to the local Bosnian Franciscans) hostility toward Tvrtko is seen in the following remark attributed to a Hungarian Franciscan in 1435 just after Tvrtko had visited the Hungarian court. "The king is only outwardly a Christian, in truth he has never been Christened; in every possible way he has interefered with the Franciscans who in his land baptize his people." (Cited by L. Thalloczy, *Poviest Jajce* . . ., p. 38.) Thalloczy gives no reference and I have not been able to find his source. I suspect it comes from some later

Franciscan chronicle and may well be entirely based upon "the slander" Tvrtko accused Jacob of spreading about him through Christendom after Jacob's angry departure from Bosnia for Hungary (see above note 128). If my suspicions are correct this remark's preservation shows that Jacob's propaganda was effective and was to cause permanent damage to Tvrtko's reputation in Franciscan sources. I believe that Tvrtko's adherence to Catholicism has been shown above. Following the quoted remark, our Hungarian Franciscan continues, "For in his land live Manichee heretics. In vain have Ragusan priests worked. They tried to convince the heretics to send envoys to Basel, but three hierarchs the djed-gost, the starac and stroinik, wild enemies of the Catholic Church, threatened the king into refusing" (Thalloczy, *op. cit.*, p. 38). These remarks too are not original but are drawn from any one of several sources about Basel. At Basel the Bosnians were accused of being Manichees (see below note 137 of this chapter). The description of the Bosnian Church hierarchy is straight from the Ragusan letter to the council (see below note 135 of this chapter) though it may well be drawn from a later historian, such as Luccari, who quotes the Ragusan letter.

131. Čremošnik, *op. cit.*, p. 33; and Wadding, Vol. X, original pagination, p. 195.

132. *AB,* p. 159.

133. *AB,* p. 163.

134. *AB,* p. 163.

135. Dinić, pp. 192-93, No. 31.

136. Monumenta conciliorum Generalium XV, Concilium Basileense, II, Vindobona, 1873, Joannis de Segovia, *Historia gestorum Generalis Synodi Basiliensis* (ed. E. Birk), lib. ix, Chap. 5, p. 750.

137. *Ibid.,* p. 750.

138. *AB,* pp. 150-51.

139. *AB,* pp. 154-55; also Jorga, II, p. 346.

140. *AB,* pp. 160-61; Jorga, II, p. 342.

141. Miklosich, pp. 377-79. Though I am accepting the Vojsalić charter as authentic, there are several things which make me suspicious about the document. First we must always be on our guard with documents about the Franciscans. Second, the document purports to have been issued from Kreševo. A Franciscan monastery had been erected there prior to the Ottoman conquest, though we do not know exactly when it was built. Kreševo is a place about which forgeries have been composed and about whose "ancient Catholic heritage" the Franciscans have actively spread tales. With no contemporary documentation to support it, they have claimed that the twelfth-century Bosnian Catholic bishop resided at Kreševo. (e.g., Luccari, pp. 34, 52, refers to two Bosnian bishops — Milovan from the time of Borić and Radigost from the time of Kulin — as Bishops of Krescevaz. Farlati, *IS,* Vol. IV, p. 46 claims that ca. 1200 Patarins destroyed both the cathedral church and the bishop's court at Kreševo. If this destruction had actually occurred, the absence of any mention of it in contemporary papal letters about Bosnia is hard to explain.) The Franciscans also insisted and possibly even forged documents to demonstrate that the Franciscan monastery at Kreševo had been erected in the fourteenth century (see Chapter II, Appendix A). Why the Franciscans have fabricated a tradition for Kreševo is not clear, but we can assume that the fabrications were made by subsequent Kreševo Franciscans who sought to build up the prestige of their monastery. The Vojsalić charter (since Kreševo was not part of his lands) might have been

forged to suggest that the vicar resided at Kreševo. In addition to the suspect combination of Franciscans and Kreševo, I am somewhat bothered by the language of the charter. I am no linguist but when reading the charter I feel that the vocabulary and syntax are more modern than that of other fifteenth-century charters. The term ''katolicaške'' (Catholic) is strange. In other Bosnian documents the Catholic Church was not called Catholic but Roman or Christian (Rimska or Kršćanska). The only way to make the term's presence plausible is to attribute it to Franciscan influence; such an explanation is possible. The expression ''vikar kon' vikara'' is also unusual; it is similar to ''did kon' dida'' used only once in a charter of 1446 (Miklosich, p. 440) about whose authenticity some doubt has been expressed. Thus it is possible that both charters are fakes worked in the same period and in the same milieu. It would be helpful if linguists would examine both of these charters for general features and also if they would make a study of this expression, find other cases of its use, and see if, for example, it was not an expression more common at some later date. However, though doubts exist in my mind, I cannot prove the Vojsalić charter a fake; the evidence for its authenticity is impressive: The forger would have to have known of Vicar Johannes (Žuvan in the charter) who held office for only a very brief term; we would expect a forger to have had as vicar Jacob de Marchia, the most famous fifteenth century vicar who, after being a visitor to Bosnia in 1432-33, became vicar in 1435. In addition the individuals named are known from other sources and the land grants are plausible. Vojsalić's Catholicism is also attested by his church building. Thus I tentatively accept the charter as authentic.

142. Miklosich, p. 440. This charter is discussed below in Chapter VI.

143. D. Mandić, *Rasprave i prilozi iz stare hrvatske povijesti,* Rome, 1963, pp. 483-88. A full discussion of this bishopric is found in J. Fine ''Mysteries about the Newly Discovered Srebrnica-Visoko Bishopric in Bosnia (1434-1441),'' *East European Quarterly,* VIII, no. 1, March, 1974, pp. 29-43.

144. Jorga, II, pp. 320-21.

145. Truhelka, ''Konavoski rat . . .,'' *GZMS,* 29, 1917, p. 210.

146. Resti, p. 264.

147. S. Delić, ''Dva stara natpisa iz Hercegovine,'' *GZMS,* 23, 1911, p. 491.

148. Delić, *op. cit.,* pp. 496, 499-500.

149. Resti, p. 264. We might point out here that in addition to his nephew and successor Stefan Vukčić who had connections with the Bosnian Church, Sandalj may have had a relative who was an ordained member of the Bosnian Church. In the village of Kosače, near Goražde was found a grave inscription which according to Bešlagić reads, ''Here lies Gospoja (B)eoka kr'stjanica daughter of Pribisav Kosača'' (S. Bešlagić, 'Nekoliko novopronadjenih natpisa na stećcima,'' *GZMS*(arch), XIV, 1959, p. 242. Bešlagić dates the inscription fifteenth century but lacked data to be more precise. M. Vego, however, disagrees with Bešlagić's reading and comes up with, ''here lies kr'stijašinov's; Beoka''; in this case instead of a monastic title we simply have the name of Beoka's husband. (M. Vego, ''Novi i revidirani natpisi iz Hercegovine,'' *GZMS,* n.s., XIX (Arh), 1964, pp. 201-02.) I have not seen the stone and thus have no basis on which to take sides. We do not know who Pribisav Kosača was, but he clearly was a member of the family, and his daughter may have been ordained as a sister in the Bosnian Church.

150. On Stefan Vukčić Kosača, see S. Ćirković, *Herceg Stefan Vukčić-Kosača i njegovo doba* (SAN, pos izd.), Beograd, 1964.

151. On this church see *Istorija Crne gore,* Titograd, 1970, II, 2, pp. 483-88.

152. Miklosich, pp. 415-17.

153. We also saw in Chapter IV that heretics and Patarins came to Catholic churches when the Franciscans were saying mass, which corroborates my belief that the Bosnian Church neither opposed church buildings nor church services.

154. M. Dinić, "Zemlje hercega sv. Save," *Glas.* (SKA), 182, 1940, pp. 218 ff., and H. Šabanović, *Bosanski pašaluk* (Djela NDBH, XIV), Sarajevo, 1959 pp. 27-31 and his "Pitanje turske vlasti u Bosni do pohoda Mehmeda II 1463 g," *GID,* VII, 1955, pp. 37-51. But though Turkish conquests in the 1430's were less extensive than has generally been thought, Bosnia was still in a most critical position. Tvrtko II, expecting the end of his kingdom, in 1441 was negotiating with Venice about finding asylum there. The pressure was not to be alleviated until the successful crusade of 1443 (on it see, Chapter VI) whose effects, of course, were short lived.

155. *AB,* p. 161.

156. Jorga, II, pp. 362-63.

157. S. Kulišić, "Tragovi Bogomila u Boki kotorskoj," *Spomenik* (SAN), 105, 1956, pp. 41-43. He quotes from the Bishop's Archive in Kotor, sveska 1, p. 27.

158. In the text we shall only note those of special interest. A complete list is found below in note 195. In the period 1433-36, there is record of only one Patarin in Dubrovnik a man named Veseochus (Veseoko?) who was given permission to carry some oil from the town (Dinić, p. 224, no. 10). We are not told where he came from; we do not even know if he had been an envoy. He may simply have been sent from his monastery to obtain oil.

159. Dinić, p. 224, No. 12.

160. Dinić, p. 225, No. 13, 14.

161. S. Ćirković, *Herceg Stefan Vukčić-Kosača i njegovo doba,* Beograd (SAN), 1964, p. 44.

162. Dinić, p. 194, No. 35.

163. Dinić, p. 194, No. 36; p. 227. No. 20, 21; and Stojanović, I, 2, p. 54.

164. Pucić, II, p. 105.

165. Dinić, p. 184, No. 10.

166. Dinić, p. 185, No. 11.

167. Dinić, p. 185, No. 12.

168. Dinić, p. 186, No. 14.

169. Dinić, p. 186, No. 15.

170. See map at end of M. Vego, *Naselje* . . . . However, there is no reason to locate Ljubskovo exactly where Vego does, since Vego's placing of Ljubskovo requires the caravan route between Višegrad and Srebrnica to take a wide looping detour that has no justification in the geography or the sources. Vego's route properly leaves Višegrad and follows the Drina north, but from the point where he correctly has the route leave the Drina, he should have drawn the route so as to have it continue straight overland to Srebrnica with no significant deviations from a basically straight path. The straight route would correspond to present day peasant paths toward Srebrnica, would be shorter, and would be more reasonable than Vego's looping detour that is not mentioned in any source

(oral communication from V. Palavestra). Since the medieval caravans most likely did follow this straight path, then it is most probable that Ljubskovo lay somewhere on the straight line between Višegrad and Srebrnica on the Bosnian side of the Drina. V. Palavestra and M. Petrić, "Srednjovjekovni nadgrobni spomenici u Žepi," *Radovi* (NDBH), XXIV, Sarajevo, 1964, p. 142. suggest that Ljubskovo may well be the modern village of Ljubomišlja in which lies the largest medieval cemetery in the area. They describe the cemetery (pp. 147-52); its stones have few motifs and no inscriptions, thus giving us no basis to connect the cemetery with a particular confession. M. Dinić, *Za istoriju rudarstva u srednjevekovnoj Srbiji i Bosni,* I deo, Beograd, 1955, p. 34 places Ljubskovo in the same general area; but rather than identify it with Ljubomišlja, he locates it in the vicinity of three settlements (Purtići, Krnić and Parabučje) in the vicinity of Osat.

171. Dinić, *Iz dubrovačkog arhiva,* III, p. 187, No. 16, 17.

172. M. Dinić, *Za istoriju rudarstva . . .,* I deo, p. 34, cites documents from the Dubrovnik archive from 1425 and 1426, one of which says ". . . usque in Glubschovo in contrata Dragisini," showing that Ljubskovo lay in the territory of Dragiša Dinjičić. On Peter Dinjičić's use of a Patarin diplomat in 1450 see below, Chapter VI.

173. Dinić, *Iz dubrovačkog arhiva,* III, pp. 187-88, No. 18.

174. See Chapter IV, note 27 for a list of villages known to have had hižas. A thorough study of the material in the Turkish defters may eventually shed further light on the size of certain specific Bosnian Church hižas.

175. Dinić, p. 194, No. 37.

176. Dinić, pp. 194-95.

177. We know Borač was still held by the Pavlovići since three months later in April 1441 Radoslav Pavlović issued a charter from Borač. Miklosich, p. 405.

178. Dinić, p. 195, No. 39.

179. Dinić, p. 196, No. 40.

180. Dinić, p. 196, No. 41.

181. Ć. Truhelka, "Natpisi iz sjeverne i istočne Bosne," *GZMS,* VII, 1895, p. 348. "(as)e lež (i Rado . . .) . . naići (na svoi) na ple(menito)i postavi na nem' kamen' novi Gost Ra( )koe i Radiv(o) sin." (Here lies Radoe (?) on his family land and over him Gost Raško (Truhelka suggest Raško, it could equally well be Ratko) and Radoe's son place a new stone.) A slightly different reading, but with no change in meaning, is found in M. Vego, *Zbornik srednjovjekovnih natpisa Bosne i Hercegovine,* IV, Sarajevo, 1970 (henceforth, Vego, *Zbornik),* p. 135, No. 312.

182. A second inscription on another gravestone from the same general area (at Batkovići, 10 km. from Bijeljina) has been announced as referring to a cleric of the Bosnian Church. The stone is quite damaged and only part of the inscription can be read. S. Bešlagić, who announced it, reads it as follows: "Here lies Ra . . . . Milorad . . . ost . . . . ." Bešlagić believes that the second line mentions a Milorad Gost. The spacing between letters would allow this. S. Bešlagić, "Novopronadjeni natpisi na stećcima," *NS,* XII, 1969, pp. 141-42. However, M. Vego comes up with a different reading for the second line "Miloje Anko," thus arriving at something very different from "gost." (Vego, *Zbornik,* IV, pp. 116-17, No. 296). Vego prints a clear photograph of the inscription which I think clearly shows the man's name was Milorad. However, the second word is totally unclear, but from what can be seen I doubt

very much it says "gost." Thus I do not believe that we have another reference to a Bosnian Church cleric in the vicinity of Bijeljina. However, we have various indefinite references in written sources to members of the Bosnian faith or heresy in the north noted in reports about Jacob de Marchia's work in Srem in the 1430's. Srem of course lies beyond the Sava at considerable distance from Bijeljina. In addition, Danilo's "Life" of Stefan Dragutin, Ban of Mačva and Usora (1284-1316), says that Dragutin converted many heretics from the Bosnian land; presumably these heretics were to be found in his realm and possibly some of them may have been adherents of the Bosnian Church. We also noted heretics and Patarins in 1376 in Mačva, see above Chapter IV.

183. Bartholomaeus of Pisa, "De Conformitate vitae . . .," *Analecta Franciscana,* IV, 1906, p. 555.

184. Č. Truhelka, *op. cit.*, p. 348. "Ase leži . . . na svoi zemli plemenitoi. Asi postavista bileg sva . . . i Trdoe a toi pisa Goičin pop."

185. J. Krajinović, "Tri revidirana nadpisa iz okolice Travnika," *GZMS,* LV, 1943, pp. 234-36. "(Se leži Dragaj), unasrajaniš (or nakrajaniš) Ase pisa Vukašin svom g(ospodi)n', koji me biše veoma s'bludi zato molu vas gospodo ne nastupajte na n'jere ćete ve (=vi) biti kako on a (o)n ne može kako v' (=vi). V ime oca i sina 'z Bogdan na Dragoja novi kameni postavih, i da viste (=znate) da ćete umriti i vsi domu poginuti pravi (veli J. K.) apustl' i slava dj(č) et (i) i sa(boru). Se pisa Gost." Krajinović projects that the beginning of the inscription must have been here lies Dragaj . . . . Then it states: "I, Vukašin, write for my lord, who greatly defended (according to J. K., deriving s'bludi from s'bjudati) me; therefore I ask you, sir, do not tread on him because you will be as he and he cannot be as you. In the name of the Father and Son, I, Bogdan over Dragoja placed a new stone and you know that you will die and your whole home (household) will die (then according to J. K.) "the apostles correctly say and the glorious group to the council." The ending is confused. It is quite possible that a thought comes to an end after "the whole house will die." In the final phrase "say" and "council" are simply projections of J. K. In 1894 (*GZMS,* VI, 780-81) Truhelka read this last line ". . . poginuti pravi a pusti slav četi . . ." Thus even the word "apostle" is not certain. Thus I do not dare try to come to any conclusions about any doctrinal ideas that might be contained. Since the inscription begins with Vukašin writing and ends with an unnamed gost writing, J. K. plausibly suggests the two are the same and the gost's name was Vukašin. Of course, we do not know how the deceased had defended Vukašin — from persecution? bandits? It is interesting to note that this is the second case of a new stone having to be erected. Does this simply signify that originally a wooden marker had been placed over the grave, which now was being replaced by a stone? Or had an earlier stone been destroyed? If so, was it a case of grave robbing — hence the reminder on this stone to leave the corpse alone, which we find on a large number of Bosnian and Hercegovinian stones — or had Catholics shown hostility toward the Bosnian Church by damaging the stones? These questions, though interesting to ponder, unfortunately cannot be answered.

186. M. Vego, "Novi i revidirani natpisi iz Hercegovine," *GZMS* (arh), XVII, 1962, p. 231, "Ase se ovoi kamene uzvuče Radovan s Kr'stieninom' Radašinom za života na se." Radovan and Krstjanin Radašin set up these stones for themselves during their lifetime. (Reprinted in M. Vego, *Zbornik,* IV, pp. 14-15, No. 209).

187. M. Vego, *Zbornik,* IV, pp. 140-41, No. 317 claims that the stone does

not say "stroinik" but "etroinik." His photograph does seem to have a line through the initial "c" (cyrilic "s"). However, when I was at Zgunje in the spring of 1967 I carefully examined the stone and it said "stroinik" as clearly as if it had been printed. The initial "c" must have acquired a scratch through it since then.

188. Š. Bešlagić, "Novopronadjeni natpisi na stećcima," *Naše starine,* IX, 1964, pp. 138-39. "Ase leži Ostoja Kr'(s)tijan na Zguno." The description of the site comes from my own visit there.

189. M. Okiç, "Les Kristians (Bogomiles Parfaits) de Bosnie d'après des Documents Turcs inédits," *Südost-Forschungen,* XIX, 1960, p. 124, from a register of 1534.

190. Lj. Stojanović, *Stari srpski zapisi i natpisi,* SKA, Beograd III, 1905, p. 10. "A se leži Dobri Gost Mišlen komu biše pr(iredio) po uredbi Avram svoe veliko gostolubstvo. Gospodine dobri, kada prideš prid Gospoda našega Isuha ednoga spomeni i nas svoih rabovi, Pisa G. M." See also M. Vego, *Zbornik,* IV, pp. 60-63, No. 249.

191. Lj. Stojanović, *Stari srpski zapisi i natpisi,* Vol. 6, SKA, Beograd, 1926, p. 63.

192. The presence of a hiža in the vicinity is suggested by the fact that Turkish defters mention lands connected with krstjani in three different villages in the later Turkish nahija of Sokol: in Podi, Suhodlak and Kunovo (see Okiç, "Les Kristians . . .," pp. 123, 127-28). In this last village at the time of Herceg Stefan (1435-66) a certain Cvatko Gost had sold some lands (see below Chapter VII). This suggests there may well have been a hiža at Kunovo. However, as we shall see from the example of Gost Radin, every gost did not have a hiža. We shall also find that Bosnian Church figures, at least in some cases, retained private property; thus we cannot assume Cvatko was disposing of church property. (On the private wealth disposed of by Gost Radin in his Testament see below Chapter VII; for 22 examples of hereditary land *(baštinas)* of specific krstjani whose names are given — including Krstjanin Ostoja of Zgunje whom we have just discussed — see Okiç, pp. 123-24, and also above in Chapter II, where I discuss *baštinas* of Bosnian Church clerics.) Also possibly pointing to the existence of a hiža in the vicinity is the treaty between Herceg Stefan and his son Vladislav signed in 1453 guaranteed by the djed and a dozen stroinici at the place Pišče on the Pivska Mountain (see below, Chapter VI). We do not know whether this agreement was signed at a hiža — like Stjepan Kotromanić's land grant to Vukoslav Hrvatinić at the Moštre hiža in 1322 — or at one of Herceg Stefan's numerous residences.

193. V. Skarić, "Grob i grobni spomenik gosta Milutina na Humskom u fočanskom srezu," *GZMS,* 46, 1934, pp. 79-82. "Va ime tvoe pričista Troice gospodina Gosti Milutina bilig rodom Cr'ničan' "; "Pogibe i on ego li milosti božiei"; "žitie a žih u časte bosanske gospode primih darove od velike gospode i vlasteo i od gr'čke gospode a vse vidomo."

194. S. Ćirković, *Herceg Stefan . . .,* p. 232, mentions Stefan's embassy to Thomas and his hospitality to Manuel Cantacuzenus.

195. The statistics include only missions sent by secular figures and do not include the few missions sent by djeds and by Gost Radin. This section also does not consider the few cases we have noted of Catholic clerics serving as diplomats for the king and for certain nobles. The missions we have taken into consideration (with source references) are as follows:

1. September, 1403, King Ostoja sent Vlatko Patarin to Dubrovnik (Dinić, p. 222, No. 3; Jorga, II, p. 98).

2. January 29, 1419, Sandalj sent Dmitar Krstjanin to Dubrovnik (Stojanović, *Stare srpske povelje i pisma,* I, 1, p. 363).

3. December 21, 1419, Sandalj sent Divac Krstjanin to Dubrovnik (Stojanović, I, 1, p. 304).

4. November 1420, Radoslav sent Krstjanin Vlatko Tu(mar)lić to Dubrovnik (Stojanović, I, 1, pp. 567-69; Miklosich, pp. 306-08).

5. July 1421, Radoslav sent Vlatko Krstjanin to Dubrovnik (Stojanović, I, 1, p. 319).

6. March 1422, Radoslav sent Raden (Radin) Krstjanin to Dubrovnik (Stojanović, I, 1, p. 324).

7. January 16, 1423, Radoslav sent a "Patarin and others" to Sandalj (Jorga, II, p. 211). January 27, 1423 we find a reference to Radoslav having sent Miaša Gost to Sandalj (Dinić, p. 189, No. 20). These two references certainly refer to the same mission.

8, 9. February 15, 1423 Radoslav sent Krstjanin Vlatko Tumr'ka (sic!) and Radin Krstjanin to Dubrovnik, and Sandalj sent Starac Dmitar to Dubrovnik. The two not only dealt with Dubrovnik but also concluded peace between the two noblemen (Miklosich, pp. 319-21).

10. In 1430 Dubrovnik expected Sandalj to send Dmitar Krstjanin thither. We do not know whether he actually came (Dinić, p. 190, No. 25).

11. June 1430, King Tvrtko sent two unnamed Patarins to Radoslav (Dinić, pp. 190-91, No. 26, 27).

12. June 1430, Radoslav sent some unnamed Patarins (including a starac) to Sandalj (Dinić, p. 191, No. 28).

13. In 1432 Radoslav sent Krstjanin Radin to Dubrovnik (Dinić, p. 224, No. 9; Jorga, II, pp. 308-09; Stojanović, I, 1, pp. 619, 628).

14. February 1432, Sandalj sent an unnamed Patarin to Dubrovnik (Dinić, p. 223, No. 7).

15. January 1437, Radoslav sent Radašin Patarin (and companions) to Dubrovnik (Dinić, p. 224, No. 11; Stojanović, I, 1, p. 633).

16. April 11, 1437, Stefan Vukčić sent Radin Starac to Dubrovnik (Dinić, p. 224, No. 12).

17. April 20, 1437, Stefan Vukčić sent Radohna Starac to Dubrovnik (Dinić, p. 193, No. 33).

18. February 1438, Stefan Vukčić sent Radašin "Christianin" to Dubrovnik (Dinić, p. 226, No. 18).

19. June 1438, Stefan Vukčić sent Starac Radin to Dubrovnik (Dinić, p. 193, No. 34 and p. 225, No. 15, 16; Stojanović, I, 2, p. 46).

20. September 1438, Stefan Vukčić sent Starac Radin and Krstjanin Radelja to Dubrovnik (Stojanović, I, 2, p. 48).

21. August 1438, Radoslav sent Radašin Krstjanin to Dubrovnik (Dinić, p. 226, No. 17).

22. November 1438, Radoslav sent Radašin Krstjanin and Juraj Krstjanin to Dubrovnik (Dinić, pp. 226-27, No. 19).

23. January 1439, Radoslav sent Krstjanin Radašin Vukčić to Dubrovnik (Stojanović, I, 1, p. 635; Miklosich, pp. 397-98).

24. August 1439, Dubrovnik requested that Radoslav send thither "one of his Patarins." We do not know whether Radoslav obliged (Dinić, p. 194, No. 35).

25. April-May 1440, Stefan Vukčić sent Radin "Starac christiano" to Dubrovnik (Dinić, p. 194, No. 36, and p. 227, No. 20, 21, Stojanović, I, 2, p. 54).

26. April 1441, Stefan Vukčić sent Starac Radin to Dubrovnik (Dinić, p. 228, No. 23).

27. August 1441, Stefan Vukčić sent Starac Radin to Dubrovnik (Dinić, p. 228, No. 24).

28. In 1443, Stefan Vukčić sent a Patarin named Radivoj to Dubrovnik (Dinić, p. 229, No. 26).

29. February 1445, Stefan Vukčić sent Starac Radin to Dubrovnik (Dinić, p. 197, No. 43, 44).

30. July 1445, Stefan Vukčić sent Starac Radin to Dubrovnik (Dinić, p. 198, No. 45; *AB,* p. 197).

31, 32. In 1450 Stefan Vukčić twice sent Gost Radin to Dubrovnik (Ćirković, *Herceg Stefan . . .,* pp. 114, 129 (Note 48)).

33. In 1450 Peter Dinjičić sent Radohna Krstjanin to Dubrovnik (Dinić, p. 229, No. 28).

34-36. In the course of 1453-54 Stefan Vukčić was represented in Dubrovnik for considerable periods of time by Gost Radin. It is impossible to tell how many trips he actually made. I estimate three)*AB,* p. 218; Dinić, p. 209, No. 64).

37. In 1454 Nikola and Peter Pavlović sent an embassy including two stroiniks (Gost Radosav Bradiević and Starac Radosav) to Dubrovnik (Miklosich, pp. 469-72).

38-41. In Februrary 1456, March 1459, March 1461 and February 1462, Stefan Vukčić sent Gost Radin to Dubrovnik (Dinić, pp. 231-33, No. 34-42).

42. In 1465 Stefan Vukčić sent Gost Radin to Dubrovnik (Dinić, p. 234, No. 46, 47).

43. In December 1465 Stefan Vukčić sent Tvrdisav Krstjanin to Dubrovnik (Stojanović, I, 2, p. 78).

44. In the summer of 1466 Stefan Vukčić sent an embassy including two Patarins (Krstjanin Tvrdisav and Krstjanin Čerenko) and an Orthodox metropolitan (David of Milesevo) to Dubrovnik (Miklosich, pp. 495-98).

We list 44 missions here and in our discussion we include only 42. The reason for this is that in the discussion we do not include numbers 10 and 24; for in these two cases our sources state that Dubrovnik was expecting or hoping for a mission that included a Patarin, but we do not know whether the missions actually came.

196. By not adding 1404, I am assuming that the krstjani and stroinici sent for Pavle Klešić in January 1404 were sent by the djed alone and not by the djed and Ostoja (as one Ragusan document states, Stojanović, I. 1, p. 435). If I am incorrect, then we must increase the number of royal Patarin missions to three. I also do not include the Spring 1438 mission to Dubrovnik sent jointly by the

djed, king and Radoslav, since we do not know who the envoys were (and thus whether any Patarins were on it).

197. In one case (Radin's second trip in 1450) we do not know whether a secular envoy participated. In my statistics I assume one did (for on the basis of statistical probability it seems most likely) however, it is possible that Radin had a secular companion on 23 rather than 24 missions.

198. See note 197; here with the "at least" I am not including Radin's second mission in 1450.

199. Examples of such joint embassies in other lands can be found in K. Jireček, *Istorija Srba,* Vol. II, pp. 28-29.

200. In what is now Hercegovina prior to 1391 all nobles whose religion we can identify were Orthodox.

201. The origins of this common opinion are a historiographical problem of little relevance to this study. It can be found expressed for example in, S. Ćirković, " 'Verna služba' i 'vjera gospodska,' " *Zbornik Filozofskog fakulteta,* Beograd Univerzitet, VI, 2, 1962, pp. 95-111; F. Šišić, *Vojvoda Hrvoje Vukčić Hrvatinić i njegovo doba,* Zagreb, 1902, p. 236; M. Dinić, "Bosanska feudalna država od XII do XV veka," in *Istorija naroda Jugoslavije,* I, 1953, p. 520, etc.

202. We also find evidence that the Pavlovići maintained peaceful relations with the Catholics. We have earlier noted Radoslav's use of a Franciscan (Brother Stephen) as a diplomat in 1422. In addition, the Franciscan monastery at Olovo (noted in 1385) was on their lands. This monastery continued to function throughout the medieval period; there is no reason to believe that Pavle Radenović or his successors the Pavlovići, despite Bosnian Church ties, interfered in any way with this monastery. Since in local folklore — which has been greatly influenced by the friars at Olovo — the Pavlovići, never called heretics, are glorified as heros against the Turks, we have reason to believe that the family had good relations with the Franciscans.

203. A. Babić, "Nešto o karakteru bosanske feudalne države," *Pregled,* V, No. 2, 1953, p. 84. Professor Babić, claiming that the Bosnian Church did not possess large estates, goes on to argue that since the Catholic Church did amass large estates wherever it established itself in Europe and since it was doing so in Bosnia (as can be seen from Bela IV's charter of 1244 as well as from grants to the church by various bans), the nobility, in the interests of keeping its estates intact, ought to have defended the Bosnian Church.

However, this theory now seems invalid; data from the Turkish defters, which have become available since Babić's article was published, give us reason to believe that the Bosnian Church like other Christian churches of its day, had considerable landed property. This problem is discussed in Chapters II and VI.

204. Unfortunately, we know nothing about the families from which Patarins came. In only a few cases are we given a last name — or patronymic — for a Patarin; these few names do not indicate connections with any of the top families. In addition none of the numerous documents about the leading families even suggests that any of their members became ordained in the Bosnian Church (excluding the disputed inscription about Beoka — see note 149 for this chapter — about which Bešlagić and Vego cannot agree whether she was or was not a krstjanica). However, it is likely that some of the Patarins were drawn from the lesser nobility; the Turkish defters' references to *baštinas* (hereditary estates) of certain krstjani indicate that some of them came from well-to-do families. Yet,

even so, the origins of the great majority of Patarins, whom we have met in the sources, remain a mystery. Therefore, since we cannot show that a significant number of Patarins were drawn from the higher classes, we cannot use family ties to argue that the Bosnian clerics should have supported the interests of the nobility.

205. Orbini, p. 354.

206. M. Dinić, "Dubrovačka srednjevekovna karavanska trgovina," *Jugoslovenski istoriski časopis*, III (No. 1-4), 1937, pp. 123-24.

CHAPTER VI

BOSNIA FROM 1443 TO 1463

I: Increased Mention of Dualism in Sources from the 1440's

Prior to the 1440's, the existence of dualism in Bosnia is suggested by only a small number of sources, most of which are Italian. Because Slavic and Hungarian sources rarely hint at it, it seems probable that this dualist current was unimportant. The material we have examined thus far on the Bosnian Church strongly argues against that church being dualist, though it may well have acquired certain practices or attitudes under dualist influence. In the period to be examined now, there are more references to dualism, some of which even try to link the Bosnian Church with dualism. These references to dualism begin appearing in documents in the late 1440's and continue until the fall of the Bosnian state, after which they were incorporated into early historical works. And thus was established the generally accepted view that both the Bosnian Church and heresy were terms for a single dualist movement existing in Bosnia throughout the Middle Ages. Yet even in this final period while some sources speak of dualists, others continue to speak of the Bosnian Churchmen in the same manner that they had throughout the earlier fifteenth century, and still others continue to speak vaguely of heretics and schismatics or even link the Bosnians with the Eastern rite. Either some of these sources are inaccurate or else the religious situation in Bosnia entered a phase so complex and varied as to be indescribable on the basis of our limited source material. No satisfactory account of these final twenty years has ever been advanced. And no hypothesis is really possible that does not make some highly speculative assumptions.

One might suggest that there was a change in the actual religious situation in Bosnia after the mid 1440's and that the dualist movement mentioned in the Italian sources (and presumably quite unimportant until then) suddenly grew and attracted a large following or possibly even influenced the Bosnian Church. Yet much evidence exists to counter such suggestions. Sources such as Gost Radin's will, which we shall examine in Chapter VII, make it clear that the mainstream of the Bosnian Church remained non-dualist. However, the possibility that a minority broke away from the mainstream and formed an actively dualist wing cannot be excluded. Such a splintering off by a dualist minority from a non-dualist parent body is not unheard of: Nina Garsoian, as we have noted above, has postulated that a dualist current developed within the ranks of the Paulicians in Constantinople,(1) and the appearance of dualist groups among the Waldensians is well documented. Such a development would give us a plausible way to explain not only the increasing number of documents calling the Bosnian Church "Manichee," but also four documents, all independent of one another, whose testimony about the church being dualist is very hard to reject. These documents, all to be discussed in this chapter, are: 1) The Ritual of Kr'stjanin' Radosav. This ritual most probably is derived from a Cathar ritual, and although there is nothing heretical in its content, it does suggest ties between some Bosnian Churchmen and dualists. 2) A treatise on trade by Benko Kotruljić. In it we find the term "Manichee" used for Bosnian Churchmen by a presumably well-informed Slavic-speaking eye-witness merchant from Dubrovnik who had no connections with the papal court or the inquisition and who had no known reason to be slanderously attacking religious beliefs in Bosnia. 3) A letter of Patriarch Gennadius of Constantinople from the 1450's referring to *Kudugers* — a strange term generally believed to mean Bogomils — being influential at the court of Herceg Stefan. 4) The documents concerning the three Bosnian noblemen, called leaders of the heresy at court, who were sent to Rome as "Manichees" by the presumably bi-lingual Bishop of Nin.

If the Bosnian Church had split into two wings, one of which exhibited dualist features, the popes in this period would have had a concrete reason to call the Bosnian Church dualist; the Catholics might have thought that dualism had infected the whole church or might, at least, have found it convenient to depict it as doing so. However, we must stress that such a split cannot be proved — the documentation is far too scanty. And one might well ask, whence did such a dualist wing receive its impetus? Although there had been a history of dualists in the area, and although the Bosnian Church, with its lack of any effective central organization as well as of a corpus of clearly defined dogma, would have been an ideal body to subvert, it seems that foreign dualist movements were dead or moribund by the 1440's. It is hard to believe that the impetus for such a direction would have come from Bosnia's non-intellectual non-speculative inhabitants.

However, if we do not want to postulate such a division within the

Bosnian Church, we are left with a unified church, which despite certain deviant practices (some of which may have seemed similar to those of dualists), remained as it had been all along — a non-dualist, more or less orthodox Slavic-rite off-shoot of a Catholic monastic order. In this case we must seek in the authors of the sources that speak of it the source for the increased references from the late 1440's to dualism in that church. And here we must turn to the papacy, and the papal legates and Franciscans in the field. We may regret that we know so little about the interaction of these three elements. Did they work out this interpretation together? Did the popes hand it down and the local Catholics find it convenient to accept it? Or did the local Catholics — legates and/or Franciscans — think it would aid their own work to pass such information on to the popes far away in Rome? Unfortunately, our limited sources allow no more than speculation about these questions. However, regardless of who initiated this interpretation, we have two choices as to how it came to the fore: a) through a misunderstanding of the situation in Bosnia, b) through a deliberate frame-up:

a) The Bosnian Church, which we have suggested originally grew out of a Catholic monastic organization with no guidance from people educated in any aspect of theology or church practices, surely did acquire various heterodox practices. Some of these may well have resembled certain dualist practices — whether they were independently arrived at or acquired under the influence of the dualists found in the area. The Franciscans and papal agents, encountering such practices, might have associated them with dualism and have concluded that the Bosnian Church was dualist. The name Patarin would have served as confirmation of that suspicion. And if these agents while in Bosnia did encounter actual dualists, they might well not have realized that these dualists had nothing to do with the Bosnian Church and instead have found in them confirmation of their suspicions. Lacking theological sophistication, the Bosnians would not have understood what the Catholics were looking for or were bothered about, and hence would have been of limited help in aiding the agents to realize that the Catholic Church had a mistaken view of the Bosnian Church. The papal agents throughout behaved in Bosnia as if "heresy" and "orthodoxy" were as meaningful and realizable to the Slavic peasant as the terms — and what they represented — were to the agents.

It is also possible that the "Manichee" view originated in Rome owing to papal confusion over the term Patarin. In this case it is probable that papal agents and Franciscans in Bosnia, who were hostile to the Bosnian Church, would have found this misunderstanding helpful to them; it was likely to result in greater papal interest in and support of their efforts in Bosnia. Thus we would not expect them to have gone out of their way to correct Rome's confused views about Bosnia.

b) Some people (whether they were papal agents, Franciscans or both, we cannot be sure) for motives of their own, decided to frame the Bosnian Church. Presumably linking the Patarins with dualism — and the valuable pejorative label Manichee — would have served to justify the destruction of a small body that was in fact only schismatic. These people would have been assisted by the fact that, throughout the Middle Ages, people had been attributing heresy to Bosnia, and by the fact that various popes in the fourteenth and early fifteenth centuries had regarded the Bosnian Church as being heretical in some way. Thus it should not have seemed difficult to convince the popes that the Bosnian Patarins were the same sort of heretics as the Italian Patarins.

Papal agents, seeking a united Bosnia loyal to Rome to participate in a crusade against the Turks, might well have felt that the destruction of the Bosnian Church and the elimination of Patarins from positions of trust at political centers would enable Catholics to occupy these positions and would insure the loyalty of these various centers to Rome.

Simply having a Catholic king had not been sufficient for this end. The rulers, except for possibly Ostoja, had regularly been Catholic since the 1340's, yet the Bosnian Church had retained its position in the state. The existence of Bosnian Churchmen at, for example, the courts of Stefan Vukčić or the Pavlovići probably prevented the permanent residence of Catholic clerics there who might have had — at least in Rome's views of matters — predominant influence on those nobles. Hence, when the Bosnian king decided on a policy pleasing to Rome, there was no guarantee that either of these magnates would support it. In fact, they could have easily opposed it and gone to war against the king. Thus to achieve a unified policy against the Turks under papal leadership, it was necessary that toleration of the Bosnian Church clergy be replaced by persecution.

The destruction of the Bosnian Church might have seemed easily realizable; its clergy, as we have argued, was not too numerous, seems to have been concentrated in monasteries, and does not seem to have forged close ties with the populace. Papal agents might well have reasoned that it would not be difficult to effect the exile of these clerics, which would in effect destroy the Bosnian Church as an institution; then the uneducated peasants, who simply considered themselves Christians, might easily be won over to Catholicism.

Because it does not assume a level of ignorance on the part of the papacy that is hard to explain, this second theory is the one toward which I lean. (Of course, we must assume a certain degree of ignorance about Bosnia on the part of Rome for such a frame-up to have worked.) However, I shall stress in the beginning that we have no documentary evidence to show such a frame-up had been plotted, and we cannot clearly link it to any particular individuals. Unsatisfactory as it may be, it is the explanation of the chaotic situation that for me best fits with all the evidence that we have examined and shall now examine which strongly argues against the Bosnian Church being dualist. And of course this hypothesis does not exclude the theory that a dualist wing broke away from the church; such a split would have been most useful to frame the non-dualist mainstream of the Bosnian Church.

II: Civil War between Stefan Tomaš and Stefan Vukčić

Shortly after Tvrtko II's death in 1443, King Ostoja's son Stefan Tomaš was chosen as king.(2) Apparently, Stefan Vukčić Kosača did not participate in his election, since he immediately refused to recognize him and announced his support of the new king's brother, Radiovj, who a decade earlier had been put forward as an anti-king by the Turks, and who had been supported by Stefan Vukčić's uncle and predecessor, Sandalj.

In 1443 the papacy was interested in mobilizing a great Christian offensive against the Turks. The papal legate in Bosnia was Thomas Tomasini, Bishop of Hvar, who, though a Venetian, must have known Slavic, both through his service in Hvar and through his long association with the Bosnians. He had been assigned as permanent papal representative to Bosnia in 1439 and was to hold that post until 1461; unfortunately the few sources we have about his mission tell us little about his personality, his politics, his activities in Bosnia, or his relations with the Franciscans. That he would have had an influential part in forming papal views on Bosnia and the Bosnian Church is certain. And if he played a role in convincing the papacy that dualism existed in Bosnia we can be certain that he acted not through ignorance or through misunderstanding matters, but deliberately. If the Bosnian Church was to be the victim of a plot, it is most probable that Thomas of Hvar played a part in it.

Resti reports that in 1443 the pope offered Stefan Tomaš a crown if the king would join the league against the Turks and if he would persecute Manichees.(3) Because Resti wrote centuries later and believed that there had been Manichees in Bosnia, we cannot take the term "Manichee" in his history as evidence of the actual situation. The pope, though, most probably would have been interested in drawing Bosnia into his league against the Turks, and Thomas of Hvar could well have approached Stefan Tomaš at this time on the subject of the league and heretics; that he may even have offered him a crown is confirmed by Stefan Tomašević's letter to Pope Pius II in 1461, a source we shall discuss below.

However, Bosnian participation at this time in any sort of league was not likely, since a civil war had broken out almost immediately upon Stefan Tomaš's accession between the new king and Stefan Vukčić. The pope could not expect the king to send troops against the Turks at such a time, and the pope had no means to make Stefan Vukčić desist from making war on the king. It may have been this situation that led papal agents to conclude that it was necessary to eliminate the Bosnian Church. They may have believed that if Radin and the Patarins were ousted from the Kosača court and replaced by Catholic clerics, the Catholic Church could then bring about peace between the two camps. And with such a peace, it might have seemed possible to create a united Catholic Bosnia, willing to refrain from separatism and civil wars and to unite behind papally-sanctioned leaders to execute papal policy, and make a sizable contribution to the crusade against the Turkish threat.

In 1443 the papal league, without any Bosnian participation, worked effectively, as Catholic armies under the valiant Hunyadi marched through Serbia and into Bulgaria, capturing Niš and Sofia, and forcing the Turks to withdraw from these regions. In June 1444 the Turks recognized these Christian gains and a ten-year truce was signed. The Balkan Christians seemed to have gained the much needed breathing space to recover and prepare their defense. Serbia also re-appeared as a state.

In Bosnia, quite oblivious to the existence of this major confrontation between Christianity and Islam and to its results, the war continued between Stefan Vukčić and the new king. The papacy clearly sympathized with the king for in 1444 Hunyadi confirmed Stefan Tomaš as king. Ivaniš Pavlović now joined the king's side and together they began to make considerable progress against Stefan Vukčić. Needing allies, Stefan Vukčić turned to Alfonso of Naples, a successor of Hrvoje's Ladislas of Naples. Alfonso dreamt of greater things than Ladislas; not satisfied with only prospects of the Hungarian throne he also fixed his sights on the Byzantine empire. As his ambassador Stefan Vukčić sent a Ragusan monk, "the Abbot of Santo Jacomo."(4) Thus in addition to Patarins, Stefan Vukčić was willing to use Catholics, and Catholics were willing to serve him. Alfonso willingly accepted Stefan Vukčić as a vassal but in February 1444 instead of a contingent of men Stefan Vukčić received from his new suzerain a belt of the Order of the Virgin Mary which also bestowed upon him any and all privileges of that order.(5) Though Stefan Vukčić would doubtless have preferred troops, at least we see that the ruler of Naples was willing to make a supporter of Patarins a member of the Virgin's order. Thus already we find Stefan Vukčić in touch with all three confessions, while, of course, maintaining cordial relations with the Turks. He was vassal of a Catholic lord and a member of the Virgin's order. The Bosnian Church was represented by Radin and other Patarin diplomats, and Orthodoxy by his wife Jelena, daughter of Balša III of Zeta. She had married him in 1424, and remained his spouse until her death in 1453.

With the bulk of Turkish troops out of the Balkans, and with momentum on the Christian side, the crusaders decided that they had a chance to finish with the Ottoman presence in Europe once and for all. Ignoring the ten-year truce that they had agreed to, they resumed the offensive.(6) This resulted in a rapid Turkish mobilization and the defeat of the crusaders at Varna in November 1444. The Serbian ruler George Branković who had wisely remained neutral reaffirmed his status as vassal of the Ottomans.

After Varna, Stefan Vukčić at war with the pope's and Hunyadi's protegé Stefan Tomaš, received help from the Turks. In addition, the Turks encouraged their vassal George Branković to provide aid. Thus Stefan Vukčić, who had lost much of his territory to the king, emerged as the stronger and, taking the offensive, he began to recover the lands he had lost. Turkish soldiers proved a better remedy for his troubles than the belt of the Virgin. The war was to continue until 1446.

During 1444 we hear of heresy but only in vague terms. Fra Fabian, the new vicar, was dispatched to root out heresy in the vicariat, in Hungary, Bosnia, Moldavia, Bulgaria, Raška, and Slavonia.(7) A month later we find him at work against the Hussites.(8) Thus, clearly, he had not gone to the kingdom of Bosnia. The next year Fra Fabian had to battle somewhere in the vicariat against the wiles of evil enchanters and enchantresses.(9) Because his chief concern seems to have been with Hussites, these fiends probably were not in Bosnia proper.

III: King Stefan Tomaš Accepts Catholicism

In May 1445 Pope Eugene IV recognized Stefan Tomaš as the legitimate king and granted him permission to end his marriage, since he had not been properly married. He had simply taken the girl according to the Bosnian custom, that she be good and faithful, without a church wedding.(10) That the pope was asked to allow the annulment suggests that Stefan Tomaš by May 1445 was already accepted by the pope as a Catholic or else as a man who had given sufficient guarantees of his intent to become one.

That the king's relations with the Catholic Church were good is also seen by the fact that in the summer of 1445 several Franciscans served him as envoys to Dubrovnik.(11) The Franciscans' service to Stefan Tomas at this time gives us strong reason to doubt the frequently advanced opinion that in 1445 the Franciscans and the king had had a serious quarrel over his toleration of heretics. No contemporary source describes such a quarrel; it is mentioned only in later Franciscan chronicles and also in the alleged letter of Pope Eugene IV given by Farlati. In Chapter II, I have discussed this letter and have explained why I do not think it is authentic. Thus having sources to suggest that relations between the king and Franciscans were good, and lacking any contemporary document to show the contrary, I conclude that this quarrel never took place.

Early in 1446 peace was concluded between Stefan Vukčić and the king. To seal it, they agreed that Stefan Tomaš should marry Stefan Vukčić's daughter, Katarina, a plan which perhaps had been envisioned as far back as early 1445 when Stefan Tomaš had requested the pope to annul his earlier marriage. A Catholic wedding was held in May 1446. We can be sure that Katarina had accepted Catholicism prior to that date. Since subsequently we find Pope Nicholas revoking privileges issued by his predecessor Eugene to Stefan Vukčić, who had expressed to Eugene his wish to become a good Christian, we can suggest that at the time his daughter accepted the Roman faith, Stefan Vukčić had announced his intention of doing the same. However, we have no evidence concerning the circumstances or date that Stefan Vukčić had expressed this wish. He could equally well have declared this intention earlier through Alfonso of Naples when he had been in difficulties over his war with Stefan Tomaš.

A papal letter of July 1446 stated that Thomas of Hvar had converted King Stefan Tomaš to Catholicism.(12) This letter did not state what the king had been prior to his conversion; we can suppose that he had been a follower of the Bosnian Church since his father Ostoja had supported that church. A letter of Pope Nicholas V from July 1452, saying the king and his wife had previously laid aside Patarin errors, confirms this supposition.(13)

The sources also do not say when Tomaš accepted Catholicism. The papal letters of 1445 seem to suggest that the king was already a Catholic. Why had the pope recognized him as king if he was not yet a Catholic? And why had the king corresponded with the pope like a Catholic about the question of annuling a non-advantageous marriage?

Yet in 1446 the papal letter states that Stefan Tomaš had just accepted Catholicism, converted by Thomas of Hvar. It is apparent that he was not baptized in 1446, for Pope Pius later states that the king followed the religion of Christ, but abstained from the sacrament of baptism until the late 1450's when he was baptized by Gioanni Cardinal di S. Angelo.(14)

Pius does not explain why the king had abstained from baptism. Some scholars have attributed his delay to a prejudice against the ceremony derived from dualists who were hostile to baptism with water. I think we can find a much simpler explanation. Most probably the king had already been baptized in a Bosnian Church ceremony and believed that he should be treated like a member of the Orthodox Church who was being received into the Catholic Church; for it was against canons to rebaptize members of the Orthodox Church. If Resti is accurate, Stefan Tomaš' father Ostoja had baptized at least one son; Resti tells us Dubrovnik had sent representatives to the ceremony in 1401.(15) And it would be likely that a father who chose to baptize one son should have baptized them all. Clearly Stefan Tomaš had not been a Catholic for he was just accepting that faith, and had he had a Catholic baptism there would have been no question of him being rebaptized. We have argued above that Ostoja was probably a member of the Bosnian Church and it is likely that he would have brought up his children in his own faith. Thus we may conclude that the ceremony of baptism Ostoja's children underwent was a Bosnian Church ceremony; and we can argue that it was a ceremony similar to the Catholic and Orthodox Church ceremonies because Dubrovnik did not hesitate to call the ceremony a baptism.

IV: The Bosnian Church: The Dragišić Charter and Radosav Ritual

There is also another odd aspect about the king's acceptance of Catholicism at this time. In August 1446, only a month after the pope stated that the king was definitely a Catholic, the supposedly Catholic king turned around and issued a land grant to the brothers Dragišić guaranteed by the djed of the Bosnian Church. Charters guaranteed by the Bosnian Church are rare; in fact up to 1446 we have come across only two authentic charters of this nature, both from the 1320's. It is curious that this custom should be revived — if indeed it had lapsed — at a time when the influence of the Bosnian Church at the royal court seems to have been essentially finished, when the king was claiming to be a loyal Catholic, and when the documents reveal Catholic successes all over Bosnia. The dating also might trouble one, for 1446 is the date given in the notable forgery about the bogus Council of Konjic. In Chapter II, I have also argued that a letter attributed to Pope Eugene, which I consider a forgery, was also dated about this time (November 1445). Thus, it is not surprising that this charter's authenticity has been questioned. Yet the language of this charter arouses no suspicion, and the witnesses all were figures active in 1446 (though one might wonder if so many dignitaries would really have been present for the granting of Ključ on the Sana to the Dragišići).(16) I accept the charter as authentic but shall not use it as a basis for conclusions unsupported by other evidence as well.

The charter begins with the usual invocation to the Trinity. Then follows the king's title and the contents of the grant. It is guaranteed that for no act of disloyalty or no sin toward the king will the grant be revoked, unless it seems just to the "Lord Did" (djed), the Bosnian Church, and the "Good Bosnians" (a term for faithful noblemen). "And this and all that is written above we deliver to the Lord Djed Miloje and to all other djeds who shall succeed him (Didu Miloju i didu kon' dida) in the hands of the church." Then follow the secular witnesses and the forces whose curse shall act upon violators of the charter.(17) Previously we have noted that the expression 'did kon' dida' means the present djed and the djeds that come after him (hence his successors); the same term had been used in the 1434 charter of Djuraj Vojsalić made before the Bosnian vicar. If authentic, this 1446 charter gives us the djed's name and shows that in 1446 the Bosnian Church still was recognized as a legitimate body in the state by the king despite his Catholicism. No doubt the grantees, the Dragišići, who presumably were members of the Bosnian Church, had suggested that the djed serve as guarantor. We cannot estimate the Bosnian Church's ability in 1446 to prevent violations of the charter's terms. This is the last reference in the sources to friendly ties between Stefan Tomaš and the Bosnian Church. If the Patarins had been present at court up to this time, they were not to remain there much longer.

At some time during the reign of Stefan Tomaš we have one more reference to a djed, namely a dedication in a Gospel manuscript which simply says that Radosav Kr'stjanin' writes this book for Goisak (? Goisav) Kr'stjanin in the days of King Tomaš and Djed Ratko. If the scribe has erred in his copying, please do not insult him and may God bless you forever, amen.(18) The manuscript, generally called the Radosav Gospel after its copyist, was copied during the reign of Stefan Tomaš (1443-61); since in the 1450's the king began to persecute the

Bosnian Church, the fact he is mentioned suggests that it was copied in the first half of that reign. From it we learn of a djed named Ratko, who is not known from any other source.

This manuscript also contains a ritual in Slavic which can accurately be described as an abridged version of the Cathar ritual of Lyons. This ritual, discussed in Chapter II, may well have been brought to Bosnia by dualists. Whatever its origins, it should be noted that it contains not one heretical thought; its contents could have been accepted by Catholics and Orthodox as well. Since the Irish Franciscan in 1322 stated that the Patarins had the same rite as the Schismatics (i.e., Slavic Orthodox),(19) and a Hungarian chronicler (we shall discuss later) calls the Bosnians Eastern rite Christians, it is possible that the Bosnian Church — if it ever used the Radosav ritual in services — utilized more than one ritual.

This would be a fitting place to note one last Gospel which scholars have been unable to date precisely. Generally assigned to the early fifteenth century, this Bosnian Gospel might have been connected with the Bosnian Church. It has the following brief entry at the end: "I wrote this by the Grace of God (a) Christian (Hrstjanin) called Tvrtko Pripković of the Gomilanin land."(20) Whether "Hrstjanin" should be read "krstjanin," making Tvrtko Pripković a Bosnian Churchman, or "hrišćanin" (or the variant form hristienin), making him an Orthodox believer is impossible to determine. The only "Gomilanin land" I have been able to locate is Gomiljani-pravica in the region of Trebinje. Since Trebinje is a region having many Orthodox believers (as well as Catholics) and since we have no evidence of Bosnian Church activities in the region, I suspect Tvrtko was Orthodox. In addition, in the vicinity of Gomiljani-pravica, Professor Radimsky turned up three church ruins; the names of two of their localities (Djurdjeva crkva and Konstantinova crkva) suggest that the original churches had been Orthodox which would indicate that previously Gomiljani had been an Orthodox place.(21)

V: Progress of Catholicism from 1446

From 1446 on when both King Stefan Tomaš and Stefan Vukčić's daughter had accepted the faith of Rome, the Catholics made steady and impressive gains. Even Stefan Vukčić seems to have expressed his willingness to become a Catholic. A Ragusan letter of 1451 mentions Pope Eugene IV (1431-1447) granting certain privileges to Stefan Vukčić who had expressed his willingness to be baptized and become a good

Christian.(22) We hear of a church being repaired at Glamoč in 1446.(23) This reflects at least a certain amount of interest in Catholicism in this town. New churches were built to the Virgin at "Bozaz" (probably Bočač, between Jajce and Krupa) and at "Rossetan" in Hum(24) (probably Raštani on the Neretva, not far from present-day Mostar.)(25) The Catholic advance thus can be seen also moving along the Neretva into Hum. The pope offered indulgences to all who visited these churches.

The king's brother, Radivoj, now at peace with Stefan Tomaš, was also Catholic; in 1446 the pope placed his lands under papal protection.(26) Djuraj Vojsalić's son and successor Peter Vojsalić was also a Catholic, since the pope stated that Vojsalić was the only prince *(princeps)* who was Catholic in the land.(27) This statement has led to some misunderstanding since some scholars have taken it to mean that Vojsalić was the only Catholic nobleman. However, the word "princeps" does not mean just any nobleman, but refers only to the greatest magnates. In Bosnia at the time only three men merited the title "princeps," Stefan Vukčić, Ivaniš Pavlović and Vojsalić. And the other two were flirting with Catholicism. Stefan Vukčić, having expressed intentions of conversion to Pope Eugene, seemed about ready to submit to Rome. Even Ivaniš Pavlović seems to have accepted Catholicism sometime in 1446, for a 1449 papal letter complained that after three years of Catholicism, Ivaniš had reverted to his vomit.(28) At about the same time that we hear of Ivaniš' acceptance of Catholicism, we hear of a priest's house (chaxa de prete) at which Ragusan caravans stopped in his capital of Borač(29) Since in no document does Dubrovnik refer to Patarins as priests, we may assume that "priest" refers to either a Catholic or an Orthodox cleric. Since this is the only reference to him, we do not know when he arrived in Borač or how long he was to remain there. However, it is plausible to consider him a Catholic and connect his presence with Ivanis's fliration with Catholicism. In 1448 Ivaniš ordered a horn from Dubrovnik upon which he wanted a picture of St. Blasius (the patron saint of Dubrovnik) holding a model of his Borač castle.(30) This shows that Ivaniš was interested in the Catholic saint; whether this interest was tied to his acceptance of Catholicism or whether he always respected saints is of course unknown.

Other nobles mentioned as Catholics were Restoje the *Protovestijar* and George Tardislavić, a Hum noble.(31) Tardislavić reflects Catholic gains in Hum.

Papal successes continued in 1447. Vojsalić's land was placed under papal protection.(32) The pope gave the king permission to have two

Franciscans at court as chaplains and to use them to help in state affairs.(33) Thereafter we find references to them at court; frequently the king was to use Franciscans as ambassadors. If Patarins were still to be found at court this late, we may suspect that the Franciscans supplanted them. The king and queen also completed several churches: St. Thomas' at Vranduk, St. George's at Jezero (near Jajce), St. Mary's at "Virben" (?) and a Holy Trinity Church at "Verlau" (?).(34) With the building of churches, Catholic influence could be expected to penetrate into the regions around them.

Ćirković points out that there were both cultural and military reasons for many nobles to accept Catholicism. In this period the Bosnian nobles, enthralled with Western culture, began to imitate it and found the values and much of the content of this culture intertwined with Catholicism. In addition, the Turkish threat increased the necessity of forging military alliances with the Catholic West; frequently their heresy would prove to be a stumbling block to an alliance's realization.

Stefan Vukčić, though, seems to have taken no steps toward fulfilling his promise. Radin was still at his side; twice in 1447 Ragusan ambassadors to his court were instructed to speak with Radin Gost.(35) Thus ten years after he had been made a starac, the Bosnian Church elevated Radin to gost, its second highest rank. Perhaps his church, in an effort to counter Catholic gains, had advanced him because he was the man among them who was most influential among the secular lords. Though other gosts presided over hižas, there is no evidence that Radin ever did. He seems to have continued to reside at Stefan's court; his influence may have helped prevent Stefan's conversion to Catholicism.

VI: Papal Mention of Manichees 1447-1453

For the first time since Pope Boniface IX's reference to Manichees in a list of foes of Sigismund in 1391, in July 1447 a pope referred to Manichees in Bosnia. Pope Nicholas V, in placing the property of King Stefan Tomaš under papal protection, stated that Stefan Tomaš, scorning the Manichee errors in which he had been ensnared, had become the first Bosnian king to accept the Catholic faith.(36) The pope had forgotten that Ban Stjepan Kotromanić from ca. 1340, Ban and later King (from 1377) Tvrtko I, and King Tvrtko II had all certainly been Catholic. Such ignorance even of recent Bosnian history in Rome leaves us little confidence that the term "Manichee" should be taken literally.

Whereas it is always possible that prior to his acceptance of

Catholicism, Stefan Tomaš had been involved with real dualists, it seems much more probable that he had been a member of the Bosnian Church. And in fact, the king's former connection with the Bosnian Church is later explicitly stated by Pope Nicholas V in 1452, when he wrote that Stefan Tomaš had laid aside Patarin errors.(37) In this case the pope, like many others before him in Western Europe, probably was using the term ''Manichee'' very loosely to mean any heretic he did not like. And while we may have reason to doubt that the Patarins were really heretical, there is little doubt that most popes thought they were. Beyond this was the continued use of the term Patarin on both sides of the Adriatic to refer to the Bosnian Church, which would easily lead Italians, already set in their belief that the Bosnian Church was heretical, to connect the Bosnian Patarins with the dualist Patarins who, until recently, were to be found in Italy. Finally, there is the virtual certainty that dualists, probably in small numbers, did exist in and around Bosnia proper. It is too much to expect that Italian clerics, concerned with religious matters on a European scale, would distinguish between different groups in Bosnia, especially since distinctions between groups there may not always have been clearly defined. The blurred distinctions between groups was further confused by the fact that the different groups seem to have influenced one another. Various pecularities sometimes found in the behavior of certain members of the Bosnian Church may well have been derived from the prejudices, ideas, or practices of real dualists, be they Bogomils or dualists from Dalmatia with ties with Italy. We have already noted in Bosnia examples of hostility toward the cross, and possibly even toward certain sacraments. We have just noted a Bosnian Church ritual which very possibly had been brought from the West to Bosnia by dualists. We may even suggest that it was links between the dualists and the Bosnian Church which started the Dalmatians calling the Bosnian Church Patarin in the first place. Thus maybe we should not be surprised that in 1447 the pope made reference to Manichee errors. But we must ask: why had the popes rarely done so earlier? Why, suddenly in the late 1440's and thereafter, does this ''heresy'' or schism about which the popes have constantly spoken in vague general terms as existing in Bosnia finally receive a specific label? Once one pope (Nicholas) used the term, however, it is not strange to find the term repeated thereafter by other popes or clerks in the papal chancellery. We also must keep in mind the possibility that Nicholas used the term ''Manichee'' because someone (a legate or Franciscan) wriging from Bosnia and trying to take advantage of the confused situation described above had used the term, thereby hoping to gain papal demands for the elimination of the Patarin.

In February 1448 Nicholas V placed the lands of Restoje, *Protovestijar* of Bosnia, under papal protection, and praised him for holding firm in his Catholicism though living among "heretics."(38) It is interesting that he does not repeat the term "Manichee" which he had used the previous July. But in any case, regardless of the vocabulary used, it seems that the papacy was beginning to exert increasing pressure on the Bosnians to eliminate the Bosnian Church.

In the period 1449-53 we have four mentions of Manichees; none give any details. The first is a letter cited in Raynaldi's eccelsiastical history, compiled in the eighteenth century. Nicholas V in 1449 when writing to Thomas of Hvar made passing reference to the latter's work against Manichees in Bosnia.(39)

The second reference appears in a letter of June 1450. Nicholas V, writing the Bosnian king, refers to the king's difficulties in waging war against both Turks and Manichees.(40)

Wadding cites a letter of Thomas of Hvar, written in 1451 to John of Capistrano. The letter speaks of the success of the Franciscans in converting heretics. The heretics, he states, are melting away like wax before a fire; there is hope that the whole kingdom would be purged of the errors of those Manichees, and will be illuminated by (the light of) the true faith. (41)

In a letter of August 1453 Nicholas V, in speaking of problems, made passing reference to teaching the true faith to heretics and Manichees.(42) The pope evidently had two groups in mind.

Regardless of what groups were being referred to or how much or how little accuracy there was in the term "Manichee," it is evident that the word was beginning to be used, and once in use we can expect to see it used with increasing frequency — if for no other reason than its pejorative value. If Wadding has accurately reproduced Thomas of Hvar's letter, and has not altered it to glorify the achievements of the Bosnian Franciscans, we also have evidence that the Franciscan mission had become very successful in gaining converts. Thomas, presumably an informed observer since he had been involved with Bosnian affairs for over a decade, was actually anticipating a time in the not too distant future when the "heresy" (presumably the Bosnian Church) would be eliminated. This then is evidence that the Bosnian Church in thse last years was becoming weaker and losing its members to Catholicism. This view is supported by Thomas' statement of February 1, 1449, already cited,(43) that there remain infected with heresy only some of the nobles and barons. Stefan Vukčić and Ivaniš Pavlović were then mentioned by name. This phrasing suggests that Thomas thought the majority of nobles and barons had been won over to Catholicism. Pope

Nicholas V in a letter of July 1452 confirms the contents of Thomas of Hvar's letter; and even if Nicholas' source of information is Thomas, it at least is confirmation that Wadding has accurately reproduced the sense of Thomas' letter. Nicholas writes that Stefan Tomas, his wife, and very many barons and nobles had laid aside Patarin errors and recognized the true light.(44) The letter when compared to Nicholas' 1447 letter shows that he was using the term Manichee and Patarin as synonyms and that they both refer to members of the Bosnian Church to which, the pope believed, prior to their conversion to Catholicism very many Bosnian barons and nobles had belonged.

VII: Stefan Vukčić's Political Affairs (1448-1451) and Religious Associations

When Stefan Vukčić had made peace with the Bosnian king in 1446, he strained his relations with his former allies, the Turks and George Branković of Serbia. In 1448 when the Turks sent a plundering expedition against Bosnia, they pillaged Stefan's lands as well. Partly to prevent a recurrence of such an attack, and partly because he and Stefan Tomaš remained mutually suspicious, Stefan Vukčić tried to improve relations with Serbia. In 1448 he abandoned his title of Vojvoda of Bosnia (which emphasize his position in the Bosnian state and suggested that he served the King of Bosnia) and took the title Herceg of Hum and the Primorje (which reflected his status as an independent ruler). By 1449 Herceg Stefan Vukčić was clearly allied to the Serbian despot in his war against Bosnia, and now began to call himself Herceg of Saint Sava — a title he retained until his death.(45)

Saint Sava had been the founder of the independent Serbian Church in 1219 and his body lay at the Monastery of Mileševo in the herceg's territories. Because Sava was one of Serbia's most popular saints, the new title can be viewed as both a sign of the herceg's re-established alliance with Serbia and of an effort to win popularity among the Orthodox. However, it should be stressed that Sava's relics were believed to work miracles without respect to race, creed, or nationality. In the Turkish period we find that Catholics, Moslems, and even Jews as well as the Orthodox honored Sava's cult and sought cures and other rewards from his relics.

That the herceg was impressed by the miracle-working ability of St. Sava's relics is seen from the charter issued by Alfonso of Naples, confirming the herceg's possessions. The charter states that at Mileševo there rests "one saint who does great miracles."(46) Clearly this information originated with Herceg Stefan. Thus the herceg may also have wished to honor this miracle-worker to gain his intervention and support in this world. That the herceg, who was almost certainly a member of the Bosnian Church, utilized in his title the name of an Orthodox saint whose relics were part of an important cult is further evidence against the Bosnian Church being dualist; dualists rejected both saints and relics.

We do not know the extent of the herceg's flirtation with the Orthodox Church. His association with Gost Radin remained as close as ever and in April 1449 Radin and Stefan had drafted letters to Dubrovnik about commercial problems.(47) Radin in the course of 1450 made two trips to Dubrovnik for the herceg.(48)

The Serbian-Bosnian war continued over Srebrnica from 1448 into 1450. At the same time military engagements were fought between Herceg Stefan and King Stefan Tomaš. Ivaniš Pavlović sided with the king. He had by now renounced Catholicism and returned to the Bosnian Church. In 1450 Ivaniš died and his holdings, much diminished largely because of Turkish attacks in 1448 and the following years, went to his brother Peter. Finally, in 1451, the Turks established themselves permanently in Vrhbosna (Sarajevo) and in the neighboring fortress of Hodidjed. With his holdings so reduced that he had little base for independent action, Peter had no choice but to become a vassal of Herceg Stefan. Peter Dinjičić, who held lands between the eastern slopes of the Romanija Mountains and the Drina, maintained a perilous existence. We know he had ties with the Bosnian Church, since in 1450 he sent a Radohna Krstjanin to Dubrovnik.(49)

In 1449 we hear of a Patarin hiža in the territory of Herceg Stefan. Complaint was made in a Ragusan court against two nobles who served the herceg. These two men had sent their retainers at night to rob some Ragusan merchants in the home (domus) of Obižen Patarin in a place called Eretva below Biograd (Eretva sub Belgrado).(50) This hiža too was used as a hostel; it was presided over by a Patarin named Obižen. "Eretva" is Neretva, and can refer at times to the town of Konjic as well as to the river. By this time Konjic was in the hands of Herceg Stefan. Professors Vego and Andjelić have convincingly located Biograd on the Neretva just a couple of kilometers above Konjic. Thus we now have a second hiža on the herceg's lands, in addition to the one near Goražde close to, if not right on, his border with the Pavlovići.(51) The hiža at Biograd also lay near the border with the king's lands and perhaps doubled as a border post or customs station like the one near Goražde. That both hižas lay near the borders of Bosnia suggests that the sources of support for the Bosnian Church as well as its sources of manpower for clergy lay within Bosnia. Except for the Kosača family and its Patarin diplomats and these two hižas near the border, we have found no Patarins in Hercegovina before 1450. However, we do find Orthodox churches and grave inscriptions which lead us to believe that most of Hercegovina had retained its Orthodox character.

VIII: The Herceg's Economic Problems — War with Dubrovnik

In these years the Bosnian king and Herceg Stefan had to raise money to pay the Turkish tribute as well as to finance their own wars and their expensive court-life, modeled on Western European courts. Both rulers spent large sums on imported textiles, tapestries, and clothing, and both hired artists and performers (both actors and musicians) to live at their residences.(52) The king, with mines on his lands, was in a far better economic position than the herceg. Yet his need for money probably explains his long struggle with the Serbs over the rich silver-mining town of Srebrnica. After the fall of Constantinople in 1453, the size of Ottoman tribute demands increased greatly. Ćirković has calculated that, between 1453-1457, Stefan Tomaš had to turn over to the sultan 160,000 ducats in tribute, and Ćirković believes that this sum must have represented the bulk of the silver extracted from Bosnia's mines.(53)

Lacking mines, the herceg was dependent on customs and tolls collected from merchants who had to pass through his territory en route from Dubrovnik or Kotor to Bosnia or Serbia. In the late 1440's he began to expend considerable money and energy toward making Novi (or Hercegnovi) into a major Adriatic port. He established trade ties with south Italy. To Novi he imported weavers to set up a weaving industry, and then he began to market salt in an effort to break Dubrovnik's near-monopoly in selling salt to the landlocked Bosnians and Serbs. He forbade his Vlachs to buy salt in Dubrovnik. Both Dubrovnik and Kotor protested, but the herceg quite practically replied that everyone had to look out for his own interests, and everyone was free to do what he wanted on his own lands.(54) The herceg clearly viewed Hercegovina as his own state. In 1450 Dubrovnik forbade its merchants to trade in his lands, accusing him of actively working against Ragusan merchants. The town claimed that the herceg was preventing Ragusan merchants from purchasing goods from the Hercegovinian merchants within Hercegovina, because he wanted these Hercegovinian merchants to go themselves to Dubrovnik and there market these goods. This suggests that the herceg had been trying to establish himself as a middle-man and force coastal merchants who wanted Hercegovinian goods to buy them through him; thus he was inviting war with Dubrovnik. Ćirković, however, argues that we should not take these Ragusan complaints for fact and points out that when negotiations for peace were initiated in 1453 between the herceg and Dubrovnik among the first things brought up by the herceg was resumption of commerce between Dubrovnik and Hercegovina, including free movement of Ragusan merchants in Hercegovina. Thus Ćirković concludes that Dubrovnik had initiated a blockade against the herceg which it tried to justify by leveling accusations against him.(55)

In early 1451 the herceg asked the sultan's permission to make war on Dubrovnik. The Ragusans sent an embassy to the Hungarian court for aid, but since Hungary was involved in a war against the Turks little help could be expected. In February 1451 Dubrovnik requested the pope to free it from its financial obligations and cited its dangerous plight. The town said its territory was surrounded by perfidious Patarins and "Manieri" (surely a slip of the pen for Manichei) with no buffer zone between them and it.(56) This is the first time Dubrovnik used the term Manichee. By saying Patarins and Manichees, the Ragusans, whose dealings with the Bosnian Church up to now clearly show that the town did not consider the Patarins heretical, were probably trying to distinguish the Bosnian Church Patarins from the heretical "Manichees." We may suspect that Dubrovnik knew the pope had been using the term "Manichee" recently and thus chose to use it as an effective way to plead its case. It is worth stressing that this reference to heretics as well as other references that follow from Ragusan writers during the next three years all came from a town at war. Previously, in 1430 when Dubrovnik had been at war with Radoslav Pavlović, it had spoken of him as a heretic. Now the town would speak the same way of the herceg. When at peace with these nobles, if Dubrovnik bothered to comment on their religions at all, it simply called them schismatic. Thus the terminology in the Ragusan sources, which we are about to examine, in no way alters the conclusion, stated earlier, that Dubrovnik dealt amicably with Bosnian Church members in peace-time, treating them as if they were schismatics, and damned them as heretical only in war-time to add strength to its appeals for aid and allies.

In May 1451 the Ragusan authorities wrote a Franciscan named Biasio de Constadino to protest to the pope against the aggressive designs of the herceg. In this letter Dubrovnik spoke of its lands as bordering on those of heretics and Patarins and particularly those of Duke Stefan.(57) It is interesting that in repeating the February letter's statement about its geographical position, it replaced the word "manieri" with "heretic." Again we may ask if Dubrovnik intended to make a distinction between heretic and Patarin. A letter written to Hungary, undated but presumably penned at about this time, spoke of Dubrovnik's suffering from infidels and heretics.(58) And in May 25, 1451 in yet another letter the town referred to Herceg Stefan's men as Patarins and Manichees.(59)

In June the herceg invaded Konavli meeting with little opposition, since Dubrovnik never was able to field an army of any quality.(60) The Catholic Celjski count supported him and had sent him cannons with engineers to man them. However, the herceg's ally, the Serbian despot, was not pleased with the attack on Dubrovnik, and relations between herceg and despot deteriorated. In July Dubrovnik spoke of the herceg as a perfidious Patarin and a public enemy of the Cross of Christ.(61) Also in July the town wrote an envoy to thank Gost Radin for the love he had always shown the town and not as a payment but as a sign of love, the town promised to pay him 200 ducats.(62) Dubrovnik presumably hoped Radin would take its side to prevent the herceg from pushing his military campaign too far. Perhaps Radin had communicated to Dubrovnik his willingness to help. In August Dubrovnik wrote to its ambassadors at the Serbian court about its war with the perfidious Patarin and heretic.(63)

In August 1451 Dubrovnik reported to its ambassador at the Bosnian court that Herceg Stefan had told Pope Eugene that he wanted to be baptized and become a good Christian, and that the pope had granted him various privileges. Since the herceg had not fulfilled his promise but had persevered in his infidelities, the (present) pope (Nicholas V) has revoked the privileges offered by his predecessor.(64) Nicholas' reaction presumably was the result of Ragusan complaints about the herceg's attack.

In August 1451 Dubrovnik became aware of dissension within the herceg's family circle(65) and opened negotiations with his son Vladislav, who agreed tht he would revolt against his father. Truhelka speculates that Radin may have had a role in instigating the revolt since a month earlier Dubrovnik had promised him 200 ducats. The great sums of money Dubrovnik expended on Radin, as well as the praise and other gifts it heaped upon him (especially after the war) must have reflected real services Radin rendered Dubrovnik.(66)

In November 1451 the Turks and Hungarians signed a peace, in which one clause condemned the herceg's war on Dubrovnik, and in December the Bosnian king agreed to aid Dubrovnik. The king's alliance with Dubrovnik was witnessed by the Bishop of Hvar (and papal legate to Bosnia) and by the Custodian Father, Chaplain Marinus.(67)

The king then called on Peter Vojsalić and Vladislav Klešić to join him; these two noblemen, not wanting to fight the herceg, attacked the king. The same two clerics then made peace between them and the king, making possible their entry into the war on his side.(68) This story illustrates the gains made by Catholicism and especially the influence it had achieved at court and in affairs of state.

Early in 1452 the herceg renewed his activities along the coast. Vladislav in March launched his well-planned revolt; immediately a whole series of Stefan's major fortresses, including Blagaj, surrendered to him. By April Vladislav held all of Hum except for the region around Ljubuški. Dubrovnik described the action of Vladislav (who almost certainly was an adherent of the Bosnian Church) against his father as being for the honor of God and the good of all Christianity.(69) The Bosnian king and Peter Vojsalić entered the fray in April by sending troops into Hercegovina against the herceg. The Turks then raided Bosnia, which caused a temporary withdrawal of Bosnian troops from Hercegovina. There then erupted a well-organized revolt against Vladislav in the Krajina which declared its loyalty to Venice. At the same time Venetian ships appeared in the mouth of the Neretva, as allies of the herceg, and temporarily occupied Drijeva. The Bosnian king almost immediately expelled the Venetians and declared Drijeva part of his patrimony.

Soon Dubrovnik's two major allies began to squabble over Blagaj which Stefan Tomaš declared was part of his patrimony as well. Vladislav, who occupied this key fortress, refused to yield it. The Bosnian king offered to confirm Vladislav in possession of all the land from the Čemerno Mountain to the sea if he yielded Blagaj, but Vladislav wisely retained Blagaj, pointing out that the whole region which the king so generously granted him was at present held by the herceg. The king then stated his intention to return to Bosnia.

Meanwhile the difficulties which Ragusan merchants had been having with their transit trade through, and commerce with, Hercegovina increased immeasurably with the outbreak of war, which closed off entirely the overland route through Hercegovina. Professor I. Voje has recently discovered several documents which illustrate the merchants' troubles. One of these speaks of Patarins in the herceg's lands. This is a contract from February 1452 between the Ragusan merchant Radosav Radohnić who had promised to deliver certain goods, now detained in Hercegovina, to another Ragusan Matko Bogosalić. The goods at present were "in the hands of certain Patarins." Radosav states that, when the war is over, he will go to Hercegovina for the goods and then make delivery of them to Matko; meanwhile the goods should be considered Matko's property.(70)

Voje believes that this contract should be interpreted to mean that Radosav had left the goods with certain Hercegovinian Patarin merchants, which in turn shows that certain Patarins in Hercegovina were involved in commerce. He suggests, however, that these Patarins might have been lay members of the Bosnian Church rather than clerics.(71) Since, excluding polemics, the Ragusans consistently used the term Patarin to denote an ordained Bosnian cleric, I think we can assume that this contract refers to clerics. In addition, the summary of the document given by Voje — he does not give us its text — does not suggest that the Patarins referred to were in any way involved in trade by profession. It seems probable to me that when war broke out Radosav's agents had simply chosen to store these goods, which they could not transport to Dubrovnik, at a Patarin monastery in Hercegovina since the monastery would have seemed a safe place to leave them. Thus all we have here would be a case of certain Bosnian monks providing secure storage for the property of certain Ragusan merchants. This of course does reflect, as Voje points out, good will between the Patarins and the Ragusans.

In 1452 Dubrovnik wrote Hungary of the evil war waged against Dubrovnik by Stefan, the perfidious heretic and Patarin, and enemy to the name of Christian, and accused him of destroying churches of God and crucifixes.(72) A second letter to Fra Blasio referred to the herceg's destruction of Konavli which Dubrovnik had held for about twenty-five years and on which Dubrovnik had spent much money and the Franciscans much labor to convert to Catholicism the population, which had been Patarin and heretical. Stefan was accused of ravaging the area, wrecking villages, cutting down trees and vines, spilling Christian blood, destroying churches and the holy objects inside them, and converting churches into stables. His men and horses trampled underfoot crucifixes and massacred many Christians, including some priests who were celebrating a mass. The letter called for an offensive against Stefan and the other perfidious heretics and Patarins in league with him.(73) A third letter sent to the Bishop of Hvar to pass on to Venice, dated September 1452, also spoke of the infidelity and heresy of Stefan. It called him a perfidious Patarin, a heretic and an enemy and persecutor of the Catholic faith. His men destroyed vineyards and orchards, burned houses, killed Christians, not sparing even the priests, and defaced the churches of God, throwing crucifixes to the ground in contempt for the divine majesty of the Holy Catholic faith.(74)

In June 1452 at the same time that Herceg Stefan was carrying out this alleged persecution of Catholics, he was approached by Pagaminus the Bishop of Ulcinj to whom the herceg held out the hope of peace and hinted that he wanted to give up heresy (probably the word heresy is that of the bishop or the pope) and become a true son of the Catholic Church.(75) The Bishop of Hvar was dispatched by the pope to see the herceg about this, but the herceg did not receive him or take any steps toward his conversion.

The herceg's conversation with Pagaminus was presumably motivated by political considerations; and if he thought it politic to speak in that way, it is unlikely that he would have directed a deliberate campaign to destroy churches which could have brought him no political profit. Thus, it is probable that if churches had been destroyed, this destruction or damage had occurred in the course of battle. Dubrovnik, however, seeking allies and aid, hoping to cause Venice to break its alliance with the herceg, and knowing that the nations it sought to influence were Catholic, naturally took the most emotional tone possible to try to turn the Catholic world against the herceg. Since Dubrovnik never called the herceg heretical in peace-time, and since the town could call his son and its ally, who was probably also a supporter of the Bosnian Church, one who worked for the honor of God and Christianity, we cannot rely too much on these accusations which Dubrovnik levelled at the herceg.

IX: Peace Treaties of 1453-54 — Role of Gost Radin

In early spring 1453 the herceg sent Radin to Dubrovnik to discuss peace terms; he even conferred with the papal legate to Bosnia, Thomas of Hvar, but without success.(76) In July Herceg Stefan came to terms with his son Vladislav and his supporters: they agreed that all should be as it had been before the rebellion. Vladislav, pleased to receive these terms, forgot his ally Dubrovnik, which was left alone still in a state of war with the herceg, to make its own peace with him as best it could. Of course, the town would have Radin to help it in negotiations, but his services were expensive.

The peace between the herceg and his family was signed at Pišče on the Pivska Mountain and announced in two documents guaranteed by the hierarchy of the Bosnian Church. The first began with a cross followed by an invocation to the Trinity. Stefan gave his titles and then stated that the charter was drawn up before the Lord living God and before all the saints. After giving the terms of the peace, he agreed not to take any action against his wife or sons unless sanctioned to do so by the Lord Djed of the Bosnian Church, and the twelve stroiniks, among which stroiniks would be Gost Radin for his lifetime, and twelve nobles . . . The herceg swore to abide by the treaty before God all powerful, the

Trinity, the Pure Mother of God and Sainted Virgin Mary, before the Cross of Our Lord, the four Evangelists, the twelve Apostles, the Seventy Chosen of God and all the Saints.(77)

The second document in general repeats the contents of the first. However, instead of referring to the twelve stroiniks, it speaks of the "twelve head krstjani" (12 poglavitijeh kr'stjan').(78)

Some scholars have assumed that the sanctioning of this treaty by the djed and his council means that they have moved to Hercegovina, and this in turn means that King Stefan Tomaš, under pressure from the Catholic Church, had initiated persecution of the Bosnian Church, causing the emigration from Bosnia of the church's hierarchy.(79) Though plausible, this hypothesis cannot be taken as fact. The treaty between the herceg and his son was an event sufficiently important to have warranted a guarantor's role for the djed and his council, regardless of where they lived.

The variation in phrasing between the two texts shows that the twelve stroiniks were the twelve heads of the church. Thus we may think of the stroiniks as the members of a council which aided the djed in administering the church. It is usually argued that the passage should be read as "the twelve" and thus the djed's council was composed of twelve stroiniks — the same number as the apostles — which would mean the whole council stood behind the treaty.(80) Whereas this theory is quite plausible, it is possible that we should read the text as "twelve *of the* stroiniks", in which case it would mean that twelve stroiniks out of a larger group stood with the djed on this occasion. After all, there were more nobles in Hercegovina than the twelve guaranteeing the treaty; perhaps the djed simply wanted the number of clerics to balance the number of laymen. That the members of the council, despite their higher ranks (i.e., gost, starac) were called "twelve head krstjani" shows that the term krstjanin could be used for any ordained cleric, including those who had achieved higher rank and was not limited to those clerics below the rank of starac.

On this occasion Gost Radin was made a stroinik. He had been a gost already in 1447, so we know that all gosts were not automatically stroiniks. Later we shall meet a starac who is a stroinik: thus men of this lower rank could be on the council while certain gosts, a rank higher than starac, might not be on it. Gost Radin was selected to be a stroinik for life. Whether this was exceptional or standard procedure, we do not know. It would be rash to try to generalize on a stroinik's tenure on the basis of one case. Nor do we know whether Herceg Stefan put pressure on the djed to appoint Radin as a stroinik; it seems, however, a sensible decision to place on the church council a churchman with so much influence in the secular world.

Finally, we might look at the theological references in the document. The curses against violators, which being magical have formularized wording and hence are almost worthless for content analysis, are absent. However, the herceg swore by a series of items which turn out to be the same figures which carried out the curses. Among these we find an all powerful God, the pure Mother of God the Sainted Virgin Mary, the strength of the revered and life-creating Cross of the Lord, and all the Saints. Since these items are regularly found in the formulated curses, they clearly had a formula base; hence we cannot place too much emphasis on them. But they do contradict the inquisition sources which allege that the ''Bosnian Patarins'' rejected all of these.

The document begins with the sign of the cross, frequently found at the beginning of charters, but by no means an essential ingredient to make a charter valid. Its presence here — like the cross at the beginning of the djed's letter in 1404 — suggests that it was accepted by the Bosnian Church, the herceg's membership in which is indicated by the fact that the djed and his council stood behind this family document. The presence of the cross also contradicts the implication in the Ragusan accusations against the herceg which stress his attacks on crucifixes. The invocation to the Trinity, as we have seen, is standard for the Bosnian Church. It is also worth noting that the herceg did swear by a variety of holy figures. This shows that laymen or followers of the Church, unlike ordained clerics in it, were allowed to take oaths.

In April 1454, after long negotiations, Dubrovnik, too, obtained peace on condition that matters be restored as they had been. The charter issued by the herceg after beginning with the sign of the cross and an invocation to the Trinity, states that ''God sent His only Son to great suffering in order to deliver His (people) from sins by the resurrection . . . .'' Near the end the herceg, his family and several secular noblemen swear to uphold the treaty. The holy items they swear by are the same ones noted in the preceding treaty.(81) Though Radin had been instrumental in bringing about this peace, neither he nor any other Bosnian Church cleric appended his name to the document. However, because Radin participated in negotiations and because the herceg almost certainly was an adherent of the Bosnian Church, we probably are justified in using the phrase I have just quoted as being representative of Bosnian Church belief. Thus we see that the Bosnian Church held perfectly orthodox views about the crucifixion and its significance. This is strong refutation of the inquisition's assertion that the Bosnian Patarins denied the resurrection, believing that Christ had only an apparent body and did not really suffer or die on the cross.(82)

The charter shows that the Bosnian Church believed that Christ suffered and died on the cross and by his resurrection enabled mankind to be delivered of sin. The charter's phrasing also shows that the Bosnian Church believed that Christ was the only son of God. It has often been stated by scholars that the Bosnian Church, like the Bogomils, believed that God had two sons Satan, the elder son, and Christ, the younger. Thus the herceg's treaty shows that this belief was not held by the Bosnians.(83) It should be stated here, however, that no source about the Bosnians ever accuses them of believing in two sons of God. This has simply been a scholarly hypothesis based entirely upon attributing beliefs found among dualists in Bulgaria and elsewhere to heretics they believe to be dualists in Bosnia.

Since the peace treaty restored matters to the ante bellum state, all the causes for tension, the issues of the trading and Novi, remained. During negotiations in August 1453 Dubrovnik, referring to the friendship between them, had paid Radin 400 ducats to maintain the armistice.(84) When Radin returned to Dubrovnik with the peace treaty, he requested and obtained still more money from the Ragusan council. Whether Radin was really a double-agent, as Truhelka and Solovjev believe, is hard to say.(85) Dubrovnik clearly thought he was working in its interests, but he may have been simply a masterful extortionist, persuading the town fathers that he was working to help them, at great personal risk, while all the time loyally serving the herceg, who certainly would not have objected to his envoy's bleeding the rich town's treasury. It is hard to believe that Radin actually ever did anything treasonable against the herceg. Dubrovnik's faith in him, though, is clearly seen in the sources: In August 1453 when Radin was given 400 ducats, Radovan Vardić, the secular negotiator, received only 100.(86) An undated letter from about this time told a Ragusan envoy to discuss matters with Radin, for the town council had much confidence in him.(87) A letter of January 1455 from Dubrovnik to its ambassador told him to speak first with Radin since the council wanted to make sure that he was favorable about an idea, for the council had hope in him.(88) In April 1455 it was suggested to the envoy that he speak secretly with Radin and the herceg's two sons about matters before he spoke to the herceg, who was unfriendly toward the town.(89) In November 1455, for the love he has shown the town, Dubrovnik presented Radin with a house for himself, his household and his servants in Dubrovnik. This shows that Radin did not live in the ascetic way that might seem fitting for an ordained "prior-gost." He was given the right to come and go as he pleased and to remain in the town as long as he wanted. If he chose to live there, it was guaranteed that he would not be forced by anyone to renounce the faith in which he believed, as it has been pleasing to him.(90)

Solovjev believes that Dubrovnik's charter was intended to enable Radin to escape and find asylum if the herceg discovered what Solovjev believes to have been Radin's "treachery." However, several noblemen, including Radič Sanković and Sandalj Hranić, had been honored with Ragusan citizenship, titles of Ragusan nobility and even houses in the town. The right of asylum was always freely granted by Dubrovnik and Radin surely could have sought it at any time even without the special grant. We have noted earlier that Pavle Radišić and Pavle Klešić both obtained asylum in Dubrovnik and we know that upon Klešić's arrival the town council voted to give him a house to live in. The grant to Radin then was a token of Dubrovnik's gratitude for his good will toward it and the services he had rendered it in achieving and maintaining peace with the herceg. If Radin did feel any need to have a place of shelter which he could escape to, we can suppose it was not to flee the herceg but to escape from Hercegovina in the event of a Turkish conquest which was an ever present possibility. The herceg sent Radin to Dubrovnik several times in the following years: in February 1456, March 1459, March 1461, February 1462.(91) It is clear that relations between him and Dubrovnik had not changed since in 1462 a Ragusan envoy was instructed not to discuss a certain matter with the herceg if Radin were not present.(92)

X: The Pavlovići and Patarins in the 1450's

The herceg was not the only magnate using Patarins to negotiate with Dubrovnik in 1454. The Pavlović brothers, Nikola and Peter, made a treaty with Dubrovnik after consulting their advisers according to custom "and with the lord stroiniks of the Bosnian Church." They sent and embassy to Dubrovnik which included two stroiniks: Gost Radosav Bradiević and Starac Radosav. Both these clerics signed the charter but did not place their names among those who swore to it. In signing, the starac wrote, "Starac Radosav called grandson by the words of his order." (unuk rekoše po svom redu ričiju.)(93)

Here we find both a gost and a starac as stroiniks. The significance of the term "grandson" is unknown; the expression is known from no other source. It is evident that it should not be related to the Cathar elder and younger sons(94) since the Cathar sons were the second and third ranking people in their church, and Radosav as a starac was clearly of lower rank than all the gosts. The Pavlovici obviously still had Patarins at their court; these Patarins were respected, since they were consulted before the treaty with Dubrovnik was drawn up. Thus it seems that after Ivaniš returned to the Bosnian Church after his three years of Catholicism, his family had remained faithful to the Bosnian Church.

They, the herceg, and possibly also the Dinjičići are the only families mentioned in the sources of the 1450's which retained ties with the Patarins.

XI: The Herceg's Dealings with Catholicism and Orthodoxy

While the negotiations between the herceg and Dubrovnik were being carried out by the Patarin gost, the herceg was flirting with both Catholicism and Orthodoxy. His old friend Alfonso of Naples wrote him in November 1454 that he would send suitable Franciscans from south Italy to teach in Hercegovina as the herceg had requested.(95) We hear no more of them, however. In 1455 the herceg initiated negotiations to marry Barbara, daughter of the Prince of Lichtenstein. She, of course, was a Catholic, and the herceg agreed to let her bring a whole entourage, including priests, with her.(96) This marriage never actually took place. Instead the herceg took as his second wife another Barbara, this one also a Catholic, daughter of the Duke of "Payro."(97) After her death in 1460, the herceg would make yet another Catholic marriage. Also in 1455 the herceg (once again) and his sons were named honorary noblemen of Venice.(98) Since the herceg had been allied with Venice in his recent war with Dubrovnik, this honor is not surprising.

The herceg's son Vladislav was also showing an interest in Catholic cults. In the spring of 1454 he requested Dubrovnik to make available to him a boat to take him to Italy so that he could make a pilgrimage to the Virgin of Loreto. His request was granted; we know he went to Dubrovnik where he was received with honor. We may assume that then he proceeded on to Italy to make his pilgrimage.(99)

In 1454 the herceg built an Orthodox church dedicated to St. George in Goražde.(100) It has been suggested that he built the church in memory of his Orthodox wife, who had died late in 1453, but it is also possible that he built it as a sign of good will toward the Serbian despot, with whom he was trying to better his relations and from whom that same year he acquired a girl (a relative of the despot's wife and of good Byzantine family, Anna Cantacuzena) to be a bride for son Vladislav. He might also have built the church to please the Metropolitan of Mileševo with whom, we shall see, the herceg maintained very cordial relations. The building of an Orthodox church in Goražde suggests that there were at least some Orthodox believers in the area. A decade earlier we noted a Patarin hiža near Goražde. Probably the region of Goražde had a mixed population of Bosnian Church adherents and Orthodox.

In addition the Kosača family (either Sandalj or Herceg Stefan) built two churches beneath their major fortress of Soko, which stands on a high rise above the junction of the Tara and Piva Rivers, and below the ridge of the Pivska Mountain on which is located village Pišče. On the mountain-side below the fortress of Soko and above Šćepan polje (at Donje Zagradje ispod Sokola) at location "Manastir" stands a church long known to the natives but recently discovered by scholars. It is basically Orthodox in style with a few architectural differences. One might attribute the Orthodox style and variants to the artisans who worked on it; thus they are not sufficient evidence to prove the church to be or not to be Orthodox. A *stećak* inside the church suggests it was built as a burial chapel.(101) Unfortunately there is no inscription on church or *stećak* to help us date the church. Popular tradition attributes the church to Herceg Stefan and this is supported by a second popular tradition referring to the paraklis as "misa." Professor Kajmaković pointed out to me that the paraklis is Catholic in style. This fact combined with the Catholic term "misa" suggests that Catholic services were held in the paraklis while services of a different confession (Orthodox or Bosnian) were held in the main church. Since the herceg's second and third wives were Catholic and we know from his negotiations with Barbara of Lichtenstein that he was willing to let his Catholic wife bring Catholic priests with her, we may suggest that at least the paraklis was built by the herceg and that the church — if not built by him — was in use in his time. Whether it was built by Sandalj or the herceg is not known and we cannot be sure whether the two Bosnian Church nobles built it as a Bosnian church for themselves or as an Orthodox church for some member of their household. However, we can make a fairly strong case that it was Bosnian, for twenty minutes walk up the mountain above the church lies a medieval cemetery which includes the stone of Krstjanin Petko, discussed in Chapter V, who was buried between 1435 and 1448 (when the herceg still bore the title vojvoda).(102)

About a kilometer below the church just described, the ruins of another church at the location "Crkvina," "Crkva" or also "Manastir" on Šćepan polje have even more recently been found. These ruins, probably from the fifteenth century, seem a bit older than the church on the mountainside.(103) This suggests that this building may have been Sandalj's (or Jelena's) work and the upper church the herceg's. Again we know nothing about the circumstances of the church's erection, or even which denomination used it. One might suggest, though, that if the upper church was Bosnian, the lower one might have been Orthodox for the Orthodox members of the Kosača household.

XII: The Question of Kudugers

The herceg's good will toward the Orthodox is also seen by the gifts he gave to some Orthodox monks on Mt. Sinai. Somewhat puzzled by the herceg's generosity, the monks had written to Gennadius, Patriarch of Constantinople (1453-59), whose reply survives in a sixteenth- or seventeenth-century Slavic translation. The patriarch began by referring to a Bishop of Bosnia who had brought many ''Kudugers'' into obedience to the Greek church. He hoped that many more Bosnians might be converted to Orthodoxy since this bishop had the love of the lord herceg. The herceg had not yet publicly separated himself from the ''Kudugers'' for he feared his nobles. But he supported the activities of the bishop and in his heart had been for a short time Orthodox. Because of this, he had sent alms to the Christians. The patriarch told the monks that they might accept charity from him, and that they might pray for him, but that they must not mention his name in the liturgy.(104)

Who was this ''Bishop of Bosnia'' mentioned? He could not have been Bosnian since there were no known Orthodox bishops there; presumably he was a Hercegovinian bishop. It is not surprising that the patriarch should confuse Hercegovina and Bosnia considering that since its annexation by Bosnia in the 14th century, the land that was to become Hercegovina had been considered part of Bosnia, and that in the fifteenth century down to 1448 the ruler of this territory bore the title Vojvoda of Bosnia. Probably the bishop was the metropolitan who resided at Mileševo, which lay in the herceg's lands. In 1466 this metropolitan helped draft the herceg's last will and testament. Perhaps the herceg's ties with that monastery were already established in the 1450's; possibly they originated in 1449 when the herceg took the title of Herceg of St. Sava. After all, Sava's relics were kept at Mileševo.

What of the strange term ''Kuduger'' that Gennadius used twice to describe certain ''Bosnians'' (i.e., people of Hercegovina)? The Byzantine historian, Laonikos Chalkokondyles, writing after the fall of Constantinople, says that the inhabitants of Sandalj's region were called ''Koudougeroi.''(105) Gennadius and Chalkokondyles are the only writers who used the term about Bosnia or Hercegovina. However, early in the fifteenth century, Symeon the Archbishop of Salonika

(1410-29), who wrote several tracts against heretics in his diocese — especially against the Latins and the Bogomils — directed one of his treatises against Bogomils and Koudougeroi, using the terms synonymously. Having mentioned Mani, Symeon states that of his followers now there still exist the Bogomils, the devil's crowd who are also called Kudugers. They existed near Salonika, pretended to pray a lot and to honor the Gospels, Acts and Epistles, but reject the remaining books, and are not evangelical in their practices. The Bogomils secretly pray to Anti-Christ, reject all the Church Sacraments, baptism with water, the Eucharist, the Cross, icons, churches, the Ten Commandments, prophets, and the saints. They are against all prayers and hymns and allow only "Our Father." Their father is the Devil whom they call "Topaka" of the Earth, which means High Lord of Sin and Dark.(106) Thus we see that according to Symeon the term Kuduger was simply another name for a Bogomil. The description he gives of the Bogomils is perfectly standard, clearly stresses dualism, and is highly polemical and pejorative. In 1926 Skarić came to the conclusion that the term Kuduger meant specifically those who hated the cross,(107) which of course coincides with Symeon's description. Thus, the generally accepted conclusion is that the Kudugers were dualists and thus both Gennadius and Chalkokondyles meant dualists when they used the word about Hercegovina.

However, we have had no evidence of any dualists in Hercegovina. Since we have found the herceg connected with the Bosnian Church until now, it is likely that Gennadius is referring to the Bosnian Church when he says "Kuduger." Yet when he says that the herceg "has not yet publicly separated himself from the Kudugers for fear of his nobles" this implies that many of the Hercegovinian nobles were members of the Bosnian Church, which does not agree with the sources, we have studied about Hercegovina; they suggest that the majority of Hercegovinian nobles were Orthodox. And if Kuduger is intended to mean dualist, we would say also that Gennadius was wrong to apply the term to the Bosnian Church.

Of course, however, we must realize that the Patriarch of Constantinople was writing as an outsider who had little correspondence about or contact with Bosnian affairs. We do not know what his source of information about Hercegovina was, though we may suspect it was correspondence (no longer extant) with the "Bosnian" bishop whom he referred to in the letter.(108) If that bishop had spoken of the Bosnians as Patarins, the patriarch might well have interpreted that term to mean dualists. Then when he wrote to the Sinai monks he could easily have dropped this Western word, which he might have expected the monks not to understand, and have substituted the term Kuduger which may still have been in use in the Greek world (and even if not too commonly used, which he could have drawn from the Archbishop of Salonika's writings.)

The same explanation could probably apply to Laonikos' use of the term. However, there is also one other place from which this term might have come. Laonikos used the term in such a general context that it need not have referred to a religious group at all. In Turkish the term "güdücü" (pronounced kuduju), attested in manuscripts of the fourteenth and fifteenth centuries, means shepherd.(109) The main occupation of the population of Hercegovina was of course raising sheep. This may well have been what Laonikos had in mind when he described the inhabitants of Hercegovina as Koudougeroi. It also would have been a more accurate generalization than one which would make the people of Hercegovina heretics since Hercegovina was a religiously mixed area with most probably an Orthodox majority. The patriarch may also have heard the term used with its Turkish meaning about the people of Hercegovina, but instead of understanding it in the intended occupational sense, related the term to the word for dualists used by the Archbishop of Salonika.

We also find a second Turkish term similar to Kuduger; in the Defter of 1515 the mahala (section of town) of Nedžar Ibrahim in Sarajevo was called "Kurudger."(110) Unfortunately the significance of this name is not known. But it does suggest that linguistic study of fifteenth and sixteenth-century Turkish documents might prove fruitful in solving the mystery of the word "Kuduger."

XIII: Catholic — Orthodox Rivalry Begins in Bosnia

In his letter Gennadius spoke of Orthodox successes in winning over the populace of "Bosnia." Orthodox proselytism was also the subject of a complaint by the Bosnian Franciscans to John of Capistrano, who wrote to Pope Calixtus III in 1455. He reported that the Raška metropolitan hindered the Catholics in their work. Many of the Bosnian heretics who held the Patarin faith, having heard the word of God and having been converted to the Roman faith, were prevented by the Raška (Mileševo?) metroplitan and others from being reconciled with the Catholic Church. As a result, many died outside the faith, preferring to die outside the faith than to accept (Serbian) Orthodoxy.(111)

Here is a Catholic writer who links the Patarins with the Bosnian heretics, and thus states that the Patarins were heretics. However, we may question whether this letter can be used to demonstrate that the

Patarins were heretics rather than schismatics since in the same letter John of Capistrano states that the Serbs (Rasciani) strayed beyond the "heresies of the Greeks."(112) The Greeks, of course, were only schismatics. Thus if he misused the term heretic once, he may well have done so twice. This letter, however, does provide a strong argument against the theory that the Bosnian-Patarin Church was really Orthodox; for John of Capistrano notes that Patarins who have accepted Catholic teaching but are prohibited from being accepted into that faith by the Serbian metropolitan, would rather die outside the faith than accept the Serbian faith. Hence it is evident that the Patarins originally had been something other than Orthodox. We may well wonder how this Raška Metropolitan had the force to prevent their being baptized as Catholics. Perhaps the "others" referred to in the passage were Orthodox members of the nobility who harassed would-be Catholics.

The letter of John of Capistrano shows that a rivalry was developing between Orthodox and Catholics in some unspecified parts of Bosnia-Hercegovina, presumably along the Orthodox-Catholic contact zones, in eastern Bosnia between Olovo and the Drina,(113) and along the Bosnian-Hercegovinian border in a region like that around Konjic. Either of these regions could well have had Orthodox priests in them under the jurisdiction of the metropolitan in Mileševo.

Although Bosnia is frequently spoken of as a meeting place and even a battle ground between the two major branches of the Christian faith, the fact is that this generalization only begins to become true in the 1440's and 1450's. In the period between the late twelfth century and 1440, we find signs of actual collision between Orthodoxy and Catholicism only in those areas where the Serbian state reached the Dalmatian coast and thus came into conflict with the Dalmatian Catholics, i.e., southern Dalmatia and inland as far as the region around Trebinje. We have noted that when Dubrovnik procured Ston and Konavli, the Ragusans persecuted the Orthodox inhabitants of these regions and forcibly reduced the populations to Catholicism. However, this conflict did not occur in Bosnia or in most of what is now Hercegovina during the medieval period.(114) Whereas inland Hercegovina was Orthodox, Bosnia was more or less a no-man's land between faiths with a weak Catholic organization, which died out in the thirteenth century only to be gradually rebuilt by the Franciscans after 1340. It is possible that Orthodox and Catholics did have contacts with one another in the region between Olovo and the Drina from the late fourteenth century; but there is no evidence of quarrels between them. For Bosnia and Hercegovina as a whole, then, the Orthodox-Catholic rivalry began only in the middle of the fifteenth century. Once begun it has lasted to the present day.

XIV: The Turkish Threat and Proposed Leagues to Meet it

In 1456 the Turks had demanded that Stefan Tomaš surrender four key towns to them; when he refused Turkish attacks became more frequent, but even so the king continued to quarrel with Herceg Stefan and with Serbia. George Branković died in 1456, and his successor in 1458. Stefan Tomaš took advantage of the disorder in Serbia in 1458 to seize eleven towns along the Drina. Then he made peace with the weak ruling family by marrying the deceased despot's daughter to his own son and heir Stefan Tomašević. As a dowry Bosnia obtained the key fort of Smederevo, which Stefan Tomašević occupied in 1459, apparently assuming the title of despot. Within a matter of months, the Turks captured Smederevo. The Hungarian king, Matyas Corvinus, immediately accused Bosnia of selling the fortress to the Turks, and rumors of Bosnian perfidy spread through Western Christendom.(115)

Papal agents had visited both Stefan Tomaš and the herceg to try to convince them to put up a united front to the Ottoman danger.(116) One of these agents, the Dominican Nicholas Barbucci, visited the Bosnian king at his court in Jajce, where the king had now established his chief residence so as to be more distant from the source of Turkish raids. Barbucci's letter — written from Jajce — about his mission is not dated, and scholars had generally ascribed it to the years 1457-60, but Professor Ćirković had convincingly dated it 1456.(117) Barbucci reported that he had broached the subject of a crusade against the Turks and found the king hesitant because "the Manichees, who were almost the majority in his kingdom, preferred the Turks to the Christians"; the king was not interested in fighting the Turks without Christian (i.e., Western) help.(118)

We cannot be sure that the king had really used the term Manichee. It is quite possible that Stefan Tomaš had spoken of Patarins or "heretics" and Barbucci, when making his report, under the influence of Italian Catholic views on Bosnia, or even under the influence of local Catholic clerics who wanted the pope to think the Patarins were Manichees, had substituted the word Manichee for whatever term the king had used. Even though there seem to have been dualist heretics in Bosnia, it is hard to believe that the dualists, so rarely specified, could have been a sizable enough movement to have been intended here. Thus the king and Barbucci are evidently referring to the Bosnian Church. Yet even so, the testimony of this 1456 letter runs counter to everything we have found in recent sources about the Bosnian Church. For the previous twenty years we have noted Catholic successes; in 1449 and 1450 Thomas of Hvar stated that there was hope that the kingdom would be purged of Manichee errors and become Catholic; heretics were disappearing before the Franciscans like wax before fire. He noted that heresy remained only among some of the barons and noted by name only Herceg Stefan and the Pavlovići.(119) Our other documents support Thomas of Hvar. Thus at a time when the Bosnian Church seems to be on the decline, we find Stefan Tomaš telling Barbucci that nearly the majority of his people are supporters of the Bosnian Church.

Thus we may suggest that the king used religion as an excuse not to fight since he feared a crusading plan would merely serve to stir the wrath of the Turks against him, and leave Bosnia to their mercy while Western aid would simply never materialize. It is certain the king wanted to save his kingdom. But we can suspect that he would have liked to see the Western armies before he risked becoming part of the venture. It is also important to note his statement that some Bosnians preferred the Turks to the Christians (i.e., foreign Christians and particularly Hungarians). This as we shall see was probably quite true. Though the Bosnians would have preferred no foreign interference at all, this was an impossible wish in the 1450's, and it was almost certain that they would be overrun either by Turks or by Hungarians. The Turks were not unknown monsters; many Bosnians had had contact with them during the preceding fifty years, and had fought beside them as allies on numerous occasions. Since the Bosnians were not "good Christians," the spectre of Islam as a foreign faith would not have seemed so important to them. Catholicism as Hungary presented it was also a foreign faith. That many would have preferred the Turks to the Hungarians, with whom Bosnia had fought wars and suffered persecution for the previous three centuries, therefore should surprise no one.

XV: Kotruljić's Evidence

In 1458 a cultured Ragusan merchant, Benko Kotruljić, who had on several occasions served as an envoy to Spain and Italy, wrote a treatise on trade. In the course of the treatise he commented: "the Bosnians, who follow Manichee customs, especially respect rich people and very hospitably receive them in their houses (i.e., hižas), while turning away the poor."(120) He evidently refers to Patarins. Excluding letters written by Dubrovnik during its war with Herceg Stefan, this is the only source written by a contemporary Slav which uses the word "Manichee" about the Bosnians. Whether he had picked up this term on a visit to Italy or whether it had been used pejoratively about Herceg Stefan during the town's 1451-53 war with him is not known. If Kotruljić was aware of the real significance of the label and was not just using it pejoratively, then we have evidence from a Slavic speaker, who had almost certainly travelled in Bosnia, that at least some Patarins had certain practices which Kotruljić felt were dualist.(121)

XVI: Catholic Progress in the Final Years of Stefan Tomaš

The scanty sources show that Catholicism continued to make progress in Bosnia, but we cannot learn anything about popular attitudes toward the Catholic Church or toward the Franciscans, or to what degree the peasants accepted Catholicism, participated in its services, or allowed its priests to influence their lives. A later report of Pius II states that King Stefan Tomaš had been a Catholic for a considerable time but had abstained from baptism until he received it from the hands of the legate Gioanni Cardinal di S. Angelo,(122) who visited Bosnia in 1457. All indications suggest that the king took his Catholicism seriously. In 1461 the pope, at the king's request, permitted him to have a portable altar on which Matins could be served.(123) Thus the king, on campaign, would not be deprived of services. Further evidence is provided by the religious upbringing of Stefan Tomaš's children. The testimony of Stefan Tomašević in 1461 to Pope Pius II about his Catholic upbringing will be discussed below. However, Orbini also describes the pilgrimage to a Benedictine church in Venice by an un-named fourteen year old son of the king in 1460. The unfortunate youngster was taken ill and died there; he was buried in the Benedictines' cemetery, clear indication of his Catholicism.(124)

In 1458 the pope, referring to heretics and schismatics in Bosnia, granted indulgences to all who visited the church of St. Catherine in Jajce,(125) presumably a church recently completed by the king and named for his queen's saint. In 1461 a papal letter reports that the relics of St. Luke were there.(126) These relics had come to Smederevo in 1453, rescued from the Constantinople disaster by one of the Serbian despot's in-laws. They became part of the dowry of Stefan Tomašević's Serbian bride, and after the fall of Smederevo were brought to Jajce. Later in 1461 they were transferred from the church of St. Catherine to a second church in Jajce dedicated to the Virgin Mary,(127) and presumably just completed at that time.

The relics clearly brought disaster with them since in 1463 Jajce in its turn fell to the Turks thus becoming the third fortress in which the relics were kept to fall within a decade. At the fall of Jajce, Tomašević's Serbian bride Jelena, who had accepted Catholicism and assumed the name of Maria, mislaid the relics in her hurry to escape. The Franciscans rescued them and fled toward Dubrovnik. At Poljice near the coast, the Franciscans were stopped by a local vojvoda named Ivaniš, a friend of the queen, who would not let the Franciscans continue with the relics without her permission. Meanwhile she was negotiating to sell the relics to Venice while Dubrovnik fired off angry letters to Ivaniš ordering him to surrender them. In August 1463 a letter of Queen Jelena-Maria showed her anger at the rich skinflints in Venice who had the nerve to question the authenticity of the relics. In the end, however, the Venetians did buy them and allowed the Bosnian Franciscans to seek shelter on Venetian territory. However, the jinx of the relics was to continue: six years later Venice exiled the Bosnian Franciscans, who had rescued St. Luke, just one more incident in a larger quarrel between the Bosnian vicariat and Dalmatian Franciscans, which had been going on for the preceding decades and which has no real relevance to our story. But after the Venetian purchase of St. Luke's relics brought them into the Italian world a new crisis arose; St. Luke was already there in the Benedictine monastery of San Giusto in Padua. The Benedictines raised a terrific howl at the Venetian claims. A law-suit that dragged on for years, more than once appearing before the pope, was instigated to determine who really owned St. Luke.(128)

In 1461 King Stefan Tomaš's brother Radivoj completed a church dedicated to St. George at Tešanj,(129) north of Vranduk. The Franciscans too had built at least eleven new monasteries in Bosnia and Hercegovina in the years since 1385, when a vicar's list had noted only four. These monasteries are noted in sources from the first seventy years of Turkish rule. Since the Turks in this period did not allow the Catholics to build new churches where they had not had them previously, it is probable that all eleven of these monasteries were built before the fall of the Bosnian kingdom. We ascribe to the years between 1385-1463 (for Bosnia), and 1385-1481 (for Hercegovina) Franciscan monasteries in: Fojnica, Kreševo, Deževica, Zvornik, Sol (Tuzla, where two may have been built), Jajce, Jezero, Rama, Konjic, Mostar, Ljubuški.(130) The last three monasteries were in Hercegovina. If Alfonso of Naples did actually dispatch the Franciscans he promised to the herceg in 1454, they were probably sent to one or more of these three Hercegovinian monasteries.

Konjic, we know, also had close by a Patarin hiža, as well as an Orthodox church at Biskup built by the Sankovići. Thus, it would have been an area of contact between all three confessions. The towns of Fojnica, Deževica, and Kreševo were all mining centers, where the Franciscans would have received support from foreign Catholic colonists. At Jajce, of course, they had the support of the king who had made this attractive town his residence.

XVII: Persecution Launched Against the Bosnian Church in 1459

In the final years of the kingdom, the Catholics prevailed on the king to engage in the active persecution of the Bosnian Church, which Rome and the Bosnian Franciscans had long wanted. Some scholars claim that the king began his persecution of the Patarins as early as 1449, using as evidence the djed and stroiniks guaranteeing the herceg's treaty in 1453. But we have argued above that their guaranteeing this treaty between the members of the herceg's family need not mean that the Bosnian Church hierarchy resided in Hercegovina. In addition, those who want to attribute Stefan Tomaš's action against the Patarins to1449 usually claim that the story found in Pius II's *Commentarii* and also in Orbini that the king gave the Manichees (i.e., Bosnian Church members) a choice of baptism or exile took place in that year. However, Pius' *Commentarii* dates this story to 1461, and Orbini dates it to 1459. I prefer to leave it dated 1459-60 as the sources state.(131)

In the passage Pius II (1458-64) described how King Stefan Tomaš gave the Manichees the choice of accepting baptism in Christ or leaving his kingdom. Pius reported that: About 2,000 were baptized and 40 (or 40,000) or a few more fled to Herceg Stefan. (duo circiter millia baptizati sunt, quadraginta aut paulo plures pertinaciter errantes ad Stephanum Bosnae ducem perfidiae secium confugere.)(132)

Many historians (and most of the popular books) have interpreted Pius' statement to mean that 40,000 fled to the herceg rather than accept baptism. Other scholars, such as Ćirković and Babić, have interpreted Pius' remark to mean that forty fled to Hercegovina.(133) The Latin is ambiguous and could mean either. However, conditions in Bosnia at the time argue strongly for the figure of forty. A successful migration of 40,000 to rocky, barren Hercegovina in the late 1450's is difficult to imagine. Where would they all have gone? How could they have supported themselves? How could such a major event have escaped attracting notice in the Dalmatian sources or in the correspondence among the papacy, Hungary and Dalmatia? And if such a large number had been adamant heretics, would the king, anticipating a Turkish invasion, ever have presented them with such a rigorous choice? Would 40,000 uneducated peasants, who clearly did not understand formal religious matters and who seem to have been relatively indifferent to formal religion, have sacrificed their lands and their homes for any creed?

The figure forty, on the other hand, is a very reasonable figure. Herceg Stefan's barren land could easily have accommodated that number. It seems likely, then, that when the Bosnian Church members were forced to choose, the larger number, 2,000, accepted Catholicism, whereas forty (presumably all or nearly all ordained clerics) preferred to keep their faith and go into exile in Hercegovina.

Whether the figure 2,000 represents just clerics or whether it includes both clerics and lay members of the church is unknown. But when we take note of the large amount of land in various parts of Bosnia associated with the krstjani in the Turkish defters, we see that it is quite possible for all 2,000 to have been clerics.(134) And if most or all of the 2,040 were clerics, it would explain why so few people were given the choice. For to direct an attack upon the clergy would have effectively destroyed the church and avoided stirring up the popular resistance that might have followed an attempt to forcibly baptize the lay population which included armed warriors. If the 2,000 who accepted Catholicism were clerics, and if they were sincere in their conversion, then presumably many of their adherents would have followed them to Catholicism.

The existence of 2,000 clerics in 1460 might at first glance suggest that the church was not as weak as we have suggested. However, when we realize that clerics concentrated in specific monasteries would have been far easier for the king's men to round up than would 2,000 secular priests living in an equal number of villages in various parts of Bosnia, we see that unless the clerics had strong popular support, their numbers alone would not have provided much strength. In addition, when we realize that roughly 2,000 — still assuming that all 2,000 were clerics — (out of roughly 2,040) preferred to accept another faith rather than go into exile, we see that the morale of the clergy was bad; clearly the church suffered from strong internal weakness.

The story about the expulsions is confirmed by two letters of Pius written on June 7, 1460, which spoke of the Bosnian king exiling Patarins and of the papal hope that Herceg Stefan could be convinced not to receive them.(135) Pius' letters thus confirm: 1) the story he reported later in *Commentarii;* 2) our dating the event 1459-60; 3) my belief that those exiled were clerics of the Bosnian Church since he referred to the exiles as "Patarins," the normal term for Bosnian Churchmen used in Dalmatia and also frequently by Bosnian Franciscans; and 4) that forty were exiled since in writing about the expulsion of the Patarins, he would presumably have made some reference to the number who had fled had the exodus been on a scale anywhere near 40,000.

The figure of forty as well as my supposition that the emigrés were ordained members of the Bosnian Church is also confirmed by a request made by Gost Radin to Venice in 1466 to allow fifty or sixty members of his sect to migrate thither in the event that they had to flee from the Turks.(136) The similarity of the two figures makes it probable that he was referring to the same people who had fled to Hercegovina, with the addition of some Hercegovinian Patarins to account for the slightly larger figure. That Radin's figure only adds ten or twenty to the number that had come from Bosnia confirms my belief that the Bosnian Church was not a major institution in Hercegovina. The djed and his hierarchy, if they had not moved to Hercegovina earlier, thus apparently came there in 1459. Their arrival was too late to establish any efficient church organization in Hercegovina.

King Stefan Tomaš apparently took advantage of the departure of the hierarchy to seize Patarin lands. N. Filipović, in a very stimulating article, argues on the basis of the defters that the krstjani had been successful and well-to-do farmers. He points out that the defters refer on a variety of occasions to "kristian" land being confiscated by the king. On the basis of this, he argues that the desire to confiscate the rich lands of the Bosnian Church may well have been a motivating factor in Stefan Tomaš's decision to persecute the Patarins.(137)We do not have enough evidence to prove this hypothesis, but the prospect of obtaining these rich lands certainly could have made the persecution of the Patarins demanded by the Catholic Church more attractive to the king. In any case, the king did obtain for the crown the rich lands that had belonged to the Bosnian Church. Filipović cites eighteen folio pages from an unpublished defter which, he claims, refer to lands that the king seized from the krstjani,(138) but he does not tell us where any of these lands were.

The material that Okiç published includes only one example of the king confiscating land. In the nahija of Neretva in village Orahovica, the defter reports the "evil king" took lands from the kristians and gave them to the villagers.(139) Orahovica is very near the Patarin hiža at Biograd. What is puzzling about this statement, though, is that this area should have belonged to Herceg Stefan; it is also clear that Patarins were not exiled from this area, since the hiža at Biograd is referred to again in 1466 after the fall of Bosnia. Thus we do not see how the King of Bosnia could have confiscated land around Konjic. One hopes the publication of further material from the defters can clarify the matter.

XVIII: Three Bosnians Abjure Fifty Manichee Errors in Rome

In August 1461, Pope Pius II wrote a letter to three Bosnian noblemen (Djuraj Kučinić, Stojšan Tvrtković, and Radmilo Večinić or Vočinić) at the time, en route home from Rome after renouncing Manichee errors and accepting the Catholic faith.(140) In writing about the event in his *Commentarii,* Pius gives more details. Calling them "three mighty leaders of heresy at the royal court", he reports that the Bishop of Nin — who doubtless knew both Latin and Slavic — had spoken with the three men and had sent them in chains to Rome, where Pius had relegated them to a monastery to learn Christian dogma. Johannes Cardinal of St. Sixtus (Torquemada) had then converted them. The three renounced their errors and went back to the King of Bosnia. Two remained firm in the new faith but the third returned like a dog to his vomit and sought refuge with Herceg Stefan.(141) Thereafter, nothing more is heard about them.

We have one more document about the trio since two texts of their renunciation have been preserved.(142) Claiming they sinned from ignorance rather than from malice, the three Bosnians renounced a list of fifty points drawn up by the aged Cardinal Torquemada who had never been to Bosnia and almost certainly knew no Slavic language. The fifty points are similar in content to the two tracts from the end of the fourteenth century, which Rački published and we have already discussed. We suspect that when Torquemada was told that Manichees were coming to study with him, he simply went to the inquisition archive and drew up a list of fifty points from the documents he found there. Thus the document itself tells us nothing about the specific beliefs of the nobles, for any given point — or most all the points — could have been drawn from the inquisition archives rather than from the beliefs they actually held.(143) And although the two fourteenth-century tracts state that they give the beliefs of "Bosnian Patarins," it is quite likely that the errors listed in them were those of Italian dualists, who, though having some ties with the Slavic world, also were greatly influenced by Western Cathars.

There is much that is puzzling in this affair and I must admit that I am not satisfied with any explanation. Because we have three documents about the trio, we must assume that three men had actually been sent from Bosnia to Rome. Then we are told that the three were powerful leaders of heresy at the royal court. Since the inquisition usually specifies if a man investigated is a heretical bishop, I think we can assume that the three were noblemen. Since no Bosnian source mentions them, and though supposedly at court they never witnessed any extant royal charters, I think we can assume that they were members of the lesser nobility. Thus, it seems probable that Pius somewhat exaggerates their importance. In addition, we have reason to believe that, prior to 1461, "heretics" had been expelled from the royal court. Many scholars believe that the Bosnian Church had been eliminated from the court in the 1440's; the last mention of a djed there is 1446. And, in any case, in 1459-60 "Manichees" were given the choice of baptism or exile. Thus it is hard to believe that this trio could have remained heretics and at court. It is possible, however, that when the others went into exile, these three had been sent to Rome and had spent about two years in Italy. Pius does not state how long they had been there.

We are faced with two alternatives: a) The three were members of the Bosnian Church, or else b) they were members of the dualist heresy. And of course in Bosnia membership in one need not exclude them from acquiring ideas or practices from the other.

a) Since they were never called Patarins, and since they had passed through the hands of the bilingual Bishop of Nin, we cannot believe that a misunderstanding had arisen over the term Patarin. Since the Bosnian Church had been referred to as Manichee with some regularity over the preceding decade, it would not have been strange for members of this church to be called Manichees. Yet it is hard to believe that the three, if they were members of the relatively orthodox Bosnian Church, could still have been sent to Rome as Manichees after being examined by this bi-lingual Bishop of Nin. Thus, we suspect that either they were framed as dualists, with the Bishop of Nin a party to the plot, and packed off to Rome in chains as Manichees, or else, despite their membership in the Bosnian Church, they held certain dualist beliefs — beliefs that may or may not have been generally accepted by the Bosnian Church — which convinced the bishop that they were dualists. Once they arrived in Rome, it is not surprising that they would have been "proved" Manichees. Rome received them as Manichees. The trio surely knew no Latin; Torquemada surely knew no Slavic. All conversations would have had to be through interpreters and surely much misunderstanding occurred. In addition, the three men were surely scared out of their wits by the inquisition procedure taking place in a foreign land in a foreign language. Thus it is probable that they would have abjured and signed anything put before them to renounce. And so they signed the Manichee renunciation presumably compiled from inquisition archives; thus we put little stock in the text of the fifty points as revealing beliefs to be found in Bosnia. And if the three were Bosnian Church members, Gost Radin's will (1466) shows the unreliability of the renunciation, since his will refutes eight to ten of the fifty points.(144) And even if a dualist wing had split off from the mainstream of the Bosnian Church to which these three men belonged, the fifty points, because they were probably based on inquisition files rather than the testimony of the three, would still be useless to determine the actual beliefs of that wing.

b) The only reason given in the sources to suspect the three of belonging to the Bosnian Church is that it is stated that the three men were leaders of the heresy *at court.* Because Rome considered the Bosnian Church a heresy, and because Patarins had long been present at court (although probably not any longer in 1461), one might argue that the three belonged to that church. However, it is not impossible that some real dualists had also been at court; Bosnia has always been a land where different faiths co-existed, probably more because of indifference about such matters than because of tolerance. In this case, the three as actual dualists would have been correctly labeled "Manichees," and we do not have to assume that the Bishop of Nin was involved in a plot. He could have interviewed them, learned of their dualism and passed them on to Rome as the dualists they were. Besides, since we suspect that Catholics working in Bosnia were trying to convince Rome that the Patarins were dangerous heretics, what better "proof" of this could they have found than to send to Rome three authentic dualists and to let the Curia think the three were Patarins. If this had been the aim of those who had sent the trio, then because a plot had been worked out, we have no reason to take statements made about the three literally. Though sent as "powerful leaders of heresy at court," the three could actually have been real dualists with no association with the court — or the Bosnian Church — at all. And although this cannot be proved, I strongly lean toward this second alternative.

In any case, be it a frame-up of Bosnian Church members, or a case of Bosnian Church members following certain dualist practices, or out and out dualist heretics, it is evident that Pius (and Torquemada, also) believed the three to be Manichees, and that Pius, at least, believed the three were members of the Bosnian Church. Prior to 1461, even though some of his predecessors had, Pius had never spoken of Bosnians as

Manichees. When he wrote letters about the exiling of Bosnian Churchmen in 1460, he called them Patarins. But when he sends the three home in August 1461, he calls them Manichees, and when he writes about the Bosnian Church in *Commentarii* and *Europa,* he without hesitation calls it Manichee. He even changes the word Patarin to Manichee in *Commentarii* when he speaks of the Bosnian Churchmen exiled in 1460. Thus we have reason to believe that it was the visit of these three noblemen and their signing the Manichee abjuration that convinced Pius of the truth of the allegation that the Bosnian Church was dualist.

XIX: The Turkish Conquest of Bosnia and Part Played by Religious Differences

In July 1461 King Stefan Tomaš died and was buried at the Franciscan monastery at Sutjeska. A story sprang up, which found its way into a variety of Franciscan chronicles, to the effect that he had been ambushed and murdered by men sent out by his brother Radivoj with the consent of his son Stefan Tomašević. However, Ćirković notes that in June 1461 Stefan Tomaš had sent to Dubrovnik for a doctor, which suggests that he had been ill.(145) In any case, he was succeeded by his son Stefan Tomašević.

If we can believe Pope Pius II's state papers in the re-worked and edited form in which they were published over a century after his death, Stefan Tomašević in 1461 sent an embassy to Rome for a crown, and requested help for his kingdom against the Turks. He referred to Pope Eugene offering a crown to his father, who refused it, fearing to provoke the Turks. His father was then a new Christian, who had not yet exiled the Manichees (probably Pius' word) from the kingdom. The wording suggests that by 1461 the "Manichees," (presumably the Bosnian Churchmen) had already been expelled, and thus were no longer a problem for the state. The new king claimed that as a boy he had been baptized, learned Latin letters and firmly grasped the Christian (i.e., Catholic) faith. Therefore he did not fear the crown as his father had. He wished it sent to him as well as holy bishops (plural). (Whether he wanted to establish a new bishopric for part of Bosnia, whether a second bishopric — e.g., possibly the Srebrnica-Visoko one noted in the papal letters dated 1434 and 1440 still existed — had been established, or whether a bishopric on peripheral territory like Duvno is intended, we do not know.) He then went on to speak of the pressing Turkish danger, and said that the Turks were promising "freedom" to the Bosnian peasants, many of whom in their simplicity believed the Turkish propaganda.(146)

Rački believes the account because it gives an accurate picture of the state of Bosnia; but Rački also argues that Pius was probably paraphrasing the royal envoy's original message.(147)

The new king's request was accepted and a papal legate crowned him at Jajce on 17 November 1461 on the Day of St. Gregory the Miracle-Worker, who ten days earlier on the request of the king, had been proclaimed Defender of Bosnia by Pope Pius II.(148) This coronation illustrates the tremendous gains made by the Catholic Church in Bosnia; in fact, in its last days the kingdom had acquired the character of a Catholic state.

In 1462 a decade after his first revolt against Herceg Stefan, his son Vladislav revolted again; unable to capitalize on an existing war as he had in 1452, he turned for aid to the willing Turks, who launched massive attacks on Bosnia and Hercegovina in 1463. This time, however, the Turks came in their own interests, and they were clearly bent on putting an end to Bosnian independence. Bosnian fortresses fell rapidly, one after the other. The king fled from Jajce in one direction, the queen in another and the Franciscans bearing Saint Luke's relics in a third. The king fled toward Croatia with a Turkish company in hot pursuit. They caught up with him at the fortress of Ključ on the Sana, persuaded him to surrender the fortress on condition that he would be allowed to escape, and then broke their promise. Stefan Tomašević was brought back to Jajce, beheaded, and buried on a near-by hill from which his castle could not be seen.

Most of Bosnia fell in a matter of weeks. The speed with which Bosnia fell surprised every one. Partly this was owing to the clever ruses used by the sultan, which made the invasion a surprise attack. However, this is far from a full explanation. Clearly, resistance was lacking. This lack can be explained in part by poor organization and lack of cooperation among the nobles and between them and the king. We may also speculate that morale was low — though this need not have been due to religious causes. After all many must have felt that it was merely a matter of time before the Turks conquered Bosnia. The rapid fall required an explanation, however, and to many that meant scape-goats. Therefore it is not surprising that, after the event, tales of betrayal were told. The most famous, reported by Pope Pius in his *Commentarii,* tells of the betrayal of Bobovac by ''Radak Manichee.'' If the story had not been added by a later editor, then apparently it was making the rounds immediately after the fall of Bosnia for Pius died the following year. This story about the Manichee who had feigned conversion to Christianity was incorporated into a variety of chronicles and histories, and even today is frequently found in serious monographs. It has also become part of the folklore of the Bobovac region, where natives tell of the betrayal and show the stone on which the Turks, evincing an inflexible attitude toward traitors of any sort, beheaded Radak. Present local tradition may of course be based on subsequent school learning derived from a variety of seventeenth-and eighteenth-century accounts; this is quite likely since no other source, written or oral, mentions anything else about Radak. Moreover, Dursum-beg and Konstantin Mihailović, who describe the conquest and clearly were more closely in touch with these events than the papal court, speak of Bobovac falling only after heavy fighting.(149)

Another passing remark about 1463 which provides material of a similar nature to the Radak story is a statement by a bishop, Nicholas Modrussienses (Modruša), that Turkish aid was being solicited in Bosnia by forcibly baptized Manichee heresiarchs.(150) Ćirković uses this source in his history; however, since elsewhere in the document Bosnia is referred to as Illyricum, a name not used, to the best of my knowledge, for Bosnia in the Middle Ages but revived as a descriptive term for Bosnia by Dalmatians at the very end of the fifteenth century, I have doubts as to whether this document really dates from the 1460's. In fact it might be a distorted version of the Radak story. However, if reliable, it suggests that somewhere in Bosnia certain Patarin clerics were favorable to the Turks and were stirring up unrest against the now Catholic state.

A third reference to betrayal, though it does not mention religious issues, is found in a letter written 27 January 1464, by King Matyas of Hungary to Pope Pius II. Matyas states that the Moslems had been invited to Bosnia by certain traitors; having noted the ease and speed with which the Turks conquered Bosnia, the Hungarian king states that the traitors suffered as much as those they betrayed.(151) Perojević finds Matyas' remarks hard to believe and suggests that the Hungarian king was trying to justify his own inactivity. Having noted that various scholars have believed that Patarins or forced converts to Catholicism had called in the Turks, Perojević stresses that there is no proof of this. The chief nobleman supporting the Patarins was Herceg Stefan, who opposed the Turks; thus, evidently, there was no Bosnian Church policy to assist the Turks.(152) If there were cases of betrayal, then most likely they were by individual choice and probably more often than not for non-religious motives.

One of the most interesting things about the sources concerning the first half of the 1460's is what is *not* found in them. Here was Bosnia, a nation supposedly rent with religious differences and presumably considerable anger at the king for the policy of persecution that he had initiated. Yet, in all of the papal, Venetian, Ragusan, and Hungarian letters that discuss plans to oppose the Turks, and which frequently mention Bosnia, only two show any concern about the loyalty of the Bosnian populace in the event of a Turkish attack. Of these two, Tomašević's reference to the success of Turkish propaganda among the peasantry and Barbucci's letter of 1456, only Barbucci's suggests religious issues might be a factor in Bosnian loyalty. Not one letter from any of the foreign secular courts — and I am excluding the papacy here — even refers to religious differences, heresy, or persecution in these years. When the Turkish attack was over, many letters were written by these same parties and in only one of their letters — the one by King Matyas which does not mention religious matters — was a betrayal spoken of; no letter mentions heretics or Bosnians siding with Turks against Catholics. The only suggestions of disloyalty beside Matyas' letter are the two stories of questionable reliability noted previously — the one of the Modruša Bishop and the other about Radak.

Although I have expressed doubt about the reliability of these two stories about ''Manichees'' aiding the Turks, I think it quite possible that there were cases of members of the Bosnian Church cooperating with the Ottomans. This would have been a natural reaction for people who had been persecuted by Catholics. However, the lack of notice given to them as a factor both before and after the conquest suggests that such Patarin cooperation with the Turks had not been on a large scale and had not been significant enough to have had any real effect on the outcome of the struggle.

The almost total absence of mention of the religious issue as a factor in plans for opposing the Turks or as a factor in the conquest, then, suggests that the religious issue had not been a very important one in 1463. A few years earlier the king had told the Bosnian Church clergy to accept Catholicism or leave; a majority accepted the new faith and a handful went to Hercegovina. Apparently the populace, indifferent about the issue, went on living as they had previously, with no great indignation against the king or Catholics for exiling a handful of monks.

XX: The Position of the Bosnian Church in the Last Years of the Kingdom

Within a year after the Turkish conquest the Catholics asked the sultan for permission to practice their religion in the new state; they received a charter from him granting them this privilege.(153) The Bosnian Church apparently never did this, scholars have wondered why. The answer, I think, is not hard to find. The Bosnian Church, consisting of only a limited number of clerics (possibly a little over 2,000 prior to 1459) who lived in monastic communities, never seems to have provided an effective ministry for the people. The only clerics in Bosnia proper from the schism in the mid-thirteenth century until the arrival of the Franciscans in the 1340's, the Patarins presumably then had had the token support of many peasants. But even in this period it is doubtful that the Bosnian Church succeeded in making itself the center of religious life for the peasantry. By the second quarter of the fifteenth century, the Franciscans had probably won over the token adherence of many of these peasants. And the remaining Bosnian Church peasants could hardly have been expected to have taken up arms to defend their clerics from persecution. By 1459-60 most of the nobility seems to have been won over to Catholicism except for Herceg Stefan in Hercegovina, the Pavlovići and possibly the Dinjičići; and the last two were greatly weakened by Turkish occupation of the bulk of their lands. Thus in 1459-60 the Bosnian Church stood alone without mass popular support and without the backing of nobility within Bosnia strong enough to oppose the king.(154) So, when the king gave the Patarin clerics a choice of becoming Catholics or going into exile, he basically destroyed the church in Bosnia. The fact that the king was able to present the Patarins with such a choice — as well as the fact that we hear of no reaction against the king as a result of his action — is evidence that the Bosnian Church organization in 1459 was weak and lacked support. That a far larger number of Patarins faced with the choice accepted baptism, rather than resist or go into exile, shows that even within the church itself loyalty and interest were lacking.(155) The few clerics (only forty of them) who felt strongly about their faith went to Hercegovina. This presumably eliminated the hierarchy, at least from the main centers of Bosnia, as well as the most important monasteries.

When the monks left or accepted Catholicism, the king undoubtedly closed their monasteries and confiscated their lands. And these confiscations have been noted in Turkish defters. In the period after 1461 we shall find Patarin monasteries within Bosnia in connection with only two places, both of which are away from the centers of the state — Uskoplje (from 1463 subject to the herceg) and possibly Seonica (near the Hercegovinian border). In addition, it is likely that some of the "kristian" lands referred to in the Turkish defters (compiled after the conquest) remained in krstjanin hands after 1460. Some of these lands, however, may have belonged to lay followers of the church rather than to clerics. And for much of the defter material published up to now, it is impossible to determine whether krstjani (be they monks or laymen) still were connected with the lands in question or whether they were simply figures from the past whose previous connections with the lands had supplied place names. We can suggest that quite possibly the community on the Drina at Zgunje, where the obelisk of Krstjanin Ostoja and the stroinik's stone (discussed in Chapter V) still stand, being beyond the borders actually controlled by the king, continued to exist. The obelisk-shape of Ostoja's stone suggests its late date — though it could perfectly well have dated from the 1450's. We may even postulate that the "stroinik" had not originally lived so far from the center of the state, but had fled thither after the king initiated persecutions.

Thus besides the one (or possibly two) known hižas existing in Bosnia after 1460, it is probable that certain other krstjani communities noted in the defters continued to exist. It is to be hoped that the publication of the defters themselves will allow scholars to decide when at least some of the krstjani linked to pieces of land were connected with those lands. We suspect such surviving communities would have been in remote regions farther away from Jajce and the center of the state.

Such scattered communities, probably lacking contacts with each other, could hardly have provided any sort of leadership or have rebuilt the church, especially since none of the Bosnian nobles apparently had risen to its defense. Thus, when the Turks assumed power, the Bosnian Church had no leadership to petition for a charter. Once the Ottoman administration took over, the Bosnian peasants were faced with problems concerned with their existence in this world under the new order. Clearly, this was no environment in which to revive a shattered church, which even in its heyday had had little more than indifferent acceptance by the peasants.

In many respects my description of the Bosnian Church resembles that of fra L. P(etrović), who argued that the krstjani were Slavic-liturgy Benedictine monks in schism with Rome.(156) And if we visualize the krstjani as a monastic order with a few monks in monasteries, it is easy to understand how their expulsion in 1459-60 would have eliminated the order — and the church — from Bosnia. It was like the situation in a country whose only Catholic priests are Jesuits; the expulsion of Jesuits in effect means the end of Catholicism there. Hence when the monks were expelled from Bosnia, the order was eliminated and with it went the "church" and whatever specific beliefs it might have had. The people who had been administered to by these monks were left without priests, and thus either had to remain priestless, accept Catholic or Orthodox priests, or accept Islam.

We last hear of the Patarins in central Bosnia on May 4, 1465, when some of them — ordained or not we do not know — attacked the Franciscan monastery of Visoko and killed five friars.(157) The

Franciscan obituary notice refers to the slayers as Patarins and not Manichees. It is ironic that the Patarins pass from the Bosnian scene on such a violent note, for throughout their history in that land and throughout the persecution that they underwent on occasions, not once prior to this do we hear of them attacking a monastery or murdering any Catholics. It is also odd that (excluding the defter material which cannot yet be dated) after this notice, and references to a hiža in Uskoplje — most probably in the Bosnian *župa* Uskoplje which was granted to the herceg in 1463 — between 1466 and 1470, the Patarins are heard of no more in Bosnia. Shortly thereafter we begin to hear of quarrels between Catholics and Orthodox in Bosnia, which were to become a common feature of Bosnian religious life in the Turkish period. The rapid disappearance of the Patarins would seem to confirm the weakness of their church and the lack of popular support for it at the time of the Turkish conquest. That they disappeared and that the Orthodox appeared in such a short period of time is one of the major arguments advanced by those who speculate that the Patarins had been more or less Orthodox all along. I disagree, although I can see why the Orthodox clerics succeeded in their proselytizing in Bosnia: Like the Patarins, the Orthodox had a vernacular liturgy; second, the Turks preferred them to the Catholics; third, the Orthodox had not seriously persecuted the Patarins earlier; fourth, the Orthodox were linked with the Serbs (fellow Slavs) and not with the hated Hungarians.

The Hungarians were clearly unpopular with large segments of the Bosnian populace. Shortly after the kingdom fell, some Bosnian nobles (apparently from the lesser nobility since they were unnamed) offered the kingdom to Venice. If Venice would not aid Bosnia, then these nobles preferred to remain under the Turks, but under no circumstances did they want to be under the Hungarians.(158)

After the Turks completed their rapid conquest of Bosnia and Hercegovina, they garrisoned the major fortresses and then withdrew the bulk of their armies. This allowed for a partial recovery of some of the lost territory. The herceg, who seems to have withdrawn to the coast with most of his troops to avoid the Turkish armies, now marched back into Hercegovina and quite rapidly recovered most of his fortresses that the Turks had conquered. Thus Hercegovina was restored as a state and was to remain independent until 1481. During these final twenty years, the Turks constantly attacked it and usually seized territory. In fact, the Turks recovered most of Hercegovina in the mid 1460's. Gost Radin remained at the herceg's side. Thus the Bosnian Church, though more or less extinct in Bosnia, was able to cling onto life for a little longer in Hercegovina. Its days were clearly numbered, however, for

Hercegovina was primarily an Orthodox region, and the Patarins had even less popular support here than they had had in Bosnia.

XXI: Mention of Religion in the Hungarian Banate of Jajce

While the herceg was restoring his state, Hungarian armies stormed in from the north and recovered a large part of northern Bosnia (including Srebrnik in Usora, Doboj, Jajce, and Ključ on the Sana). This northern territory was incorporated into a Bosnian banate under Hungarian tutelage. Herceg Stefan and his son Vladislav had sent armies to help the Hungarians recover Jajce. The Hungarians also recovered the regions of Završje and part of the Krajina which included most of the land between the west bank of the Neretva and the Adriatic coast.

In December 1463 the herceg and his son visited Matyas (probably at Jajce), and Matyas issued them a charter granting them the *župa* of Rama, with Prozor, the *župa* or Uskoplje on the Vrbas, Livno, and the town of Vesela straža.(159) Thus the herceg was able briefly to add considerable territory to his holdings.

Gost Radin's will, drawn up in 1466, left, among other things, a fur robe he received from King Matyas.(160) It seems probable that Matyas had given the robe to Radin on this occasion. Thus we may suggest that Gost Radin had accompanied the herceg to this important meeting with the Hungarian king. That the herceg received this extensive territorial grant and the gost this gift indicate that Matyas, the zealous defender of the Catholic faith, did not find Radin's presence offensive. It is hard to believe that Matyas did not know who and what Radin was; thus we may conclude that he did not consider Radin a heretic.

The following year the Turks returned and laid siege to Jajce, but the Hungarians were able to hold out. This Hungarian Banate of Bosnia, though constantly losing territory to the Turks, was able to survive until 1527.

Now that Catholic Hungary had occupied this Bosnian territory we might expect to find reference to persecution of heretics or schismatics, or action to convert the populace to Catholicism. However, one looks in vain through the letters about the banate by King Matyas Corvinus, and through letters between him and his governors in the banate, for reference to any of these matters.

The lack of Hungarian concern about the religious issue confirms my belief that the Bosnian Church in 1463 was no longer an institution of significance in Bosnia. This absence of interest also confirms my contention that the Bosnian Church was not dualist. For were there dualists around — even if their numbers were small — we would expect the Catholic Church in characteristic fashion to have carried out a major campaign to reduce them to Catholicism. In fact, the one piece of information from Hungarian sources that does exist, instead of depicting the Bosnians as dualists, presents them as orthodox in rite.

Peter Ranzanus, who describes the Hungarian conquest of northern Bosnia in 1463, gives no hint of any heresy among the people. He states that the inhabitants of Bosnia worshipped Christ according to the rite of the Eastern Church.(161) Clearly, Ranzanus was describing Bosnia and not Hercegovina. It is also obvious that the majority of people about whom he had information were not Catholics. And since there is no evidence of any Orthodox penetration this early into the area that was made into the Hungarian Banate of Bosnia, it seems that Ranzanus must have been speaking of the Bosnian Church. Thus this testimony from the final year of the kingdom re-enforces my view that the Bosnian Church had been a Slavic liturgy church,relatively orthodox in theology, that was derived from the Catholic organization in the thirteenth century. A Hungarian finding Slavs using a Slavic liturgy would naturally have identified their faith with the Eastern Church.

Ranzanus also shows that the Bosnian Church had not died out completely in Bosnia. Perhaps a few Patarin priests still existed in this region, more distant from the centers of the old state and thus more immune from persecution. If so, their days were obviously numbered; their hierarchy was in exile, and after the establishment of the Hungarian banate they would have found themselves under hostile foreign rule.

FOOTNOTES TO CHAPTER VI

1. See above, Chapter III, note 9.

2. The narrative portions of this chapter unless otherwise noted follow Ćirković's *Istorija . . .*, and his monograph *Herceg Stefan . . .* Any matter that is disputed or controversial will receive a special footnote; otherwise the reader may assume that an event or fact mentioned is generally accepted in the historical literature. Matters related to religious questions will in all cases be documented.

3. Resti, p. 290.

4. Jorga II, p. 399. Stefan Vukčić had also sent a Franciscan as an envoy to Genoa in the course of 1443 (S. Ćirković, *Herceg Stefan . . .*, p. 74).

5. L. Thalloczy, *Studien zur Geschichte Bosniens . . .*, pp. 357-58.

6. Fairly recently O. Halecki, *The Crusade of Varna,* New York, 1943, has advanced the view that the ten-year peace had never been ratified and thus the crusaders should not be accused of breaking the treaty. G. Ostrogorsky, *History*

of the Byzantine State, New Brunswick, 1969 and many other scholars have not been convinced by Halecki's arguments. I do not want to enter the controversy and have stuck to the traditional view.

7. *AB,* p. 183.
8. *AB,* p. 184.
9. *AB,* p. 187.
10. Theiner, *MSM,* I, p. 388; *AB,* p. 198.
11. S. Ćirković, *Herceg Stefan . . .,* p. 90.
12. Theiner, *MSM,* I, p. 396; *AB,* p. 202.
13. Theiner, *MH,* II, p. 264.
14. *Europa Pii Pontificis Maximi nostrorum temporum varias continens historias,* p. xxiii.
15. Resti, p. 188.
16. The Dragišić of the 1446 charter (Pavle, Marko, and Iuri sons of Ivaniš) should not be confused with Stjepan, Radosav, and Ostoja Dragišić Kosače. On the latter see, Lj. Stojanović, *Stare srpske povelje i pisma,* I, 2, pp. 94, 96, 97. However, we do have confirmation on the existence of our 1446 charter Dragišići; in 1461 Knez Marko Dragišić with brothers witnessed a royal charter (Stojanović, *op. cit.,* I, 2, p. 164).
17. Miklosich, pp. 438-40.
18. Lj. Stojanović, *Stari srpski zapisi i natpisi,* SKA, Beograd, I, 1902, p. 47.
19. See Chapter IV.
20. V. Vrana "Književna nastojanja u sredovječnoj Bosni," in *Poviest B i H,* Sarajevo, 1942, p. 814.
21. On Gomiljani-pravica see, Radimsky's *Topografija: Arheološki leksikon* (arranged alphabetically). This work exists in manuscript only and is to be found only at the Zavod za zaštitu spomenika kulture in Sarajevo.
22. Dinić, p. 204, No. 59.
23. G. Čremošnik, "Ostaci arhiva bosanske franjevačke vikarije," *Radovi* NDBH, III, Sarajevo, 1955, p. 38.
24. Jorga II, p. 420.
25. M. Vego, "Patarenstvo u Hercegovini u svjetlu arheoloških spomenika," *GZMS* (arh), n.s. 18, 1963, p. 207.
26. Jorga II, p. 420. Radivoj is called Count of Vranduk. Thus it seems probable that he had been granted lands in this region after he had made peace with his brother the king.
27. Theiner, *MH,* II, p. 230; *AB,* p. 203.
28. *AB,* p. 208.
29. Cited from the Dubrovnik archive by M. Dinić, *Za istoriju rudarstva . . .,* I deo, Beograd, 1955, p. 38. Also amidst the ruins of the Pavlović court at Borač is a location called "Ciža" which Mazalić believes is derived from "chiesa"; he suggests that once a church had stood on that spot. Since an Italian word is used, we may assume that the church, if in fact there had been one, had been Catholic. Quite possibly it was built during Ivaniš' Catholic period and our priest had served in it. On the location and the word derivation see, Dj. Mazalić, "Kraći članci i rasprave," *GZMS,* n.s. IV-V, 1950, pp. 223-24.
30. Jorga, II, p. 426, note 1.
31. Jorga, II, p. 420.
32. *AB,* p. 205.

33. *AB,* p. 205.

34. *AB,* pp. 204-05, Nos. 863-866; Theiner, *MH,* II, pp. 233-34. M. Vego, *Naselja bosanske . . .*, p. 136, suggests that "Verlau" refers to Vrili near Kupres. Dj. Basler, "Kupres," *GZMS,* n.s. VIII, 1953, p. 340, agrees.

35. Dinić, p. 198, Nos. 46-47.

36. Theiner, *MH,* II, p. 237. Shortly prior to this letter of Nicholas is a letter attributed to Eugene IV dated November 1445 which refers to Manichees in Bosnia and gives details similar to those of the late fourteenth-century Italian inquisition tracts. I am convinced that this letter published by Farlati (*IS,* Vol. 4, pp. 256-57) just before he gives an account of the bogus Council of Konjic, is a forgery. See the discussion about this letter in Chapter II.

37. Theiner, *MH,* II, p. 264.

38. *AB,* p. 207.

39. Raynaldi, *AE,* 1449, No. 9.

40. Theiner, *MH,* II, p. 255.

41. Wadding XII, original pagination pp. 111-112.

42. Theiner, *MH,* II, p. 267.

43. *AB,* p. 208; Raynaldi also cites this letter (1449 No. 9).

44. Theiner, *MH,* II, p. 264.

45. S. Ćirković has pointed out that this title passed through these two stages; see his *Herceg Stefan . . .*, pp. 106-08.

46. L. Thalloczy, *Studien zur Geschichte Bosniens . . .*, p. 361.

47. Dinić, p. 229, no. 27.

48. S. Ćirković, *Herceg Stefan . . .*, pp. 114, 129 (note 48).

49. Dinić, p. 229, no. 28.

50. Cited from the Dubrovnik archive by P. Andjelić, "Srednjevjekovni gradovi u Neretvi," *GZMS,* n.s. XIII (arh.), 1958, p. 180.

51. See Chapter V.

52. See S. Radojcić, "Reljefi bosanskih i hercegovačkih stećaka," *Letopis Matice srpske* (Novi Sad), Jan. 1961, knj. 387, No. 1, pp. 1-15. Besides giving much interesting material on the cultural and artistic level of the courts of the nobility (Ragusan artists hired by Sandalj, and a goldsmith from Bruges in Belgium by Herceg Stefan), Radojčić's article is one of the most stimulating and solid about the meaning of *stećci* motifs. On actors and musicians, see the interesting article of A. Babić, "Fragment iz kulturnog života srednjovjekovne Bosne," *Radovi* (Filozofski fakultet u Sarajevu), II, 1964, pp. 325-336.

53. Ćirković, *Istorija . . .*, p. 312.

54. Cited from the Dubrovnik archive by Ćirković, *Herceg Stefan . . .*, p. 131. For a full discussion of these events the reader is referred to Ćirković's excellent account in Chapter VI of his work.

55. *Ibid.*, p. 135.

56. J. Gelcich and Thalloczy, *Diplomatarium . . .*, p. 483 and J. Radonić, *Acta et diplomata Ragusina,* Beograd, 1934, p. 523. Radonić has rendered "manieri" as "manicei." Though his rendering may well be what the Ragusan author intended to write, the form "manieri" is what he did write. I have checked the original document.

57. Dinić, p. 201, No. 53.

58. Dinić, p. 202, No. 54.

59. Lettere e Commissioni di Levante (1449-1453) Fol. 164, 25 May 1461. "Patarinos et manicheos." In all other cases my citations to documents in the

Dubrovnik archive have been to works in which the material has been published for that is more useful for reference. This one reference, however, I have not seen cited anywhere else.

60. The best account of the herceg's war with Dubrovnik is to be found in S. Ćirković, *Herceg Stefan . . .*, Chapters seven and eight.

61. Dinić, p. 203, No. 55.

62. Dinić, p. 203, No. 57.

63. Dinić, pp. 203-04, No. 58.

64. Dinić, p. 204, No. 59. The content of this message is also given in G. Gondola (Gundulić) *Croniche . . .*, p. 321. This chronicle has a detailed account of the whole war and the concomitant diplomatic activity.

65. Some merchants, knowing that the herceg's elest son Vladislav sought a bride, brought a girl of low-reputation but of great beauty from Siena to the herceg's court. They claimed she was of noble birth, and hoped to be richly rewarded. Vladislav liked her but the herceg immediately made her his mistress. Angry words were exchanged between father and son; Vladislav was imprisoned for a few days until some nobles freed him, after which he and his mother, jealous over the Sienese beauty, left the court and joined Dubrovnik in war against the herceg (S. Ćirković, "Vesti Brolja da Lavelo kao izvor za istoriju Bosne i Dubrovnika," *Istorijski časopis,* XII-XIII, 1963, pp. 169-70). In addition to the contemporary Italian account summarized above and recently discovered by Ćirković a variety of early chronicles discuss the girl, and she is also mentioned in various contemporary documents. A 1451 Ragusan protocol, sending greetings to the herceg's court, includes a "Domina Helisabeta" who most probably is our girl. We might note here that the Sienese concubine was to remain at the herceg's court even after peace was concluded between the herceg and his wife in 1453. Ćirković found mention of her in a Ragusan document from December 1453 (*Herceg Stefan . . .*, p. 163, note 78). For additional material on the girl and the family quarrels she stirred up, see J. Radonić, "Herceg Stipan Vukčić Kosača i porodica mu u istoriji i narodnoj tradiciji," in *Zbornik u slavu Vatroslava Jagića,* Berlin, 1908, pp. 406-14.

66. Ć. Truhelka, "Testament gosta Radina — prinos patarenskom pitanju," *GZMS,* XXIII, 1911, p. 362.

67. Miklosich, pp. 447-50.

68. Theiner, *MH,* II, p. 265.

69. J. Gelcich, and Thalloczy, *Diplomatarium . . .*, p. 525.

70. I. Voje, "Sitni prilozi za istoriju srednjovjekovne Bosne," *GID,* XVI, 1965, Sarajevo, 1967, p, 281.

71. *Ibid.,* pp. 281-82.

72. Dinić, p. 199, No. 48.

73. Dinić, pp. 205-06, No. 60.

74. Dinić, pp. 206-07, No. 61.

75. Theiner, *MH,* II, p. 264.

76. *AB,* p. 218.

77. Miklosich, pp. 457-60; Stojanović, I, 2, pp. 66-69.

78. Miklosich, pp. 460-63; Stojanović, I, 2, pp. 69-72.

79. For example this view can be found expressed in A. Solovjev, "Nestanak bogomilstva i islamizacija Bosne," *GID,* I, 1949, p. 47, and in A. Babić, *Bosanski heretici,* Sarajevo, 1963, p. 156.

80. For example this belief can be found in F. Rački, *Bogomili i patareni,* p. 522. D. Mandić, *Bogomilska crkva,* pp. 216-18.

81. On negotiations see Dinić, pp. 208-10; Gondola, *Croniche . . .*, pp. 333-46. For the text of the treaty see Miklosich, pp. 465-69.

82. For these inquisition charges, see document given as Appendix A for this chapter, article numbers 14, 15, 44.

83. Pavle Radenović, most probably also a supporter of the Bosnian Church, also refers to God's only son in an invocation to the Trinity in a charter he issued in 1397. See Miklosich, p. 229.

84. Dinić, p. 209, No. 64.

85. Ć. Truhelka, *op. cit.*, p. 362; and A. Solovjev, "Gost Radin i njegov testament," *Pregled,* II, 1947, pp. 310-318.

86. Dinić, p. 209, No. 64.

87. Dinić, p. 211, No. 70.

88. Dinić, p. 211, No. 72.

89. Dinić, p. 212, No. 73.

90. Miklosich, pp. 472-73.

91. Dinić, pp. 231-33, Nos. 34-42.

92. Dinić, p. 233, No. 43.

93. Miklosich, pp. 469-72.

94. The organization of each Western Cathar church was headed by a bishop, beneath whom was a filius maior (usually rendered into English as elder son) and a filius minor (usually rendered into English as younger son). When the bishop died, the elder son became bishop and the younger son became elder son. Specific cases of this system in practice can be found in the tracts published by A. Dondaine in "La Hierarchie Cathare en Italie," *Archivum Fratrum Praedicatorum,* XIX, 1949, pp. 280-312, and XX, 1950, pp. 234-324. The best work on the Cathar church and its organization is A. Borst, *Die Katharer,* Stuttgart, 1953.

95. Thalloczy, *Studien zur Geschichte Bosniens . . .*, p. 401.

96. *Ibid.*, pp. 175-76.

97. S. Ćirković, *Herceg Stefan . . .*, p. 218.

98. Ljubić, *Listine,* X, pp. 75-76.

99. S. Ćirković, *Herceg Stefan . . .*, p. 209.

100. Dj. Mazalić, "Hercegova crkva kod Goražda i okolne starine," *GZMS,* LII, 1940, pp. 27-43.

101. Z. Kajmaković, *Zidno slikarstvo u Bosni i Hercegovini,* Sarajevo, 1971, pp. 56-57. Much of the description I give here, however, comes from my visit to Šćepan polje and environs with Professors Kajmaković and Palavestra.

102. Further evidence of Patarin activity in the vicinity of Soko also suggests the church might have been Patarin. On this activity see the discussion in Chapter V and especially note 192. In addition on the Pivska Mountain, a seven hour walk from the church, is village Pišče at which the treaty between the herceg and his family, and guaranteed by the Bosnian Church, was signed.

103. Oral communication from Z. Kajmaković.

104. L. Kovačević, (ed.) "Odgovor carigradskog patriarha Genadija na pitanja sinajskih kaludjera," *Glasnik Srpskoga učenoga društva,* LXIII, 1885, pp. 12-13.

105. L. Chalkokondyles, Bonn ed., 1843, p. 249.

106. A. Solovjev, "Fundajajiti, paterini, i kudugeri u vizàntiskim izvorima," *ZRVI* (Beograd), I, 1952, pp. 130-33.

107. V. Skarić, "Kudugeri," *Prilozi*, VI, 1926, pp. 107-110.
108. However, if the "Bosnian bishop" were Gennadius' source of information, we may wonder why he does not get his title straight.
109. Türk Dil Kurumu, *Türkiye Türkçesinin Tarihi Sözlüğü Hazirliklarindan XIII. Asirdan Günümüze Kadar Kitaplardan Toplanmiš Taniklariyle Tarama Sözlüğü*, I, Istanbul, Cumhuriyet Basimevi, 1943; II, A-I, Istanbul, Cumhuriyet Matbaasi, 1945. Vol. I, p. 336: Güdücü: çoban. Gives a quotation using the term from a ms. of the fourteenth century. For dating of ms. see Vol. I, p. XVI. The term is used again in Vol. II, p. 472, and a second passage using it is given; this time it is taken from a fifteenth-century ms. For dating of that ms. see Vol. II, p. XV.
110. H. Šabanović, "Postanak i razvoj Sarajeva," *Radovi* (NDBH), XIII, 1960, p. 107.
111. *AB*, pp. 224-26.
112. *AB*, p. 225.
113. By this time it is possible that Orthodox were settled as far west as Vrhbosna. There is a tradition that Angelo Zvijezdović, a prominent Catholic cleric in the second half of the fifteenth century, had been born into an Orthodox family of Vrhbosna (he was born in the 1420's). See, M. Batinić, *Franjevački samostan u Fojnici od stoljeća XIV-XX*, Zagreb, 1913, p. 129. This tradition also illustrates the attempts — here successful — of Catholics to win over Orthodox believers.
114. In the earlier medieval period we even noted signs of cooperation between the Orthodox and Catholics. For example in the 1290's the Orthodox ruler of Mačva and Usora, Stefan Dragutin, requested the pope to send him Franciscans.
115. *AB*, p. 240; Theiner, *MH*, II, p. 330.
116. Papal legates visited the herceg in 1457, 1458 and 1462 to discuss projects of a crusade against the Turks (see S. Ćirković, *Herceg Stefan . . .*, pp. 235, 246). Such embassies clearly show that the pope considered the herceg part of the Christian world.
117. Ćirković, *Istorija . . .*, p. 381.
118. Thalloczy, *Studien zur Geschichte Bosniens . . .*, pp. 415-16.
119. Wadding, XII, original pagination pp. 111-112; *AB*, p. 208.
120. B. Cotrugli (Kotruljić), *Della Mercatura et del Mercante perfetto*, Brescia, 1602. On this work see M. Vujić, "Prvo naučno delo o trgovini Dubrovčanina Benko Kotruljića," *Glas* (SKA), 80, 1909, pp. 25-123. Our merchant author is frequently also called Kotruljević.
121. Further evidence of hostility being felt by Ragusans toward the Bosnian Church (Patarins) is provided by a court case of 5 February 1457 in which a Ragusan woman sues a relative for insulting her; he had called her a whore and a "wet-nurse of Patarins" (babiza de Patarinis). Cited by K. Jireček, *Istorija Srba*, II, p. 278, note 129.
122. Pius, *Europa*, xxiii.
123. Theiner, *MH*, II, pp. 374-75.
124. Orbini, p. 370.
125. *AB*, p. 238.
126. *AB*, p. 243.
127. *AB*, p. 244.
128. For the amusing adventures of these relics, see Ć. Truhelka, *Kraljevski*

grad Jajce, povijest i znamenosti, Sarajevo, 1904, pp. 55-56.

129. *AB,* p. 242. Previously we noted Radivoj as lord of Vranduk (see above note 26) and suggested that he had been granted this territory as his own personal holding by the king.

130. D. Mandić, *Hercegovački spomenici franjevačkog reda iz turskoga doba,* Mostar, 1934, pp. 7-21. Of course 1481 was the date of the conquest of the last parts of Hercegovina. Most of Hercegovina fell to the Turks in the earlier part of 1465-81.

131. One could connect the initiation of persecution with the king's desire to prove his Catholicism to the pope and papal allies in the face of Hungarian slander about Bosnia's disloyalty to Christendom after the fall of Smederevo to the Turks in 1459. We know that in 1459 the king sent an embassy to justify himself to the pope. Perhaps, he initiated persecutions to strengthen the case of his envoy, or perhaps the envoy returned with a papal order to persecute the Patarins, which the king, in his present difficulties and needing Western aid, felt unable any longer to resist.

132. Pius, *Commentarii,* (Ed. 1584) V, p. 227; Orbini, p. 369.

133. Ćirković, *Istorija . . .,* p. 320, dating the expulsion 1459-61; A. Babić, *Bosanski heretici,* Sarajevo, 1963, dating the expulsion 1449-50.

134. Or possibly a combination of clerics and laics (servants and peasants) serving at the Bosnian Church monasteries.

135. Theiner, *MH,* II, pp. 358-59.

136. M. Šunjić, "Jedan novi podatak o gostu Radinu i njegovoj sekti," *GID,* XI, 1960, pp. 265-68.

137. N. Filipović, "Osvrt na položaj bosanskog seljaštva u prvoj deceniji uspostavljanja osmanske vlasti u Bosni," *Radovi* (Filozofski fakultet u Sarajevu), III, Sarajevo, 1965, p. 66. This view about the king's motives was already suggested by Orbini who said that Stefan Tomas acted to prove his religion "or maybe, as many believe, he was motivated by greed" since the lands of those who left went to the state. (Orbini, p. 369). That the king may also have needed to prove his religion, as Orbini suggests, confirms the suggestion made above in note 131 to this chapter that the king found himself in serious difficulties as a result of Hungarian claims that the Bosnians were disloyal to Christianity and had not tried to defend Smederevo.

138. *Ibid.,* p. 66. For the benefit of those with access to the defters I quote Professor Filipović's footnote: "Defter 0-76, page 16v, 17v, 26r, 29v, 30r, 30v, 32v, 35r, 39r, 46r, 47r, 48v, 61v, 82r, 107 r-v, 138v, 155v." The original defter is in the "Belediye Kütüphanesi" in Istanbul. The Oriental Institute in Sarajevo has a microfilm of it.

139. M. T. Okiç, "Les Kristians (Bogomiles Parfaits) de Bosnie d'après des documents Turcs inédits," *Südost-Forschungen,* XIX, 1960, p. 128.

140. Theiner, *MH,* II, pp. 363-64.

141. Pius, *Commentarii,* V, p. 227.

142. D. Kamber, "Kardinal Torquemada i tri bosanska bogomila (1461)," *Croatia Sacra,* III, 1932, pp. 27-93; a second manuscript in slightly abbreviated form had been published previously by F. Rački, "Dva nova priloga za povijest bosanskih patarena," *Starine,* XIV, 1882, pp. 1-21. See Appendix A, Chapter VI, for the 50 points as extracted by Kniewald.

143. In addition in a renunciation prepared for dualists, one would expect the inquisition to make the heretics renounce not only beliefs they were known to

hold but also other beliefs which were associated with dualism which they could be suspected of having. And thus the inquisition could obtain at one shot a comprehensive renunciation of "Manicheeism" which it could hold against the three should they reappear the following year with some new and different "Manichee" ideas. If this should be the case, we have even less cause to use the document to show beliefs actually found in Bosnia.

144. In Appendix A I list the 50 points and then discuss those that clearly do not have relevance for Bosnia, as well as those that could have relevance.

145. Ćirković, *Istorija . . .*, p. 323.

146. Pius, *Commentarii* (1584, ed.), XI, p. 547.

147. F. Rački, *Bogomili i patareni*, pp. 476-77.

148. Theiner, *MH*, II, p. 371.

149. Prof. Andjelić, who has excavated the Bobovac ruins, found signs of fire which also suggests that warfare had occurred (cited by I. Bojanovski, "Stari grad Bobovac i njegova konservacija," *Naše starine*, VIII, p. 88). Bojanovski, however, points out (*ibid.*, p. 87) that the accounts of fighting and the Radak betrayal need not be mutually exclusive; Radak could well have betrayed the town after some severe fighting had occurred.

150. G. Mercati, "Note varie sopra Niccolo Modrussiense" in *Opera Minori*, IV, *Studi e Testi*, LXXIX, 1937, p. 218.

151. L. Thalloczy and Horvath, *Codex Diplomaticus partium Regno Hungariae adnexarum* (Banatus, castrum et oppidum Jajcza): Monumenta Hungariae Historica Diplomataria, vol. XL, Budapest, 1915, p. 14. Excluding the remarks attributed to the Bishop of Modruša, the only reference in our sources to someone in 1463 seeking aid from the Turks is to the herceg's son Vladislav. However, it is doubtful that Matyas was thinking of him: first, Vladislav did not suffer ill effects from the Turkish campaigns of 1463; and second, Matyas would not have spoken thus of one who was in his good graces — the herceg and Vladislav, having helped Matyas recover Jajce, had been confirmed by Matyas a month earlier in December 1463 in possession of territory recovered from the Turks (on this confirmation, see below, Section XXI).

152. *Poviest BH*, Sarajevo, 1942, p. 570.

153. "Turski dokumenti o Bosni iz druge polovine XV stoljeća," *Istorisko-pravni zbornik*, Sarajevo (pravni fakultet), 2, 1949, pp. 200-04.

154. This view is confirmed by Raynaldi's rendering of Thomas of Hvar's 1449 letter, see Raynaldi, *AE*, 1449, No. 9., which states that only some of the barons and nobles (noting by name Herceg Stefan and Ivaniš Pavlović) remain infected with heresy, as (are) others who are called "religiosi." Thus in 1449, the papal legate to Bosnia felt that only some of the nobility and the (Patarin) clergy remained heretics. Thomas of Hvar certainly suggests that the position of the Bosnian Church was considerably weakened and anticipates that it would become even more so in the years that followed.

155. This argument, of course, assumes that 40 and not 40,000 fled to Herceg Stefan. This issue has been discussed above. I am absolutely convinced that Pius did mean 40.

156. Fra L. P. *Kršćani bosanske crkve*, Sarajevo, 1953.

157. J. Jelenić, "Necrologium Bosnae Argentinae - prema kodeksu franjevačkog samostana u kr. Sutjeskoj," *GZMS*, XXVIII, 1916, p. 357.

158. S. Ćirković, "Vlastela i kraljevi u Bosni posle 1463," *Istoriski glasnik*,

1954, pp. 123-31.

159. On the herceg's assistance to the Hungarians see, S. Ćirković, *Herceg Stefan . . .*, p. 258; and V. Stefanović, "Ratovanje kralja Matije u Bosni," *Letopis Matice srpske*, 332, 1932, pp. 195-213, especially pp. 202-03.

160. See below, Chapter VII, Section III.

161. L. Thalloczy and Horvath, *Codex Diplomaticus partium Regno Hungariae adnexarum (Banatus, castrum et oppidum Jajcza);* Monumenta Hungariae Historica Diplomataria, XL, Budapest, 1915, p. 7.

APPENDIX A FOR CHAPTER VI

The 50 Points Renounced before Cardinal Torquemada and Their Irrelevance for the Bosnian Church

The fifty points renounced by the three Bosnian noblemen before Torquemada in 1461: from Kniewald, "Vjerodostojnost . . .," pp. 178-81.

1. There are two Gods, the one supremely good, the other supremely evil.

2. There are two principles: one spiritual and bodiless, the other corruptible, with a body or visible. The first is the God of light, the second of darkness.

3. Certain angels are evil by nature, and cannot help but sin.

4. Lucifer ascended to heaven, fought with God there and brought down many angels.

5. Souls are demons imprisoned in bodies.

6. Evil angels imprisoned in bodies, by baptism and purification and repentance may return to heaven.

7. They damn and reject the Old Testament, saying that it originated with the principle of darkness.

8. They say the Angel appearing to Moses on Mt. Sinai was evil.

9. They do not accept all the New Testament, but only certain parts of it. They do not believe that Christ was born of woman and reject Christ's geneology.

10. They condemn the patriarchs and prophets of the Old Testament.

11. They condemn John the Baptist, saying that there is no devil in hell worse than he.

12. The tree of knowledge of good and evil was a woman. Adam knew her, that is, sinned with her and was expelled from paradise.

13. The blessed Mary was not a woman or female creature but an angel.

14. The Son of God did not assume a real body but a fantastic (i.e., apparent) one.

15. Christ did not really suffer, die, descend to hell, or ascend to heaven. Everything he did was only seemingly done.

16. Their church is that of God.

17. They are the successors of the Apostles; their heresiarch is Bishop of the Church and successor of Peter.

18. The Roman Church is condemned and excommunicated.

19. All the popes from Peter to Silvester were of their faith; and that Silvester was the first to apostatize from it.

20. They condemn material churches, saying they are synagogues of

Satan, and that those who worship in them are idolaters.

21. That the use of images in churches is idolatry (i.e., against worship of images).

22. The sign of the cross is a sign of the devil.

23. They condemn the mass and church singing as being opposed to Christ's Gospel and doctrine.

24. They ridicule and damn the veneration of holy relics.

25. They ridicule and damn the worship of saints done in church, saying that only God should be worshipped and adored.

26. Their leaders allow the people to adore them, saying they are holy and without sin, and that they have the Holy Spirit within them.

27. They damn ecclesiastical sacraments.

28. They renounce baptism with water, saying it is John's baptism, and by it no one can be saved.

29. They maintain Christ's way of baptizing without water, by placing the Gospel on the chest and with the laying-on of hands.

30. Through their baptism, anybody may have his sins remitted and become as holy as was St. Peter.

31. That a boy before the age of discretion cannot be saved.

32. Complete holiness and the power of baptism comes to the baptized only through the merit of the baptizer.

33. As often as the baptizer sins, the souls of those whom he has baptized fall,from a state of blessedness in heaven, to hell.

34. As often as the baptizer sins, all those baptized by him must be remitted by re-baptism.

35. They condemn the sacrament of confirmation.

36. They deny the sacrament of the eucharist, saying that the body of Christ cannot be made into bread, and if it could, we should not eat it.

37. They condemn the sacrament of repentance, they say sins are remitted by re-baptism.

38. They condemn the sacrament of extreme unction.

39. They condemn the sacrament of ordination (for a priest).

40. Bodily marriage is adultery.

41. All sins are mortal.

42. They deny all authority to the Church, saying no one can be excommunicated.

43. They condemn the eating of meat, saying that whoever eats meat, cheese or milk cannot be saved unless re-baptized.

44. They deny resurrection, saying that no body that dies now will ever be resurrected; it is the spirit that will be resurrected.

45. They deny purgatory, saying there is no middle road between heaven and hell.

46. They condemn church prayers for the deceased.

47. It is a mortal sin to kill animals, birds, or to break eggs.

48. They condemn capital punishment by secular powers.

49. They condemn all oath-taking.

50. They prohibit all acts of charity and mercy.

All these points could well concern Western Cathars, which makes it possible that they were drawn from inquisition records about Cathars.

The following of the 50 points are contradicted in Bosnian sources: (The documents most frequently cited are the herceg's treaty in 1453 guaranteed by the djed and 12 stroiniks (Chapter VI, Section IX), Gost Radin's will of 1466 (Chapter VII, Section III), and the Gospel manuscript of Krstjanin Hval from 1404 (Chapter V, Section II).

1,2) There are two Gods or two principles: Radin's will and the herceg's 1453 treaty both refer to omnipotent God.

7, 8, 10) They damn the Old Testament, Moses' Law, the Prophets and the Patriarchs: Hval's Gospel contains the Ten Commandments, the Psalms and Songs. We also find a favorable reference to the Patriarch Abraham on the gravestone inscription of Gost Mišljen (Chapter V, Section XI).

9, 13) They do not accept all the New Testament, they do not believe Christ was born of woman; Mary was an angel: Hval's Gospel manuscript contains the complete New Testament. Herceg Stefan in 1453, swears before Mary, Mother of God.

11) They condemn John the Baptist: Hval's Gospel has a complimentary picture of John.

20) They condemn material churches: We have presented a number of reasons to suggest that the Bosnian Church did have churches. Here we shall simply note that in 1472 money had been taken, according to wishes specified in Radin's will, to build him a sepulchre and chapel (Chapter VII, Section IV).

22) They condemn the cross: Though it can be argued that the Bosnian Church may have rejected the cross in the thirteenth or fourteenth century, this certainly was not true in the fifteenth century — and the 50 points were compiled in 1461 — the djed's letter of 1404 begins with a cross (Chapter V, Section III), so does the herceg's 1453 treaty, and Radin's will. Radin's will also has a large cross down the left margin. The 1453 treaty also has the herceg swearing before the life-

giving and venerable cross. This last phrase suggests that he and the djed and stroiniks accepted Christ's actual resurrection and also bodily resurrection for mankind. Thus we probably have good reason to also reject point No. 15 which says that Christ only apparently participated in the events of his passion, and No. 44 which says they deny bodily resurrection for mankind. And if Christ's passion becomes real, then we eliminate No. 14 which denies Christ had a real body. The herceg's 1454 treaty issued to Dubrovnik (Chapter VI, Section IX), stating "God sent His only Son to great suffering in order to deliver His (people) from sins by resurrection" also contradicts points 14, 15, 44. Though no Patarins witnessed the document, the herceg's adherence to their church probably is sufficient justification to cite this treaty as evidence for the Bosnian Church's position.

24) They condemn relics of saints: We find Sandalj's Orthodox widow leaving relics to her grand-nephew, almost certainly an adherent of the Bosnian Church (Chapter V, Section VIII). We also find Herceg Stefan leaving relics in his will (Chapter VII, Section II). The herceg also accepted the cult of St. Sava (Chapter VI, Section VII).

25) They ridicule the worship of Saints: The herceg in 1453 swore by all the saints. Radin's will requests prayers for his soul on the Days of St. Peter, St. Paul, St. Stephan, and St. George. He even notes that he celebrated St. George's Day as his *slava.*

41, 46) All sins are mortal, they condemn church prayers for the deceased: Radin requests prayers by both krstjani and by Catholics for his soul after death, and speaks of God forgiving sins and pardoning us at the Last Judgment. Thus he clearly believed that sins could be remitted after death and that prayers for the dead were of value. Thus by extension we can eliminate No. 37 which says sin can only be forgiven through re-baptism.

50) They condemn all charity: Radin's will not only leaves money to be distributed to poor and crippled of his own faith, but also to poor and crippled Catholics.

Thus, immediately 15 of the 50 points can be shown not to pertain to the Bosnian Church in the fifteenth century. And if we make a logical extension to No. 22, we add three more to that figure, and by extending Nos. 41, 46 we add a fourth. Thus it is reasonable to say that 19 of the 50 points are irrelevant.

The following points seem somewhat inaccurately expressed in the 50 points, but we cannot be certain:

18) The Roman Church is condemned and excommunicated. The Bosnian Church clearly had broken away from Rome, and was condemned by Rome. How strongly the Bosnians attacked Rome is

unknown. Relations certainly were on the whole friendly between Patarins and Catholic Dubrovnik, and in his will Radin asks for prayers in Catholic churches and leaves money for charity to crippled Catholics. Thus this point seems somewhat overstated.

21) The use of images in churches is idolatry. This point seems to refer to worship of images. Serbian anathemas state the krstjani refuse to bow down to images. The Bosnian Church clearly did not reject religious pictures since its Gospels were illustrated with them. Various lay members of the Bosnian Church also possessed icons. It is possible, though, that the church while allowing religious pictures, did prohibit the worship of them.

23) They condemn the Mass and church singing. Once again, we cannot be sure. We have one ritual (Radosav's Chapter II, "The Gospel Manuscripts"), which is similar to (but shorter than) a Lyons Cathar ritual. Thus the Bosnian Church may have had its own service that varied from the Roman. Whether this means they condemned the Roman is unknown, as are their views on church singing. However, we also have two sources suggesting they had a rite similar to the Serbian Orthodox, which would include a Mass and singing: The Franciscan Symeon in 1322 (Chapter IV, Section I) and a Hungarian writer in 1463 (Chapter VI, Section XXI). And Radin in his will requests Catholics to light candles for his soul on Sundays and holidays which needless to say are times when the mass was being served.

28) They renounce baptism with water: In 1404, King Ostoja, almost certainly a member of the Bosnian Church, having conferred with the djed, wrote Dubrovnik to recall Klešić. He dated the letter, the Day of the Hallowing of the Waters (Chapter V, Section III). Ostoja also in 1401 had one son baptized, and Dubrovnik's calling the ceremony a baptism suggests it was an orthodox (i.e., with water) baptismal ceremony. (Chapter V, Section III). Radin's will refers to baptized (kršteni) krstjani, though we do not know how they were baptized. Though sufficient evidence to prove the point is lacking, the above shows the Bosnian Church had some sort of baptismal rite and suggests it was with water.

39) They condemn the sacrament of ordination. While it is likely that the Patarins had a somewhat different procedure of ordination from the Catholics, they clearly had some rite of ordination for their clergy. This is seen from Radin's will which refers to krstjani (i.e., the ordained) who were "baptized in the correct way." And since the krstjani seem to have been a continuation of a Catholic monastic order, it would not be surprising if their ceremony was quite similar to the Catholic.

40) Bodily marriage is adultery: Though the Bosnians did not look

upon marriage as a sacrament, there is no reason to believe they condemned marriage or sex. It seems the krstjani were celibate; however, Catholic priests were too. Lay members married.

43) They condemn eating of meat, cheese and milk. We know that the Bosnian Church clerics kept some sort of fasts since Radin mentions fasting in his will, and speaks of fasters and "mrsni" (non-fasting) people. However, we do not know whether the ordained Patarins kept fasts all the time, or only long ones during certain periods of the year. We also do not know if they eliminated all the items noted in No. 43 from their diet. It is also evident that only the ordained kept the strict fasts; lay member's diets, as far as we know, were not regularly restricted.

Only the following four points accurately describe some Bosnian phenomenon. The first two are not doctrinal or in themselves heretical and would apply to any Christian Church of that day.

16) Their church is that of God. Every church believes that of itself.

17) They are the successors of the Apostles: Radin's will refers to krstjani of the Apostolic faith.

26) The (heretical) leaders allow the people to adore them. The pope in 1373 quoting Franciscans (Chapter IV, Section V) states some group in the vicariat had believers adoring heretics. It is not stated that this refers to the Bosnian Church. We do not know if the adoration was similar to that of the Cathars. The 1373 letter never states that it was believed that the heresiarchs were in possession of the Holy Spirit. We also find a visiting Catholic bishop more or less adored by Bosnians in the 1430's (Chapter V, Section VI).

49) They condemn all oath-taking. It is clear that the ordained Patarins did not swear oaths (Chapter V, Section III). However, on occasions they were able to endorse a document or testify in court according to their own customs, whatever that means. Lay members of the Church freely took oaths.

Thus of the 50 points we find 15 do not apply to the Bosnian Church, and if we make logical extension of the information found in our sources, we can raise this number to 19; 7 more most probably do not (though one may want to question point No. 43 on fasting); and only 4 (or 5 if we place the fasting point here), 2 of which are of a non-doctrinal nature, apply to some Bosnians; 3 or these 4 clearly are true for the Bosnian Church. And since all 4 (or 5) of the relevant points also pertain to the Western Cathars, we cannot be sure that they were placed on the abjuration as a result of any knowledge about Bosnia.

The other 20 points are not mentioned one way or the other in non-inquisitional sources. Thus, in addition to the reasons given in Chapter

II, for rejecting the details of these inquistion or polemical documents, the irrelevance of the contents of the 50 point renunciation — the most detailed of these documents and the only one about which it could be argued that its author had had contact with some real Bosnians — supports my contention that these documents cannot be used as sources for beliefs of the Bosnian Church.

But though we believe that all evidence indicates that the Bosnian Church was an orthodox church, we believe, as we have said before, that these inquistion and polemical documents are not pure fantasy. There were certainly dualists in Bosnia, but their movement, probably very small in size, was distinct from the Bosnian Church. Could the 50 points give an accurate picture of the beliefs of these Bosnian dualists? For after all, we have suggested in the text, the three noblemen who renounced them were members of this heresy and not of the Bosnian Church. Unfortunately, we cannot come up with any sort of answer to that question. We simply do not have any detailed sources other than inquistional and foreign polemical sources about Bosnian dualists. And we have already in our discussion of inquisition sources (in Chapter II) as well as in our discussion of the three noblemen in 1461 (in Chapter VI) given sufficient reasons why we should be wary of the contents of this inquisition material, much — if not most — of which was certainly drawn from the beliefs of Western heretics. Thus we can only use this material to demonstrate the presence of dualists in Bosnia. Of the specific beliefs of these dualists we must sadly admit our ignorance until such time as some sources providing reliable information about them should be discovered.

CHAPTER VII
HERCEGOVINA FROM 1463 TO 1481

I: Patarins in Hercegovina After 1463

After the Patarins who were expelled from Bosnia in 1459-60 took refuge with Herceg Stefan, we begin to find increased numbers of Patarins in Hercegovina in the 1460's. This, however, was to be a phenomenon of short duration. Even if the Turks had not annexed Hercegovina over the next twenty years, we would expect to see the number of Patarins in Hercegovina decline when the generation of exiles died out. For without popular support in Hercegovina, there would not have been many Hercegovinians ready to be ordained to replace them. A Slavic liturgy church, linked only to a state that had fallen, could not have expected much success in a region the majority of whose populace, being Orthodox, already had the Slavic liturgy.

In the defters we find considerable land associated with the krstjani in certain limited regions of Hercegovina; for most of Hercegovina no source indicates their presence at all. Lands of krstjani were chiefly found in the regions of Konjic and Goražde, both near the Bosnian border, and both areas having hižas in the 1440's. Presumably, some of these lands mentioned by the defters were connected with these hižas. Doubtless when the Patarin exiles left Bosnia, they migrated to regions of Hercegovina where their co-religionists were already established. They may have joined the existing hižas or have formed new ones in the vicinity. Also, quite possibly, they were granted lands by the herceg. Thus, from our limited sources, we cannot be sure any krstjan land in Hercegovina (other than some of the land around Goražde and Biograd-Konjic) belonged to the krstjani prior to the late 1450's.

In addition to the hižas that we have met and those that we shall meet in written sources about Hercegovina, a Turkish defter (Istanbul No. 5) written shortly after 1588 reports that part of the čiftlik of Mahmud and his father Radko in the village of Kunovo, nahija Soko (i.e., Sokolgrad at the junction of the Tara and Piva Rivers) had been bought by the family during the reign of the herceg from Cvatko Gost.(1) We do not know whether these lands had been Cvatko's private estates or lands attached to his hiža.

The entry also shows that the family who bought the land had retained it under the Turks even though they had not converted to Islam until after the middle of the sixteenth century. The father Radko was clearly a Christian while the other owner, son Mahmud, had obviously accepted Islam. We meet Cvatko Gost in no other source, thus we cannot date his sale of the land more precisely than the period between 1435 and 1466, the years of Herceg Stefan's rule.

II: The Herceg's Relations with Different Faiths in His Last Years

The herceg maintained his relations with the different confessions. In 1465 Catholic Venice awarded him, "for the good of Christendom," the land it had held near the mouth of the Neretva, as well as part of the Krajina which it did not feel strong enough to defend in the event of a Turkish attack.(2) Gost Radin remained at the herceg's side and served him as envoy to Dubrovnik in 1465.(3) In December another Patarin, Tvrdisav Krstjanin, accompanied by a secular envoy, represented the herceg in Dubrovnik.(4) In that same year the herceg drew up and deposited his will, drafted by the Orthodox Metropolitan of Mileševo, David, before a Ragusan court. He left, among other things, some relics and a gold and silver icon decorated with pearls to his son Stefan. Whether these items had any religious meaning for the herceg or whether he kept them for their monetary value, we cannot say. However, it is worth noting that these items were among the effects of a man considered to have been a supporter of the Bosnian Church. The herceg also left a sum of money for the good of his soul, without specifying which church should receive it. The will shows that the herceg was still keeping contacts with different confessions. Both Gost Radin and the Orthodox metropolitan — classified together as ordained monks *(redovnici)* — witnessed the will.(5) That an Orthodox metropolitan was willing to participate in this venture with the Patarin gost seems to me further evidence that Radin could not have been dualist.(6)

The following summer (1466) Herceg Stefan deposited a large sum or money in Dubrovnik for safekeeping in the event that the Turks might conquer his lands. This transaction was also witnessed by Metropolitan David, this time in the company of various people including two Patarins, Krstjanin Tvrdisav and Krstjanin Čerenko.(7) Once again the Orthodox bishop seems to have been satisfied with the company in which he was included. This same Krstjanin Tvrdisav also came to

Dubrovnik in 1466 as an envoy for Gost Radin.(8) Thus we see that the gost (even if he did not have a hiža) had in his service Patarins whom he could utilize to perform tasks for him.

The herceg died in 1466. Had he become a Catholic prior to his death, Dubrovnik, the papacy, or the early chroniclers presumably would have mentioned it. Unreliable popular traditions noted down in the late nineteenth and early twentieth centuries in various parts of Hercegovina would have it that he was Orthodox. A more reliable story is recorded by the traveler Ramberto in the 1530's, who reports that the inhabitants around Ključ in Hercegovina pointed out a small church where the herceg had attended services.(9) This church was probably Orthodox (possibly the same one later converted into a mosque and said by tradition to have been built by Sandalj's Orthodox wife Jelena). Yet even if the herceg occasionally did attend this church, we cannot be sure he belonged to its rite. And it is possible that in the herceg's day the church had been Patarin. The published defters do not speak of Patarin lands here; but a court Patarin could well have performed services for the herceg in this church.

Orbini calls Herceg Stefan "schismatic,"(10) which, in his history, usually means Orthodox. However, Orbini also speaks of Gost Radin as the herceg's confessor and a monk of St. Basil (i.e., Orthodox).(11) Since we know Radin was not a Basilian monk, Orbini is not reliable here. It is probable that Orbini's source for these remarks is the herceg's will; Orbini, seeing Radin's name next to Metropolitan David's, could well have assumed that the herceg and Radin belonged to the same faith as the Metropolitan of Mileševo.

Lacking evidence to suggest Herceg Stefan was received into either the Catholic or Orthodox Church, I conclude that he died a member (not necessarily a devout one) of the Bosnian Church. However, this affiliation did not affect Dubrovnik which took, from the money the herceg had deposited there for safekeeping, the 10,000 ducats he had left for his soul and presented the money to the pope. Thus if the herceg had had any doubts about which church to leave the money to, Dubrovnik efficiently resolved them for him.

III: The Testament of Gost Radin

On January 5, 1466, Gost Radin deposited his will in Dubrovnik.(12) This document has a large cross down the left margin of the first page; this is not standard form, for no other will in the 1467-71 folio where Radin's will is placed has such a cross. Thus Radin did not reject the cross. At the head of the text is a second cross, followed by the statements that Radin wrote before omnipotent God and the Ragusan nobles, that he was of sound mind, and that certain Ragusan nobles (i.e., Catholics) would serve as his executors. Radin's reference to omnipotent God suggests he believed in one God, and hence was not a dualist.

He left 600 gold ducats for his soul in the service of God, i.e., to help the unfortunate. Half of these 600 ducats should be turned over to his nephew Gost Radin Seoničan; his name shows he had some connection with Seonica, a village not far north of the Neretva above Konjic. Whether Seonica was his birthplace or the location of a hiža he administered is not known. The nephew with pure motives was to distribute this money to properly baptized correct krstjani-peasants of both sexes, who were of the correct apostolic faith (dobriem' načinom' kršteniem koi su prave vere apostolske praviem' krst'janom' kmetem' i pravem' kmeticam' krstjanicam'). The word "kršteniе" is a regular term for baptism; but whether it refers here to a rite of baptism, initiation or ordination is unknown. Since Radin called these individuals "krstjani" evidently they were ordained clerics. In every other Slavic source about the Bosnians, the term krstjani has been strictly used to designate only ordained clerics. And later in the will, when Radin left money to ordained individuals, he uses the term "krstjani" in the standard way as a title for the ordained; and when he speaks of people — including some other peasants *(kmeti)* — who are obviously lay-members of the church, he does not call them krstjani. Thus, since Radin here is speaking of ordained churchmen, we can assume that these individuals had first undergone some rite to become members of the church, and subsequently a second rite of ordination. The term "krštenie" could therefore refer to either rite. If Kniewald is correct in deriving the term for ordained clerics "krstjani" from the verb "krstiti" (to baptize),(13) then the fact that not laymen but only ordained clerics are called krstjani, suggests that "krštenie" must refer to their rite of ordination. We also see that the Patarins traced their faith and practices back to the apostles. The will states that both men and women were ordained as monks and nuns, and that some of them worked the land like peasants *(kmeti).* It is not surprising to find both nuns and monks. The text does not say they co-habited the same hižas, as they had at Bilino polje. In fact, the will tells us nothing about their living arrangements. We find that certain monks and nuns farmed the land themselves, and that a regular term for peasant *(kmet)* was used for them. The reference to these peasant monks agrees with material in the Turkish defters about kristian lands. In a peasant society it is not surprising to find peasants after ordination continuing to work in the manner to which they were accustomed.(14) In a poor and barren area like Hercegovina, where the majority of the populace was poor and probably also belonged to the Orthodox faith, we could hardly expect the Patarin monks and nuns to live off charity.

These krstjani, on each holiday, holy Sunday, and holy Friday, were to pray for Radin's soul, kneeling on the ground and reciting the Holy Lord's prayer that God forgive us our sins and pardon us at the Last Judgment at the end of time. This shows that the Patarins considered Fridays and Sundays holy, that they believed there was value in praying for the souls of the dead, that they prayed in a kneeling position and particularly valued the Lord's prayer, that they believed that God forgave sins, and that they believed in a Last Judgment.

Radin further prescribes that the gost, his nephew, is to give money to elderly peasants who have led good lives and to krstjani and krstjanice who do not like sin, and to those of our law (presumably those following the prescriptions for ordained members of the Bosnian Church) who are blind, lame and crippled in different ways. This shows that Radin believed in charity and alms. He also mentions old peasants *(kmeti)* who have lived good lives, i.e., lay-members of the Bosnian Church. Since he does not call these peasants who are lay members of the church krstjani, we find confirmation for the view that the "krstjani kmeti" were ordained. He makes reference to "our law" which shows that there was a rule for the order (which contradicts what was said in the 1433 letter sent by Dubrovnik to the Council of Basel).

Radin orders next that money be given to unfortunates among the "mrsni ljudi" (i.e., people who do not fast).(15) This passage in the will generally has been interpreted to mean that ordained Patarins kept strict dietary habits not followed by ordinary believers, who are therefore referred to as "mrsni."

Radin directs that the alms be given at the great holidays: Christmas, the Annunciation, Easter, St. George's Day — his *slava* — (na dan svetoga Georgija moga krsnoga imena), Ascension Day, St. Peter's Day, St. Paul's Day, St. Stephan the first martyr's Day, Mikhail the Holy Archangel's Day, the Day of the Virgin, and All Saints. Clearly, the Patarins honored all these holidays; thus they must have believed in and honored the saints. Like the Orthodox they observed both a St. Paul's Day and a St. Peter's Day as opposed to the Catholics who celebrated the two saints on the same holiday. We also see that Radin in Serbian style had a *krsno ime.* I think this means that he celebrated his *slava* on St. George's Day, and thus had that saint as his patron. If we want to follow M. Filipović's interpretation of the significance of this term in medieval Bosnia, then George would have been the church name Radin received at his baptism.

Radin then goes on to instruct the Ragusan nobles to give these 300 ducats to Radin's nephew, the Seoničan gost, for the needs of the faith that he believes in and the fasts which he observes. This suggests that differences did exist between Patarin and Ragusan Catholic Christianity. A similar expression was used in the Ragusan privilege to Radin in 1455, which stated that no one would force him to give up the faith which he believed in. Once again reference is made to the Patarin fasts. Radin then calls on his nephew to carry out his last requests if he wants to go, at his death, with peaceful soul before the Lord High God and the Holy Indivisible Trinity. Thus we see a solemn adjuration akin to an oath, although no oaths appear in the will. Radin surely believes in a personal judgment by God and in an indivisible Trinity.

The other 300 gold ducats are to be distributed by his Ragusan executors for remembrances for his soul to the blind, the crippled, and other unfortunates of Dubrovnik, who will light candles for Radin's soul in the temples *(hram)* of God on the great holidays named above, and on each holy Friday and each holy Sunday. The Ragusan poor who would light candles for Radin's soul were Catholics, and the temples *(hram)* in which they were to remember him were clearly Catholic churches. There were no non-Catholic churches in Dubrovnik. Yet even though the Ragusan poor were Catholics, Radin leaves alms to them as well as to the Bosnian Church poor. Radin clearly believed that prayers for the dead and lighting candles for the deceased, even by those of another faith, would have a positive influence on the fate of his soul in the other world. That a Patarin gost should request this suggests that his church had beliefs and practices similar to or even identical with the Catholic Church in regard to the dead. If the Bosnian Church had grown out of a Catholic monastic order in the thirteenth century, as we have suggested, such similarities in beliefs should not seem surprising. That Radin wanted candles lit in Catholic churches also shows that he was not opposed to church buildings.

In the next clause of the will Radin disposes of money to a whole series of relatives. Truhelka thought that certain of them had been Radin's children and thus apparently Radin had been married prior to ordination.(16) Solovjev originally agreed with this interpretation, but subsequently has concluded that these supposed children were nephews, nieces and cousins.(17) If Solovjev in his recent article is correct, then it is accurate to say that we do not have one example of a child of an ordained Patarin. Thus, unless further evidence appears, we probably should follow the universally accepted opinion that Bosnian Churchmen were celibate. As members of a monastic organization that originally had been Catholic, this too would not be surprising. Among the relatives receiving money are: Vukava Krstjanica, the daughter of his cousin; his nephew Gost Radin Seoničan; another Vukava Krstjanica, a servant; Vukše Krstjanin, Radoe Krstjanin and Mil'sava Krstjanica; Radan Krstjanin (whose title is repeated twice but this surely was a slip of the pen); his cousin Gost Radivoj; Vuk Gost ''Uspopalski.''

All of this shows how closely Radin's family was tied to the Bosnian Church, supplying several ordained members, including four gosts (if we count Radin himself) all known to be alive in 1466.

''Uspopalski'' is usually interpreted as an error for Uskopalski (i.e., Uskoplje). Scholars have located this gost at the Uskoplje close to the town of Hum near Trebinje. But since we have no (other) evidence to suggest that there were Patarins in this area, and since Trebinje had both Catholic and Orthodox organizations, and lay under the scrutiny of coastal bishops, I doubt the gost lived here. I would suggest the gost's hiža was in the *župa* Uskoplje north of Rama near Bugojno. This region is much nearer Bosnia and earlier Patarin centers. In addition, the material from the Turkish defters published by Okiç refers to kristian land in four different villages in the Turkish nahija of Uskoplje.(18) Because of the region's remoteness, the clerics of this hiža may well have escaped being exiled in 1459-60. It is also possible that the gost had been exiled from Bosnia and only came to, or returned to, Uskoplje after December 1463, when that *župa* was given to Herceg Stefan by the Hungarian king.

Finally Gost Radin leaves to ''our friend Knez Tadiok Marojević'' (a Ragusan noble Thadeus Marini de Nale) an expensive fur robe given to him by King Matijas (i.e., Matyas Corvinus) and 140 ducats for a ''temple'' *(hram)* and grave for his bones. The gift to Radin from the zealous Hungarian Catholic king further argues against Radin being a dualist. We have suggested above that King Matyas had given this robe to Radin in December 1463 when the herceg and his son had met with Matyas and been awarded with a considerable portion of the territory recovered from the Turks. The presence of this robe among Radin's effects is the only evidence we have that Radin had attended this historic meeting.

What Radin had in mind by a *hram* (temple) is a matter of dispute among scholars. Some would have it an elaborate mausoleum and others a funeral chapel. Earlier in the will Radin used the word *hram* to refer to a Catholic church. A 1472 source, which we shall examine, refers to a sepulchre and chapel built for Radin, as ordered in his will. Since his will says temple and grave, and since in 1472 we hear of a chapel and sepulchre, it seems evident that Radin by *hram* meant a chapel. And if it was a chapel, then we have one more piece of evidence in favor of the Patarins having churches. In any case, Radin was leaving a large sum of money to erect an elaborate monument for his remains, which is not what we would expect from a dualist who should have scorned the body as material and valued only the soul and the other world. However, it is just what we would expect from a man who clearly valued the riches and glory of this world as Radin did.

Truhelka has calculated that in his will Radin left a total of 5,640 ducats, which clearly made him one of the richest men of his times in Bosnia or Hercegovina.

IV: Patarins in Hercegovina After 1466

Quite recently, Professor Šunjić, while working in Venice, came across another document on Radin from 1466. On March 10, 1466, the Venetian senate voted on a request from Gost Radin that he and fifty or sixty of his sect and law be allowed to settle on Venetian territory without fear of unpleasantness. Radin describes himself as a leading baron and councillor of Herceg Stefan. Sixty Venetian senators voted to accede to the request, with seven opposed and fifteen abstaining,(19) but as far as we know, no Patarins ever took advantage of this permission. We can assume that the request was made for the purpose of finding asylum in the event the Turks conquered the rest of Hercegovina and Dalmatia. That the Venetian Senate voted so heavily in favor of the request argues strongly against Radin's group being dualist. We have speculated that fifty or sixty included the majority of ordained clerics in the church. Most of them, by now, presumably lived in Hercegovina, where they probably had come under the influence of the leading Patarin there, Radin. Radin's request confirms my view that the Patarin priesthood's numbers were by this time small, and that the Bosnian Church was nearing its end. The death of the herceg surely dealt the church a further blow. And when, upon the herceg's death in 1466, Radin left Hercegovina and went to Dubrovnik, we suspect that the organization of the church more or less broke down. We hear no more about stroiniks, or church administration, or organizational ties between hižas. In these years we can picture the members of the exiled Patarin hierarchy from Bosnia, now in Hercegovina, with little or no general leadership, living out their sheltered lives in a few hižas — farming their lands and occasionally taking in travelers who sought a bed for the night.

In October 1466 Radisav Patarin, a member of the society of Belosav Patarin in the Neretva region of Bosnia (i.e., the herceg's lands) on this (the southwest) side of that river below ''Belgrad'', appeared before a Ragusan court to ask a layman Rusko Tvrdović to return a sum of money deposited with him. Radisav testified that he had deposited twenty perperos with Rusko, who denied that he had ever received such a deposit. The judges asked Radisav if he had any witnesses and he said that he did, a certain Cvjetko (Zvietchus) one of his Patarin colleagues from the above named society. When interrogated by the judges,Cvjetko swore by the oath that was their practice, as was their custom, that he knew nothing, had seen nothing, and had heard nothing about the said matter. Since his witness could not support Radisav's accusation, the judges said they would free Rusko, unless Radisav would support his accusation with an oath. Since Radisav was unwilling to swear it, the case against Rusko was dropped.(20)

This case refers to Patarins of the Bosnian Church monastery at Biograd on the Neretva, where in 1449 the merchants had been robbed. At that time the monastery had been under Obižen Patarin; now in 1466 Belosav was leader. Since on neither occasion, do the Ragusan sources call the leaders gosts, we suspect that this hiža was headed by a man of lower rank than gost; possibly even by an ordinary krstjanin. Two other monks of the monastery (Radisav and Cvjetko) are named. We still find that ordained Patarins bore popular names. We also see that at least some ordained Patarins in hižas had private money. From a 1470 case (not involving Patarins) we learn that Rusko Tvrdović was a citizen of Dubrovnik.(21) Hence we see why Radisav had brought his complaint before a Ragusan court.

The matter of oaths in this passage is important: Radisav throughout refused to take an oath, even to recover a serious financial loss. This agrees with our earlier evidence that Patarins refused to swear oaths. However, his witness, also a Patarin, swore some sort of oath that was according to Patarin practice and custom. We do not know what sort of oath this was, but we recall that in 1403-05 Ragusan envoys were told to get oaths from secular figures but to let the Patarins subscribe to treaties in any way they chose, or according to their custom.(22) We may suspect that by the fifteenth century the Patarins, in order to have practical dealings with the world, had had to find a substitute for the oath which they rejected. Thus a prescribed way of giving one's word, other than the usual oath, had been found. Whatever this compromise was, Ragusan envoys had accepted it earlier in the century, and in 1466 we even find a Ragusan court willing to accept Patarin testimony given this way. This shows that Ragusan courts must have tried more Patarin cases than we have records of, and that Ragusan courts, in accepting testimony according to Patarin religious custom rather than under traditional oaths, were surprisingly tolerant. However, it seems that all Patarins did not accept this compromise-oath, for Radisav even to recover his lost money refused to swear it.

Gost Radin died in 1467, initiating a series of court cases over his large estate. In July 1467 Gost Radin Seoničan, his nephew, collected a sum of money deposited in Dubrovnik by Radin. With him was a Radiz (Radič?) Krstjanin.(23) In March 1470 two gosts, Radivoj Priljubović and Vochus (Vuk) Radivojević testified that prior to Gost Radin's death, they had entrusted sums of money to him; now they sought to recover this money from his estate. These two gosts sent letters to Dubrovnik by the same two envoys, a secular figure named Tvrtko Brajanović and a Patarin, Cvjetko Krstjanin Radinović. Gost Radivoj Priljubović dated his letter from Bielo, and Gost Vuk Radivojević from Uscopia (Uskoplje).(24) These gosts are presumably the Vuk Gost ''Uspopalski,'' and Gost Radivoj mentioned in Radin's will. Gost Radivoj maintained a hiža at Bijela, not far south of Konjic. Thus in 1470 we hear for the first time of a Patarin monastery at Bijela.

Thus in the vicinity of Konjic in the fifteenth century we find three (and possibly four) hižas: 1) In 1418 not far north of Konjic in Bradina in Bosnia. 2) In 1449 and 1460 at Biograd on the Neretva just above Konjic in Hercegovina. 3) In 1470 in Bijela also in Hercegovina. It is possible that there was also a fourth hiža at Seonica which lay in Bosnia not far north of Konjic and the Neretva. Gost Radin's nephew was called ''Seoničan''; which could be a surname indicating his birthplace, or a nickname indicating the locality of a hiža he administered. Since we do not hear anything about the Bradina hiža after 1418, it may no longer have existed in the 1460's. Thus other than Seonica (if a hiža existed there) and excluding the undated defter data all of the monasteries which sources mention in the period after the 1459-60 exile were in the herceg's lands.

In 1472 a lay nephew of Gost Radin named Tvrtko testified that he had received the dinars deposited by Radin for the ''sepulchre and chapel'' (pro sepultura . . . et pro capella) provided for in his will.(25) We can assume that these two structures had been built. From this testimony in 1472 we see that the gost in his will by the word ''hram'' had intended to have a chapel. Thus we see that beside his tomb Radin wanted (and presumably received) a small chapel, which I think is evidence that the Bosnian Church did not reject church buildings.

After the 1472 notation about Gost Radin's burial arrangements, and despite the rash of references to various gosts and krstjani during 1466-70, we hear no more about Bosnian Church clerics. However, since we have no evidence about any persecution, we may assume that Patarin monasteries continued to exist in Hercegovina at least until 1481, when the last parts of Hercegovina were conquered by the Turks.

We may suggest that the Bosnian Church throughout the 1470's was declining in numbers and influence. We have no direct evidence that Patarins were to be found as advisors and diplomats at the courts of the herceg's sons and successors, Herceg Vlatko and Vladislav. Vlatko was closely tied to Venice; all his hopes to maintain what was left of his state or to recover what had been lost depended on Venice. Yet a 1481 Ragusan document refers to him as a Patarin.(26) Thus it seems that up to 1481 Vlatko had not joined either the Catholic or Orthodox Church, that the Bosnian Church still existed in his lands, and that quite possibly some Patarins could still be found at his court. After the fall of Hercegovina in late 1481, Vlatko retired to Venice.

The third son of Herceg Stefan, also named Stefan, who had been left holy relics and an icon in the herceg's will, went to Istanbul in 1473 or 1474 and became a Moslem, taking the name Ahmed. Known as Ahmed Hercegović, he rose rapidly in the Turkish administration and twice would serve as grand vizier under Selim I (from 1503-06, and 1510-14). Unlike many of Selim's grand viziers, Ahmed died a natural death in 1519. Stefan-Ahmed not only exemplifies the phenomenon of "Christians" accepting Islam, but also the fact that Bosnians of the highest families at the time the Turks were conquering their land had so little animosity against the Turks that they willingly accepted the Turk's religion, joined them, and served the sultan loyally.(27)

FOOTNOTES TO CHAPTER VII

1. Okiç, "Les Kristians . . .," *Südost-Forschungen,* 19, 1960, pp. 127-28.
2. Ljubić, *Listine,* X, p. 346.
3. Dinić, p. 234, No. 46, 47.
4. Stojanović, I, 2, p. 78.
5. Pucić, II, pp. 124-30; Stojanović, *op. cit.,* I, 2, pp. 87-92.
6. We recall a similar event back in 1305 when Djed Miroslav witnessed a grant to a Kotor monastery in the presence of Orthodox and Catholic bishops.
7. Miklosich, pp. 495-98.
8. Dinić, p. 235, No. 54.
9. P. Matković, "Putovanja po balkanskom poluotoku XVI vieka," *Rad* (JAZU), CXXIV, 1895, p. 82.
10. Orbini, p. 385.
11. Orbini, p. 388. Orbini calls him here Rasi gost but it is evident that Radin is intended since Orbini goes on to tell us that Rasi gost took the herceg's testament to Dubrovnik, and we have noted above that Radin was involved in

the drafting of Herceg Stefan's testament.

12. L. Stojanović, *op. cit.*, I, pt. 2, pp. 153-56.

13. D. Kniewald, "Hierarchie und Kultus Bosnischer Christen," *Accademia Nazionale dei Lincei, Problemi Attuali di Scienza e di Cultura*, Rome, 1964, pp. 592-93.

14. Bosnian Orthodox monks, at least in the sixteenth century, also farmed the land. The traveler Zeno reports that the Orthodox monks of Mileševo worked in the fields after services in addition to making their own bread and wine. See, P. Matković, "Putovanja po balkanskom poluotoku XVI vieka," *Rad* (JAZU), Vol. LXII, 1882, p. 98.

15. *Mrsni* is the antonym of the word "fast." To the present day in Jugoslavia meals on religious holidays are either "postna" — fasting, no meat, but fish cooked with oil — or "mrsna" — meat, lard used in cooking, unrestricted diet.

16. Č. Truhelka, "Testament gosta Radina . . .," *GZMS*, 23, 1911, pp. 366-67.

17. A. Soloviev, "Le Testament du Gost Radin," *Mandićev zbornik*, Rome, 1965, pp. 141-156, esp. 148-55. His arguments are intricate and I cannot judge what terms of relationship, complicated enough in the twentieth century, meant in the fifteenth century. This problem I shall not even go into, but shall simply follow Solovjev's re-evaluation.

18. Okiç, "Les Kristians . . .," *Südost-Forschungen*, 19, 1960, p. 125 (fortress Novi near Vinčac), p. 127 (village Skrobovići), p. 129 (village Zdrmci), p. 123 (village Kuti).

19. M. Šunjić, "Jedan novi podatak o gostu Radinu . . .," *GID*, XI, 1960, p. 267.

20. Dinić, pp. 215-16, No. 81.

21. Dinić, p. 153, No. 306.

22. See above, Chapter V.

23. Dinić, pp. 216-17, No. 82.

24. Dinić, p. 219, No. 84. It is not known whether this Krstjanin Cvjetko is the same Cvjetko who testified before the Ragusan court in 1466. In these documents we learn that the famous Gost Radin bore the last name of Butković.

25. Dinić, p. 216, No. 82.

26. J. Tadić, "Nove vesti o padu Hercegovine pod tursku vlast," *Zbornik Filozofskog fakulteta, Beograd univerzitet*, VI, No. 2, 1962, p. 144.

27. A second example of this was Sigismund, the son of Stefan Tomaš and the herceg's daughter Katarina — thus a younger half-brother of Stefan Tomašević. He was taken by the Turks in 1463 to Istanbul; he also converted to Islam. Under the name of Ishak Kral Ogli (Isaac the King's son), he served in 1487 as sandžak beg for Karasa in Asia Minor.

CHAPTER VIII
RELIGION IN BOSNIA AFTER THE TURKISH CONQUEST

I: Patarins and Other Heterodox Christians in the Turkish Period

References to Patarins or other heterodox Christians (excluding mixtures between Christianity and Islam) in sources other than the defters (whose information cannot yet be dated) after 1463 in Bosnia and after 1481 in Hercegovina are amazingly few. After the murder of the five Franciscans by Patarins in Visoko in 1465, and references to a gost in Uskoplje in 1470, we do not meet the term Patarin again in references to Bosnia.(1) After the Ragusan reference to Herceg Vlatko of Hercegovina as a Patarin in 1481, no known source uses the word again until the seventeenth century. No sixteenth century traveler mentions any religious group in Bosnia or Hercegovina other than the Orthodox, Catholics, and Moslems.

Then, in 1605, a Croatian prince in Vinodol wrote that some Krmpočani Vlachs (from the region of Krmpota near Lika) made the short migration into the coastal region of Vinodol, just south of the junction of the mainland with Istria. The Vlachs came as Christian believers (i.e., Catholics), allegedly at the summons of St. John, who had appeared to them in the night. The Vlach leader stated that they had come from Patarin land to Christianity.(2)

Because the Vlachs had made a short migration within an area which had long had Vlach shepherds, we have no reason to connect them with Bosnia. Their Catholicism also argues against their being Bosnian; by the seventeenth century, most of Bosnia's Vlachs were Orthodox. It is also worth noting that no medieval source mentions Patarins or heretics in this region. Thus we do not know how Krmpota could have come to be called Patarin land.

In 1692 a Venetian priest visiting Dalmatia noted the settlements of Bosnians and Hercegovinians in Dalmatia. He said that these people were of three types. The majority were Latins or Greeks (i.e., Catholics or Orthodox), but there were also Christians of a third sort who were in mournful condition, so blinded in matters of faith that it is only the fact that they were not circumcized which made them Christians (i.e., that they were not Moslems or Jews). They lived in full darkness and lacked all correct belief.(3) Besides the fact that these people belonged to a third group of Christians of Bosnian-Hercegovinian origin, we have no compelling evidence to connect them with remnants of the Bosnian Church. The description tells us nothing about their faith, except that they were ignorant of all correct belief.(4) These people might well have been descended from medieval Bosnian Churchmen or heretics, but they could equally well have been descended from any group of Christians who, subsequent to the Turkish conquest, had fallen into error — possibly owing to the lack of priests or to the influence of Islam.

In 1703 a visitation report from the region of Trebinje says that the schismatics there were Patarins,living according to the customs of the Greeks. They followed the rules of their priests and Basilian monks, who were ignorant and illiterate. Their rite was Greek Patarin, and their errors consisted of rejecting the authority of the pope, the process of the Holy Spirit, Purgatory, and consecration with azymes.(5) It is evident that the "errors" of these "Patarins" consisted of those matters that distinguished the Orthodox from the Roman Church. This fact, combined with the fact that their clergy included Basilian (i.e., Orthodox) monks, shows that these people were Orthodox believers. This has been used as an argument that medieval Patarins had been more or less Orthodox as well;(6) however, it could equally well mean that these people were descendants of medieval "Patarins" who converted to Orthodoxy but, despite their acquisition of new doctrines, had somehow retained their former name. In any case, by 1703 so much time had passed since the Turkish conquest of Bosnia that we cannot rely upon a report such as this to give us an accurate picture of the beliefs and practices of medieval Patarins.

The time interval also detracts from a visitation report from 1751 for Trebinje and western Hercegovina by S. Tudisić, which says that many of the Greek schismatics also practiced errors of the ancient Manichees. Their errors consist in: affirming that Hell's sufferings are not eternal, and that Jesus Christ did not die on the Cross to save humanity but that Archangel Michael did. They believe in two types of creed, hate the Catholics greatly and re-baptize those who accept their faith. They apparently professed the Greek rite but had their own special rites.(7)

The beliefs of these schismatics, not surprisingly, were a synthesis of a variety of beliefs. Tudisić's accuracy may also be questioned, since presumably he drew his information not from the schismatics themselves but from Catholic peasants in the area who could not be expected to have understood or even cared about the beliefs of their schismatic neighbors. Presumably Tudisić had read that the medieval Bosnians were Manichees, and he may well have assumed that deviations in the beliefs of these un-educated schismatics would have been derived from this former heresy. The beliefs he describes do not seem dualist. The reference to the archangel is not related to any known medieval belief, though Archangel Michael is a popular figure in South Slavic folklore and has long been connected with taking human souls upon death. His Day was also an important Orthodox Church festival and on it was celebrated the *slava* of many families. Hatred for Catholics after the centuries of rivalry in the Trebinje region between the Orthodox and Catholics is not odd at all. Visitation reports from the seventeenth and eighteenth centuries do, from time to time, report Orthodox priests and Franciscans re-baptizing converts from the other faith. Thus this practice, though against canons, does not necessarily reflect heresy. Unfortunately, we are not told how the rite of these schismatics deviated from that of the Greeks (i.e., Orthodox). Thus this report can only be used to show heterodox beliefs existing in the eighteenth century.

Finally, in 1737 and 1739, the Bosnian chronicler Fra Nikola Lašvanin mentions old believers *(starovir'cy)* in the Vrbas region.(8) Unfortunately, since he tells us nothing whatsoever about their beliefs, we do not know from which, if any, of our medieval groups the old believers might have been descended.

These, then, are the references to Patarins and heterodox sects in Bosnia-Hercegovina after the Turkish conquest. I have not included descriptions of syntheses between forms of Christianity and Islam. In the nineteenth century traditions about Bogomil families begin to be reported. These traditions were discussed in Chapter II. Since there we showed that these traditions were not actual medieval survivals — the term Bogomil was not used in medieval Bosnia — but were based on tales told by priests who had learned about Bogomils from books and schooling, we do not discuss these traditions here.

From the above we would argue that the Bosnian Church and any other dissident movements that might have existed disappeared very rapidly after the fall of the medieval states of Bosnia and Hercegovina. Orbini dates the end of the Patarins in 1520,(9) but since we have shown that he obtained this date from the date of the conversions of Paulichiani around Nicopolis during the Austro-Turkish war, we maintain that he was not speaking about the Bosnian Church.

II: Main Trends in Bosnian Religious History 1463-1600

Briefly, I shall now point out some of the main trends in Bosnian religious history from 1463 to 1600, for no study of the Bosnian Church would be complete without it.

In the course of this study I have postulated that by 1460 the Bosnian Church had become weak and vulnerable. Its clergy, living monastic lives and having failed to create a preaching order, had not forged close ties with the populace, which remained fairly indifferent to the fate of the church. This picture is confirmed by the ease with which the king was able to expel the church's leadership. Having been hopelessly crippled by this expulsion, the Bosnian Church was therefore more or less destroyed prior to the Turkish conquest. The few Bosnian Patarins who preferred to flee than accept Catholicism went to Hercegovina, an Orthodox land whose populace also seems to have been quite indifferent to the few Patarins who had already been living there in monasteries near Hercegovina's Bosnian borders. Under the protection of Herceg Stefan, the Bosnian Church continued to exist in Hercegovina — particularly in the regions of Konjic and the upper Drina (between Goražde and Soko) — and enjoyed a brief flowering in the 1460's. However, we must emphasize that it always remained a small-scale operation, limited to a few monasteries in these border regions. In Hercegovina the church was already on the wane by 1481 when Hercegovina fell to the Turks. The picture summarized above is not contradicted by any source, and serves to explain the following facts drawn from our limited sources:

1) After the Turkish conquest of Bosnia in 1463, we hear of Patarins in Bosnia only in 1465 when they raided a Franciscan monastery, and up to 1470 at a hiža in Uskoplje.(10) After 1470, excluding the defters whose information cannot yet be dated, no source mentions their presence in Bosnia. 2) Not one extant document discussing problems related to the administration of the Hungarian Banate of Jajce (formed in 1463 and lasting to 1527) mentions the existence of heresy or the need to reform or persecute the religious beliefs of the populace. 3) In Hercegovina, again excluding the undated defter information, we find Patarins in two hižas from 1460 up to 1470 at Biograd and Bijela (both near Konjic); most of these references concern Gost Radin and his relatives — particularly in connection with Ragusan court action about Radin's will. However, the fact that the same few Patarins are mentioned in many documents at this time does not mean that there were many Patarins then in Hercegovina. In fact, Radin's request that fifty or sixty of his sect be allowed to seek asylum on Venetian territory in the event of a Turkish invasion of Hercegovina suggests that their numbers were few. Since we hear no more about hižas or Patarin clerics in Hercegovina after 1470, this is probably an accurate supposition. In the final twelve years of independent Hercegovina, the term Patarin — as well as any synonym for it — is used only once again, i.e., in 1481 about

Herceg Vlatko. After that no source uses the term about people in Hercegovina until 1703, when evidently the term had come to mean an Orthodox believer.

It is possible that the weakness of the Bosnian Church by and after 1460 has been slightly overstated in the course of this study. We may find that when the defters are published themselves — as opposed to extracts from them by scholars — certain references in them to "kristians" may pertain to monks and Bosnian Church land still active after the Turkish conquest. Because the defters may contain information about how long "kristian" land was held by krstjani, publication of all the defter texts is a necessary prerequisite to determine how weak the Bosnian Church had actually become by the time of the Ottoman conquest. Hence I realize that I may be forced to modify my conclusions slightly if the defters should show the existence of various other Patarin communities or significant Patarin land-holding after 1463. However, even if the Patarins did retain certain other monasteries and lands in various places in Bosnia and Hercegovina after 1463, it is still plain from the evidence of all the sources examined in this study that by the 1460's the Bosnian Church had become a feeble organization, lacking in popular support and not fulfilling any vital needs for the society or state.

We have noted that Orthodox believers seemingly were not present in medieval Bosnia, except in the regions near the Drina and the Hercegovinian border. Most of the Serbs who migrated west after Kosovo (1389) probably settled in the Orthodox regions of Hercegovina and Zeta (Montenegro). Some may have penetrated into the border regions of Bosnia where there already were Orthodox. The Orthodox may have been settled in Bosnia as far west as Sarajevo by the 1420's; this is indicated by a tradition that Angelo Zvijezdović, the Catholic who obtained the privileges for his church from the sultan in 1463, had been born to an Orthodox family of Vrhbosna.(11) Toward the middle of the fifteenth century, after the second Turkish conquest of Serbia and the Turkish annexation of eastern Bosnia, it is probable that further migration of Orthodox believers into Bosnia occurred.

In the 1450's we noted that the Catholics and Orthodox were competing in some unspecified place in Bosnia to convert Bosnian Church believers and heretics.

This rivalry continued after the Turkish conquest. And it is in the period after 1463 that we begin to find mention of Orthodox believers and priests in many parts of Bosnia where we had not found them earlier. This fact has led various scholars to suggest that the Patarins had been more or less Orthodox all along, and that after the fall of the

Bosnian state and the diminution of the Bosnian Church clergy, the Serbian hierarchy, preaching the same — or almost the same — doctrines, was accepted by the lay members of the Bosnian Church. We have argued that, although the Bosnian Church may not have differed greatly in rite from the Orthodox Church, it was not Orthodox: the Serbian sinodiks anathematized it and John of Capistrano's letter refers to Orthodox trying to convert Patarins.(12) It also should be stressed that references to Orthodoxy in Bosnia after 1463 only gradually become more and more frequent. We find references to it occasionally in the 1480's, and only with real frequency in the sixteenth century. The gradualness with which Orthodoxy appeared in Bosnia at this time confirms my belief that prior to the Turkish conquest there had been few Orthodox believers in Bosnia, and thus argues against equating the Bosnian Church with Orthodoxy. In fact, Orthodoxy appears at the rate we would expect if its appearance had been caused by gradual migration into Bosnia and by conversions from other denominations.

Although immediately upon the Turkish conquest the Franciscans requested and obtained from the sultan permission to exist and practice their religion, it is evident that the Turks preferred the Orthodox to the Catholics. The reasons for this preference are not hard to find. The Serbian-Orthodox Church hierarchy resided in Ottoman-occupied territory. The Catholic Church leadership did not. Besides, the pope and the Hungarians (who had recovered a part of northern Bosnia in 1463 and held it into the sixteenth century) were the leaders of Christian anti-Turkish crusading ventures. Thus the Catholic Church was viewed as a fifth-column with close ties to a foreign and dangerous enemy.

By the sixteenth century we find Orthodox believers in most regions of Bosnia. Their spread is seen by the number of Orthodox monasteries erected in the sixteenth century: e.g., Tavna, Lomnica, Paprača, Ozren, and Gostović. In 1533 the Orthodox also received permission to build a church in Sarajevo. The privileges that allowed the Orthodox to build new churches in these years contrasts sharply with the prohibitions against Catholic church building, which were strictly enforced. Only occasionally did the Catholics receive permission to repair or rebuild a ruined or damaged church that had existed in the same spot previously, and then on condition that the restored church not exceed its earlier dimensions.

Particularly interesting are the numbers of Orthodox found in the region of Bosnian Krajina between western Bosnia and Dalmatia (in the region of Glamoč and Duvno)(13) where there was no evidence of Orthodox believers in the medieval period. Ethnological research and investigations of the defters have shown that most of these Orthodox had come from Hercegovina, and that the bulk of newcomers were nomadic Vlach shepherds whose migration routes had followed the path of their seasonal journeys with their herds.

After the Turkish conquest of Bosnia in 1463, the earliest Turkish cadastral surveys (defters) in 1468 show a large number of abandoned villages, and others which were greatly depopulated. The former populace had either been killed or captured by the Turks, or had fled. Many Catholics did flee from Bosnia into the Hungarian Jajce banate, into Slavonia and other Croatian lands, or into Dalmatia. These events, of course, led to a reduction in the number of Catholics in Bosnia. The defters from the 1480's show many of these villages repopulated. In some cases the new populace was Moslem. We do not know whether these were Moslems brought from outside Bosnia and settled there by the Turks or whether they were Christian converts transferred from elsewhere in Bosnia or Hercegovina to the abandoned villages.(14) In some cases the new populations were Christians; the defters do not usually state whether Catholic or Orthodox. Again we cannot be sure whether these newcomers were chiefly from internal migrations within Bosnia, whether they had freely come from Hercegovina or Serbia, or whether they had been forcibly transferred from Serbia by the Turks.

For most people to move legally within the Ottoman areas of the Balkans it was necessary to receive "Vlach" classification or privilege. Thus a Serb peasant could receive the right to migrate only if he received Vlach status. After receiving this status, the peasant had to pay a tax of a ducat in cash instead of the usual payments in kind. Most of the legal migrants and new settlers in Bosnia and Hercegovina in the first century and a half of Turkish rule were Orthodox "Vlachs." Thus we find that the term Vlach has acquired new meaning;(15) hence care must be exercised before drawing any conclusions about "Vlachs" referred to in the sources. These "Vlach" migrations occurred gradually and steadily throughout this century and a half, and had a large role in repopulating villages. There is little evidence of massive migrations occurring at single times. However, it must be stressed that many, if not the great majority, of these "Vlachs" were real Vlachs (i.e., shepherds), since in many regions such as the Krajina we find a great increase in the number of shepherds compared with the number of settled farmers. Serbs, who had acquired Vlach privilege, presumably upon their arrival at their new homes would have worked the land. The majority of Vlachs coming into Bosnia and the Krajina were Orthodox, and we have noted earlier that the great majority of Hercegovinian Vlachs in the Middle Ages had been Orthodox. The large number of actual Vlachs that must have migrated is also shown by the general picture we obtain of Bosnia in the seventeenth and eighteenth century. Then we find Moslems, Jews, and other Balkan and Anatolian people living in towns, Moslems and Catholics in the valleys and flat farmland, and Orthodox in the mountains. Because the populace of the mountains is chiefly pastoral, this strongly suggests that many Orthodox immigrants into Bosnia had been pastoral people.

Along with these "Vlachs", a number of Moslems also settled. Most of these settlers — be they Moslem or Orthodox — replaced Catholics who had fled to Catholic regions not under the Turks. Catholics, certainly, would not have come in any significant numbers from outside Bosnia at this time. However, presumably, a certain amount of internal movement of Catholics from one place to another within Bosnia did occur.(16)

There is no evidence that the Turks pressured people to convert (except for specific short periods under particular fanatical pashas). Whenever a Christian accepted Islam, the Turks suffered the loss of the special "infidel" taxes collected from Christians. Nor did Christians need to accept Islam in order to retain their lands. Many Bosnian-Hercegovinian Christians retained their estates and their Christianity. A prime example is the Orthodox Vlach family of Miloradović from the Stolac region in Hercegovina. Not only did this family retain its enormous estates in this area and its leaders serve the Turks at great personal gain, but its leaders even built Orthodox churches, including the famous monastery of Žitomislić.

In the earliest censuses of the 1460's we still find references to krstjani. It is not always evident whether they still retained the lands with which their names were linked or whether, as former owners, their names had become place names for the lands in question. It is probable that both of these circumstances did occur. In 1469, a defter extract which Okić published lists eleven villages inhabited exclusively by krstjani.(17) Since the families are listed as so many households and so many unmarried males, it is evident (as we have argued earlier) that "families" refers to married couples. If they were all ordained monks, everyone would have been a celibate male (or female). Thus it is evident that here the defters refer to believers in the church as opposed to monks. Possibly in some cases these villages belonged to the Bosnian Church. Yet the overall picture obtained from the defters of the 1460's (and more so from the 1480's) is that Bosnia was by then a land of Christians (Catholics and Orthodox). And since the land had long been Catholic (as opposed to Orthodox) and since we have no evidence of large-scale Orthodox migrations yet, we may presume that the majority of these Christians was still Catholic.

In these years we see the beginning of the conversion of part of the population to Islam. Frequently in court and legal documents from the first century of Turkish rule we see the father having a Christian or popular Slavic name, showing that he was a Christian, and the son bearing a Moslem name, showing that he had accepted Islam. The defters, too, sometimes illustrate conversions by referring to fathers with Slavic names being succeeded by sons bearing Moslem names; the defters also on occasions refer to people as "New Moslems", indicating recent conversions.(18)

The defter statistics published by Šabanović on confessions around Lepenica in central Bosnia also show the gradual acceptance of Islam by the populace there.(19) In 1468 we find 279 householders and 61 adult single males. All of them were Christian. In 1485 we find 329 Christian householders and 92 Christian adult unmarried males, with 18 Moslem householders and 26 unmarried Moslem males. This shows both the increased population of the area through resettlement and the beginning of the acceptance of Islam. Of course, we do not know whether the Moslems mentioned were recent converts or long-time Moslems brought in from elsewhere. If they were converts, we see that the rate of accepting Islam over these first twenty years of Turkish rule was slow. The 1489 census notes 165 Christian families and 65 Moslem. This shows that there was either a decline of population, possibly owing to epidemics, or else that the Turks had re-classified the villages in the area, assigning many former free villages to timars. In 1509 we find that the population had risen again and by now the Moslems are the majority: 393 Moslem families to 160 Christian. However, because the number of Christian families was essentially unchanged, it is possible that most of these Moslems appearing between 1489 and 1509 had been brought in from elsewhere. Since the Christian figure has decreased by only five families, we can argue that the number of actual conversions during these twenty years among the local populace of Lepenica was small. Even so, though, Bosnia was clearly becoming more Moslem in character.

We also find that the ratio of Moslems to Christians varied from place to place. In some areas we find many Moslems and in others almost none. This is clearly shown by the 1485 defters for central Bosnia.(20) Once again we do not know how many of these Moslems were old-timers converted to Islam and how many were new settlers. However, when we find an early Moslem cemetery on the site of a medieval cemetery, then we are justified in concluding that the buried Moslems were relatives of the medieval deceased; in such cases, then, we are faced with conversion rather than migration. It would be an interesting project to determine whether the villages noted in the defters as containing Moslem families have old cemeteries, and if so, whether *stećci* and *nišani* are found at the same sites. The only time a migration has obviously occurred is when we see a great increase in the actual population. Unfortunately, however, the defters rarely tell us from where the new settlers have come. Migration is a most important subject that requires a great deal more work. It is necessary for Turkologists who have access to the defters to extract all the relevant material, so that their results may be incorporated with the work done by Jugoslav anthropologists on popular traditions about migrations and about place of origin of families.

The defters, then, show that conversions to Islam occurred but that the rate of conversion varied from place to place and that in many places the rate of accepting Islam was slow. The defters of 1528/29 give the following figures for the number of Christian (including both Catholic and Orthodox) and Moslem *households:* for the Sandžak of Bosnia, 19,619 Christian and 16,935 Moslem; for the Sandžak of Zvornik, 13,112 Christian and 2,654 Moslem; for the Sandžak of Hercegovina (probably not the whole Sandžak) 9,588 Christian and 7,077 Moslem. One should probably multiply these figures by five or six to arrive at the total number of individuals.(21) The defter data is supplemented by the contents of certain Catholic visitations.(22) Peter Masarechi states that in Bosnia in 1624 there were about 900,000 Moslems, 300,000 Catholics, and 150,000 Orthodox. His figures, though of course approximations, do not include Hercegovina.(23) Athanasius Georgijević, writing in 1626 does not agree. He states that there were 250,000 Catholics, that the number of Orthodox exceeded that of the Catholics, and together the two Christian faiths outnumbered the Turks. The fact he had the Orthodox exceeding the Catholics probably can be attributed to the fact that he included Hercegovina. His smaller estimate for Moslems is interesting. In 1655 Marijan Maravić stresses the conversions that had occurred and states that the majority were Moslems. As a result of conversions and flight (some 2,000 Catholics had fled to Croatia in the last decade), the number of Catholics in Bosnia, he says, had fallen to 73,000. He mentions the presence of Jews in Sarajevo and Gypsies everywhere.

Although we find conversions to Islam in this period, we find evidence in seventeenth century visitations that changing religion was a multi-directional phenomenon. Catholics abandoned their faith to both rivals. We learn of Catholics accepting Islam or Orthodoxy in the region of Trebinje in the 1620's.(24) The main cause for abandoning Catholicism noted by the visitor was the shortage of priests, the ignorance of the existing priests, and the indifference of the local bishop. Bernard Pomazanić, a Capuchin, traveling in Gacko in 1529-30, found many former Catholics, who, owing to the lack of priests of their faith, had converted to Orthodoxy.(25) Thus we see the importance of having an effective clergy to retain believers and keep them from errors.

The dying Bosnian Church lost all its believers to other faiths. We have seen in Patriarch Gennadius' letter that some Bosnian Churchmen accepted Orthodoxy. The reports of Pius II clearly show that other Bosnian Church believers accepted Catholicism. It is only when we try to show Bosnian Churchmen accepting Islam that we hit a blank. Professor N. Filipović, having observed that for years scholars have referred to adherents of the Bosnian Church accepting Islam en masse, notes that the Turkish sources do not show such a mass conversion. He stresses that in all the defters he could not find a single case of a krstjanin or a son of a krstjanin converting to Islam. He could only find cases of people, living in villages in which krstjani live, converting,(26) but there is no way of knowing whether or not these converts had been adherents of the Bosnian Church. In the general atmosphere of changing religion, we can be certain that there were many Bosnian Church members converting to Islam; but it is ironic — in the face of the generally accepted opinion that this church supplied the largest number of new Moslems — that Bosnian Church members are the one group which cannot be shown on the basis of our sources converting to Islam.

The Orthodox, despite their successes, also lost believers. The visitations not only speak of Catholics accepting Orthodoxy but also show Orthodox believers turning to Catholicism. The Orthodox also converted to Islam; we find an inscription from a Psalter in the Holy Trinity Church in Plevlje, dated Sarajevo 1517, which states that in these days in this land (Bosnia?, the Sarajevo region?) there is a great increase in the number of Moslems, while the Orthodox Christian faith in this land has become greatly reduced in numbers.(27) Thus we find that in some parts of Bosnia, the Orthodox religion, which by migration and conversion had recently increased its membership, was now in the early sixteenth century suffering losses to Islam.

Thus we see that changes of religion were a general and common occurrence at this time, and that the widely discussed "Islamization" was only one aspect of this broader phenomenon. The only direction that conversion did not go openly was from Islam to Christianity. The reason for this was the frightful penalty for apostatizing from Islam. However, the visitations do report that Moslems frequently became secret Christians.(28) Thus, on the sly, converts to Islam or their descendants also reverted to Christianity.

Why did so many changes in religious confession occur in Bosnia and Hercegovina and not elsewhere in the Balkans (excluding Albania)? The reason I suggest is not hard to find and has nothing to do with the content of beliefs of the former heresy, even though such a view has frequently been advanced. The other Balkan states had one dominant form of Christianity. They had had and continued to preserve under the Turks fairly efficient and territorially organized church administrations. Bosnia, unlike them, had had competing faiths. And as a result of its religious history no faith in Bosnia was able to establish an efficient territorially based organization that could bind believers to its church — be it through belief or through a sense of community. Thus Bosnia's Christians, of whatever confession, had had little contact with any church, and few Bosnians were deeply attached to any religious community. In the 1450's and early 1460's many Bosnians had been forcibly brought to Catholicism. These converts certainly had not had time to become strong believing Catholics, they probably lacked interest in Catholicism, and many may have resented being forced to accept that faith. Thus many Bosnians were more or less between faiths — having renounced an earlier faith and not yet committed to the new one — with no deep belief in any.

After 1463, Islam — a dynamic and well-preached new religion — appeared. It had the advantage of being the religion of the conquering state, which gave its members all sorts of worldly advantages. In the early years of Turkish rule the Orthodox had not yet had time to build an effective organization in much of Bosnia and Hercegovina; the Catholic organization in this period was effective only sporadically. Thus, in a locality where Christianity was poorly organized and generally ineffectively preached, it is not surprising to find people without any strong religious attachment accepting a new faith. And because the Bosnians had long been shaky Christians who had dealt with the Turks for half a century before the conquest, they had no strong prejudices against Islam as did people from most other Christian lands. This general situation was surely intensified by their general feelings of hatred toward Hungary which many probably identified with Catholicism. In addition, we can be certain that frequently religious motives were not the major ones which determined one's acceptance of a new faith. And finally, acceptance is a more accurate term than conversion for what occurred in Bosnia. Probably few Bosnians in accepting Islam underwent any deep changes in patterns of thought or way of life. Most of those who became Moslems probably lived as they always had, retaining most of their domestic customs as well as many Christian practices. They adopted now a few Islamic practices, which quickly would acquire great symbolic value and which would soon come to be viewed as the essentials of Islam.

Thus the Bosnian Church simply passed away. Probably only in the days of the nativistic reaction in the middle of the thirteenth century, when Bosnians fought to maintain their independence from the Hungarians, had the Bosnian Church had much significance for sizable numbers of Bosnians. Throughout the fourteenth and fifteenth centuries, the Bosnian Churchmen — excluding those who lived and served at secular courts — seem to have spent most of their time in their monasteries; they did not build a territorial organization or attempt to establish close ties with the peasant population. Thus when other faiths began to proselytize and put pressure on the rulers to persecute, there were few who cared enough about the Bosnian Church to defend its clerics. With the Turkish conquest and the new order established by it, the irrelevance of the Bosnian Church to Bosnian life became complete, and the Bosnian Church became extinct. Historical selection parallels natural selection.

Throughout this study we have stressed that the Bosnian Church exerted relatively little influence on political developments or upon society. And as an inefficient religious organization existing in the middle of a peasant society quite indifferent to religious matters, its religious and moral influence was also small. Thus the legacy of the Bosnian Church is nil. And though frequently historians have used the Bosnian Church to explain the Islamization of Bosnia, it is more accurate to explain that phenomenon by the absence of strong Catholic, Orthodox, or even Bosnian Church organizations.

At times, subsequent historians have magnified the importance of this medieval church, giving it an influential role in the development of the medieval state.(29) Others have made it into a romantic symbol of Bosnian independence in the face of its predatory neighbors. Apparently its contemporaries did not usually find it so.

FOOTNOTES TO CHAPTER VIII

1. And of course Uskoplje, in the years following 1463, had become part of the territory belonging to the ruler of Hercegovina.

2. E. Laszowski, "Urbar Vinodolskih imanja knezova Zrinskih," *VZA*, 17, 1915, p. 107.

3. G. Stanojević, "Jedan pomen o kristjanima u Dalmaciji iz 1692 godine," *GID*, XI, 1960, p. 273.

4. The ignorance of the Bosnians about matters of faith is also illustrated by the fact that after 1590 when the Catholic Church introduced the Gregorian calendar in Bosnia, many Bosnians, thinking the change in calendar signified a

change in faith, preferred to stick with the old calendar than the sacraments and converted to Orthodoxy (D. Mandić, *Etnička povijest Bosne i Hercegovine*, Rome, 1967, p. 475).

5. Pandžić, appendix, p. 132, presenting text of 1703 visitation of A. Righus.

6. M. Vego, "Patarenstvo u Hercegovini u svjetlu arheoloških spomenika," *GZMS* (arh), n.s.18, 1963, pp. 209-11.

7. Pandžić, appendix, p. 147, presenting text of visitation of 1751 of S. Tudisić.

8. *Lètopis fra Nikole Lašvanina* (J. Jelenić ed.) Sarajevo, 1916, pp. 76, 78.

9. Orbini, p. 353.

10. And Uskoplje had been under the ruler of Hercegovina since December 1463. Thus, since we only hear of a hiža there between 1466 and 1470 we must consider the possibility that the hiža did not date from the Bosnian kingdom. It is quite possible that Bosnian Churchmen, under the protection of the herceg, had moved to there after Uskoplje had been granted to Herceg Stefan.

11. M. Batinić, *Franjevački samostan u Fojnici od stoljeća XIV-XX*, Zagreb, 1913, p. 129. He, of course, would be an example of an Orthodox believer being converted to Catholicism in the period prior to the collapse of the medieval state.

12. That the Bosnian Church was not a branch of the Serbian is also suggested by the great difference in character and spirit between the two churches. The Bosnians, though they formed their own church, never attempted to have a Bosnian saint with his own cult like the Serbs' Sava. The Bosnians were not church oriented; churches of all denominations in Bosnia were small and not at all like the magnificent monastery churches built in Serbia and supported by extensive land grants and costly gifts from the rich who saw these donations as a cultural and religious duty. The Bosnians do not seem to have felt such an obligation particularly strongly. In addition we find the Serbs giving gifts to foreign centers of Orthodoxy, particularly Mount Athos. If the Bosnian Church had been Orthodox we would expect to find ties between it or its rich adherents and some of those foreign centers. Excluding the herceg's gifts to the Sinai monks (and the source referring to the gift specifically states that he was not yet Orthodox) we have no evidence of any Bosnian giving to any foreign Orthodox center. The Bosnian Church was monastic, but if it were an outgrowth of the Serbian Church we would expect it to have ties with Serbia and to acquire some of the characteristics of Serbian monasticism. That it did not and that the Serbian Church anathematized the krstjani seems to me sufficient evidence that the Bosnian Church, though more or less orthodox in its theology, was not Orthodox.

13. M. Vasić, "Etničke promjene u Bosanskoj krajini u XVI v," *GID*, 13, 1962, pp. 233-49.

14. On a few occasions the defters do inform us, in the manner they name individuals holding land, that a Moslem landholder was a newcomer from beyond the borders of Bosnia (e.g., Ilyas of Florina). Okiç, "Les Kristians . . .," p. 119 gives a fairly long list of such examples. Unfortunately the names are of individuals, thus we cannot say whether they had arrived individually or whether they came as parts of a large migration from one place (e.g., in the cited case from Florina).

15. Though the term Vlach originally referred to a specific ethnic and linguistic group of people, the limited evidence we have about Vlachs in Bosnia and Hercegovina in the Middle Ages suggests that they were already Slavic speakers, and already intermarried with Slavs. Thus in medieval sources for

Bosnia and Hercegovina the term "Vlach," regardless of the origin of the word and the fact that the majority of Vlachs were of non-Slavic origin, had come to designate "shepherds"; the new significance gained under the Ottoman's "Vlach" status was the right to resettle and pay a set tax in cash rather than in kind. This then would add to the "Vlach" category some non-shepherds. On the Vlachs in Bosnia and Hercegovina, see *Simpozijum o srednjovjekovnom katunu,* edited by M. Filipović, Sarajevo, Poseb. izd. NDBH, II, Sarajevo, 1963. The fact that the Vlachs became Slavicized in language and racially mixed with Slavs did not eliminate cultural differences between the Vlach shepherds and the settled Slavs. Thus, probably owing to the fact that many if not most of the medieval "Vlachs" were of common ethnic Vlach origin, "Vlachs" have preserved to the present day a variety of practices not found among other peoples in the Balkans, some of which may well date back to classical times. Some examples can be found in M. Wenzel, "The Dioscori in the Balkans," *Slavic Review,* 26, 1967, pp. 363-381.

16. On the problem of migrations I am much indebted to the fruitful conversations I was able to have with the anthropologists at the Zemaljski muzej in Sarajevo, particularly with Professors M. Petrić and V. Palavestra. See also M. Petrić's brief but excellent article on migration, "O migracijama stanovništva u Bosni i Hercegovini," *GZMS,* 18, 1963 (etn), pp. 5-16. See also the pioneer works of J. Dedijer, "Porijeklo bosanskog i hercegovačkog stanovništva," *Pregled,* I, 1911, pp. 420-30, and V. Skarić, "Porijeklo pravoslavnoga naroda u sjeverozapadnoj Bosni," *GZMS,* XXX, 1918, pp. 219-265. Also the volumes of the Cvijić, *Naselja i poreklo stanovništva* series, particularly J. Dedijer's excellent work on Hercegovina and M. Filipović on Visoko. I also greatly profited from my discussions about material in the Turkish defters and migrations in the early Turkish period with H. Šabanović at the Oriental Institute in Sarajevo. See also note 13.

17. Okiç, "Les Kristians . . .," *Südost-Forschungen,* 19, 1960, p. 122.

18. *Ibid.,* pp. 118-19.

19. H. Šabanović, "Lepenica u prvom stoljeću turske vladavine," in NDBH, poseb. izd. vol. 3, *Lepenica,* Sarajevo, 1963, pp. 193-207, esp. pp. 195-96. The breakdown of individual villages, pp. 197-207.

20. H. Šabanović, "Bosansko krajište 1448-1463," *GID,* IX, 1958, pp. 177-220. I here choose examples from 1485 of certain larger villages that well illustrate the great variation in the ratios between Christian and Moslem households: Dolac (near Travnik): 84 Christian, 13 Moslem. Hodidjed (the fortress above Sarajevo): 35 Christian, 9 Moslem. Glavogodina: 3 Christian, 29 Moslem. Several villages which together compose Doljani: 19 Christian, 16 Moslem. Butmir: 21 Christian, 14 Moslem. Otes: 3 Christian, 17 Moslem. and Presjenica: 38 Christian, 39 Moslem.

21. Ö. L. Barkan, "Essai sur les Données Statistiques des Registres de Recensement dans l'Empire Ottoman aux XVe et XVIe siècles," *Journal of Economic and Social History of the Orient,* I, 1, 1957, p. 32.

22. For references to all the visitations that follow, see Chapter II, note 64.

23. Masarechi also specifically notes conversions from Catholicism to Islam, mentioning that recently some 7,000 Catholics from the vicinity of Sutjeska in central Bosnia had accepted Islam.

24. For example, see the interesting article and texts presented by K. Draganović, "Tobožnja stjepanska biskupija u Hercegovini," *Croatia Sacra,* 7,

1934, pp. 29-58, esp. pp. 35, 40, 42, 50-51.

25. D. Mandić, *Etnička povijest Bosne i Hercegovine,* Rome, 1967, p. 460. Mandić emphasizes the fact that in the seventeenth century Catholics all over Bosnia and Hercegovina converted to Orthodoxy. He cites as reasons Turkish favor and the shortage of Catholic priests.

26. N. Filipović, "Napomene o islamizaciji u Bosni i Hercegovini u XV vijeku," *Godišnjak,* Knj. VII, *Centar za balkanološkog ispitivanja,* Knj. 5 (AN i U, B i H), 1970, pp. 151, 161.

27. Pop Stjepa and V. Trifković, *Sarajevska okolina, I: Sarajevo polje* (SKA, SEZ, vol. XI, Naselja srpskih zemalja vol. V, general editor J. Cvijić), Beograd, 1908, p. 65.

28. K. Draganović, "Tobožnja stjepanska biskupija . . .," p. 35. Also illustrating the desire of Moslems to return to Christianity is the fact that during and after Austrian-Turkish wars, Moslems as well as Catholics sought asylum in Austrian territory; these Moslems upon arrival accepted Catholicism. We also have data on Moslems migrating to Dalmatia to accept Catholicism.

29. The tendency to exaggerate the importance of religious issues in medieval Bosnia — as well as in many other medieval states — is a natural one since such a large proportion of our sources were written by churchmen who were, of course, concerned with these issues.

INFORMATION ABOUT THE ORGANIZATION OF THE BIBLIOGRAPHY

Under Primary Sources (A: Written) are included all documentary sources from the medieval period. Since certain early modern histories include documents now lost (and therefore primary material), we have had to draw an arbitrary dateline between works classified as primary and secondary. Thus as primary sources we have included all post-medieval histories from 1463 up to and including the work of Junius Resti (1669-1735). The eighteenth century historians, such as Farlati and Raynaldi, despite their presentation of certain documents not found elsewhere — some of which are of doubtful authenticity — are included with secondary material.

Archaeological works have presented a problem. Straight descriptions of cemeteries, gravestone motifs, excavations of medieval churches, or texts of inscriptions are clearly primary sources. However, rarely is archaeological material presented in a clear-cut manner. Frequently the archaeologist, in addition to providing descriptive data about a site, will devote considerable space to theorizing about the meaning of motifs. We also find new inscriptions or gravestones announced in more general secondary works. Therefore, while certain archaeological works may clearly be categorized as "primary" or "secondary," there are a vast number of works which fit both categories. Thus, if one wants to avoid listing many titles twice, one again must be arbitrary about classifying works under Primary Sources (B: Archaeological and Epigraphical) or Secondary Sources. The bulk of these difficult to classify works have been placed under the Primary heading, hence the reader interested in the views and scholarly judgments of able archaelogists who have worked on questions connected with medieval Bosnia are referred to Section B of Primary Sources.

For descriptions of sources, and what texts appear in particular collections, the reader is referred to Chapter II of this study which is devoted to problems connected with the sources.

Under the secondary literature are included works on a variety of topics. These include works on Bosnian history, religion, ethnography, folklore, and culture as well as recent travel accounts. Also included are a variety of works on related fields that I followed up in the hopes of discovering links between Bosnian and foreign phenomena (e.g., works on Eastern and Western dualism). I have also included a variety of Anthropological works, whose contents have nothing about Bosnia, but which were read for methodological reasons — to find questions to put to the Bosnian sources.

Works on Bosnia fall into three categories — and once again distinctions between the categories are frequently blurred: works of quality and relevant, works of quality on various topics of Bosnian history, but which did not contribute specifically to this study, and works which are chiefly polemical. Of course, lines between these categories are frequently hard to draw; a polemical work of little scholarly value may occasionally have an important insight. Since judgments on the merits or demerits of many of these works are put forward in the course of the book and its footnotes, the bibliography makes no attempt to evaluate the secondary works.

BIBLIOGRAPHY

KEY TO ABBREVIATIONS USED IN THE BIBLIOGRAPHY

AFP, *Archivum Fratrum Praedicatorum* (Publication of Instituto Storico Dominicano, Rome, I, 1930-.

AHID, *Anali Historijskog Instituta u Dubrovniku,* Dubrovnik, I, 1952-.

BAN, B'lgarska Akademija na Naukite.

CEC, *Cahiers d'Études Cathares,* Arques, I, 1946-.

FFS, Filozofski Fakultet u Sarajevu. Have a publication, *Radovi.*

Fontes: Fontes, Pontificia Commissio ad Redigendum codicem iuris canonici orientalis, Rome.

GEMB, *Glasnik Etnografskog muzeja u Beogradu,* Beograd, I, 1926-.

GID, A Journal originally entitled, *Godišnjak Istoriskog društva Bosne i Hercegovine* and now entitled, *Godišnjak Društva istoričara Bosne i Hercegovine,* Sarajevo, I, 1946-.

GNČ, *Godišnjica Nikole Čupića,* Beograd, I, 1877-1941.

GSUD, *Glasnik Srpskog učenog društva,* Beograd, I-LXXV, 1847-1892 (first 18 volumes entitled *Glasnik Društva srbske slovesnosti).*

GZMS, *Glasnik Zemaljskog muzeja,* Sarajevo, I, 1889-1943; new series, I, 1946-.

IČ, *Istorijski časopis,* Beograd, I, 1948-.

IG, *Istoriski glasnik,* Beograd, begun in 1948; for some reason with volume numbers only within a year and not by year.

IJKSN, Istorija, jezik i književnost srpskog naroda. A sub-title for a periodical and a series of documents issued by a department of the Serbian Academy of Sciences. The document series is Zbornik za IJKSN; and the periodical, *Prilozi.*

JAZU, Jugoslavenska Akademija znanosti i umjetnosti. Zagreb. The Academy of Science and Art in Zagreb. Has many publications: Two periodicals, *Starine,* and *Rad.* Also publishes collections of documents, see MSHSM. In addition see also the ethnographic publication, ZNŽOJS.

JIČ, *Jugoslovenski istoriski časopis,* Beograd. A periodical that began in 1935 and expired before World War II began.

MSHSM, Monumenta spectantia historiam Slavorum Meridionalium. The source publication series of JAZU, initiated in 1868.

NDBH, Naučno društvo, Bosne i Hercegovine. Institute in Sarajevo. Publications include the periodical *Radovi;* it also publishes two series of monographs: Djela, and Posebna izdanja.

NS, *Naše starine,* Organ of the Zavod za zaštitu spomenika kulture for the Republic of Bosnia and Hercegovina. Sarajevo, I, 1953-.

Pos. izd. Posebna izdanja. Individual Monographs. The heading for monograph series of both SAN and NDBH.

Prilozi. *Prilozi IJKSN;* Periodical issued by SAN, Beograd, I, 1921-.

SAN, Srpska Akademija Nauka i Umetnosti. Serbian Academy of Sciences and Arts. Has many publications. The monograph series, pos. izd. Collections of sources Zbornik za IJKSN. As well as journals *Prilozi IJKSN,* and the Zbornik radova series of the various institutes. See also, SEZ (below). Before World War II the Academy was the Serbian Royal Academy, abbreviated SKA.

SEZ, *Srpski etnografski zbornik.* A multi-volume publication. In addition to monographs, it publishes the series on migration and settlements of populations on an area to area basis: Naselja i poreklo stanovništva . . . as well as the volumes of SEZ devoted to ethnographic studies of particular customs or particular areas: Život i običaji narodni.

SKA, Srpska kraljevska Akademija. The pre-revolutionary name for SAN.

SNSBH, Srednjevjekovni nadgrobni spomenici Bosne i Hercegovine. The monograph series of which each volume is devoted to medieval cemeteries of a particular location. Begun in 1950. Originally published by the Zemaljski muzej in Sarajevo, publication has recently been under the auspices of the Zavod za zaštitu spomenika kulture BH.

VV, *Vizantijskij vremennik,* Moscow, 1894-1927. New Series, I, 1947-.

VZA, *Vjesnik zemaljskog arkiva,* Zagreb, I-XXII, 1899-1920. Resumed under title *Vjesnik državni arhiv* 1925-45.

ZNŽOJS, *Zbornik za narodni život i običaje južnih Slavena.* The ethnographic and folklore publication of JAZU, Zagreb, I, 1896-.

ZREI, *Zbornik radova Etnografskog Instituta,* SAN, Beograd. Journal of the Ethnographic Institute of SAN. Beograd, I, 1950-.

ZRFFB, *Zbornik radova Filozofskog fakulteta u Beogradu.* Journal of the Filozofski fakultet of the University of Beograd, I, 1948-.

ZRVI, *Zbornik radova Vizantološkog Instituta,* SAN, Beograd. Journal of the Byzantine Institute of SAN. Beograd, I, 1952-.

I: PRIMARY SOURCES

A. Written

Amati, G. "Processus contra Waldenses in Lombardia Superiori anno 1387," *Archivio Storico Italiano* (Firenze), Ser. III, Vol. II, pt. 1, 1865, pp. 3-61.
Anonymi Spalatensis, in *Scriptores Rerum Hungaricarum, Dalmaticarum, Croaticarum et Slavonicarum veteres ac Genuini*, III, Vindobona (Vienna), 1748 (ed. I. Schwandtner).
Bajraktarević, F. "Turski dokumenti manastira Sv. Trojice kod Plevlja," *Spomenik* (SKA), LXXIX, 1935, pp. 25-85.
Bartholomaeus de Pisa, "De Conformitate vitae B. Francisci," *Analecta Franciscana*, IV, 1906.
Benedict XI, see Delorme and Tautu (eds.).
Benedict XII, see Tautu (ed.).
Bertrand de la Broquière, *Le Voyage d'Outremer* (M. Rajičić, V. Rabotin eds.), Beograd, 1950.
Boniface IX, see Monumenta Vaticana Historiam Regni Hungariae . . . .
Chalkokondyles, Laonikos, *De Rebus Turcicis*, Bonn Corpus, Vol. 48, Bonn, 1843.
La Chanson de la Croisade Albigeoise (ed. E. Martin-Chabot), 3 vols., Paris, 1961.
Chronica XXIV generalium Ordinis Minorum, *Analecta Franciscana*, III, 1897.
Cledat, L. *Le Nouveau Testament, traduit au XIIIe siècle en langue Provençale, suivi d'un rituel Cathare*, Paris, 1887.
Clement V, see Delorme and Tautu (eds.).
Clement VI, see Tautu (ed.).
Çotrugli, B. *Della Mercatura et del Mercante perfetto*, Brescia, 1602.
Čremošnik, G., ed. "Nekoliko dubrovačkih listina iz XII-XIII stoleća," *GZMS*, XLIII, 1931, pp. 25-54.
Čremošnik, G., ed. *Istoriski spomenici dubrovačkog arhiva*, "Kancelariski i notariski spisi 1278-1301"; (SKA, Zbornik za istoriju jezik i književnost srpskog naroda, Ser. III, 1 sv.), Beograd, 1932.
Čremošnik, G., ed. *Spisi dubrovačke kancelarije*, knj. I, "Zapisi notara Tomazine de Savere 1278-82," Zagreb (JAZU), 1951.
Čremošnik, G., ed. "Ostaci arhiva bosanske franjevačke vikarije," *Radovi* NDBH, III, 1955, pp. 5-56.
Daničić, G., ed. "Hvalov rukopis," *Starine*, III, 1871, pp. 1-146.
Danilo, Arhiepiskop. *Život Kraljeva i Arhiepiskopa Srpskih* (L. Mirković, ed.), Beograd, 1935.
Davis, G. *The Inquisition at Albi, 1299-1300*, New York, 1948.
Delorme, F., and Tautu, A., eds. *Acta Romanorum Pontificum ab Innocentio V ad Benedictum XI (1276-1304)*, in Fontes, Series III, Vol. V, Tom. II, Rome, 1954.
Delorme, F., and Tautu, A., eds. *Acta Clementis PP, V (1303-1314)*, in Fontes, Ser. III, Vol. VII, Tom. I, Rome, 1955.
Dinić, M., ed. *Odluke veća Dubrovačke Republike*, I-II (SAN, Zbornik za IJKSN, 15, 21), Beograd, 1951, 1964.
Dinić, M., ed. *Iz dubrovačkog arhiva*, I-III (SAN Zbornik za IJKSN), Beograd,

1957, 1963, 1967.
Dinić, M., ed. "Nekoliko ćiriličkih spomenika iz Dubrovnika," *Prilozi*, XXIV (1-2), 1958, pp. 94-110.
Dlugosz, J. *Historiae Polonicae*, XI, in Opera Omnia, XIII, Cracow, 1877.
Döllinger, I. *Beiträge zur Sektengeschichte des Mittelalters*, I (sources), Munich, 1890.
Dondaine, A., ed. *Un traité Néo-Manichéen du XIIIe siècle. Le liber de duobus principiis*, Rome, 1939 (This volume also includes Rayner Sacchoni's *Summa*, pp. 64-78).
Dondaine, A., ed. "Les Actes du Concile Albigeois de St. Felix de Caraman," *Miscellaneae G. Mercati*, V (Studi e testi, 125), The Vatican, 1946, pp. 324-55.
Dondaine, A., ed., "Le Manuel de l'Inquisiteur 1230-1330," *AFP*, XVII, 1947, pp. 85-194
Dondaine, A., ed. "La Hiérarchie Cathare en Italie," *AFP*, XIX, 1949, pp. 280-312; XX, 1950, pp. 234-324. The first contains "De Heresi Catharorum in Lombardia" (ca. 1210), the second, Anselm of Alexandria's "Tractatus de hereticis" (ca. 1270).
Draganović, K. "Tobožnja Stjepanska Biskupija u Hercegovini," *Croatia Sacra*, VII, 1934, pp. 29-58.
Draganović, K. "Izvješće fra Tome Ivkovića Biskupa Skradinskog iz godine 1630," *Croatia Sacra*, VII, 1934, pp. 65-78.
Draganović, K. "Izvješće apostolskog vizitatora Petra Masarechija o prilikama Katoličkih naroda u Bugarskoj, Srbiji, Srijemu, Slavoniji i Bosni g. 1623 i 1624," *Starine*, XXXIX, 1938, pp. 1-48.
Djurdjev, B. "Kanun-nama Bosanskog sanžaka iz godine 1530," *GZMS*, n.s. III, 1948, pp. 189-200.
Esposito, M. "Un Auto da Fé à Chieri en 1412," *Revue d'Histoire Ecclesiastique*, XLII (3-4), 1947, pp. 422-32.
Fejér, G. *Codex diplomaticus Hungariae Ecclesiasticus ac Civilis*, IX, pts. 5-7; X, pts. 1-8, XI, pt. 1. Budapest, 1829-44. (Since Smičiklas, *CD*, incorporated all relevant material from Fejer up to 1378 (through Fejer, IX, pt. 4), I began to use Fejer at the year at which Smičiklas' collection stopped.)
Fermendžin, E., ed. "Chronicon Observantis Provinciae Bosnae Argentinae Ordinis S. Francisci Seraphici," *Starine*, XXII, 1890, pp. 1-67.
Fermendžin, E., *Acta Bosnae*, JAZU MSHSM, 23, Zagreb, 1892 (Abbreviated throughout as *AB*).
Filipović, N. "Jedna kanun-nama Zvorničkog Sandžaka," *GZMS*, n.s. III, 1948, pp. 223-34.
Gelcich, J. and Thalloczy, L. *Diplomatarium relationum Reipublicae Ragusinae cum Regno Hungariae*, Budapest, 1887.
Glassenberger, Chronica Fratris Nicolae. *Analecta Franciscana*, II, 1887, pp. 178-225.
Gregory IX, see Tautu (ed.).
Gregory XI, see Tautu (ed.).
Grigorovič, V. *O Serbii v' eja otnošenijah k sosednim deržavam preimuščestvenno v XIV-XV stoletijah*, Kazan, 1859.
Gui, B. *Manuel de L'Inquisiteur*, 2 vols., Paris, 1926-27.
Gundulić, J. (J. Gondola). *Croniche ulteriori di Ragusa* (Nodilo ed.), JAZU

MSHSM 25, Scriptores 2, Zagreb, 1893.
Hadžibegić, H. "Bosanska kanun-nama iz 1565," *GZMS*, n.s. III, 1948, pp. 201-222.
Haluščynskyj, P. T., ed. *Acta Innocentii PP. III (1198-1216)*, in Fontes, Ser. III, Vol. II, Rome, 1944.
Hamm, J. "Apokalipsa Bosanskih krstjana," *Slovo*, IX-X, 1960, pp. 43-104.
H. Hoberg, *Taxae pro Communibus Servitiis ex Libris Obligationum ab anno 1295 usque ad annum 1495 confectis*, Vatican (Studi e Testi, 144), 1949.
Honorius III, see Tautu (ed.).
Horvat, K. "Monumenta Historica Nova Historiam Bosnae et Provinciarum Vicinarum Illustrantia," *GZMS*, XXI, 1909, pp. 1-104; 313-424; index, 505-18.
Innocent III, see Haluščynskyj (ed.).
Innocent V, see Delorme and Tautu (eds.).
Innocent VI, see Tautu (ed.).
Jagić, V. "Analecta Romana," *Archiv für slavische Philologie*, XXV, 1903, pp. 1-47.
Jelenić, J., ed. "Necrologium Bosnae Argentinae — prema kodeksu franjevačkog samostana u Kr. Sutjesci," *GZMS*, XXVIII, 1916, pp. 337-57.
Jelenić, J., ed. "Dva letopisa Bosne Srebrene," *GZMS*, XXX, 1918, pp. 115-28.
Jelenić, J., ed. "Spomenici kulturnoga rada bosanskih franjevaca (1437-1878)," *Starine*, XXXVI, 1918, pp. 81-162.
Jelenić, J., ed. "Ljetopis franjevačkog samostana u Kr. Sutjesci," *GZMS*, XXXV, 1923, pp. 1-30; XXXVI, 1924, pp. 1-26.
Jelić L., ed. "Discorso del priorato della Wrana di Giovanni Marnavich Bosnese Canonico di Sibenico," *GZMS*, XVIII, 1906, pp. 279-305.
Jireček, K. *Spomenici Srpski*, published in *Spomenik*, SKA, XI, 1892.
Joannis de Segovia. *Historia gestorum Generalis Synodi Basiliensis* (ed. E. Birk), in Monumenta Conciliorum Generalium XV, Concilium Basileense, II, Vindobona (Vienna), 1873.
John XXII (Joannis), see Tautu (ed.).
Jorga, N. *Notes et Extraits pour servir à l'Histoire des Croisades au XVe siècle*, II, Paris, 1899.
Kajer, P., Vuletić-Vukasović, V. "Listina Stefana Tomaša od 1450," *GZMS*, XVII, 1905, pp. 253-56.
Kamber, D. "Kardinal Torquemada i tri bosanska Bogomila," *Croatia Sacra*, II, 1932, pp. 27-93.
Kosanović, S. "Beleška o Bogomilima," *GSUD*, XXXVII, 1873, pp. 179-88.
Kovačević, L. "Nekoliko priloga za crkvenu i političku istoriju južnih slovena," *GSUD*, LXIII, 1885, pp. 1-40. (Includes "Odgovor Carigradskog Patrijarha Genadija (1453-59) na pitanja Sinajskih kaludžera," pp. 11-14.)
Kritovoulos. *History of Mehmed the Conqueror* (trans-ed. C. Riggs), Princeton, 1954.
Kuripešić, B. *Putopis kroz Bosnu, Srbiju, Bugarsku i Rumeliju 1530* (trans-ed., Dj. Pejanović), Sarajevo, 1950.
Lasić, D., "Fra Bartholomaei de Alverna Vicarii Bosnae 1367-1407 quemdam scripta hucusque inedita," *Archivum Franciscanum Historicum*, LV (1-2), 1962, pp. 59-81.

Lašvanin, fra Nikola. *Letopis* (ed. J. Jelenić). Sarajevo, 1916.
Laszowski, E. "Urbar Vinodolskih imanja knezova Zrinskih," *VZA*, XVII, 1915, p. 107.
Ljubić, S. *Listine o odnošajih izmedju Južnoga Slavenstva i Mletačke Republike*, I-X, JAZU, MSHSM, 1-5, 9, 12, 21, 22. Zagreb, 1868-91.
Luccari, Jacob. *Copioso ristretto degli annali di Ragusa di Giacomo di Pietro Luccari* (1st ed., 1605), I used and cite from the 1790 edition published in Ragusa.
Lučić, J. "Prilog pitanju nestanka bosanskih bogumila (patarena)," *Historijski zbornik*, XIV, 1961, pp. 239-42.
Lucius, I. *De Regno Dalmatiae et Croatiae*, in Scriptores Rerum Hungaricarum, *Dalmaticarum, Croaticarum et Sclavonicarum Veteres*, III (ed. I. Schwandtner), Vindobona (Vienna), 1748.
Dr. F. M. (Zagreb). "Dva savremena izvještaja o Bosni iz prve polovine XVII stoljeća," *GZMS*, XVI, 1904, pp. 251-66.
Mandić, D. *Hercegovački spomenici franjevačkog reda iz Turskog doba*, I (1463-99), Mostar, 1934.
Manuel di S. Giovanni, G. "Un Episodio della Storia del Piemonte nel secolo XIII," *Miscellanea di Storia Italiana*, XV, Torino, 1874, pp. 7-84.
Martene-Durand. *Amplissima collectio veterum scriptorum*, I, Paris, 1724.
Martene-Durand. *Thesaurus novus anecdotarum*, I, Paris, 1717.
Matasović, J. "Fojnička Regesta (Acta Turcica Bosnensia et Latina ex archivio conventus Fratrum Minorum de Observantia Spiritus S. Fojnicae Bosnae Argentinae)," *Spomenik* (SKA), LXVII, Beograd, 1930, pp. 61-432.
Matković, P., "Spomenici za Dubrovačku povjest u vrieme Ugarsko-hrvatske zaštite," *Starine*, I, 1869, pp. 141-210.
Matthaeus Parisiensis, *Chronica Majora* (ed. H. Luard), Rerum Brit. Medii aevi scriptores, III, 1876.
Matthias Corvinus, *Epistolae Matthiae Corvini Regis Hungariae ad Pontifices, Imperatores, Reges, Principes, aliosque . . .* (ed. S. Vida), Cassovia, 1744.
Mercati, G. "Note Varie sopra Niccolo Modrussiense," in Mercati, *Opere Minori*, IV, published as volume 79 of *Studi e Testi*, Vatican, 1937, pp. 205-58.
Mihailović, Konstantin iz Ostrovice. *Janičareve uspomene ili Turska hronika*, ed.-tr. Dj. Živanović, *Spomenik* (SAN), 107, Beograd, 1959.
Miklosich, F. *Monumenta Serbica Spectantia Historiam Serbiae, Bosnae, Ragusii*, Vienna, 1858 (photostat of 1858 edition done in Graz, 1964).
Monumenta Ragusina. Libri Reformationum, I-V, JAZU, MSHSM, 10, 13, 27-29, Zagreb, 1879-97.
Monumenta Turcica Historiam Slavorum Meridionalium Illustrantia, I (Kanuni i kanun-name za Bosanski . . . etc. Sandžak), Orientalni Institut, Sarajevo, 1957.
Monumenta Vaticana Historiam Regni Hungariae Illustrantia, Ser. I, Vol. I (Rationes collectorum Pontificorum in Hungaria 1281-1375), Budapest, 1887; Ser. I, Vol. 3 (Bullae Bonifacii IX 1389-96), Budapest, 1888.
Mošin, V. "Serbskaja redakcija Sinodika v Nedelju Pravoslavija," *VV*, XVII, 1960, pp. 278-353.
Nodilo, S., ed. *Annales Ragusini Anonymi item Nicolai de Ragnina*, JAZU, MSHSM, 14, Scriptores, 1, Zagreb, 1883.
Novaković, R., ed. *Brankovićev letopis* (SAN, pos. izd.), Beograd, 1960.

Orbini, M. *Il Regno degli Slavi hoggi corrottamente detti Schiavoni Historia di Don Mauro Orbini Rauseo Abbate Melitense*, Pesaro, 1601. A Serbian translation by Z. Šundrica with notes by S. Ćirković entited *Kraljevstvo Slovena* was published in Beograd in 1968.
Pandžić, B. *De Dioecesi Tribunensi et Mercanensi*, Rome, 1959.
Pfeiffer, N. *Die ungarische Dominikanerprovinz von ihrer Gründung 1221 bis zur Tatarenverwüstung 1241-42*, Zurich, 1913. (Dissertation from Freiburg, No. 86) (Documentary appendix)
Pope Pius II (Enea Silvio de Piccolomini), *Europa Pii Pontificis Maximi nostrorum temporum varias continens historias*, Basel, 1501.
Pope Pius II. *Commentarii rerum memorabilium, quae temporibus suis contigerunt*, Rome, 1584.
Popruženko, M. "Sinodik carja Borila," *B'lgarski starini*, VIII, Sofija, 1928.
Popruženko. "Kozma Presviter bolgarskij pisatelj X veka," *B'lgarski starini*, XII, Sofija, 1936.
Pucić, M. *Spomenici srpski*, I, Beograd, 1858; II, Beograd, 1862.
Puech, H. & A. Vaillant. *Le traité contre les Bogomiles de Cosmas le Prêtre*, Paris, 1945.
Rački, F. "Prilozi za poviest bosanskih patarena," *Starine*, I, 1869, pp. 109-38.
Rački. F. "Notae Joannis Lucii," *Starine*, XIII, 1881, pp. 211-68.
Rački, F. "Dva nova priloga za poviest bosanskih Patarena," *Starine*, XIV, 1882, pp. 1-29.
Rački, F. "Prilog za povjest slavenskoga juga god. 1448," *Starine*, XXVII, 1895, pp. 226-38.
Radojičić, Dj. "Odlomak bogomilskog jevandjelja Tepačije Batala iz 1393 godine," *B'lgarska Akademija na Naukite, Izvestija na Institut za Istorija*, XIV-XV, 1964, pp. 495-509.
Radonić, J. *Dubrovačka Akta i povelje* (Acta et Diplomata Ragusina), SKA, Zbornik IJKSN, Ser. I, Vol. I, sv. 1-2, Beograd, 1934.
Resti, J. *Chronica Ragusina ab origine urbis usque ad annum 1451* (ed. S. Nodilo), JAZU, MSHSM, 25, Scriptores, 2, Zagreb, 1893.
Ristić, S. *Dečanski spomenici*, Beograd, 1864.
Roger de Wendover, *Chronica sive flores historiarum* (ed. H. Coxe), IV, London, 1842.
Sabellicus, M. *Rhapsodiae historiarum ab orbe condito Enneades*, 2 vols., Venice, 1494-1504.
Skok, P. *Tri starofrancuske hronike o Zadru u godini 1202*, Zagreb, 1951.
Smičiklas, T. *Codex diplomaticus Regni Croatiae, Dalmatiae et Slavoniae*, II-XV, Zagreb, 1904-34 (completed up to 1378; cited throughout this study as *CD)*.
Stanojević, G. "Jedan pomen o krstjanima u Dalmaciji 1692 godine," *GID*, XI, 1960, pp. 273-74.
Stojanović, Lj. *Miroslavljevo jevandjelje* (facsimile edition), Beograd, 1897.
Stojanović, Lj. *Stari srpski zapisi i natpisi* (SKA), I-VI, Beograd, 1902-26.
Stojanović, Lj. *Stari srpski rodoslovi i letopisi* (SKA), Beograd, 1927.
Stojanović, Lj. *Stare srpske povelje i pisma* I, pts. 1-2, SKA (Zbornik za IJKSN 19, 24), Beograd, 1929, 1934.
Itineraria Symonis Simeonis et Willelmi de Worcestre (ed. J. Nasmith), Cambridge, 1778.

Šišić, F. "Letopis Pavla Paulovića patricija zadarskoga," *VZA,* VI, 1904, pp. 1-59.
Šišić, F. "Nekoliko isprava iz početka XV st.," *Starine,* XXXIX, 1938, pp. 129-320.
Šundrica, Z. *Acta Consiliorum Reipublicae Ragusinae,* Reformationes, XXVIII; and *Acta Consilii Minoris,* Reformationes, XXIX (unpublished typescript — to be published).
Šunjić, M. "Jedan novi podatak o gostu Radinu i njegovoj sekti," *GID,* XI, 1960, pp. 265-68.
Šurmin, Dj. *Hrvatski spomenici,* I, JAZU, Monumenta Historico-juridica Slavorum meridionalium VI, Zagreb, 1898.
Tadić, J. *Pisma i uputstva Dubrovačke Republike* (Litterae et Commissiones Ragusinae), SKA, Zbornik za IJKSN, Ser. 2, Vol. I, Beograd, 1935.
Tautu, A., ed. *Acta Honorii III (1216-1227) et Gregorii IX (1227-1241),* in Fontes, Ser. III, Vol. III, Rome, 1950.
Tautu, A., ed. *Acta Ioannis XXII (1317-1334),* in Fontes, Ser. III, Vol. VII, Tom II, Rome, 1952.
Tautu, A., ed. *Acta Benedicti XII (1334-1342),* in Fontes, Ser. III, Vol. VIII, Rome, 1958.
Tautu, A., ed. *Acta Clementis PP VI (1342-1352),* in Fontes, Ser. III, Vol. IX, Rome, 1960.
Tautu, A., ed. *Acta Innocentii PP VI (1352-1362),* in Fontes, Ser. III, Vol. X, Rome, 1961.
Tautu, A., ed. *Acta Urbani PP V (1362-1370),* in Fontes, Ser. III, Vol. XI, Rome, 1964.
Tautu, A., ed. *Acta Gregorii PP XI (1370-1378),* in Fontes, Ser. III, Vol. XII, Rome, 1966.
Thallóczy, L. *Codex Diplomaticus Comitum de Blagaj,* Budapest, 1897.
Thallóczy, L. "Istraživanja o postanku bosanske banovine sa naročitim obzirom na povelje Körmendskog arhiva," *GZMS,* XVIII, 1906, pp. 401-444.
Thallóczy, L. *Studien zur Geschichte Bosniens und Serbiens im Mittelalter,* Munich, 1914.
Thallóczy, (with Horvath). Monumenta Hungariae Historica, Diplomataria, XL, *Codex Diplomaticus partium Regno Hungariae adnexarum* (Banatus, Castrum et Oppidum Jajcza), Budapest, 1915.
Theiner, A. *Vetera monumenta historica Hungariam sacram illustrantia,* I-II, Rome, 1859-62 (cited throughout as Theiner, *MH).*
Theiner, A. *Vetera monumenta Slavorum Meridionalium historiam illustrantia,* I, Rome, 1863, II, Zagreb, 1875 (cited throughout as Theiner, *MSM).*
Thomas Archidiaconus. *Historia Salonitana* (F. Rački ed.), JAZU, MSHSM 26, Zagreb, 1894. A new edition has now appeared edited by N. Klaić, SAN, Beograd, 1967.
de Thurocz, Joannes (Janos Thuroczi). *Chronica Hungarorum* (1488), Monumenta Hungarica, 1, Budapest, 1957.
Tkalčić, I. *Monumenta Historica Episcopatus Zagrebiensis saec. XII et XIII,* I-II, Zagreb, 1873-74.
Truhelka, Ć. "Liječništvo po narodnoj predaji bosanskoj i po jednon starom rukopisu," *GZMS,* I, No. 4, 1889, pp. 95-116.
Truhelka, Ć. "Iz starih rukopisa," *GZMS,* VI, 1894, pp. 449-63.

Truhelka, Ć. (signed Dr. T.). "Jedan zanimiv zapis pisan bosančicom," *GZMS*, XVIII, 1906, pp. 349-54.
Truhelka, Ć. (Dr. T.). "Opet jedan zapis pisan bosančicom," *GZMS*, XVIII, 1906, pp. 540-51.
Truhelka, Ć. (ed.). "Fojnička kronika," *GZMS*, XXI, 1909, pp. 443-57.
"Turski Dokumenti u Bosni iz druge polovine XV stoljeća," *Istorisko-pravni zbornik*, I, No. 2, 1949, pp. 179-208. Published by Pravni fakultet, Sarajevo.
Urban V, see Tautu (ed.).
Venanzio da Fabriano, *La Vita di S. Giacomo della Marca* (ed. M. Sgatonni), Zadar, 1940.
Volaterrano, R. *Commentariorum urbanorum*, Basel, 1530.
Wadding, L. *Annales Minorum seu trium ordinum a S. Francisco institutorum*, 25 volumes, Rome 1731-1886. I utilized, however, the third revised edition, Quaracchi, 1932. Volumes 7-13 cover the relevant period for our study (1325-1471). Citations are to original pagination which is also given in the third edition.
Zlatović, S. "Bribirski nekrolog XIV-XV vieka," *Starine*, XXI, 1889, pp. 83-85.
Zlatović, S. "Izvještaj o Bosni god. 1640 od Pavla iz Rovinja," *Starine*, XXIII, 1890, pp. 1-38.

B. Archaeological and Epigraphical

Andjelić, P. "Dva srednjevjekovna nalaza iz Sultića kod Konjica," *GZMS* (arh), n.s. XIV, 1959, pp. 203-14.
Andjelić, P. "Iskopavanje na srednjevekovnom gradu Bobovcu," *Arheološki pregled*, I, 1959, pp. 159-61.
Andjelić, P. "Arheološka ispitivanja," chapter in *Lepenica*, Pos. izd. III, NDBH, Sarajevo, 1963, pp. 151-91.
Andjelić, P. "Revizija čitanja Kulinove ploče," *GZMS* (arh), n.s. XV-XVI, 1961, pp. 288-308.
Andjelić, P. "Grobovi bosanskih kraljeva u Arnautovićima kod Visokog," *GZMS* (arh), n.s. XVII, 1962, pp. 165-71.
Andjelić, P. *Srednjovjekovni pečati iz Bosne i Hercegovine*, Akademija nauka i Umjetnosti, BH, Djela, knj. 38, odj, Društv. nauk, 23, Sarajevo, 1970.
Anonymous. "Epigrafske crtice iz Bosne i Hercegovine," *GZMS*, I, No. 1, 1889, pp. 65-77.
Babić, S. "O starim grobovima u okolini Grahova," *GZMS*, IV, 1892, pp. 273-75.
Barjaktarević, M. "O grobljima i grobovima u gornjem Polimlju," *GEMB*, XXII-XXIII, 1960, pp. 217-23.
Basler, Dj. "Ivanjsko polje," *GZMS*, n.s. VII, 1952, pp. 411-20.
Basler, Dj. "Kupres," *GZMS*, n.s. VIII, 1953, pp. 335-43.
Basler, Dj. "Kreševo-Kiseljak-Fojnica," *GZMS*, n.s. IX, 1954, pp. 299-306.
Basler, Dj. "Dolina Neretve od Konjica do Rame," *GZMS* (arh), n.s. X, 1955, pp. 219-29.
Baum, M. "Nekoliko interesantnih stećaka sa područja istočne Bosne," *Članci i gradja za kulturnu istoriju istočne Bosne*, II, Tuzla, 1958, pp. 53-63.

Benac, A. *Radimlja,* SNSBH, 1, Sarajevo, 1950.
Benac, A. *Olovo,* SNSBH, 2, Beograd, 1951.
Benac, A. *Široki Brijeg,* SNSBH, 3, Sarajevo, 1952.
Benac, A. "Srednjevekovni stećci od Slivna do Čepikuća," *AHID,* II, 1953, pp. 59-82.
Benoit, F. "Le Sarcophage de Lurs en Provence," *Cahiers Archéologiques,* X, Paris, 1959, pp. 27-70.
Bešlagić, Š. "Proučavanje i zaštita stećaka," *NS,* I, 1953, pp. 167-75.
Bešlagić, Š. *Kupres,* SNSBH, 5, Sarajevo, 1954.
Bešlagić, Š. "Stećci u dolini Neretve," *NS,* II, 1954, pp. 180-212.
Bešlagić, Š. "Stećci Kod Raške Gore," *NS,* III, 1956, pp. 253-60.
Bešlagić, Š. "Mastan Bubanjić," *GID,* VII, 1955, pp. 67-80.
Bešlagić, Š. "Stari krstovi u Drežnici," *NS,* III, 1956, pp. 179-88.
Bešlagić, Š. "Stećci Duvanjskog polja," *Starinar,* VII-VIII, 1956-57, pp. 375-96.
Bešlagić, Š. "Stećci na Nekuke kod Stoca," *GEMB,* XXI, 1958, pp. 155-75.
Bešlagić, Š. *Stećci na Blidinju,* Zagreb (JAZU), 1959.
Bešlagić, Š. "Uredjenje nekropole stećaka u Boljunima," *NS,* VI, 1959, pp. 135-42.
Bešlagić, Š. "Selo Kosače i njegovi spomenici," *NS,* VI, 1959, pp. 243-46.
Bešlagić, Š. "Nekoliko novopronadjenih natpisa na stećcima," *GZMS* (arh), n.s. XIV, 1959, pp. 239-47.
Bešlagić, Š. "Stećci u Gornjem Hrasnu," *NS,* VII, 1960, pp. 91-112.
Bešlagić, Š. "Stećci u Opličićima," *NS,* VII, 1960, pp. 145-54.
Bešlagić, Š. "Boljuni srednjovjekovni nadgrobni spomenici," *Starinar,* n.s. XII, 1961, pp. 175-205.
Bešlagić, Š. "Stećci u Brotnjicama," *AHID,* VIII-IX, 1962, pp. 65-83.
Bešlagić, Š. *Kalinovik,* SNSBH 7, Sarajevo, 1962.
Bešlagić, Š. "Srednjovekovni nadgrobni spomenici - stećci - u Dolini Trebišnjice," *NS,* VIII, 1962, pp. 17-36.
Bešlagić, Š. "Novopronadjeni natpisi na stećcima," *NS,* IX, 1964, pp. 133-44.
Bešlagić, Š. "Stećci u Bitunjoj," *NS,* IX, 1964, pp. 79-100.
Bešlagić, Š. "Ćirilski natpis iz doba Kulina bana," *NS,* X, 1965, pp. 203-07 (written with Z. Kajmaković and F. Ibrahimpašić).
Bešlagić, Š. "Ljubinje — srednjovjekovni nadgrobni spomenici," *NS,* X, 1965, pp. 113-63.
Bešlagić, Š. "Stećci u Ziemlju," *Starinar,* n.s. XV-XVI, 1965, pp. 279-92.
Bešlagić, Š. *Popovo,* SNSBH, 8, Sarajevo, 1966.
Bešlagić, Š. *Stećci Centralne Bosne,* SNSBH, 9, Sarajevo, 1967.
Bešlagić, Š. "Trnovo — Srednjovjekovni nadgrobni spomenici," *NS,* XI, 1967, pp. 101-35.
Bešlagić, Š. "Nekoliko novopronadjenih natpisa na stećcima," *NS,* XI, 1967, pp. 41-50.
Bešlagić, Š. "Novopronadjeni natpisi na stećcima," *NS,* XII, 1969, pp. 133-48.
Bešlagić, Š. "Stećci okoline Kladnja," *NS,* XII, 1969, pp. 155-76.
Bešlagić, Š. *Stećci: Kataloško - Topografski pregled,* Sarajevo, 1971.
Bihalji-Merin, and A. Benac. *Bogomil Sculpture,* Beograd, 1962.
Bijelić, M. "Dva starobosanska natpisa u Vlahovićima," *GZMS,* II, 1890, pp. 225-28.

Bijelić, M. "Natpis iz Miljanovića," *GZMS*, III, 1891, pp. 54-55.
Bogunović, P. "Stari nadgrobni spomenici u selu Vrućici," *GZMS*, XLII, 1930, pp. 177-82.
Bojanovski, I. "Stari grad Bobovac i njegova konservacija," *NS*, VIII, 1962, pp. 71-96.
Bojanovski, I. "Arheološki spomenici u dolini Trebišnjice," *NS*, VIII, 1962, pp. 7-14.
Bojanovski, I. "Zaštita spomenika kulture i prirode u području buduće akumulacije na Perućcu na Drini," unpublished typed paper in his possession.
Čremošnik, I. "Iskopavanje crkvine u Zgošći," *GZMS*, n.s. IV-V, 1950, pp. 411-16.
Čremošnik, I. "Značaj arheologije u istraživanju srednjeg vijeka," *Pregled*, V, 1953, pp. 85-86.
Čremošnik, I. "Izveštaj o iskopinama u Rogačićima kod Blažuja," *GZMS*, n.s. VIII, 1953, pp. 303-15.
Čremošnik, I. "Dva srednjevjekovna grada u okolici Grahova," *GZMS*, n.s. VIII, 1953, pp. 349-51.
Čremošnik, I. "Izveštaj o iskopavanjima na crkvini u Lisičićima kod Konjica," *GZMS*, n.s. IX, 1954, pp. 211-23.
Čremošnik, I. "Novi srednjevjekovni nalazi kod Prijedora," *GZMS*(arh), n.s. X, 1955, pp. 137-45.
Čremošnik, I. "Crkvina u Golubiću," *GZMS*(arh), n.s. XI, 1956, pp. 127-34.
Čremošnik, I. "Narodna simbolika na rimskim spomenicima u našim krajevima," *GZMS* (arh), n.s. XII, 1957, pp. 217-34.
Ćorović, V. "Jedan srpski natpis iz XII vijeka," *GZMS*, XXIII, 1911, pp. 549-51.
Ćorović, V. "Izvještaj o iskopavanjima u Brezi 1913 godine," *GZMS*, XXV, 1913, pp. 409-20.
Ćorović, V. "Obnova manastira Banje," *Starinar*, n.s. IV, 1926-27, pp. 223-33.
Ćorović, V. "Iskopavanja u Petrovom manastiru kod Trebinja," *JIČ*, I (3-4), 1935, pp. 750-51.
Ćorović-Ljubinković, M. "Nekropole i grobni belezi," *Gradja*, IX, Arheološki Institut, II, Arheološki spomenici i nalazišta u Srbiji I, Zapadna Srbija, Beograd, 1953, pp. 169-98.
Delić, S. "Dva stara natpisa iz Hercegovine," *GZMS*, XXIII, 1911, pp. 485-502.
Dragičević, T. "Starobosanski natpis iz Vlasenice," *GZMS*, IV, 1892, pp. 248-49 (with V. Vuletić-Vukasović).
Dvorniković, V. "Tri starinska groblja u Studeničkom kraju," *GEMB*, XVII, 1954, pp. 5-39.
Filipović, M. "Natpis na stećku u Tičićima," *GZMS*, XXXVIII, 1926, pp. 79-80.
Filipović, M. "Starine u Bakićima kod Olova," *GZMS*, XL, No. 2, 1928, pp. 69-78.
Filipović, M. "Nadgrobni spomenik iz Puste reke iz 14 v," *IČ*, I (1-2), 1948, p. 254.
Filipović, M. "Starinska groblja u predelu Takovo — prilog proučavanju

stećaka," *NS*, VIII, 1962, pp. 153-62.
Hörmann, K. "Starobosanski natpis iz XV vijeka," *GZMS*, III, 1891, pp. 48-53.
Hőrmann, K. "Ošanić kod Stoca," *GZMS*, IV, 1892, pp. 40-49 (with V. Radimsky).
Horvat, A. "O stećcima na području Hrvatske," *Historijski zbornik*, IV, 1951, pp. 157-62.
Ivanović, V. "Srednjovjekovni nadgrobni spomenici u Podrinju," *GEMB*, XVII, 1954, pp. 221-71.
Jurišić, K. "Stari hrvatski natpisi Makarskog primorja (XV-XVIII st.)," *Starine*, LIII, 1966, pp. 89-133.
Kajmaković, Z. "Značajniji likovni i arhitektonski spomenici u dolini Trebišnjice," *NS*, VIII, 1962, pp. 39-51.
Kajmaković, Z. "Živopis u Dobrunu," *Starinar*, XIII-XIV, 1962-63, pp. 251-60.
Kajmaković, Z. "Natpisi i krstače sa Gatačkog polja," *NS*, IX, 1964, pp. 145-65.
Kajmaković, Z. "Jedno starije kultno mjesto," *NS*, XI, 1967, pp. 175-80.
Katić, L. "Stećci u Imotskoj Krajini," *Starohrvatska prosvjeta*, Ser. III, sv. 3, 1954, pp. 131-69.
Korošec, J. *Uvod u materijalno kulturo Slovanov zgodnjeg srednjega veka*, Ljubljana, 1952.
Korošec, P. "Srednjevjekovne nekropole okoline Travnika," *GZMS*, n.s. VII, 1952, pp. 375-407.
Krajinović, J. "Tri revidirana nadpisa iz okolice Travnika," *GZMS*, LV, 1943, pp. 227-36.
Mandić, M. "Patarensko groblje pod Osmačom planinom," *GZMS*, XIV, 1902, pp. 560-62.
Mandić, M. "Turbe kod Travnika," *GZMS*, XXXVI, 1924, pp. 83-90.
Mandić, M. "Starine kod Fojničkog Kiseljaka," *GZMS*, XXXVII, 1925, pp. 61-66.
Mandić, M. "Arheološke crtice iz Bosne," *Starinar*, n.s. III-IV, 1928, pp. 9-13.
Mazalić, Dj. "Starine po okolini Sarajeva," *GZMS*, LI, No. 1, 1939, pp. 15-35.
Mazalić, Dj. "Hercegova crkva kod Goražda," *GZMS*, LII, No. 1, 1940, pp. 27-43.
Mazalić, Dj. "Borač, Bosanski dvor srednjeg vieka," *GZMS*, LIII, 1941, pp. 31-94.
Mazalić, Dj. "Starine u Dobrunu," *GZMS*, LIII, 1941, pp. 101-123.
Mazalić, Dj. "Gradac kod Hadžića," *GZMS*, LIV, 1942, pp. 193-206.
Mazalić, Dj. "Stara crkva u Sarajevu," *GZMS*, LIV, 1942, pp. 241-70.
Mazalić, Dj. "Travnik i Toričan," *GZMS*, n.s. III, 1948, pp. 145-66.
Mazalić, Dj. "Semizovac i okolina," *GZMS*, n.s. IV-V, 1950, pp. 403-10.
Mazalić, Dj. "Stari grad Jajce," *GZMS*, n.s. VII, 1952, pp. 59-100.
Mazalić, Dj. "Tešanj," *GZMS*, n.s. VIII, 1953, pp. 289-303.
Mazalić, Dj. "Visoki,bosanski grad srednjeg vjeka," *GZMS*, n.s. IX, 1954, pp. 227-53.
Mazalić, Dj. "Zvonik (Zvornik) stari grad na Drini," *GZMS* (ist. etn), n.s. X, 1955, pp. 73-116; XI, 1956, pp. 243-277.
Mazalić, Dj. "Hrišćanski nišani u okolini Travnika - evidencija važnijih

spomenika i predlozi za zaštitu,'' *NS*, IV, 1957, pp. 97-118.
Miletić, N. ''Crkva sv. Klimenta u Mostaćima,'' *GZMS*, n.s. IX, 1954, pp. 281-96.
Miletić, N. ''Nekropole u selu Mihaljevićima kod Rajlovca,'' *GZMS* (arh), n.s. XV-XVI, 1961, pp. 249-57.
Miletić, N. ''Izveštaj o zaštitnom iskopavanju u Potocima kod Mostara,'' *GZMS* (arh), n.s. XVII, 1962, pp. 153-57.
Palavestra, V. ''Srednjovjekovni nadgrobni spomenici u Žepi,'' *Radovi*, NDBH, XXIV, 1964, pp. 139-180 (with M. Petrić).
Palavestra, V. ''Nekropola 'Deminov krst' u Burmazi kod Stoca,'' *NS*, X, 1965, pp. 195-201.
Radimsky, V. *Arheološki leksikon*, unpublished ms. kept at Zavod za zaštitu spomenika kulture in Sarajevo, existing in a German handwritten edition and a typed Serbian translation. It was written in the last decade of the nineteenth century.
Radimsky, V. ''O materijalu bogumilskih stećaka,'' *GZMS*, I, No. 1, 1889, pp. 59-60.
Radimsky, V. ''Bišće-polje kod Mostara,'' *GZMS*, III, 1891, pp. 159-92.
Radimsky, V. ''Ostaci rimskih naseobina u Šipragi i Podbrgju, za tim starobosanski stećci u Šipragi i uz Vrbanju u Bosni,'' *GZMS*, IV, 1892, pp. 75-80.
Radimsky, V. ''Arheološke crtice,'' *GZMS*, IV, 1892, pp. 117-25, 221-22.
Radimsky, V. ''Crkvena razvalina kod Dabravine u kotaru Visočkom u Bosni,'' *GZMS*, IV, 1892, pp. 372-87.
Radimsky, V. ''Razvaline crkve na Carevoj luci kod Rmanja u Bosni,'' *GZMS*, V, 1893, p. 487.
Sergejevski, D. ''Srednjevjekovna groblja u Stuparima i Rastiku,'' *GZMS*, LIII, 1941, pp. 95-99.
Sergejevski, D. ''Putne bilješke iz Glamoča,'' *GZMS*, LIV, 1942, pp. 113-71.
Sergejevski, D. ''Arheološki nalazi u Sarajevu i okolici,'' *GZMS*, n.s. II, 1947, pp. 13-50.
Sergejevski, D. ''Putne bilješke sa Nevesinjskog polja,'' *GZMS*, n.s. III, 1948, pp. 239-50.
Sergejevski, D. ''Japodske urne,'' *GZMS*, IV-V, 1950, pp. 44-93.
Sergejevski, D. *Ludmer*, SNSBH, 4, Sarajevo, 1952.
Sergejevski, D. ''Staro-hrišćanska bazilika u Klobuku,'' *GZMS*, n.s. IX, 1954, pp. 189-207.
Skarić, V. ''Grobni natpis braće Radilović u Čadovini,'' *GZMS*, XXXIX, 1927, pp. 193-96.
Skarić, V. ''Grob i grobni spomenik Gosta Milutina na Humskom u Fočanskom srezu,'' *GZMS*, (ist-etn) XLVI, 1934, pp. 79-82.
Slijepčević, P. ''Staro groblje po Gacku,'' *GZMS*, XL, No. 2, 1928, pp. 57-63.
Truhelka, Ć. ''Dva heraldička spomenika iz Bosne,'' *GZMS*, I, No. 2, 1889, pp. 73-76.
Truhelka, Ć. ''Bosančica,'' *GZMS*, I, No. 4, 1889, pp. 65-83.
Truhelka, Ć. ''Starobosanski mramori,'' *GZMS*, III, 1891, pp. 368-87.
Truhelka, Ć. ''Stari bosanski natpisi,'' *GZMS*, III, 1891, pp. 86-95.
Truhelka, Ć. ''Katakombe u Jajcu,'' *GZMS*, IV, 1892, pp. 57-68.
Truhelka, Ć. ''Nekoliko hercegovačkih natpisa,'' *GZMS*, IV, 1892, pp. 24-32.

Truhelka, Ć. "Stari hercegovački natpisi," *GZMS,* IV, 1892, pp. 107-16, 215-20; V, 1893, pp. 93-97.
Truhelka, Ć. "Iskopine u dolini Lašve 1893," *GZMS,* V, 1893, pp. 685-89.
Truhelka, Ć. "Starobosanski pismeni spomenici," *GZMS,* VI, 1894, pp. 771-82.
Truhelka, Ć. "Natpisi iz Sjeverne i Istočne Bosne," *GZMS,* VII, 1895, pp. 337-56.
Truhelka, Ć. "Starobosanski natpisi," *GZMS,* VII, 1895, pp. 259-82, 567-71.
Truhelka, Ć. "Stari natpisi iz Hercegovine," *GZMS,* VIII, 1896, pp. 325-28.
Truhelka, Ć. "Natpis Kulina Bana," *GZMS,* X, 1898, pp. 617-22.
Truhelka, Ć. "Grobnica bosanskog Tepčije Batala, obretena kod Gornjeg Turbeta (Kotor Travnik)," *GZMS,* XXVII, 1915, pp. 365-73.
Vego, M. "Srednjevjekovni bihaćki latinski spomenici XVI vjeka," *GZMS,* n.s. IX, 1954, pp. 255-69.
Vego, M. *Ljubuški,* SNSBH, 6, Sarajevo, 1954.
Vego, M. "Nadgrobni spomenici porodice Sankovića u selu Biskupu kod Konjica," *GZMS*(arh), n.s. X, 1955, pp. 157-65; XII, 1957, pp. 127-39.
Vego, M. "Humačka ploča," *GZMS*(arh), n.s. XI, 1956, pp. 41-61.
Vego, M. "Ćirilski natpisi iz Hercegovine," *GZMS*(arh) n.s. XIII, 1958, pp. 169-77.
Vego, M. "Crkva u Razićima kod Konjica," *GZMS*(arh), n.s. XIII, 1958, pp. 159-67.
Vego, M. "Arheološko iskopavanje u Zavali," *GZMS*(arh), n.s. XIV, 1959, pp. 179-88.
Vego, M. "Novi i revidirani ćirilski natpisi iz župe Broćno u Hercegovini," *GZMS* (arh), n.s. XIV, 1959, pp. 221-35.
Vego, M. "Arheološko iskopavanje u Viništima kod Konjica," *Arheološki pregled,* I, 1959, pp. 179-82.
Vego, M. "Novi i revidirani natpisi iz Hercegovine," *GZMS*(arh), n.s. XV-XVI, 1961, pp. 259-86; XVII, 1962, pp. 191-243; XIX, 1964, pp. 173-211.
Vego, M. *Zbornik srednjovjekovnih natpisa Bosne i Hercegovine,* I-IV, Sarajevo, 1962-70 (further volumes will be forthcoming)
Vidović, D. "Srednjevekovni nadgrobni spomenici u okolini Zvornika," *NS,* III, 1956, pp. 221-38.
Vuković, J. "Jedan stari bosanski nadgrobni spomenik i natpis," *GZMS,* n.s. II, 1947, pp. 51-68 (with A. Kučan).
Vuletić-Vukasović, V. "Drevni bosanski božak," *GZMS,* II, 1890, pp. 363-66.
Vuletić-Vukasović, V. "Stari manastirski pečat iz Duvna," *GZMS,* III, 1891, pp. 213-14.
Vuletić-Vukasović, V. "Starobosanski natpis u Kalesiji," *GZMS,* III, 1891, pp. 193-95.

II: SECONDARY WORKS

Alphardery, P. "La Gnosticisme dans les Sectes Medievales Latines," *Revue d'Histoire et de Philosophie Religieuses,* VII, 1927, pp. 395-411.

Alphardery, P. "Traces de Manichéisme dans le Moyen Age Latin (VI-XII siècles)," *Revue d'Histoire et de Philosophie Religieuses,* IX, 1929, pp. 451-67.
Andjelić, P. "Srednjovjekovna kultna mjesta u okolini Konjica," *GZMS*(arh), n.s. XII, 1957, pp. 185-99.
Andjelić, P. "Srednjovjekovni gradovi u Neretvi," *GZMS* (arh), n.s. XIII, 1958, pp. pp. 179-229.
Andjelić, P. "Trgovište, varoš, i grad u Srednjovjekovnoj Bosni," *GZMS* (arh), n.s. XVIII, 1963, pp. 179-193.
Andjelić, P. "Neka pitanja bosanske heraldike," *GZMS* (arh), n.s. XIX, 1964, pp. 157-72.
Andjelić, P. "Doba srednjovjekovne bosanske države," in A. Benac (et al.), ed., *Kulturna Istorija* . . ., pp. 405-536.
Andjelić, P. "Originalni dijelovi dviju bosanskih povelja u falsifikatima Ivana Tomke Marnavića," *GZMS,* n.s. XXVI (arh), 1971, pp. 347-60.
Andjelić, P. "Stara bosanska župa Vidogošća ili Vogošća," *GZMS,* n.s. XXVI (arh), 1971, pp. 332-46.
Andonovski, H. "O Dragovitskim Bogomilima Koste Racina," *Pregled,* VI, No. 2, 1954, pp. 108-9.
Angelov, D. *Bogomilstvoto v B'lgarija*(2nd ed.), Sofija, 1961.
Angelov, D. Review of Obolensky's *The Bogomils,* in *Byzantinoslavica,* X, 1949, pp. 303-12.
Angelov, D. "Le Mouvement Bogomile dans les pays Balkaniques et son Influence en Europe Occidentale," *Actes du Colloque International de Civilisations Balkaniques* organized by the Commission nationale Roumaine pour UNESCO, 1962, pp. 173-82.
Angelov, D. "Le Mouvement Bogomile dans les pays Slaves Balkaniques et dans Byzance," *Accademia Nazionale dei Lincei, Problemi Attuali di Scienza e di Cultura,* Rome, 1964, pp. 607-16.
Aničkov, E. "Les Survivances Manichéennes en Pays Slave et en Occident," *Revue des Études Slaves,* VIII, 1928, pp. 203-25.
Aničkov, E. "Maniheji i Bogumili," *Glasnik Skopskog naučnog društva,* V, 1929, pp. 137-56.
Apocryphal Tale, "O prepiranju Isusa Hrista sa djavolom," 2 texts in *Starine,* XVI, 1884, pp. 86-89; and XXII, 1890, pp. 200-03.
Asbóth, J. *An Official Tour Through Bosnia and Herzegovina,* London, 1890.
Atom (Pseudonym). "Kakva je bila srednjevekovna 'crkva bosanska' 'vjera bosanska,' " *Bosanska vila,* VII, 1892, and IX, 1894.
Babić, A. "Nešto o karakteru bosanske feudalne države," *Pregled,* V, No. 2, 1953, pp. 83-86.
Babić, A. "Noviji pregled u nauci o pitanju srednjevjekovne crkve bosanske," *Pregled,* VI, No. 2, 1954, pp. 101-07.
Babić, A. "O odnosima vazaliteta u srednjevjekovnoj Bosni," *GID,* VI, 1954, pp. 29-44.
Babić, A. "O pitanju formiranja srednjevekovne bosanske države," *Radovi,* NDBH, III, 1955, pp. 57-79.
Babić, A. "Diplomatska služba u srednjevjekovnoj Bosni," *Radovi,* NDBH, XIII, 1960, pp. 11-70.
Babić, A. "Struktura srednjovjekovne bosanske države," *Pregled,* XIII, No. 1,

1961, pp. 1-8.
Babić, A. *Bosanski heretici,* Sarajevo, 1963.
Babić, A. "Fragment iz kulturnog života srednjovjekovne Bosne," *Radovi,* FFS, II, 1964, pp. 325-36.
Bagby, P. *Culture and History,* Berkeley, 1958.
Bamberger, B. *Fallen Angels,* Philadelphia, 1952.
Bakula, P. *Schematismus topographico-historicus Custodiae Provincialis et Vicariatus Apostolici in Hercegovina,* Split, 1867.
Banović, S. "O porijeklu slave krsnog imena," *GZMS,* XXIV, 1912, pp. 265-73.
Banović, S. "Mašta prema istini u našim narodnim pjesama," *ZNŽOJS,* XXVI, 1928, pp. 193-256.
Banović, S. "Praznovjerica našeg naroda o prekopavanju starih grobova," *ZNŽOJS,* XXIX, 1933, pp. 88-89.
Barada, M. "Šidakov problem 'bosanske crkve,' " *Nastavni vjesnik,* XLIX, 1940-41, pp. 398-411.
Barada, M. "Osvrt na odgovor J. Šidaka," *Nastavni vjesnik,* L, No. 2, 1941, pp. 17-20.
Baring-Gould, S. *Curious Myths of the Middle Ages* (2nd Ser.), London, 1868.
Barišić, N. "Franjevački samostan i crkva u Sutjeskoj," *GZMS,* II, 1890, pp. 28-38.
Barišić, N. "Posušje," *GZMS,* IV, 1892, pp. 276-77.
Barišić, N. "Tihaljina i njezine znamenitosti," *GZMS,* V, 1893, pp. 558-60.
Barkan, Ö.L. "Essai sur les Données Statistiques des Registres de Recensement dans l'Empire Ottoman auz XVe et XVIe siècles," *Journal of Economic and Social History of the Orient,* I, 1, 1957, pp. 9-36.
Baroja, J.C. *The World of Witches,* Chicago, 1964.
Bašagić, S. *Kratka uputa u prošlosti Bosne i Hercegovine od 1463-1850,* Sarajevo, 1900.
Bašagić, S. "Bošnjaci i Hercegovci u islamskoj književnosti," *GZMS,* XXIV, 1912, pp. 1-88.
Basler, Dj. "Kasnoantičko doba," in A. Benac (et al.), ed., *Kulturna Istorija . .* , pp. 303-375.
Batinić, M. *Djelovanje franjevaca u Bosni i Hercegovini,* I, Zagreb, 1881; II, Zagreb, 1883.
Batinić, M. "Nekoliko priloga k bosanskoj crkvenoj poviesti," *Starine,* XVII, 1885, pp. 77-150.
Batinić, M. *Franjevački samostan u Fojnici od stoljeća XIV-XX,* Zagreb, 1913.
Benac, A. "Jedan značajan problem naše Istoriografije," *Pregled,* V, No. 4, 1953, pp. 285-87.
Benac, A., et al., editors, *Kulturna istorija Bosne i Hercegovine,* Sarajevo, 1966.
Bjelovučić, N. *Povijest poluotoka Rata (Pelješca),* Split. 1921.
Bjelovučić, N. *Bogomilska vjera od XII-XVI v,* Dubrovnik, 1926.
Bodenstein, G. "Tri turske listine Monastira Duža," *GZMS,* XX, 1908, pp. 113-116.
Bogičević, V. "Vlasteoska porodica Miloradovića Hrabrenih u Hercegovini," *GZMS* (arh), n.s. VII, 1952, pp. 59-100.
Bonfitto, P. "Eretici in Piemonte al tempo del gran scisma (1378-1412),"

Studi e documenti di Storia e diritto, XVIII, 1897, pp. 381-431.
Borst, A. *Die Katharer*, Stuttgart, 1953. (Schriften den Monumenta Germaniae Historica Deutsches Institut für Erforschung des Mittelalters, 12.)
Borst, A. "Neue Funde und Forschungen zur Geschichte der Katharer," *Historische Zeitschrift* 174, 1952, pp. 17-30.
Bošković, Dj. "Prilog proučavanju Tračkog konjanika," *Starinar*, n.s. VII-VIII, 1956-57, pp. 159-62.
Božić, I. "O jurisdikciji kotorske dijeceze u srednjevekovnoj Srbiji," *Spomenik* (SAN), 103, 1953, pp. 11-16.
Božić, I. "Mlečani prema naslednicima Hercega Stevana," *ZRFFB*, VI, No. 2, 1962, pp. 113-128.
Božičević, J. "Postanak jedne priče o vragu," *ZNŽOJS*, XXIII, 1918, pp. 1-4.
Božitković, J. *Kritički ispit popisa bosanskih vikara i provincijala 1339-1735*, Beograd, 1935.
Brajković, C. "Vatikanska politika u prošlosti prema Bosni i Hercegovini," *Pregled*, V, 1953, pp. 280-85.
Brandt, M. "Susret Viklifizma s Bogomilstvom u Srijemu," *Starohrvatska Prosvjeta*, III Ser., sv. 5, 1956, pp. 33-64.
Brandt, M. "Wyclifitism in Dalmatia in 1383," *Slavonic and East European Review*, XXXVI, No. 86, 1957, pp. 58-68.
Brandt, M. "Utjecaji patristike u ranom Bogomilstvu i Islamu," *Rad* (JAZU), 330, 1962, pp. 57-83.
Brlek, M. *Rukopisi knjižnice Male Braće u Dubrovniku*, I, Zagreb (JAZU), 1952. Further volumes never published.
Bru, C. "Notes en vue d'une interpretation sociologique de Catharisme Occitan," *Annales Institut d'Études Occitanes*, XII, 1952, pp. 57-80.
Brückner (Brikner), A. "O paganstvu kod starih Slovena," *Knjiga o Balkanu*, II, Beograd, 1937, pp. 51-61.
Buconjić, N. *Život i običaji Hrvata katoličke vjere u Bosni i Hercegovini*, Sarajevo, 1908.
Budimir, M. Review of Ante Kučan, "Jedan stari balkanski nadgrobni spomenik i natpis," *IČ*, I (1-2), 1948, pp. 261-63.
Budimir, M. "Trikleti babuni i babice patarenske," *GID*, X, 1958, pp. 73-86.
Bušetić, T. "Verovanja o djavolu u okrugu Moravskom," SEŽ, 32, Beograd, 1925, pp. 400-05.

The letter C, shall be followed by Č and then by Ć.

Campbell, J., (ed.). *The Mysteries*, Bollingen Series XXX, Papers from the Eranos Yearbooks, 1942.
Campbell, J. "Primitive Man as Metaphysician," in S. Diamond, *Primitive Views . . .*, pp. 20-32.
Campbell, J., (ed.). *Pagan and Christian Mysteries*, New York, 1955.
Cannon, W. *History of Christianity in the Middle Ages*, New York, 1960.
Carcopino, J. "Les symbolismes de l'Ascia," *Actes du Ve Congres International d'Archéologique Chrétienne*, Rome-Paris, 1957, pp. 551-66.
Cermanović-Kuzmanović, A. "Kult i spomenici tračkog herosa u balkanskim zemljama," *ZRFFB*, VI, No. 2, 1962, pp. 71-91.

Serbian ch's (Č or Ć) shall follow all the c's.

Challet, J. "Bogumili i simbolika stećaka," *NS*, X, 1965, pp. 19-37.
Charles, R. *A Critical and Exegetical Commentary on the Revelation of St. John*, New York, 1920.
Chaumette-des-Fossés, M. A. *Voyage en Bosnie dans les années 1807 et 1808*, Paris, 1816.
Clark, J. *The Dance of Death*, Glasgow, 1950.
Coulton, G. *Medieval Village, Manor and Monastery* (1925), ed., New York, 1960.
Coulton, G. *Medieval Faith and Symbolism*, Oxford, 1928.
Coulton, G. *Life in the Middle Ages*, Cambridge (Eng.), 2nd ed., 1928.
Coulton, G. *Inquisition and Liberty*, London, 1938.
Cvijić, J. "Studies in Jugoslav Psychology," *Slavonic Review*, IX, 1930-31, pp. 375-90, 662-81.
Čajkanović, S. "Srpska pravoslavna crkva Svetih Arhangela u Sarajevu," *GZMS*, I, No. 4, 1889, pp. 1-11.
Čajkanović, V. "Nekolike opšte pojave u staroj srpskoj religiji," *GNČ*, XLI, 1932, pp. 167-228.
Čajkanović, V. "Starinska religija u našim dnevnim listovima," *GNČ*, XLVIII, 1939, pp. 164-83.
Čajkanović, V. *O srpskom vrhovnom bogu*, Beograd (SKA, Pos. Izd. 132), 1941.
Čarić, A. "Rudica kćer Poganica," *GZMS*, IX, 1897, pp. 329-31.
Čremošnik, G. "Prodaja bosanskog primorja Dubrovniku god. 1399 i kralj Ostoja," *GZMS*, XL, No. 2, 1928, pp. 109-26.
Čremošnik, G. "Oko Bogomilstva u srednjevekovnoj Bosni," *Prosveta*, XXI, 1937, pp. 10-16.
Čremošnik, G. "Pravni položaj našeg roblja u srednjem veku," *GZMS*, n.s. II, 1947, pp. 69-73.
Čremošnik, G. "Bosanske i humske povelje srednjega vijeka," *GZMS*, n.s. III, 1948, pp. 103-43 and *GZMS*, n.s. X, 1955, pp. 137-46.
Čremošnik, G. "Zadnja raziskovanja o bosanskih Bogomilih" (1952 oral paper) printed in *Zgodovinski časopis*, 1965, pp. 229-36; with an addenda bringing it up to date by B. Grafenauer, pp. 236-38.
Črnja, Z. *Cultural History of Croatia*, Zagreb, 1962.
Čubrilović, V. "Poreklo muslimanskog plemstva u Bosni i Hercegovini," *JIČ*, I (1-4), 1935, pp. 368-403.
Čubrilović, V. "Srpska pravoslavna crkva pod Turcima od XV do XIX veka," *ZRFFB*, V, No. 1, 1960, pp. 163-88.
Ćatić, A. "Božić kod Muslimana (Prozor u Bosni)," *ZNŽOJS*, XXVI, sv. 2, 1928, pp. 379-80.
Ćirković, S. "Dve godine bosanske istorije 1414-1415," *IG* (3-4), 1953, pp. 29-41.
Ćirković, S. "Vlastela i kraljevi u Bosni posle 1463," *IG* (3), 1954, pp. 123-31.
Ćirković, S. "Ostaci starije društvene strukture u bosanskom feudalnom društvu," *IG* (3-4), 1958, pp. 155-64.
Ćirković, S. "Jedan prilog o banu Kulinu," *IČ*, IX-X, 1959, pp. 71-77.
Ćirković, S. " 'Verna služba' i 'vjera gospodska,' " *ZRFFB*, VI, No. 2, 1962, pp. 95-111.
Ćirković, S. "O Djakovačkom ugovoru," *IG* (1-4), 1962, pp. 3-10.

Ćirković, S. "Vesti Brolja da Lavelo kao izvor za istoriju Bosne i Dubrovnika," *IČ*, XII-XIII, 1963, pp. 167-88.
Ćirković, S. *Istorija srednjovekovne bosanske države*, Beograd, 1964.
Ćirković, S. "Die Bosnische Kirche," *Accademia Nazionale dei Lincei, Problemi Attuali di Scienza e di Cultura*, Rome, 1964, pp. 547-75.
Ćirković, S. *Herceg Stefan Vukčić-Kosača i njegovo doba*, Beograd (SAN pos. izd. 376), 1964.
Ćirković, S. Review of A. Babić's *Bosanski heretici* in *GID*, XIV, 1964, pp. 288-92.
Ćirković, S. "Sugubi venac," *ZRFFB*, VIII, 1964, pp. 343-69.
Ćorović, V. "Hercegovački manastiri," *GZMS*, XXIII, 1911, pp. 505-33; XXIV, 1912, pp. 545-54.
Ćorović, V. "Ban Kulin," *GNČ*, XXXIV, 1921, pp. 13-41.
Ćorović, V. "Srebrnik za vlade Despota Stevana (1413-27)," *Prilozi*, II, 1922, pp. 61-77.
Ćorović, V. "Hercegovački manastiri," *Starinar*, n.s. I, 1922, pp. 209-30; II, 1923, pp. 69-77.
Ćorović, V. "Sandalj Hranić u Dubrovniku 1426," *Bratstvo*, XVII, 1923, pp. 102-07.
Ćorović, V. "Jedan novi izvor za srpsku historiju iz početka XIV veka," *Prilozi*, IV, 1924, pp. 67-74.
Ćorović, V. *Kralj Tvrtko I Kotromanić*, Beograd (pos. izd. SKA, LVI), 1925.
Ćorović, V. *Mostar i njegova srpska pravoslavna opština*, Beograd, 1933.
Ćorović, V. "Teritorialni razvoj bosanske države u srednjem veku," *Glas* (SKA), 167, 1935, pp. 5-47.
Ćorović, V. "Iz prošlosti Bosne i Hercegovine," *GNČ*, XLVIII, 1939, pp. 133-45.
Ćorović, V. *Historija Bosne*, Beograd (Pos. izd. SKA), 1940.
Ćorović, V. "Prilog proučavanju načina sahranjivanja i podizanja nadgrobnih spomenika u našim krajevima u srednjem vijeku," *NS*, III, 1956, pp. 127-47. This unfinished article was found among his papers after his death.
Ćurčić, V. "Starinsko oružje u Bosni i Hercegovini," *GZMS*, LV, 1943, pp. 21-225.

Serbian Dj's (or Đ) shall follow all the D's.

Davidović, S. *Srpska pravoslavna crkva u Bosni i Hercegovini od 960 do 1930*, Sarajevo, 1930.
Dawkins, R. M. "The Crypto-Christians of Turkey," *Byzantion*, VIII, 1933, pp. 247-75.
Dedijer, J. *Hercegovina*, Beograd, 1908. SEZ, XII (SKA, Naselja Srp. zem. 6).
Dedijer, J. "Porijeklo bosansko-hercegovačkog stanovništva," *Pregled*, I (7-8), 1911, pp. 420-30.
Delić, S. "Samobor kod Drine," *GZMS*, IV, 1892, pp. 255-69.
Delić, S. "Kod Djavola babica," *GZMS*, V, 1893, pp. 352-53.
Delić, S. "Gatačko polje," *Školski vjesnik*, XI, 1904, pp. 549-554, 655-665; XIII, 1906, pp. 162 ff.
Delić, S. "Petrov manastir kod Trebinja," *GZMS*, XXIV, 1912, pp. 275-82; XXV, 1913, pp. 129-32.
Delmas-Boussagol, B. "Les monuments funeraires Bogomiles à l'Exposition,"

CEC, II, No. 5, spring, 1950, pp. 31-37.
Deroko, A. "Crkva sv. Apostola Petra u Bijelom polju," *Glasnik Skopskog naučnog društva,* VII-VIII, 1929-30, pp. 141-46.
Diamond, S. (ed.). *Primitive Views of the World,* New York, 1964.
Dinić, M. "Jovan Angel 'dominus Syrmiae,' " *Glasnik Istoriskog društva u Novom sadu,* IV, 1931, p. 301.
Dinić, M. *O Nikoli Altomanoviću,* Beograd (pos. izd., SKA, 90), 1932.
Dinić, M. "O krunisanju Tvrtka prvog za kralja," *Glas* (SKA), CXLVII, 1932, pp. 135-45.
Dinić, M. "Srebrnik kraj Srebrnice," *Glas* (SKA), 161, 1934, pp. 185-96.
Dinić, M. "Vesti Eberharda Vindekea o Bosni," *JIČ,* I (3-4), 1935, pp. 352-67.
Dinić, M. "Dubrovački tributi," *Glas*(SKA), 168, 1935, pp. 205-57.
Dinić, M. "Dubrovačka srednjevekovna karavanska trgovina," *JIČ,* III, 1937, pp. 117-46.
Dinić, M. "Trg Drijeva i okolina u srednjem veku," *GNČ,* XLVII, 1938, pp. 109-47.
Dinić, M. "Zemlje Hercega Svetoga Save," *Glas* (SKA), 182, 1940, pp. 151-257.
Dinić, M. "Jedan prilog za istoriju Patarena u Bosni," *ZRFFB,* I, 1948, pp. 33-44.
Dinić, M. "Bosanska feudalna država od XII do XV veka," in *Istorija naroda Jugoslavije,* I, Beograd, 1953, pp. 515-64.
Dinić, M. *Državni sabor srednjevekovne Bosne,* Beograd (pos. izd. SAN 231), 1955.
Dinić, M. *Za istoriju rudarstva u srednjevekovnoj Srbiji i Bosni,* I deo, Beograd (pos. izd., SAN 240), 1955; II deo, Beograd (pos. izd., SAN, 355), 1962.
Dinić, M. "Odnos izmedju kralja Milutina i Dragutina," *ZRVI,* III, 1955, pp. 49-80.
Dinić, M. "Dubrovčani kao feudalci u Srbiji i Bosni," *IČ,* IX-X, 1959, pp. 139-48.
Dinić, M. "Comes Constantinus," *ZRVI,* VII, 1961, pp. 1-10.
Dinić, M. "Jedna dubrovačka arhivska knjiga XV veka," *IČ,* XII-XIII, 1961-62, pp. 15-30.
Dinić, M. "Relja Ohmučević — istorija i predanje," *ZRVI,* IX, 1966, pp. 95-118.
Dinić, M. *Humsko-Trebinjska vlastela,* Beograd (pos. izd. SAN, 397), 1967.

NB: Serbian Ð (or Dj) shall follow all the D's.

Döllinger, I. *Fables Respecting the Popes in the Middle Ages* and *Prophetic Spirit and the Prophecies of the Christian Era* (trans. by A. Plummer and H. Smith), New York, 1872.
Dollinger, I. *Beiträge zur Sektengeschichte des Mittelalters* (I, sources, II, his discussion), Munich, 1890.
Dondaine, A. See also under sources since each text he publishes is accompanied by a long critical and interpretative essay.
Dondaine, A. "Nouvelles sources de l'histoire doctrinale du néo-Manichéisme au Moyen âge," *Revue des Sciences et Théologie,* XXVIII, 1939, pp. 465-88.

Dondaine, A. "L'Origine de l'heresie Mediévale," *Rivista di Storia della Chiesa in Italia,* VI, 1952, pp. 47-78.
Dondaine, A. "St. Pierre martyr-Études," *AFP,* XXIII, 1953, pp. 66-162.
Dondaine, A. "Durand de Huesca et la polémique anti-Cathare," *AFP,* XXIX, 1959, pp. 228-76.
Draganović, K. See also under sources since he publishes several visitation texts and accompanies the texts with interesting and valuable essays.
Draganović, K. "Katolička crkva u Bosni i Hercegovini nekad i danas," *Croatia Sacra,* IV, 1934, pp. 175-216. At the same time he published individually a useful map entitled "pregledna karta katol. crk. u Bosni."
Draganović, K. "Katolička crkva u sredovječnoj Bosni," in *Poviest BH*-1942, pp. 685-766.
Draganović, K. "Katalog katoličkih župa XVII vieka u Bosni i Hercegovini," *Croatia Sacra,* XIII-XIV (no. 22-23), 1944, pp. 89-126.
Dragičević, P. "Narodne praznovjerice," *GZMS,* XIX, 1907, pp. 311-313.
Dragić, N. "Oboljenje, smrt i pogrebni običaji u okolini Tavne," *ZREI* (SAN), II, Beograd, 1951, pp. 119-128.
Dragomanov, M. *Notes on the Slavic Religio-Ethical Legends: The Dualistic Creation of the World* (written ca. 1890) trans. E. Count, Indiana University Publications-Russian and East European Series, V. 23, Bloomington, 1961.
Duda, P. Bonaventure. *Joannis Stojković de Ragusio O.P. (+1433) Doctrina de Cognoscibilitate Ecclesiae,* Rome, 1958.
Dujčev, I. "I Bogomili nei Paesi Slavi e la loro Storia," *Accademia Nazionale dei Lincei, Problemi Attuali di Scienza e di Cultura,* Rome, 1964, pp. 619-41.
Dujčev, I. " Dragvista-Dragovita," *Revue des Études Byzantines,* XXII, 1964, pp. 215-221.
Durham, M. E. *Some Tribal Origins, Laws and Customs of the Balkans,* London, 1928.
Duvernoy, J. "Un traité Cathare du début du XIII siècle," *CEC,* XIII, 2nd Ser. II, Spring 1962, pp. 22-54.
Djivanović, N. "Flagelanti u našem primorju," *Glasnik Skopskog naučnog društva,* VII-VIII, 1929-30, pp. 193-96.
Djordjević, T. *Naš narodni život,* I-VI, Beograd, 1930-32; of particular interest are the following studies: "Konj i oružje kao znaci položaja i starešinstva," I, pp. 129-36; "Svatovska groblja," II, pp. 106-16; "Preislamski ostaci medju Jugoslovenskim Muslimanima," VI, pp. 26-56.
Djukić, T. "Tri legende iz oblasti Komova," *Rad Kongresa Folklorista,* V. Zaječar, 1958, pp. 185-89.
Djurdjev, B. "Uloga srpske crkve u borbi protiv osmanske vlasti," *Pregled,* V, No. 1, 1953, pp. 35-52.
Djurdjev, B. "Srpska crkva pod turskom vlašću do 1557," *Historijski pregled,* IV, No. 1, 1958, pp. 20-28.
Djurdjev, B. *Uloga crkve u starijoj istoriji srpskog naroda,* Sarajevo, 1964.
Djurić, J. "Jevandjelje Divoša Tihoradića," *ZRVI* (SAN), VII, 1961, pp. 153-61 (with R. Ivanišević).
Djurić, V. "Minijature Hvalovog rukopisa," *IG*(1-2), 1957, pp. 39-52.
Elezović, G. "Turski izvori za istoriju Jugoslovena," *Bratstvo,* XXVI, 1932, pp. 51-125.

Elezović, G. "Je li kral' k'zi 'kultno' lice sa više grobova," *IČ*, I (1-2), 1948, pp. 297-300.
Elezović, G. "Iz putopisa Evlije Čelebije njegov put iz Beograda u Hercegovinu," *IČ*, I (1-2), 1948, pp. 105-31.
Eliade, M. *Patterns in Comparative Religion*, Cleveland, 1963.
Eterović, K. "U kojim su mjestima Franjevci Zaostroga vršili dušobrižničku službu u 16 i 18 vijeku," *Nova Revija* (Makarska), V, 1926, pp. 492-93.
Evans, Arthur. *Through Bosnia and Hercegovina on Foot*, London, 1876.
Evans, Athur. *Illyrian Letters*, London, 1878.
Evans, A. "Social Aspects of Medieval Heresy," in *Persecution and Liberty: Essays in Honor of G. L. Burr*, New York, 1931, pp. 93-116.
Evans-Pritchard, E. *Nuer Religion*, Oxford, 1956.
Evans-Pritchard, E. *Theories of Primitive Religion*, Oxford, 1965.
Faber, M. "Zur Entstehung von Farlati's 'Illyricum Sacrum,'" *Wissenschaftliche Mitteilungen aus Bosnien und der Herzegowina*, III, 1895, pp. 388-95.
Farlati, D. *Illyricum Sacrum*, I-IX, Venice, 1751-1819 (especially Vol. IV).
Ferdon, E. (jr.). "Agricultural Potential and the Development of Cultures," *Southwestern Journal of Anthropology*, XV, Spring, 1959, pp. 1-19.
Ferri, R. "Prilog poznavanju ilirske mitologije," *AHID*, II, 1953, pp. 419-28.
Fiala, F. "Crtice sa Glasinca," *GZMS*, IV, 1892, pp. 336-40.
Filipović, M. "Dubrovnik u Bosni," *GZMS*, XXXVI, 1924, pp. 101-04.
Filipović, M. "Manastir Udrim ili Gostović," *GZMS*, XXXVI, 1924, pp. 109-112.
Filipović, M. *Visočka nahija*, Beograd, 1928. SEZ, XLIII (SKA naselja i porek. stanovništva, 25).
Filipović, M. "Slava ili služba: (Visočki gornji kraj u Bosni)," *ZNŽOJS*, XXVI, sv. 2, 1928, pp. 329-41.
Filipović, M. "Varošica Olovo s okolinom," *Franjevački vijesnik*, 1934, pp. 231-47, 270-81, 301-12.
Filipović, M. "Tragovi Perunova kulta kod Južnih Slovena," *GZMS*, n.s. III, 1948, pp. 63-80.
Filipović, M. "Stare srpske knjige i rukopisi po Severo-istočnoj Bosni," *GZMS*, n.s. III, 1948, pp. 251-60.
Filipović, M. "Kršteni muslimani," *ZREI*(SAN), II, 1951, pp. 120-28.
Filipović, M. "Folk Religion among the Orthodox Population in Eastern Jugoslavia," *Harvard Slavic Studies*, II, 1954, pp. 359-74.
Filipović, M. "Društvene i običajno-pravne ustanove u Rami," *GZMS*, n.s. IX (ist-etn), 1954, pp. 169-80.
Filipović, M. "Još o tragovima Perunova kulta kod Južnih Slovena," *GZMS*, n.s. IX (ist-etn), 1954, pp. 181-82.
Filipović, M. *Rama u Bosni*, Beograd, 1955. SEZ (SAN), LXIX (naselja i porek. stanovništva, 35).
Filipović, M. "Beleške o narodnom životu i običajima na Glasincu," *GZMS*, n.s. X (ist-etn), 1955, pp. 117-36.
Filipović, M., ed. *Simpozijum o srednjovjekovnom Katunu*, Sarajevo, 1963 (pos. izd. NDBH, II).
Filipović, M. "Sklapanje i razvod hrišćanskih brakova pred kadijama u tursko doba," *Radovi* (NDBH), XX, 1963, pp. 185-95.
Filipović, M. "Studije o slavi, službi i krsnom imenu," *Zbornik za društvene nauke* (Matica Srpska - Novi Sad), XXXVIII, 1964, pp. 51-76.

Filipović, N. "Osvrt na položaj bosanskog seljaštva u prvoj deceniji uspostavljanja osmanske vlasti u Bosni," *Radovi,* FFS, III, 1965, pp. 63-75.

Filipović, N. "Napomene o Islamizaciji u Bosni i Hercegovini u XV Vijeku," *Godišnjak,* knj. VII, Centar za Balkanološka Ispitivanja, knj. 5, Sarajevo, 1970, pp. 141-67.

Fine, J., Jr. "Aristodios and Rastudije - A Re-examination of the Question," *GID,* XVI, 1965 (Sarajevo, 1967), pp. 223-229.

Fine, J., Jr. Review of S. Ćirković's *Istorija . . .*, in *Speculum,* XLI, 1966, pp. 526-29.

Fine, J., Jr., "The Dead Man's Hand and the Hand of Glory," *Narodno stvaralaštvo* (Beograd), VI (sv. 21), January, 1967, pp. 23-25 (Serbian translation, pp. 20-22).

Fine, J., Jr. "Was the Bosnian Banate Subjected to Hungary in the Second Half of the Thirteenth Century," *East European Quarterly,* III, no. 3, June, 1969, pp. 167-77.

Fine, J., Jr. Review of N. Garsoian, *The Paulician Heresy,* in *Speculum,* XLIV, no. 2, April 1969, pp. 284-88.

Fine, J., Jr. "Uloga Bosanske crkve u javnom životu srednjovekovne Bosne," *GID,* XIX, 1970-71 Sarajevo, 1973 , pp. 19-29.

Fine, J., Jr. "Mysteries about the Newly Discovered Srebrnica-Visoko Bishopric in Bosnia (1434-1441)," *East European Quarterly,* VIII, no. 1, March, 1974, pp. 29-43.

Firth, R. *Human Types,* New York, 1958.

Fliche, A. *Histoire de l'Église,* 20 vols., Paris, 1941- (with V. Martin).

Florinskij, T. D. "K voprosu o bogomilah," *Sbornik statej po slavjanovedeniju sostavlennyj i izdannyj učenikami v čest' V. I. Lamanskogo po slučaju 25-letija ego učenoj i professorskoj dejatel'nosti.* St. Petersburg, 1883, pp. 33-40.

Folz, R. "Le Catharisme d'après un livre recent" (Review of Borst, *Die Katharer . . .*), *Revue d'histoire et de philosophie religieuses,* XXXIII, 1953, pp. 382-88.

Franknoi, V. "Kardinal Carvajal u Bosni," *GZMS,* II, 1890, pp. 9-12.

Gabričević, B. "Sarajevski medaljon s prikazom tračkog konjanika," *GZMS,* n.s. IX, 1954, pp. 41-46.

Gaković, P. *Prilozi za istoriju i etnografiju Bosne - nestanak Bogumila, islamizacija Bosne, porijeklo Srba na sjevero-zapadu,* Sarajevo, 1933.

Gaković, P. "Testament gosta Radina," *Razvitak* (Banja Luka), VII, No. 5, 1940, pp. 329-333.

Gaković, P. "Putopis Benedikta Kuripešića," *Razvitak,* VII, No. 5, 1940, pp. 147-50.

Garsoian, N. *The Paulician Heresy,* The Hague, 1967.

Gaster, M. *Ilchester Lectures on Grecko-Slavonic Literature,* Oxford, 1887.

Gaster, M. *Rumanian Bird and Beast Stories,* London, 1915.

Gecić, M. "Prilog bosanskoj istoriji 1397-99," *IG,* (1-2), 1953, pp. 55-63.

Gil'ferding, A. *Bosnija putevyja zametki pis'mo k A.S. Homjakovu,* Moscow, 1859.

Glik, L. "Tetoviranje kože kod Katolika u Bosni i Hercegovini," *GZMS,* I, No. 3, 1889, pp. 81-88.

Glik, L. "O urocima," *GZMS,* I, No. 4, 1889, pp. 58-65.

Glušac, V. "Srednjovekovna 'bosanska crkva,' " *Prilozi*, IV, 1924, pp. 1-55.
Glušac, V. "Problem Bogomilstva," *GID*, V, 1953, pp. 105-38.
Gouillard, J. "L'Hérésie dans l'Empire Byzantin des Origines au XIIe siècle," *Travaux et Memoires*, I, 1965, pp. 299-324.
Grabar, A. *Recherches sur les influences Orientales dans l'Art Balkanique*, Paris, 1928.
Gračev, V. "Terminy 'župa' i 'župan' v Serbskih istočnika XII-XIV vv i traktovka ih v istoriografii," *Istočniki i istoriografija slavjanskogo srednevekov'ja* (AN Institut slavjanovedenija), Moscow, 1967, pp. 3-52.
Grant, R. *Gnosticism and Early Christianity* (rev. ed.), New York, 1966.
Grdjić-Bjelokocić, L. " 'Kozi-grad', 'zvoni grad', 'gradina' u kotaru fojničkom," *GZMS*, IV, 1892, pp. 91-94.
Grégoire, H. "Cathares d'Asie Mineure, d'Italie, de France," *Mémorial Louis Petit*, Mélanges d'histoire et d'Archéologie. Archives de l'Orient Chrétien. I, Paris, 1948, pp. 142-51.
Gruber, D. "Bertold Meranski Ban Hrvatski i Nadbiskup Kaločki," *VZA*, V, 1903, pp. 18-41.
Gruber, H. "Les Tombes Bogomiles et leur Significance," *CEC*, XIV, Spring 1963, pp. 31-40.
Grujić, M. "Crkveni elementi krsne slave," *Glasnik Skopskog naučnog društva*, VII-VIII, 1929-30, pp. 34-74.
Grujić, M. "Kosmološki problemi po našim starim rukopisima," *Godišnjak Skopskog filozofskog fakulteta*, I, 1930, pp. 177-204.
Grujić, M. Jedno jevandjelje bosanskoga tipa XIV-XV veka u južnoj Srbiji," in *Belićev Zbornik*, Beograd, 1937, pp. 263-77.
Grujić, R. "Konavli pod raznim gospodarima od XII do XV v," *Spomenik* (SKA), LXVI, 1926, pp. 3-121.
Grumel, V. *Les Regestes des Actes du Patriarcat de Constantinople*, I (1-3), Istanbul, 1932.
Grundmann, H. "Neue Beiträge zur Geschichte der Religiosen Bewegungen im Mittelalter," *Archiv für Kulturgeschichte*, XXXVII, 1955, pp. 129-82.
Gržetić-Gaspičev, N. *O vjeri starih Slovena*, I deo Mostar, 1900 (2nd part never published).
Guldescu, S. *History of Medieval Croatia*, The Hague, 1964.
Gunjača, Š. "Prinos poznavanju porijekla i načina prijevoza stećaka," *IČ*, V, 1954, pp. 140-47.
Hadžijahić, M. "O islamizaciji bosanskih krstjana," *Obzor*, Zagreb, 31 December 1937.
Hadžijahić, M. "Kaimija o Bogomilima," *Život*, 1952, No. 5, pp. 125-26.
Hallowell, A. "Ojibwa Ontology, Behavior and World View," in S. Diamond, *Primitive Views . . .*, pp. 49-82.
Hammel, E. *Alternative Social Structures and Ritual Relations in the Balkans*, Englewood Cliffs, New Jersey, 1968.
Hamp, E. P. "Notes on Medieval Inscriptions of Bosnia and Hercegovina," *Zbornik za filologiju i lingvistiku* (Novi Sad), XII, 1969, pp. 83-91.
Hand, W. "Status of European and American Legend Study," *Current Anthropology*, VI, No. 4, Oct. 1965, pp. 439-46.
Handžić, H. *Islamizacija Bosne i Hercegovine i porijeklo bosansko-hercegovačkih Muslimana*, Sarajevo, 1940.

Hardie, M. "Christian Survivals among Certain Moslem Subjects of Greece," *Contemporary Review* (London), CXXV, Jan.-June, 1924, pp. 225-32.
Haskins, C. H. "The Spread of Ideas in the Middle Ages," *Speculum*, I, 1926, pp. 19-30.
Hasluck, F. W. *Christianity and Islam under the Sultans*, 2 volumes, Oxford, 1929.
Heer, F. *The Middle Ages*, London, 1962.
Hefele, C. J. (Le Clerque), *Histoire des Conciles*, VII, pt. 2 (On Basel), Paris, 1916.
Hilmi, M. "Prilog istoriji Sarajeva," *GZMS*, I, No. 1, 1889, pp. 17-20.
Hofer, J. *Johannes von Capestrano*, Innsbruck, Vienna, Munich, 1936.
Hoffer, A. "Stara crkva na Bukovici kod Travnika," *GZMS*, XIII, 1901, p. 468.
Hoffer, A. "Dva odlomka iz povećeg rada o kršćanskoj crkvi u Bosni," *Spomen knjiga iz Bosne*, Zagreb, 1901, pp. 59-142.
Hofstadter, R. "The Paranoid Style in American Politics," in a collection of essays under the same title, Vintage pb. ed., New York, 1967, pp. 3-40.
Hörmann, K. "Olovo," *GZMS*, I, No. 3, 1889, pp. 63-74.
Hrabak, B. "Prilog datovanju hercegovačkih stećaka," *GZMS*, n.s. VIII, 1953, pp. 325-28.
Hrabak, B. "Uticaj primorskih privrednih centara na društveno-ekonomsku istoriju Bosne i Hercegovine u srednjem veku," *Pregled*, V, No. 5, 1953, pp. 382-85.
Hrabak, B. "Herak Vraneš," *GID*, VII, 1955, pp. 53-64.
Hrabak, B. "Prošlost Pljevlja po dubrovačkim dokumentima do polovine XVII v," *Istorijski zapisi*, VIII, 1955, pp. 9-12.
Hrabak, B. "O hercegovačkim vlaškim katunima prema poslovnoj knjizi Dubrovčanina Dživana Pripčinovića," GZMS, (ist. etn), n.s. XI, 1956, pp. 29-38.
Huizinga, J. *The Waning of the Middle Ages* (1924), New York, 1954.
Ilić, J. *Die Bogomilen in ihrer geschichtlichen Entwicklung*, Sremski Karlovci, 1923.
Inalcik, H. "Ottoman Methods of Conquest," *Studia Islamica*, II, 1954, pp. 103-30.
Ivančević, P. "Kula," *GZMS*, III, 1891, pp. 43-45.
Ivanka, E. "Gerardus Moresanus, der Erzbischof Uriel und die Bogomilen," *Orientalia Christiana Periodica*, XXI, 1955, 143-46.
Ivanov, J. *Bogomilski knigi i legendi* (BAN), Sofija, 1925.
Ivanović, R. "Srednjovekovni baštinski posedi humskog eparhiskog vlastelinstva," *IČ*, IX-X, 1959, pp. 79-95.
Ivić, A., "Kadje i od koga je Stepan Vukčić dobio titulu 'herceg od Sv. Save,' " *Letopis Matice srpske*, 230, 1905, pp. 80-94.
Ivić, A., "Kad je i od koga je Stepan Vukčić dobio titulu 'herceg od Sv. Save,' " *Letopis Matice srpske*, 230, 1905, pp. 80-94.
Jablanović, J. *Bogomilstvo ili Patarenstvo*, Mostar, 1936.
Jackson, A. "The Doctrine of Metempsychosis in Manichaeism," *Journal of the American Oriental Society*, XLV, 1925, pp. 246-68.
Jagić, V. "Historija književnosti naroda hrvatskoga i srpskoga" (written 1867), in *Djela Vatroslava Jagića*, IV, Zagreb, 1953, pp. 207-364.
Jagić, V. "Opisi i izvodi iz nekoliko južnoslovenskih rukopisa: sredovječni liekovi, gatanja i vračanja," *Starine*, X, 1878, pp. 81-126.

Jagić, V. "Nekoliko riječi o bosanskim natpisima na stećcima," *GZMS,* II, 1890, pp. 1-9.

Jagić, V. "Ein neu entdeckter urkundlicher Beitrag zur Erklärung des Bosnischen Patarenentums," *Archiv für slavische Philologie,* XXXIII, 1912, pp. 585-87.

Jagić, V. *Izabrani kraći spisi,* Zagreb, n.d. (clearly compiled after World War II).

James, E. *Christian Myth and Ritual*(1933), Cleveland, 1965.

Jelenić, J. "Križarska vojna proti bosanskim Patarenima," *Franjevački glasnik,* 1901, pp. 200-04.

Jelenić, J. *Kraljevsko Visoko i samostan sv. Nikole,* Sarajevo, 1906.

Jelenić, J. *Kultura i Bosanski Franjevci,* I, Sarajevo, 1912.

Jelenić, J. "Problem dolaska Franjevaca u Bosni i osnutka Bosanske vikarije," *Nove Revije*(Makarska), VII (No. 3-4), 1926, pp. 1-27.

Jireček, K. "Vlastela humska na natpisu u Veličanima," *GZMS,* IV, 1892, pp. 279-85.

Jireček, K. "Glasinac u srednjem vijeku," *GZMS,* IV, 1892, pp. 99-101.

Jireček, K. "Das Gesetzbuch den Serbischen Caren Stephan Dušan," *Archiv für slavische Philologie,* XXII, 1900, pp. 144-214.

Jireček, K. *Trgovački drumovi i rudnici Srbije i Bosne u srednjem vijeku* (1879), Sarajevo, 1951; also included in *Zbornik Konstantina Jirečeka,* Beograd, 1959 (SAN pos. izd. 326), pp. 205-304.

Jireček, K. *Istorija Srba,* 2 vols., Beograd, 1952, translated and revised by J. Radonić.

Joannou, P. "Les Croyances Démonologiques au XIe siècle à Byzance," *Actes du VIe Congrès International d'études Byzantines,* I, Paris, 1950, pp. 245-60.

Jokanović, V. "Prilog izučavanju bosanskog feudalizma," *Radovi* NDBH, II, Sarajevo, 1954, pp. 221-67.

Jonas, H. *The Gnostic Religion*(rev. ed.), Boston, 1963.

Jovanović, Lj. "Ratovanje Hercega Stjepana s Dubrovnikom 1451-54," *GNČ,* X, 1888, pp. 87-198.

Jovanović, Lj. "Stjepan Vukčić Kosača," *Glas*(SKA), XXVIII, 1891, pp. 1-88.

Jugie, M. "Phoundagiagites et Bogomiles," *Echos d'Orient,* XII, 1909, pp. 257-262.

Jung, L. *Fallen Angels in Jewish, Christian and Mohammedan Literature,* Philadelphia, 1926.

Kajmaković, Z. *Zidno Slikarstvo u Bosni i Hercegovini,* Sarajevo, 1971.

Karaman, Lj. "O bosanskim srednjovekovnim stećcima," *Starohrvatska prosvjeta,* Ser. III, sv. 3, 1954, pp. 171-182.

Karanović, M. "Jedan zanimljiv mramor kod Skender-vakufa," *GZMS,* XL, No. 2, 1928, pp. 135-39.

Karanović, M. "Grobna crkva grafički izražena po bosanskom srednjevekovnom spomeniku," *Novosti iz bosansko-hercegovačkog muzeja,* XI, Sarajevo, 1934.

Karanović, M. "Problem Kulina bana," *Novosti iz bosansko-hercegovačkog muzeja,* XII, Sarajevo, 1935.

Karanović, M. "Granice srednjevekovne bosanske župe Zemljanik," *GZMS,* XLVIII, 1936, pp. 27-36.

Karanović, M. "Istorijsko-etnografske crtice o župama Rami i Skoplju," *GZMS,* L, 1938, pp. 73-93.

Karanović, M. "O mramoru vojvode Momčila," *GZMS,* LII, No. 1, 1940, pp. 69-74.

Katona, Stephano, *Historia critica Regum Hungariae,* Vol. XIV, Kalocza, 1792.

Kemp, P. *The Healing Ritual: Studies in the Technique and Tradition of the Southern Slavs,* London, 1935.

Kemura, S. "Stara hrišćanska crkva u Sarajevu," *GZMS,* XXIII, 1911, pp. 297-302.

Kemura, S. *Bilješke iz prošlosti bosanskih katolika i njihovih bogomolja po turskim dokumentima,* Sarajevo, 1916.

Klaić, V. *Poviest Bosne do propasti kraljevstva,* Zagreb, 1882.

Klaić, V. *Bribirski knezovi od plemena Šubića do god. 1347,* Zagreb, 1897.

Klaić, V. "Crtice o Vukovskoj županiji i Djakovu u srednjem vijeku," *VZA,* II, 1900, pp. 98-108.

Klaić, V. "Herceg Hrvoje i Hval krstjanin," *Nastavni vjesnik,* XXXV, 1927, pp. 83-88.

Klaić, V. " 'Bosanska crkva' i Patareni" included in a collection of essays by him entitled, *Crtice iz hrvatske prošlosti,* Zagreb, 1928, pp. 69-82.

Klarić, I. "Porod, ženidba, smrt," *ZNŽOJS,* XXVII, 1930, pp. 166-75.

Knežević, S. "Lik zmije u narodnoj umetnosti i tradiciji Jugoslovena," *GEMB,* XXII-XXIII, 1960, pp. 57-97.

Kniewald, D. "Origine française du plus ancien sacramentaire de Zagreb," *Annales de l'Institut français de Zagreb,* 1938, pp. 194-200.

Kniewald, D. "Vjerodostojnost latinskih izvora o bosanskim krstjanima," *Rad* (JAZU), 270, 1949, pp. 115-276.

Kniewald, D. "Hierarchie und Kultus Bosnischen Christen" *Accademia Nazionale dei Lincei, Problemi Attuali di Scienza e di Cultura,* Rome, 1964, pp. 579-605.

Kočo, D. "L'Ornamentation d'un Vase à Mesure du Musée Cluny et les Stećci Bosniaques," *Artibus Asiae,* XV, Ascona, 1952, pp. 195-201.

Korać, V. "Crkve s prislonjenim lukovima u staroj Hercegovini i dubrovačko graditeljstvo XV-XVI veka," *ZRFFB,* VIII, 1964, pp. 561-99 (with V. Djurić).

Korać, V. *Trebinje - istorijski pregled,* Trebinje, 1966.

Kovačević, B. " 'Bogomili,' crkva bosanska," *Srpski književni glasnik,* LIV, Beograd, 1938, pp. 32-41, 217-25.

Kovačević, B. "O Takozvanim Bogomilima," *Naša prošlost* (Kraljevo), IV-V, 1970, pp. 1-27.

Kovačević, D. *Trgovina u srednjovjekovnoj Bosni,* Sarajevo, 1961, (NDBH, Djela XVIII).

Kovačević, D. "Pad bosanske srednjovjekovne države po dubrovačkim izvorima," *GID,* XIV, 1964, pp. 205-220.

Kovačević, J. "Beleške za proučavanje Miroslavljevog jevandjelja i materijalne kulture XI-XII veka," *IČ,* I (1-2), 1948, pp. 218-33.

Kovačević, J. "Nadgrobni natpis i reljef Kaznaca Nespine," *GZMS* (arh) n.s. XV-XVI, 1961, pp. 317-22.

Kovačević, J. "Prvi klesari ćirilskih potpisa na Balkanu," *GZMS* (arh) n.s. XV-XVI, 1961, pp. 309-316.
Kovačević, K. "Selo Praščijak i njegova pećina," *GZMS*, I, No. 1, 1889, pp. 21-23.
Kovačević, K. "Selo Gorijevac i njegove starine," *GZMS*, II, 1890, pp. 25-27.
Kovačević, Lj. "Znamenite srpske vlasteoske porodice srednjega veka," *GNČ*, X, 1888, pp. 199-214.
Kovijanić, P. "Bosanski protovestijar Zore," *Istoriski zapisi*, VIII (1-2), 1955, pp. 353-56 (with I. Stjepčević).
Krekić, B. *Dubrovnik i Levant (1280-1460)*, Beograd, 1956, (SAN pos. izd. 261).
Krekić, B. "Vuk Bobaljević," *ZRVI*, IV, 1956, pp. 115-40.
Kreševljaković, H. "Odakle su i što su bosansko-hercegovački Muslimani?," *Danica* (Zagreb), 1916, pp. 317-28.
Kreševljaković, H. "Stari bosanski gradovi," *NS*, I, 1952, pp. 7-47.
Kreševljaković, H. "Stari hercegovački gradovi," *NS*, II, 1954, pp. 9-22 (with H. Kapidžić).
Kristić, A. *Bosanske crkve i njihovo obskrbljivanje za Otomanske vlade*, Sarajevo, 1936.
Kučan, A. "Kameni stolac iz Bukovica," *GZMS*, (arh) n.s. X, 1955, pp. 41-48 (with Dj. Mazalić).
Kukuljica, L. "Ivan Stojković," *Dubrovnik* za 1867, Split, 1866, pp. 280-99.
Kulišić, Š. "Zmija kao životinjski pretstavnik duha žita," *Razvitak* (Banja Luka), VII, No. 5, 1940, pp. 143-47.
Kulišić, Š. "Razmatranja o porijeklu Muslimana u Bosni i Hercegovini," *GZMS*, n.s. VIII, 1953, pp. 145-58.
Kulišić, Š. "Tragovi bogomila u Boki Kotorskoj," *Spomenik*, SAN, 105, 1956, pp. 41-43.
Kulišić, Š. "Starije etničke formacije i etnički procesi u obrazovanju naroda Bosne i Hercegovine," *Pregled*, XIII, No. 1, 1961, pp. 9-16.
Kulišić, Š. "Etnička prošlost stanovništva Livanjskog polja," *Pregled* (between 1962-64), pp. 211-16 (I read as an undated off-print).
Kuljbakin, S. *Paleografska i jezička ispitivanja o Miroslavljevom jevandjelju*, Sremski Karlovci, 1925.
Kurtz, L. *The Dance of Death and the Macabre Spirit in European Literature*, New York, 1934.
Lanternari, V., *The Religions of the Oppressed* (Mentor pb edition), New York, 1963.
Lea, H. C. *A History of the Inquistion of the Middle Ages* (1887), 3 Vols., New York, 1922.
Leff, G. *Heresy in the Later Middle Ages*, 2 Vols., Manchester, 1967.
Leger, L. "L'heresie des Bogomiles en Bosnie et en Bulgarie au moyen âge," *Revue des Questions Historiques*, VIII, 1870, pp. 479-517.
Leovac, O. "Bogomili u Bosni i Hercegovini," *Glasnik Jugoslovenskog profesorskog društva*, XVIII, 1937, pp. 81-83.
Leslie, C. *Anthropology of Folk Religion*, New York, 1960.
Lessa, W. *Reader in Comparative Religion* (2nd ed.), New York, 1965 (with E. Vogt).

Levy-Strauss, C. *The Savage Mind,* Chicago, 1966.
Liepopili, A. *Ston u srednjim vijekovima,* Dubrovnik, 1915.
Lilek, E. "Riznica porodica Hranići (nadimak Kosače)," *GZMS,* I, No. 2, 1889, pp. 1-25.
Lilek, E. "Vjerske starine iz Bosne i Hercegovine," *GZMS,* VI, 1894, pp. 141-66, 259-81, 365-88, 631-74.
Lilek, E. "Bilješke o zadružnim i gospodarstvenim prilikama u Bosni i Hercegovini," *GZMS,* XII, 1900, pp. 213-25.
Linton, R. "Nativistic Movements," *American Anthropologist,* XLV, 1943, pp. 230-40. Reprinted in W. Lessa and E. Vogt, *Reader in Comparative Religion,* 2nd ed., New York, 1965, pp. 499-506.
Ljubić, S. "Poviestnička iztraživanja o Hrvoju velikom bosanskom vojvodi i spljetskom hercegu," *Rad* (JAZU), XXVI, 1874, pp. 74-92.
Ljubinković, R. "Humsko eparhisko vlastelinstvo i crkva Svetog Petra u Bijelom polju," *Starinar,* IX-X, 1959, pp. 97-123.
Lockwood, W. "The Market Place as a Social Mechanism in Peasant Society," *Kroeber Anthropological Society Papers,* 1965, pp. 47-67.
Loos, M. "Le prétendu temoinage d'un traité de Jean Exarque intitulé Šestodnev et relatif aux Bogomiles," *Byzantinoslavica,* XIII, 1952-53, pp. 59-67.
Loos, M. "Le Mouvement Paulicien à Byzance," *Byzantinoslavica,* XXIV, No. 2, 1963, pp. 258-86; XXV, No. 1, 1964, pp. 52-68.
Lopašić, R. *Bihać i Bihačka Krajina,* Zagreb, 1890.
Lovrić, A. "Glagoljski natpis na kamenu iz okolice banjalučke," *GZMS,* XLIX, 1937, pp. 31-36.
Machal, H. "Slavic Mythology," in *Mythology of all Races,* III, Boston, 1918, pp. 227-314.
Machek, V. "Essai comparatif sur la Mythologie Slave," *Revue des Études Slaves,* XXIII, 1947, pp. 48-65.
Maksimović, J. "Ilustracije mletačkog zbornika i problem minijatura u srednjevekovnoj Bosni," *IG* (1-2), 1958, pp. 117-30.
Malinowski, B. *Magic, Science and Religion and Other Essays* (collection of essays written between 1916-26), published by Free Press, Glencoe, Indiana, 1948. Citations in my study to the Anchor edition.
Mandelbaum, D. "Social Trends and Personal Pressure," (India) in C. Leslie, *Anthropology of Folk Religion . . .,* pp. 221-55.
Mandić, D. *Bogomilska crkva bosanskih krstjana,* Chicago, 1962 (which is Vol. II of his *Bosna i Hercegovina).*
Mandić, D. *Rasprave i prilozi iz stare hrvatske povijesti,* Rome, 1963.
Mandić, D. *Etnička povijest Bosne i Hercegovine,* Rome, 1967 (which is volume 3 of his *Bosna i Hercegovina).*
Mannheim, K. *Ideology and Utopia,* New York, 1936.
Marić, R. *Antički kultovi u našoj zemlji,* Beograd, 1933.
Marić, R. "Bind ilirski bog izvora," *GNČ,* XLVIII, 1939, pp. 146-49.
Marić, R. *Studije iz srpske numizmatike,* Beograd, 1956 (SAN pos. izd. 259).
Marjanović, M. "Aristodije zadranin hereziarha bosanski," *Zadarska revija,* I, 1952, No. 1, pp. 3-10; No. 2, pp. 9-15; No. 3, pp. 13-21; No. 4, pp. 12-16.
Marković, M. "Dve narodne pripovetke," *GZMS,* IV, 1892, pp. 203-04.
Marković, Z. "Drvene grobnice u okolini Rudog," *GZMS,* N.S. VIII, 1953,

pp. 317-19.
Marriott, M. "Little Communities in an Indigenous Civilization" (India), in C. Leslie, *Anthropology of Folk Religion,* pp. 169-218.
Matanić, A. "De duplici activitate S. Iacobi de Marchia in regno et vicaria Franciscali Bosnae," *Archivum Franciscanum historicum,* 53, 1960, pp. 111-127.
Matasović, J. "Tri humanista o Patarenima," *Godišnjak Skopskog filozofskog fakulteta,* I, 1930, pp. 235-251.
Matasović, J. "Rerum Bogomilicarum scriptores novi aevi antiquiores," *Godišnjak Skopskog filozofskog fakulteta,* III, 1934-38, pp. 167-99.
Matković, P. "Prilozi k trgovačko-političkoj historiji republike dubrovačke," *Rad* (JAZU), VII, 1869, pp. 180-266.
Matković, P. "Spomenici za dubrovačku povjest u vrieme Ugarsko-hrvatske zaštite," *Starine,* I, 1869, pp. 141-210.
Matković, P. "Dva taljanska putopisa po balkanskom poluotoku iz XVI vieka," *Starine,* X, 1878, pp. 201-56.
Matković, P. "Putovanja po Balkanskom poluotoku XVI vieka," *Rad*(JAZU), XLII, 1878, pp. 56-184; XLIX, 1879, pp. 103-64; LV, 1881, pp. 116-84; LVI, 1881, pp. 141-232; LXII, pp. 45-133; LXXI, 1884, pp. 1-60; LXXXIV, 1887, pp. 45-99; C, 1890, pp. 65-168; CV, 1891, pp. 142-201; CXII, 1892, pp. 154-243; CXVI, 1893, pp. 1-112; CXXIV, 1895, pp. 1-102; CXXIX, 1896, pp. 1-89; CXXX, 1897, pp. 86-188; CXXXVI, 1898, pp. 1-96.
Mazalić, Dj. "Kratki izvještaj o ispitivanju starina crkava manastira Ozrena, Tamne, Papraće, Lomnice," *GZMS,* L, 1938, pp. 95-109.
Mazalić, Dj. "Kraći clanći i rasprave," *GZMS,* n.s. IV-V, 1950, pp. 215-42.
Mažuranić, V. "Izvori dubrovačkog historika Jakova Lukarevića," *Narodna starina,* VIII, 1924, pp. 121-31.
Meggers, B. "Environmental Limitation on the Development of Culture," *American Anthropologist,* LVI, October 1954, pp. 801-24.
Meščerskij, N. "K istorii teksta slavjanskoj knigi Enoha," *VV,* XXIV, 1964, pp. 91-108.
Migne, A. *Dictionnaire des Apocryphes,* I, Paris, 1856.
Mihajlović, H. "Popovo u Hercegovini," *GZMS,* I, No. 1, 1889, pp. 15-17.
Mihajlović, H. "Vjetrenica pećina u Zavali," *GZMS,* I, No. 4, 1889, pp. 18-21.
Mihajlović, H. "Manastir Zavala i Vjetrenica pećina," *GZMS,* II, 1890, pp. 130-43.
Mihajlović, H. "Crkva sv. Petra u Zavali," *GZMS,* II, 1890, pp. 271-73.
Mijušković, J. "Humska vlasteoska porodica Sankovići," *IČ,* XI, 1960, pp. 17-54.
Milas, M. "Bogumilska riječ 'nežit' u Stonu i Dalmaciji," *GZMS,* VIII, 1896, p. 539.
Miletić, N. "Ranoslovensko doba," in A. Benac (et al.), *Kulturna istorija . . .,* pp. 379-402.
Millet, G. "La Religion Orthodoxe et les Heresies chez les Yugoslaves," *Revue de l'Histoire des Religions,* LXXV, Jan.-June, 1917, pp. 277-94.
Milobar, F. "Ban Kulin i njegovo doba," *GZMS,* XV, 1903, pp. 351-72; 482-528.
Milošević, A. "Petao u htonskom kultu kod antičkih Grka i u srpskom narodu," *GEMB,* XVII, 1954, pp. 106-113.

Mirković, P. "Manastir Panagjur," *GZMS*, I, No. 1, 1889, pp. 12-15.
Mirković, P. "Vojvoda Toma," *GZMS*, I, No. 2, 1889, pp. 29-32.
Mirković, P. "Jankovića razbojište ili svatovsko groblje na Brezovcu," *GZMS*, I, No. 4, 1889, pp. 21-23.
Mirković, P. "Bilaj," *GZMS*, II, 1890, pp. 152-57.
Mitrović, A. "Krsno ime nije ni patarenskog ni srpskog ni pravoslavnog postanka," *GZMS*, XXIV, 1912, pp. 391-96.
Mladenović, M. "Osmanli Conquest and the Islamization of Bosnia," *Études Slaves et Est Européennes*, III, 1959, pp. 219-226.
Močul'skij, V. "O mnimom dualizme i mifologiji Slavjan," *Russkij filologičeskij vestnik, XXI, 1889, pp. 153-204.*
Momirović, P. "Stari rukopisi i štampane knjige u Čajniču," *NS*, III, 1956, pp. 173-77.
Moore, B., Jr. *Social Origins of Dictatorship and Democracy*, Boston, 1966.
Mošin, V. Review of G. Čremošnik's "Bosanske i humske povelje . . .," in *Historijski zbornik*, II, 1949, pp. 315-21.
Mošin, V. "Rukopis pljevaljskog Sinodika Pravoslavlja," *Slovo*, VI-VIII, 1957, pp. 154-76.
Mošin, V. "Serbskaja redakcija Sinodika v Nedelju Pravoslavija," *VV*, XVI, 1959, pp. 317-94.
Mujezinović, M. "Groblja nad Kovačima u Sarajevu," *NS*, IX, 1964, pp. 123-32.
Murray-Aynsley, H. *Symbolism of the East and West*, London, 1900.
Nelli, R. *Le Phenomène Cathare*, Presses universitaires de France, 1964.
Nikolajević-Stojković, I. "Skulptura srednjevekovnih crkava Bosne i Hercegovine," *ZRVI*, V, 1958, pp. 111-122.
Nikolajević-Stojković, I. "La Sculpture ornamentale au XIIe Siècle en Bosnie et en Hercegovine," *ZRVI*, VIII, No. 2, pp. 295-309.
Njegoš, P. P. "The Ray of the Microcosm," *Harvard Slavic Studies*, III, 1957 (trans. with interesting introduction on dualistic ideas in his work by A. Savić-Rebac).
Novak, V. "Slavonic Latin Symbiosis in Dalmatia During the Middle Ages," *Slavonic and East European Review*, XXXII, No. 72, Dec. 1953, pp. 1-28.
Novaković, S. "Burkard i Bertrandon de-la Brokijer o balkanskom poluostrvu XIV-XV veka," *GNČ*, XIV, 1894, pp. 1-66.
Obolensky, D. *The Bogomils*, Cambridge, England, 1948.
Obolensky, D. "Bogomilism in the Byzantine Empire," *Actes du VIe Congres International d'Etudes Byzantines*, I, Paris, 1950, pp. 293 ff.
Obolensky, D. "Le Christianisme Oriental et les Doctrines Dualistes," *Accademia Nazionale dei Lincei, Problemi Attuali di Scienza e di Cultura*, Rome, 1964, pp. 643-51.
Obradović, D. *The Life and Adventures of Dositej Obradović* (ed. and trans. by G. Noyes), Berkeley, 1953.
Oeconomos, L. *La Vie Religieuse dans l'Empire Byzantin au Temps des Comnènes et des Anges*, Paris, 1918.
Okiç, M. T. "Les Kristians (Bogomiles Parfaits) de Bosnie d'après des Documents Turcs inédits," *Südost-Forschungen*, XIX, 1960, pp. 108-33.
Okiç, M. T. "Les Kristians (Bogomiles) de Bosnie," *CEC*, XIV, winter 1963-64, pp. 3-6.
Origo, I. "The Domestic Enemy: The Eastern Slaves in Tuscany in the

Fourteenth and Fifteenth Century," *Speculum*, XXX, 1955, pp. 321-66.
Palavestra, V. "Komparativno istraživanje narodnih pripovedaka kao pomoćno sredstvo za proučavanje etničkih odnosa," *Rad Kongresa folklorista Jugoslavije*, VI, at Bled, 1959, Ljubljana, 1960, pp. 117-21.
Palavestra, V. "Narodna predanja o kamenim krstovima," *Narodno stvaralaštvo*, XV-XVI, Beograd, 1965, pp. 1191-1196.
Palavestra, V. "Nekoliko nepoznatih stihova o hercegu Stjepanu," *GZMS* (etn), n.s. XXIII, 1968, pp. 133-37.
Panofsky, E. *Tomb Sculpture*, New York, 1964.
Paškvalin, V. "Kultovi u antičko doba na području Bosne i Hercegovine," *GZMS* (arh), n.s. XVIII, 1963, pp. 127-53.
Pecarski, B. "Byzantine Influences on Some Silver Book Covers in Dalmatia," *Actes du XIIe Congrès International d'Études Byzantines* (Ohrid, 1961), Beograd, 1964, III, pp. 315-20.
Perćo, Lj. "Običaji i verovanja iz Bosne," *SEZ*(SKA), XXXII, 1925, pp. 359-86.
Peisker, J. "Koje su vjere bili stari Slaveni prije krštenja?" *Starohrvatska prosvjeta*, II (1-2), 1928, pp. 53-88.
Pejanović, Dj. *Stanovništvo Bosne i Hercegovine*, Beograd, 1955 (SAN pos. izd. 229).
Perojević, M. The historical chapters in *Poviest BH*, 1942, pp. 196-592.
Pervan, M. "Prilog rješenju problema Bosanske crkve ili Bogumila u Bosni," *Dobri Pastir*, IV-V, 1955, pp. 95-102.
Pervan, M. "Vjerske prilike za vrijeme Kulina bana," *Dobri pastir*, VIII, 1957, pp. 128-42.
Petković, P. "Motiv arkada i stolova na stećcima," *Starinar*, n.s. VII-VIII, 1956-57, pp. 195-205.
Petrić, M. "O migracijama stanovništva u Bosni i Hercegovini i doseljavanja i unutrašnja kretanja," *GZMS*(etn), n.s. XVIII, 1963, pp. 5-16.
P(etrović), L. fra. *Kršćani bosanske crkve*, published in *Dobri pastir*, 1953, and also as a separate book, Sarajevo, 1953.
Petrović, P. "Motiv ljudskih očiju kod balkanskih slovena," *GEMB*, XXII-XXIII, 1960, pp. 33-55.
Pfeiffer, N. *Die ungarische Dominikanerprovinz von ihrer Gründung 1221 bis zur Tatarenverwustung 1241-2*, Zurich, 1913 (dissertation from Freiburg, No. 86).
Pilar, I. *Bogomilstvo kao religijozno povjestni - te kao socijalni i politički problem*, Zagreb, 1927.
Pilar, I. "O dualizmu u vjeri starih Slovjena i o njegovu podrijetlu i značenju," *ZNŽOJS*, XXVIII, No. 1, 1931, pp. 1-86.
Polenaković, H. "Hagada u Makedonskoj pisanoj i usmenoj književnosti," *Rad kongresa folklorista Jugoslavije*, VI, at Bled, 1959, Ljubljana, 1960, pp. 333-39.
Popović, C. "Da li su Bosanci u XVII veku znali za Bogomile," *Zivot*, 1952, No. 3, pp. 187-89.
Popović, C. "Reč 'Bogomil' u Bosni," *Život*, 1952, No. 7, pp. 273-75.
Popović, C. "Bosansko-hercegovačke preslice i vretena," *GZMS*, n.s. VIII, 1953, pp. 159-86.
Popović, C. "Manji prilozi za pitanje Bogumila u Bosni," *GZMS* (arh), n.s. XII, 1957, pp. 235-39.

Poviest hrvatskih zemalja Bosne i Hercegovine, collective work issued by "Napredak," Sarajevo, 1942 (abbreviated as *Poviest BH).*
Preveden, F. *A History of the Croatian People,* 2 volumes, New York, 1955, 1962.
Primov, B. "Medieval Bulgaria and the Dualist Heresies in Western Europe," *Études Historiques à l'occasion du XIe Congrès international des Sciences historiques* held at Stockholm, August, 1960, publication, Sofija, 1960, pp. 79-106.
Primov, B. "Rayner Sakoni kato izvor za vr'zkite meždu Katari, Pavlikjani i Bogomili," *Izsledvanija v čest na Marin S. Drinov,* Sofija, 1960, pp. 535-70.
Prohaska, D. "Bogomilstvo i Boljševizam," *Jugoslavenska njiva,* III, 1919, pp. 281-83.
Puech, H. C. See also under Sources, since his French edition of Kosmas' Tract against the Bogomils, also contains a long and valuable study about dualism.
Puech, H. C. "Catharisme Mediéval et Bogomilisme," *Convegno di Scienze morali, Storiche e Filologiche,* Rome, 1957, pp. 56-84.
Pulver, M. "Jesus' Round Dance and Crucifixion according to the Acts of St. John," Bollingen Series XXX, Papers from the Eranos Yearbooks (J. Campbell, Ed.), *The Mysteries,* 1942, pp. 169-193.
Rački, F. "Prilozi za zbirku srbskih i bosanskih listina," *Rad*(JAZU), I, 1867, pp. 124-163.
Rački, F. *Bogomili i Patareni* (originally published serially in *Rad,* JAZU, 1869-70; re-issued in book form (SKA pos. izd. 87), Beograd, 1931. All citations in my study are drawn from the 1931 edition.
Rački, F. "Dubrovački spomenici o odnošaju dubrovačke obćine naprama Bosni i Turskoj godine razspa bosanske kraljevine," *Starine,* VI, 1874, pp. 1-18.
Radcliffe-Brown, A. *Structure and Function in Primitive Society,* New York, 1965, particularly the essay "Religion and Society," pp. 153-77.
Radić, F. "Odlomak starog slovenskog rukopisnog evangelijara s predgovorima i s blagoslovima, gotskim pismom napisana, u knjižnici franovačkog samostana na Otoku (Badia) kod grad Korčule," *GZMS,* II, 1890, pp. 254-62.
Radin, P. *Primitive Religion* (1937), New York, 1957.
Radin, P. *Primitive Man as a Philosopher*(rev. ed.) New York, 1957.
Radojčić, N. "O zemlji i imenima Bogomila," *Prilozi,* VII, 1927, pp. 147-59.
Radojčić, N. *Obred krunisanja bosanskoga kralja Tvrtka I,* Beograd (SAN, pos izd. 143), 1948.
Radojčić, N. "O jednom naslovu velikoga Vojvode Hrvoja Vukčića," *IČ,* I, 1948, pp. 37-53.
Radojčić, N. *Srpska istorija Maura Orbinija,* Beograd, 1950 (SAN pos. izd. 152).
Radojčić, N. "Dva odlomka Ilariona Ruvarca o Kosačama," *GZMS,* n.s. IV-V, 1950, pp. 200-12.
Radojčić, N. "Die Wichtigsten Darstellungen der Geschichte Bosniens," *Südost-Forschungen,* XIX, 1960, pp. 146-63.
Radojčić, S. "Crkva u Konjuhu," *ZRVI,* I, 1952, pp. 148-65.
Radojčić, S. "O nekim zajedničkim motivima naše narodne pesme i našeg starog slikarstva," *ZRVI,* II, 1953, pp. 159-77.
Radojčić, S. "Reljefi bosanskih i hercegovačkih stećaka," *Letopis Matice srp-*

ske, CXXXVII, knj. 387, Novi Sad, Jan. 1961, pp. 1-15.
Radojičić, Dj. "Grdeša (Grd) trebinjski župan XII veka," *Prilozi,* XI, 1931, pp. 155-56.
Radonić, J. "Der Grossvojvode von Bosnien Sandalj Hranić Kosača," *Archiv für slavische Philologie,* XIX, 1897, pp. 387-456.
Radonić, J. "O knezu Pavlu Radenoviću - priložak istoriji Bosne krajem XIV i početkom XV veka," *Letopis Matice srpske,* 211, 1902, pp. 39-62; 212, 1902, pp. 34-61.
Radonić, J. "Herceg Stipan Vukčić Kosača i porodica mu u istoriji i narodnoj tradiciji," in *Zbornik u slavu Vatroslava Jagića,* Berlin, 1908, pp. 406-14.
Radonić, J. "Donaldo da Lezze i njegova 'Historia Turchesca,'" *GNČ,* XXXII, 1913, pp. 302-38.
Raynaldi, O. *Annales Ecclesiastici,* Continuatio Baronii, 15 vols., Lucca, 1747-56.
Redfield, R. *Folk Culture of the Yucatan,* Chicago, 1941.
Redfield, R. *The Little Community* and *Peasant Society and Culture* (two books published together under one cover), Chicago, 1960.
Redfield, R. "Thinker and Intellectual in Primitive Society," in S. Diamond, *Primitive Views . . . ,* pp. 33-48.
Rendić-Miočević, D. "Da li je spelaeum u Močićima služio samo mitrijačkom kultu," *GZMS,* n.s. VIII, 1953, pp. 271-76.
Rendić-Miočević, D. "Ilirske pretstave Silvana na kultnim slikama sa područja Dalmata," *GZMS* (arh), n.s. X, 1955, pp. 1-40.
Rengjeo, I. "Novci bosanskih banova i kraljeva," *GZMS,* LV, 1943, pp. 237-91.
Rešetar, M. "Početak kovana dubrovačkoga novca," *Rad* (JAZU), 266, 1939, pp. 149-70.
Rice, D. T. "The Leaved Cross," *Byzantinoslavica,* XI, 1950, pp. 72-81.
Ristić, M. *Bosna od smrti bana Matije Ninoslava do vlade sremskoga Kralja Stevana Dragutina, 1250-1284,* Beograd, 1910.
Rother, A. "Johannes Theutonicus (von Wildeshausen)," *Römische Quartalschrift für christliche Alterthumskunde und für Kirchengeschichte,* IX, 1895, pp. 139-70.
Rovinskij, P. "Material dlja istorii Bogomilov v Serbskih zemljah," *Žurnal Ministerstva Narodnogo Prosveščenija,* 220, Mar. 1882, pp. 32-51.
Runciman, S. *The Medieval Manichee,* Cambridge (Eng.), 1947.
Runeberg, A. *Witches, Demons and Fertility Magic: Analysis of Their Significance and Mutual Relations in Western European Folk Religion,* Societas Scientiarum Fennica Commentationes Humanarum Litterarum (Finska Vetenskaps Soccetatem), 1947.
Rupčić, B. "Povodom članka dra N. M—a gde je bilo sijelo Biskupije Sarsitenensis," *Dobri Pastir,* VIII, 1957, pp. 69-82.
Russell, J. "Interpretations of the Origins of Medieval Heresy," *Medieval Studies,* XXV, 1963, pp. 26-53.
Russell, J. *Dissent and Reform in the Early Middle Ages,* Berkeley, 1965.
Ruvarac, I. "Nešto o Bosni, Dabarskoj i Dabro-bosanskoj episkopiji i o srpskim manastirima u Bosni," *GNČ,* II, 1878, pp. 240-61.
Ruvarac, I. "Miroslav brat Stefana Nemanje velikoga župana srpskog," *GNČ,* X, 1888, pp. 65-68.
Ruvarac, I. "O natpisu na crkvi hercega Stefana u Goraždu," *Glas* (SKA),

XVI, 1889, pp. 1-30.
Ruvarac, I. "Draga, Danica, Resa," *GZMS*, III, 1891, pp. 225-38 (reprinted in *Zbornik I. Ruvarca*, I, Beograd, 1934 (SKA pos. izd. 103), pp. 331-347.
Ruvarac, I. "Katarina kći Tvrtka I," *GZMS*, IV, 1892, pp. 205-11 (reprinted in *Zbornik I. Ruvarca*, I, pp. 351-58).
Ruvarac, I. "Banovanje Tvrtka bana 1353-77," *GZMS*, VI, 1894, pp. 225-40, 611-20.
Ruvarac, I. *O humskim episkopima i hercegovačkim mitropolitima do godine 1766*, Mostar, 1901.
Rycaut, P. *Present State of the Ottoman Empire*, Cologne, 1676.

NB. The Serbian letter Š(sh) shall follow all the S's.

Sadnik, L. "Religiöse und soziale Reformbewegungen bei den Slavischen Völkern," part II, "Die 'Bosnische Kirche,' " *Blick nach Osten*, II, no. 4, March 1952, pp. 261-66.
Sapir, E. *Culture, Language and Personality*, Berkeley, 1949 (especially the two essays, "Language" and "The Meaning of Religion").
Savić-Rebac, A. "O narodnoj pesmi 'car Dukljan i krstitelj Jovan,' " SAN, *Zbornik radova*, X, *Institut za proučavanje književnosti*, 1, Beograd, 1951, pp. 253-71.
Schmaus, A. "Der Neumanichäismus auf dem Balkan," *Saeculum*, II, 1951, pp. 271-91.
Schneweiss, E. "Glavni elementi samrtnih običaja kod Srba i Hrvata," *Glasnik Skopskog naučnog društva*, V, 1929, pp. 263-82.
Schwoebel, R. *The Shadow of the Crescent: The Renaissance Image of the Turks 1453-1517*, Nieuwkoop, 1967.
Sergejevski, D. "Slike pokojnika na našim Srednjevjekovnim nadgrobnim spomenicima," *GZMS*, n.s. VIII, 1953, pp. 131-39.

NB. The Serbian letter Š, sh, shall follow all the S's.

Shannon, A. *The Popes and Heresy in the Thirteenth Century*, Villanova, Pa., 1949.
Shannon, A. Review of C. Thouzellier, *Une Somme anti-Cathare . . .*, in *Speculum*, XLI, 1966, p. 182.
Shimkin, D. "Culture and World View: A Method of Analysis Applied to Rural Russia," *American Anthropologist*, LV, No. 3, 1953, pp. 329-48 (with P. Sanjuan).
Simić, M. "Groblja i nadgrobni spomenici u okolini Novog Brda," *GEMB*, XVII, 1954, pp. 160-67.
Sjoberg, G. " 'Folk' and 'feudal' Societies," *American Journal of Sociology*, LVIII, no. 3, November 1952, pp. 231-39.
Skarić, V. "Porijeklo pravoslavnoga naroda u sjeverozapadnoj Bosni," *GZMS*, XXX, 1918, pp. 219-65.
Skarić, V. "Župa i grad Borač u Bosni," *Prilozi*, II, 1922, pp. 184-88.
Skarić, V. "Kudugeri," *Prilozi*, VI, No. 1, 1926, pp. 107-110.
Skarić, V. "Trzan," *GZMS*, XL, No. 2, 1928, pp. 127-33.
Skarić, V. "Jedan slovenski uzor bosanskih mramorova (stećaka), *GZMS*, XL, No. 2, 1928, pp. 141-44.

Skarić, V. "L'attitude des peuples Balkaniques à l'égard des Turc," *Revue Internationale des Études Balkaniques,* Year 1, Volume 2, Beograd, 1935, pp. 235-46.
Skarić, V. "Širenje Islama u Bosni i Hercegovini," *Gajret kalendar za 1940,* Sarajevo, 1939, pp. 28-33.
Skendi, S. "Crypto-Christianity in the Balkan Area under the Ottomans," *Slavic Review,* XXVI, June 1967, pp. 227-46.
Skok, P. "Bogomili u svjetlosti lingvistike," *JIČ,* I, 1935, pp. 462-72.
Söderberg, H. *La Religion des Cathares,* Uppsala, 1949.
Solovjev, A. "Svetosavski nomokanon," *Bratstvo,* XXVI, 1932, pp. 19-43.
Solovjev, A. "Trgovanje bosanskim robljem do god 1661," *GZMS,* n.s. I, 1946, pp. 139-62.
Solovjev, A. "Pravni položaj seljaka u srednjevekovnoj Bosni," *Pregled,* II, 1947, pp. 244-50.
Solovjev, A. "Gost Radin i njegov testament," *Pregled,* II, 1947, pp. 310-18.
Solovjev, A. "Postanak i pad bosanske crkve," *Prosvjetni Radnik,* Sarajevo, Nos. 6-7, 1947.
Solovjev, A. "O natpisu na grobu velikog Kaznaca Nespine," *GZMS,* n.s. II, 1948, pp. 235-37.
Solovjev, A. "La Doctrine de l'église de Bosnie," *Académie de Belgique* - Bulletin - lettres - 5e Ser. XXXIV, 1948, pp. 481-534.
Solovjev, A. Review of V. Glušac's, *Istina o Bogomilima,* in *IČ,* I (1-2), 1948, pp. 263-73.
Solovjev, A. "Les Bogomiles veneraient-ils la Croix," *Académie de Belgique - Bulletin* - lettres - 5e Ser. XXV, 1949, pp. 47-62.
Solovjev, A. "Saint Grégoire, Patron de Bosnie," *Byzantion,* XIX, 1949, pp. 263-79.
Solovjev, A. "Vlasteoske povelje bosanskih vladara," *Istorisko-pravni zbornik,* I, Sarajevo, 1949, pp. 79-105 (published by Pravni fakultet, Sarajevo).
Solovjev, A. "Nestanak Bogomilstva i islamizacija Bosne," *GID,* I, 1949, pp. 42-79.
Solovjev, A. Review of Runciman's, *The Medieval Manichee,* in *Byzantinoslavica,* X, 1949, pp. 96-102.
Solovjev, A. "Le Messe Cathare," *CEC,* III, No. 12, 1952, pp. 199-206.
Solovjev, A. "Fundajajiti, Patarini i Kudugeri u vizantiskim izvorima," *ZRVI,* I, 1952, pp. 121-47.
Solovjev, A. "Autour des Bogomiles," *Byzantion,* XXII, 1952, pp. 81-104.
Solovjev, A. "Le témoinage de Paul Rycaut sur les restes des Bogomiles en Bosnie," *Byzantion,* XXIII, 1953, pp. 73-86.
Solovjev, A. "Svedočanstva pravoslavnih izvora o Bogomilstvu na Balkanu," *GID,* V, 1953, pp. 1-103.
Solovjev, A. "Novi podaci za istoriju neomanihejskog pokreta u Italiji i Bosni," *GZMS,* n.s. VIII, 1953, pp. 329-334.
Solovjev, A. "Prinos za bosansku i ilirsku heraldiku," *GZMS,* n.s. IX, 1954, pp. 87-135.
Solovjev, A. "Le Tatouage symbolique en Bosnie," *CEC,* V, No. 19, 1954, pp. 157-62.
Solovjev, A. "Le Symbolisme des Monuments funeraires Bogomiles," *CEC,* V, No. 18, 1954, pp. 92-114.
Solovjev, A. "Bogumili," *Enciklopedija Jugoslavije,* I, Zagreb, 1955, pp. 641-49.

Solovjev, A. "Broj grobnih spomenika u Bosni i Hercegovini," *GZMS* (arh), n.s. X, 1955, pp. 217-18.
Solovjev, A. "Simbolika srednjevekovnih spomenika u Bosni i Hercegovini," *GID*, VIII, 1956, pp. 5-67.
Solovjev, A "Le Testament du gost Radin," *Mandićev zbornik*, Rome, 1965, pp. 141-56.
Sommariva, L. "Studi recenti sulle eresie Medievali," *Rivista Storica Italiana*, LXIV, 1952, pp. 237-68.
Speranskij, M. "Ein bosnisches Evangelium in der Handschriftensammlung von Srećković," *Archiv für slavische Philologie*, XXIV, 1902, pp. 172-83.
Srejović, D. "Jelen u narodnim običajima," *GEMB*, XVIII, 1955, pp. 231-37.
Srejović, D. "Les Anciens Eléments Balkaniques dans la Figure de Marko Kraljević," *Živa Antika*, VIII, 1958, Skopje, pp. 75-97.
Stanojević, S. "Bogomili i Husiti u Sremu i u Bačkoj," *Glasnik Istorijskog društva* (Novi Sad), I, 1928, pp. 114-15.
Stanojević, S. "Dogadjaji 1253 i 1254 godina," *Glas* (SKA), CLXIV 1935, pp. 191-98.
Stefanović, V. "Ratovanje Kralja Matije u Bosni," *Letopis Matice srpske*, 332, 1932, pp. 195-213.
Stjepčević, L. "Hranići-Kosače u kotorskim spomenicima," *IČ*, V, 1954-55, pp. 311-21 (with R. Kovijanić).
Stojanović, Lj. "Jedan prilog poznavanju bosanskih Bogumila," *Starine*, XVIII, 1886, pp. 230-32.
Stojković, I. "Jonski impost-kapiteli iz Makedonije i Srbije," *ZRVI*, I, 1952, pp. 169-78 (see also Nikolajević-Stojković).
Stratimirović, G. "Srednjevjekovno groblje kod Zgošće," *GZMS*, III, 1891, pp. 122-40.
Stratimirović, G. "Rimsko kamenje kod Gostilje," *GZMS*, III, 1891, pp. 286-91.
Stratimirović, G. "Opis polja Glasinca," *GZMS*, III, 1891, pp. 323-34.
Stratimirović, G. "O crkvi Ozrenskoj," *GZMS*, IV, 1892, pp. 68-70.
Stratimirović, G. "Opet o crkvi Ozrenskoj," *GZMS*, V, 1893, pp. 309-10.
Stratimirović, G. "Starinarske bilješke," *GZMS*, XXXVII, 1925, pp. 85-86.
Stratimirović, G. "Zgošćanski stećak," *GZMS*, XXXVIII, 1926, pp. 45-46.
Stričević, Dj. "Majstori minijatura Miroslavljevog Jevandjelja," *ZRVI*, I, 1952, pp. 181-203.
Strohal, R. "Patareni i hrvatska glagolska knjiga," *VZA*, XVII, 1915, p. 348.
Strohal, R. "Vrag u crkvi," *ZNŽOJS*, XXVII, 1930, pp. 180-82.
Strukić, I. *Povjestničke crtice Kreševa i franjevačkoga samostana*, Sarajevo, 1899.
Strukić, I. "Katolička crkva u Bosni," *Spomen knjiga iz Bosne*, Zagreb, 1901, pp. 151 ff.
Subotić, G. "Kraljica Jelena Anžujska - ktitor crkvenih spomenika u primorje," *IG*, (1-2) 1958, pp. 131-48.
Swanson, G. *The Birth of the Gods*, Ann Arbor, 1960.
Šabanović, H. "Pitanje turske vlasti u Bosni do pohoda Mehmeda II 1463 g," *GID*, VII, 1955, pp. 37-51.
Šabanović, H. "Bosansko krajište 1448-1463," *GID*, IX, 1958, pp. 177-220.
Šabanović, H. *Bosanski pašaluk*, Sarajevo, 1959 (NDBH, Djela, 14).

Šabanović, H. "Postanak i razvoj Sarajeva," *Radovi* (NDBH), XIII, 1960, pp. 70-114.
Šabanović, H. "Lepenica u prvom stoljeću turske vladavine," chapter in *Lepenica,* pos. izd. III, NDBH, Sarajevo, 1963, pp. 193-207.
Šabanovic, H. "Žepa i njena okolina u prvim decenijama turske vlasti," *GZMS,* n.s. XIX, 1964, pp. 39-44 (with V. Palavestra).
Šajnović, I. "Sveta Petka i Nedjelja: Kola u Bosni," *ZNŽOJS,* XXVI, sv. 2, 1928, pp. pp. 378-79.
Šamić, M. *Les voyageurs français en Bosnie à la fin du XVIIIe siècle et au debut du XIXe et le pays tel qu'ils l'ont vu,* Paris, 1960.
Šerić, H. "Katoličke crkve i samostani u Dubici i njezinoj okolici u srednjem vieku," *Croatia Sacra,* XIII-XIV (no. 22-23), 1944, pp. 72-88.
Šidak, J. "Problem 'bosanske crkve' u našoj historiografiji od Petranovića do Glušca," *Rad* (JAZU), 259, 1937, pp. 37-182.
Šidak, J. " 'Bosanska crkva' i problem bosanskog 'bogumilstva' i sredovječna 'crkva bosanska,' " *Savremenik,* XXVI, no. 11, 1937, pp. 385-93.
Šidak, J. "Pravoslavni istok i 'crkva bosanska,' " *Savremenik,* XXVII, no. 9, 1938, pp. 769-93.
Šidak, J. "Samostalna crkva bosanska i njezini redovnici," *Nastavni vjesnik,* L, No. 2, 1941, pp. 1-16.
Šidak, J. "Bosna i zapadni dualisti u prvoj polovici XIII stoljeca," *Zgodovinski časopis* VI-VII, 1952-53, pp. 285-300.
Šidak, J. "Pitanje 'crkve bosanske' u novijoj literaturi," *GID,* V, 1953, pp. 139-60.
Šidak, J. "O vjerodostojnosti isprave bosanskog bana Tvrtka Stjepanu Rajkoviću," *Zbornik radova, Sveučilište u Zagrebu,* II, 1954, pp. 37-48.
Šidak, J. "Današnje stanje pitanja crkve bosanske," *Historijski zbornik,* VII, 1954, pp. 129-42; VIII, 1955, pp. 219-220.
Šidak, J. "Franjevačka 'Dubia' iz 1372/3 kao izvor za povijest Bosne," *IČ,* V, 1955, pp. 209-220.
Šidak, J. "Eccelsia Sclavoniae i misija Dominikanaca u Bosni," *Zbornik filozofskog fakulteta u Zagrebu,* III, 1955, pp. 11-40.
Šidak, J. "Problem Bogumilstva u Bosni," *Zgodovinski časopis,* IX, 1955, pp. 154-62.
Šidak, J. "Kopitarevo bosansko jevandjelje u sklopu pitanja 'crkve bosanske,' " *Slovo,* IV-V, 1955, pp. 47-63.
Šidak, J. "Marginalija uz jedan rukopis 'crkve bosanske' u mletačkoj Marciani," *Slovo,* VI-VIII, 1957, pp. 134-53.
Šidak, J. "Bogumilstvo i heretička 'crkva bosanska,' " *Historijski pregled,* IV, No. 2, 1958, pp. 101-14.
Šidak, J. "Problem popa Bogumila u suvremenoj nauci," *Slovo,* IX-X, 1960, pp. 193-97.
Šidak, J. "Dva priloga o minijaturima u rukopisima 'crkve bosanske,' " *Slovo,* IX-X, 1960, pp. 201-06.
Šidak, J. "O nekim prilozima B. Primova problemu Bogomilstva," *Historijski zbornik,* XIV, 1961, pp. 311-17.
Šidak, J. "Heretički pokret i odjek Husitizma na Slovenskom jugu," *Zbornik za društvene nauke* (Matica srpska, Novi Sad), XXXI, 1952, pp. 5-24.
Šidak, J. "O autentičnosti i značenju jedne isprave bosanskog 'djeda' (1427)," *Slovo,* XV-XVI, 1965, pp. 282-97.

Šidak, J. "O pitanju heretičkog 'pape' u Bosni 1223 i 1245," *Rasprave v Hauptmannov Zbornik*, Slovenska Akademija znanosti in umetnosti, Razred za Zgodovinske in Družbene vede, Ljubljana, 1966, pp. 145-59.

Šišić, F. "Kako je Justinijan postao Slaven," *Nastavni vjesnik*, IX, 1901, pp. 390-415.

Šišić, F. *Vojvoda Hrvoje Vukčić-Hrvatinić i njegovo doba*, Zagreb, 1902.

Šišić, F. "Pad Mladena Šubića, Bana hrvatskoga i bosanskoga," *GZMS*, XIV, 1902, pp. 335-66.

Šišić, F. "Studije iz bosanske historije," *GZMS*, XV, 1903, pp. 319-49.

Šišić, F. "Važan prilog za poviest ugarsko-bosanskih ratova," *VZA*, VI, 1904, p. 135.

Šišić, F. "Nešto o Bosansko-Djakovačkoj Biskupiji i Djakovačkoj Katedrali," *GNČ*, XLIV, 1935, pp. 54-70.

Šišić, F. "Hrvatska historiografija od XVI do XX stoljeća," *Jugoslovenski istoriski časopis*, II, 1936, pp. 16-48.

Šišić, F. "Izvori bosanske povijesti," in *Poviest BH*, 1942, pp. 1-38.

Škrivanić, G. "Vojničke pripreme srednjovekovne bosanske države pred propast 1463," *GID*, XIV, 1964, pp. 221-28.

Štefanić, V. "Glagolski zapis u Čajničkom jevandjelju i u Radosavljevom rukopisu," *Zbornik Historijskog instituta* (JAZU), II, 1959, pp. 5-15.

Šunjić, M. "Prilozi za istoriju bosansko-venecijanskih odnosa 1420-63," *Historijski zbornik*, XIV, 1961, pp. 119-45.

Tadić, J. "Nove vesti o padu Hercegovine pod tursku vlast," *ZRFFB*, VI, No. 2, 1962, pp. 131-52.

Taškovski, D. *Bogomilskoto dviženie*, Skopje, 1949.

Taškovski, D. *Bogomilstvoto i njegovoto istorisko značenje*, Skopje, 1951.

Taškovski, D. "Koja je prava domovina Bogomila," *Pregled*, V, No. 3, 1953, pp. 379-82.

Telebaković-Pecarski, B. "Novi pogledi na problematiku stećaka," *Starohrvatska prosvjeta*, III Ser., sv. 8-9, 1963, pp. 217-20.

Thallóczy, L. See also under sources, since frequently his publications of source material contain historical and interpretative essays.

Thallóczy, L. "Vojvoda Hrvoja i njegov grb," *GZMS*, IV, 1892, pp. 170-87.

Thallóczy, L. "Prilozi k objašnjenju izvora bosanske historije," *GZMS*, V, 1893, pp. 1-87.

Thallóczy, L. "Beiträge zur Kenntnis der Bogomilenlehre," *Wissenschaftliche Mitteilungen aus Bosnien und der Herzegowina*, III, 1895, pp. 350-71.

Thallóczy, L. *Studien zur Geschichte Bosniens und Serbiens im Mittelalter*, Munich, 1914.

Thallóczy, L. *Povijest (banovine, grada, i varoši) Jajce, 1450-1527*, Zagreb, 1916.

Thouzellier, C. "Heresie et Croisade au XIIe siècle," *Revue d'Histoire Ecclesiastique* (Louvain), XLIX, No. 4, 1954, pp. 855-72.

Thut-Weitnaur, M. "Symboles en Grafiti de Provence," *CEC*, XIII, Spring, 1962, pp. 16-21.

Tkalčić, I. "Pavlinski samostan u Dubici," *Viestnik Hrvatskog arheološkog družtva*, I, 1895, pp. 189-202.

Tomasic, D. *Personality and Culture in Eastern European Politics*, New York, 1948.

Tomić, P. *Pravoslavlje u Bosni i Hercegovini,* Beograd, 1898.
Traljić, S. "Da li su Bosanci u XVIII vijeku znali za Bogomile," *Historijski zbornik,* V (3-4), 1952, pp. 409-10.
Trifković, V. *Sarajevska okolina* (I. Sarajevsko polje), Beograd, 1908, SEZ, XI (SKA Naselja Srp. zem. 5) (with Pop Stjepa).
Trpković, V. "Kad je Stepan II Kotromanić prvi put prodro u Hum," *IG* (1-2), 1960, pp. 151-54.
Trpković, V. "Branivojevići," *IG* (3-4), 1960, pp. 55-84.
Trpković, V. "Humska zemlja," *ZRFFB,* VIII, 1964, pp. 225-59.
Truhelka, Ć. "Ko je bio slikar fojničkog grbovnika," *GZMS,* I, No. 2, 1889, pp. 86-90.
Truhelka, Ć. "Tetoviranje katolika u Bosni i Hercegovini," *GZMS,* VI, 1894, pp. 241-57.
Truhelka, Ć. *Kraljevski grad Jajce, povijest i znamenitosti,* Sarajevo, 1904.
Truhelka, Ć. "Dubrovačke vesti o godini 1463," *GZMS,* XXII, 1910, pp. 1-24.
Truhelka, Ć. "Tursko-Slovjenski spomenici dubrovačke arhive," *GZMS,* XXIII, 1911, pp. 303-49, 437-74.
Truhelka, Ć. "Testament gosta Radina - prinos patarenskom pitanju," *GZMS,* XXIII, 1911, pp. 355-75.
Truhelka, Ć. "Još o testamentu gosta Radina i o Patarenima," *GZMS,* XXV, 1913, pp. 363-81.
Truhelka, Ć. "Osvrt na sredovječne kulturne spomenike Bosne," *GZMS,* XXVI, 1914, pp. 222-52.
Truhelka, Ć. "Vuk Banić-Kotromanić - povijestima dubrovačke arhive," *GZMS,* XXVII, 1915, pp. 359-63.
Truhelka, Ć. "Konavoski rat 1430-33," *GZMS,* XXIX, 1917, pp. 145-211.
Truhelka, Ć. "Larizam i krsna slava," *Glasnik Skopskog naučnog društva,* VII-VIII, 1929-30, pp. 1-33.
Truhelka, Ć. *O porijeklu bosanskih Muslimana,* Sarajevo, 1934.
Truhelka, Ć. *Studije o podrijetlu,* Zagreb, 1941.
Truhelka, Ć. "Sredovječni stećci Bosne i Hercegovine," *Poviest BH,* 1942, pp. 629-41.
Truhelka, Ć. "Bosanska narodna (patarenska) crkva," *Poviest BH,* 1942, pp. 767-93.
Turdeanu, S. "Apocryphes Bogomiles et Apocryphes Pseudo-Bogomiles," *Revue de l'Histoire des Religions,* 138, 1950, pp. 22-52, 176-218.
Uglen, S. "Četvrtak kao narodni opći praznik u okolici jajačkoj," *GZMS,* IV, 1892, pp. 270-71.
Uglen, S. "Odakle je došlo ime 'šokac'," *GZMS,* IV, 1892, pp. 272-73.
Uglen, S. "Priče iz Foče," *GZMS,* IV, 1892, pp. 452-53.
Vaillant, A. "Un Apocryphe Pseudo-Bogomile-La Vision d-Isaïe," *Revue des Études Slaves,* XLII, 1963, pp. 109-21.
Vasić, M. "Etničke promenu u Bosanskoj Krajini u XVI vijeku," *GID,* XIII, 1962, pp. 233-50.
Vasiliev, A. "Jorg of Nurenberg," *Byzantion,* X, 1935, pp. 205-09.
Vego, M. *Povijest humske zemlje,* part I, Samobor, 1937, (part II was never written).
Vego, M. Review of fra L. P.(etrović)'s *Kršćani bosanske crkve,* in GZMS, n.s. IX (ist-etn), 1954, pp. 183-85.

Vego, M. *Naselja bosanske srednjevjekovne države,* Sarajevo, 1957 (contains the best map of Medieval Bosnia).
Vego, M. *Historija Broćna od najstarijih vremena do turske okupacije,* Sarajevo, 1961.
Vego, M. "Patarenstvo u Hercegovini u svjetlu arheoloških spomenika," *GZMS* (arh) n.s. XVIII, 1963, pp. 195-215.
Vego, M. *Bekija kroz vijekove (Područje općine Grude i Posušje),* Sarajevo, 1964.
Vego, M. Review of S. Ćirković, *Istorija . . .,* in *GZMS* (arh) n.s. XX, 1965, pp. 295-304.
Vego, M. "Postanak imena Bosna i Hercegovina," *Pregled,* LXII, No. 1, Jan. 1972, pp. 109-18.
Verner, E. See Werner.
Vidović, D. "Pretstava kola na stećcima i njihovo značenje," *GZMS,* n.s. IX 1954, pp. 275-78.
Vidović, D. "Simbolične predstave na stećcima," *NS,* II, 1954, pp. 119-36.
Vinaver, V. "Trgovina bosanskim robljem tokom XIV veka u Dubrovniku," *AHID,* II, 1953, pp. 125-46.
Vitanović, T. "Manastir Ozren," *GZMS,* I, No. 2, 1889, pp. 32-38.
Vladić, J. *Uspomena o Rami i Ramskom franjevačkom samostanu,* Zagreb, 1882.
Voje, I. "Sitni prilozi za istoriju srednjovjekovne Bosne," *GID,* XVI, 1965, Sarajevo, 1967, pp. 277-82.
Vračko-Korošec, P. "La Necropole de Han-Hreša," *GZMS,* LII, No. 1, 1940, pp. 45-47.
Vrana, V. "Književna nastojanja u sredovječnoj Bosni," in *Poviest BH,* 1942, pp. 794-822.
Vryonis, S. "Byzantium and Islam, Seventh-Seventeenth Century," *East European Quarterly,* II, 1968, pp. 205-40.
Vujić, M. "Prvo naučno delo o trgovini Dubrovčanina Benka Kotruljića," *Glas* (SKA), LXXX, 1909, pp. 25-123.
Vukanović, T. "Srebrnica u srednjem veku," *GZMS,* n.s. I, 1946, pp. 51-80.
Vukičević, M. "Iz starih Srbulja," *GZMS,* XIII, 1901, pp. 31-70, 289-350.
Vulcanescu, R. "Les Signes Juridiques dans la Region Carpato-Balkanique," *Revue des Études Sud-est Européenes* (Bucharest), II (1-2), 1964, pp. 17-69.
Vuletić-Vukasović, V. "Bilješka o Evangeliju Manojla Grka," *Starinar,* XI, 1894, pp. 90-94.
Vuletić-Vukasović, V. *Bilješka o kulturi Južnijeh Slavena,* Dubrovnik, 1897.
Vuletić-Vukasović, V. "Funeral Customs and Rites Among the Southern Slavs in Ancient and Modern Times" (paper delivered at International Folklore Congress at Chicago, 1893; published in *Archives of the International Folklore Association,* Vol. I, Chicago, 1898, pp. 70-82.
Vygotsky, L. *Thought and Language* (trans. E. Hanfmann, G. Vakar), Cambridge, 1962.
Wach, J. *The Sociology of Religion,* Chicago, 1944.
Wach, J. *The Comparative Study of Religion,* New York, 1958.
Wallace, A. *Culture and Personality,* New York, 1964.
Weber, M. *The Sociology of Religion* (Eng. trans. of 4th ed.), Boston, 1963.
Welter, G. *Histoire des Sectes Chrétiennes des Origines à nos jours,* Paris, 1950.

Wenzel, M. "A Medieval Mystery Cult in Bosnia and Hercegovina," *Journal of the Warburg and Courtauld Institute,* XXIV, 1961, pp. 89-107.
Wenzel, M. "Some Notes on the Iconography of St. Helen," *Actes du XIIe Congrès - International d'Études Byzantines* (held, Ohrid, 1961), pub. Beograd, 1964, III, pp. 415-21.
Wenzel, M. "Some Reliefs outside the Vjetrenica Cave at Zavala," *Starinar,* n.s. XII, 1961, pp. 21-34.
Wenzel, M. "O nekim simbolima na dalmatinskim stećcima," *Prilozi povijesti umjetnosti u Dalmaciji* (Split), XIV, 1962, pp. 79-94.
Wenzel, M. "Lisnati krst na stećcima s područja Neretve," *Muzej primenjene umetnosti - Zbornik* (Beograd), VIII, 1962, pp. 39-49.
Wenzel, M. "Graveside Feasts and Dances in Yugoslavia," *Folklore,* 73, Spring 1952, pp. 1-12.
Wenzel, M. "Bosnian and Herzegovinian Tombstones - Who Made Them and Why," *Südost-Forschungen,* XXI, 1962, pp. 102-43.
Wenzel, M. *Ukrasni motivi na stećcima,* Sarajevo, 1965 (gives excellent maps of the geographical distribution of each motif. With bi-lingual Serbian and English text).
Wenzel, M. "Štitovi i grobovi na stećcima," *Vesnik* (Vojni muzej), Beograd, XI-XII, 1966, pp. 89-109.
Wenzel, M. "The Dioscuri in the Balkans," *Slavic Review,* XXVI, 1967, pp. 363-81.
Werner, E. "Bogomilstvoto i Rannosrednovekovnite Eresi v Latinskija Zapad," *Istoričeski Pregled,* XIII, No. 6, 1957, pp. 16-31.
Whorf, B. *Language, Thought and Reality,* Cambridge, 1956.
Wild, G. " 'Bogu mili' als Ausdruck des Selbstverständnisses der Mittelalterlichen Sektenkirche," *Kirche im Osten* (Göttingen, Vandenhoeck), VI, 1963, pp. 16-33.

The Serbian letter Ž (zh) shall follow all the Z's.
Zajcev, V. "K voprosu o vzaimootnošenii Bogomilstva i nekotoryh zapadnyh eretičeskih dviženij," *Vestnik Leningradskogo Universiteta,* 14 Ser. Ist., Jaz., Lit., III, 1964, pp. 109-16.
Zaninović, A. "Pogled na apostolsko-znanstveni rad Dominikanaca u hrvatskim zemljama," *Bogoslovska Smotra,* VIII, 1917, pp. 262-85.

NB. The Serbian letter Ž (Zh) shall follow all the Z's.

Zovko, I. "Narodne priče - kako su postali stećci," *GZMS,* I, No. 3, 1889, pp. 97-98.
Zovko, I. "Svetica Nedelja," *GZMS,* II, 1890, p. 116.
Zovko, I. "Nekoliko narodnih pripovedaka," *GZMS,* III, 1891, p. 99.
Zovko, I. "Vjerovanja iz Herceg-Bosne," *ZNŽOJS,* IV, 1899, pp. 132-50; VI, pt. 1, 1901, pp. 115-60; VI, pt. 2, 1901, pp. 292-311.
Žuljić, M. "Crtice o Bogomilima uopće a napose u Bosni i Humu," *Školski vjesnik,* XIII, 1906, pp. 119-30, 373-78, 531-37, 600-07, 697-710, 782-90, 891-942.
Županić, N. "Šišano kumstvo kod jugoistočnih Slovenaca i ostalih Slovena," *Glasnik Etnografskog instituta* (SAN), I (1-2), 1952, pp. 521-27.

REGISTER OF PERSONAL AND PLACE NAMES

In the course of this study many people appear once in lists (e.g. of those being anathematized or of those witnessing charters) or as diplomats, and about whom nothing else is known. Such people are usually not listed in this register. A complete list of Patarin diplomats can be found in note 195 for Chapter V (pp. 290-92). Other major lists of people can be found on pp. 128, 173-74, 213-16, 227. In the same way various places mentioned once about which little is known and which are not important to this study (e.g. lists of villages being granted in charters) are also not listed in this register. Scholars are listed only where their views are being discussed but are not listed for simple reference citations. Economic reasons have necessitated a shorter index; thus topics are not included. See Table of Contents for headings of chapter sub-divisions for major topics discussed. Bosnia, since it appears on almost every page, does not appear in this register.

Acontius (papal legate), 135, 161
Albania, 48, 102-03, 386
Albi, 117
Albrecht (King of Hungary, 1437-39), 255
Alexander III (pope), 114
Alfonso of Naples, 300-01, 302, 310, 322, 331
Altoman (Serbian nobleman) 208
Altomanović, Nikola (Serbian nobleman), 191, 192 208
Ambrosius, fra. (Franciscan), 76
Anatolia, 115, 117
Andjelić, P., 85, 122, 175, 204-05, 311, 353
Anselm of Alexandria (Inquisitorial author) 54, 117, 119-20, 200
Apulia, 118
Aristodios (Dalmatian heretic) 118, 217
Athos, Mt., 44, 388
Atom (pseudonym of a Serbian clerical scholar), 3
Babić, A., 2, 212, 272, 293, 332
Bakšić (author of 17th century visitation among Paulichiani) 101-02
Bakula (19th century clerical compiler) 85
Balšići of Zeta, 208. (Djuraj B., 208, Djuraj Stracimirović B. 233, Balša III B. 301)
Banja Monastery, 191, 277
Banja Luka, 37
Bar (Antibari), 146, 148, 155, 210
Barbara (Queen of Hungary), 234
Barbara of Lichtenstein, 322, 323
Barbara of "Payro", 322
Barbucci, Nicholas (papal envoy to Bosnia), 328-29, 341.
Bartholomaeus (a Cathar leader), 74-75
Bartholomaeus of Pisa (Franciscan vicar), 60, 106, 192-94, 260, 269, 284
Bašagić, Ibrahim beg, 85

Basel (Council of), 51, 52, 73, 195, 248-50, 285, 367
Batkovići (village, with stećci), 288
Batalo, Tepčija (Bosnian nobleman) 81, 109, 156, 201, 215-18, 220, 268, 279
Bela IV (King of Hungary 1235-70), 140, 142, 144-46, 153, 164, 205, 293
Beloslav, Patarin, 371
Benedict XII (pope), 179, 180
Benedict (Archbishop of Kalocsa), 146
Benedict Ovetarius, 53
Bernard, (Archbishop of Dubrovnik), 122, 123
Bernard (Archbishop of Split), 118, 124-26
Bešlagić, Š., 261, 286, 288, 293
Bijela (village with Patarin hiža), 205, 372, 378
Bijeljina, 196, 207, 260, 288-89
Bileća, 92, 282
Bilino polje, 11, 15, 51, 126-34, 138, 150, 174, 214, 217, 262, 277, 366
Biograd (village with hiža), 205, 311-12, 334-35, 363, 370, 372, 378
Biskup (near Glavatičevo, village with cemetery), 191, 332
Bladostus (Bladosius), Bosnian Catholic monk, 128, 217
Blagaj, 29, 253, 315
Blasius, St., (Patron saint of Dubrovnik), 240, 306
Bobali, Domagna di Volzo (Dubrovnik cleric, advisor to Stjepan Kotromanić), 66-67, 68, 71, 167, 182, 202, 206
Bobovac, 65, 106-07, 229, 247, 340, 353
Bočac, 306
Bogutovo (village with cemetery), 196, 260
Bohemia, 18, 59, 60, 110-11, 168, 247
Boljuni (village with stećci), 91
Boleslav (early Bosnian Church leader), 156, 157, 215
Boniface VIII (pope), 154
Boniface IX, (pope), 51, 198-99, 224, 307
Borač, 29, 205, 220, 239, 241, 258, 282, 283, 288, 306, 347
Borić, Ban, 285
Boril (Bulgarian Tsar), 17, 155
Bosigchus (Bosnian swindler), 258
Bradina (village with Patarin hiža), 204, 257, 277, 372
Branivojevići (noble family of Hum), 171
Branković, George (Serbian despot, 1427-56), 248, 301, 310, 328
Bratislav (Catholic Bishop of Bosnia), 148
Brdo (site of church near Vrhbosna), 140
Bribir, 154, 179
Buda, 221, 232-33
Bulgaria, 1, 5, 17-18, 55, 59, 72, 75, 83-84, 101, 115-17, 118, 119, 137, 149, 153, 155, 158-59, 168, 174, 192, 193, 232, 300, 301, 320
Bugojno, 369
Butko, pop (Catholic priest, made a Gospel), 82, 220, 224, 268, 280
Byzantine Empire, 113, 115, 116, 117, 161, 232

Calixtus III, (pope), 326
Caloiannes (Cathar Bishop of Mantua), 118, 119
Campanus (papal secretary), 64
Cantacuzena, Anna (wife of Vladislav Kosača), 263, 322
Cantacuzenus, Manuel, 263
Cernica (village in Hercegovina), 252, 263
Charles Robert (King of Hungary, 1301-42), 177
Chieri (Italian town, residence of 'Bosnian heretics') 56, 98, 199
Clement VI (pope), 48, 65, 203
Coloman (Duke of Croatia), 121, 138-39, 141, 142, 143, 144, 163
Conrad (papal legate to France), 74-75, 137
Constantinople, 117, 119, 120, 136,

158, 159, 232. 296, 312, 324, 330
Croatia, 17, 59, 75, 89, 119, 135, 137, 138, 154, 155, 160, 170, 173, 179, 180, 198, 199, 221, 223, 232, 233, 235, 279, 381, 384
Cvatko gost, 290, 363-64
Cvjetko Patarin, 371, 374
Čačak, 88
Čemerno (mountain), 315
Čihorić family, 191
Čremošnik, G., 45, 164, 245, 246
Ćirković, S., 2, 124, 150-51, 202, 211, 225, 256, 277, 278, 307, 312, 313, 328, 332, 338, 340
Ćorović, V., 2, 201

Dabiša (King of Bosnia, 1391-95), 211, 216, 219-21, 277
Dalmatia, 10, 17, 31, 42, 45, 52, 55, 57, 62, 75, 76, 89, 90, 91, 113, 114, 117, 118-21, 124-26, 129, 131-33, 135-37, 147, 154, 160-63, 177-78, 185, 197-200, 203, 218, 220-25, 232, 234, 249, 255, 308, 327, 331, 333, 334, 340, 345, 375, 380, 381, 390
Danilo (Serbian Archbishop and author), 43, 154, 196, 289
Danube, 66, 72, 248
David (Metropolitan at Mileševo), 71, 153, 156, 166, 292, 364, 365
D'bar, 44
Debar, 44
Deževica, 185, 331, 332
Dinić, M., 42, 45, 211, 230, 255, 266, 281, 288
Dinjičić family (Bosnian nobility) 257, 262, 264, 265, 270, 271, 276, 288, 322, 342
Dinjičić, Peter, 270, 292, 311
Dioclea (Zeta), 124
Djakovo, 46, 95, 148, 150, 153, 169, 185-189, 192, 200, 207, 233, 245
Dmitar, starac (Patarin diplomat), 238-40, 249, 264, 265, 272, 283, 291
Doboj, 235, 345
Dobor (in Usora), 232
Dobrovojević, Vlah (Bosnian noble), 214, 268, 278
Dobrun, 191, 220, 282
Dondaine, A. 119-20
Donji kraji, 106, 154, 170, 172, 185, 188-90, 198, 212, 219-20, 223-24, 251
Dorothea (of Vidin, Tvrtko I's wife), 192
Dorothy (Tvrtko II's wife), 242, 247, 250, 284
Drageta (Brageta, Bergela, Dražeta; Bosnian Catholic monk), 128, 133, 216-17
Dragice (Dragite: Bosnian Catholic monk), 128, 216-17
Dragišić brothers (Bosnian noblemen: Pavle, Marko, Iuri), 270-71, 275, 303-04
Dragohna (Catholic bishop of Bosnia), 134-35
Dragovica, 55, 117, 118, 119
Draživojević family (Hum nobility), 171
Drežnica, 85
Drijeva, 237, 315
Drina, 44, 136, 191, 196; 201, 203, 208, 220, 241, 250, 254, 258, 260, 261, 278, 281, 287-88, 311, 327, 328, 343, 378, 379
Dubočani, 85
Dubrovnik, 2, 6, 10, 13, 29, 41, 42, 44, 45, 52, 62, 63, 67, 68, 70, 71, 73, 74, 78, 114-15, 121-26, 128, 134-36, 140, 142-43, 145-50, 152-53, 163, 166-68, 171-72, 182, 187, 190, 199-203, 208, 210, 212, 222-32, 236, 238-44, 248, 249, 251-60, 263-67, 269-72, 274, 277-79, 282-84, 291-92, 296, 302, 303, 305-07, 311-17, 319-22, 327, 329, 331, 338, 340, 349, 351, 359, 364-68, 370-73, 375, 378
Dursum beg (Turkish historian) 340
Duvno, 170, 172, 220, 226, 227, 269, 338, 380

Eger, 232
Elena (Stefan Vukčić's daughter), 253
Elizabeth (Sigismund of Hungary's daughter), 233
Emanuel the Greek (author of a chronicle), 41, 73, 94
Endre (King of Hungary, 1205-35), 135, 136, 160
Eugene IV (pope), 54, 59, 63, 251, 301, 302, 304, 305, 306, 314, 338
Evans, Arthur, 88
Evans-Pritchard, E., 22-23
Fabian, fra., (Franciscan missionary), 177, 178, 179, 207
Fabian, fra (Franciscan vicar), 301
Farlati, 1, 53, 54, 98, 302
Fatnica, 282
Fejer, G., 60
Fermendžin, E., 60, 76, 246
Filipović, M., 203-04, 367-68
Filipović, N., 334, 385
Foča, 13, 253, 258, 263
Fojnica, 75, 76, 185, 331, 332
France, 5, 9, 43, 50, 55, 57, 74-75, 99, 117, 119-20, 125, 137-38
Friuli, 66

Gacko, 15, 203, 208, 385
Garsoian, N., 296
Gennadius (Patriarch of Constantinople), 296, 324-26, 385
Georgijević, Athanasius (17th century visitor), 384
Gerard Odinis (Franciscan general), 180-82
Gioanni, Cardinal di S. Angelo, 63, 70, 303, 330
Glamoč, 170, 171, 172, 207, 220, 226, 227, 269, 306, 380
Glasinac, 241, 273
Glavatičevo, 191
Glaž, 192, 207
Glušac, V., 2
Gobellinus (papal copyist), 64
Goičin, pop, 260
Goislava of Hum (Serbian noblewoman), 208
Gojisav gost, 78, 258-60, 261, 264, 273, 276, 277
Gomilanin land, 305
Gomiljani pravica (site Orthodox church ruins near Trebinje), 305
Gondola, Giovanni di Marino (chronicler), 74
Goražde, 205, 253, 258-59, 261, 270, 276-77, 286, 311, 322-23, 363, 378
Gorica, (Beški on Lake Skadar), 253
Gorjanski, Ivan (Johannes Garai), 242
Gostović (Orthodox monastery), 380
Gradješa (Bosnian judge), 122
Greece (and Greeks), 55, 66, 72, 119, 153, 263, 327
St. Gregory (patron saint of Bosnia), 173, 174, 200, 203-04, 339
Gregory IX (pope), 136, 137-43
Gregory XI (pope), 11, 51, 59, 183, 192-97, 200-01
Gregory XII (pope), 232
Grigorovič, V., 44
Gundulić (Ragusan ambassador to Bosnia), 236, 277

Hercegnovi (Novi), 312, 320
Hercegovina, 172, 190-91, 208, 220, 223, 231, 270-71, 278, 293, 312, 313, 315-16, 318, 322-27, 331-34, 341, 342, 344-45, 352, 363-74, 375-76, 378-79, 381, 382, 386-88
Hlivno (Livno), 170, 172, 345
Hodidjed (fortress above Sarajevo), 255, 311
Hofstadter, R., 50
Honorius III (pope), 135, 136, 161
Hrvatinić (family), 188, 189, 198, 223, 237. (See also Stjepanići-Hrvatinići)
Hrvatinić, Balša (son of Hrvoje Vukčić), 237
Hrvatinić, Vuk (Brother of Hrvoje), 222, 279
Hrvatinić, Vukac (or Vlkac), 107, 189, 198, 216, 220
Hrvatinić, Vukoslav, 173, 269, 290
Hrvoje Vukčić, 41, 73, 81, 82, 108-09, 198, 219-26, 228-30, 232-37, 240, 243, 248, 251, 264, 268-69, 271, 279, 281, 282

Hum (region of), 13, 43, 48, 67, 82, 114, 117, 122, 124, 136, 140, 142, 155, 158, 160, 163, 164, 171, 172, 173, 188, 198, 208, 212, 220, 225, 278, 306, 315
Hum (town of), 369
Humsko (near Foča; site of important stećak), 263
Hungary (and Hungarians), 45-46, 48, 50, 59, 63, 75, 90, 95, 96, 106-07, 114, 121, 123, 124, 126, 128, 133-52, 160-65, 170, 173, 177-81, 186, 188-90, 197-99, 202, 207, 212, 220, 221, 224-25, 228-30, 232-33, 235-36, 242, 243, 246, 249-50, 255, 267, 295, 301, 305, 313, 314, 316, 328, 329, 333, 340, 341, 344-46, 352, 378, 380, 386, 387
Hunyadi, 300, 301
Hval Krstjanin, 81, 82, 214, 218-19, 224, 231, 268, 357
Hvar, 299

Imjani, 87-88
Imota, 170, 207
Imre, (King of Hungary, 1196 1204), 124, 133, 160
Innocent III (pope), 51, 97, 124, 125, 126, 128, 132
Innocent IV (pope), 145-48
Innocent VI (pope), 48, 95, 189
Isak Beg (Turkish governor of Skopje), 235, 236, 237
Istria, 154, 375
Italy, 5, 9, 43, 51-52, 54, 56-58, 60, 62, 65, 98, 99, 117-20, 124, 178, 186, 199-200, 209-10, 295, 308, 322, 329, 331, 335, 336

Jacob Bech (Italian heretic), 56, 57, 98-99, 199-200
Jacob de Marchia (Franciscan vicar), 18, 59-60, 70, 110-11, 207, 244-48, 250, 286, 289
Jacob of Preneste (papal legate), 138
Jagić, V., 44
Jajce, 106, 207, 220, 234, 237, 245-46, 248, 306, 307, 328, 330-31, 332, 339, 343, 345, 353, 378, 381
Janjići (village associated with djed), 184, 204, 227, 256, 262
Jelena (also known as Gruba, Queen of Bosnia, 1395-98), 221
Jelena (Hrvoje's wife; later wife of King Ostoja), 223, 224, 237, 269, 282
Jelena (Hrvoje's niece; Sandalj's first wife), 221, 231, 233
Jelena (Knez Lazar's daughter; Sandalj's second wife), 233, 239, 241, 252, 253-54, 283, 323, 358, 365
Jelena Balšić (Herceg Stefan's first wife), 301, 322
Jelena (wife of King Stefan Tomašević, taking name Maria), 330-31, 339
Jelenić, J., 60
Jelisaveta (Daughter of Stj. Kotromanić), 188
Jelisaveta (Daughter of Stefan Dragutin), 154, 167
Jelisaveta (Duchess of Mačva), 153
Jezero, 307, 331
Jireček, K., 17, 241
Johannes de Casamaris (papal legate), 48, 125, 126-34, 138
Johannes Horvat, 196
Johannes of Korčula (Bosnian vicar), 245, 251, 286
Johannes von Wildeshausen (Catholic Bishop of Bosnia), 139, 140
John XXII (pope), 51, 76, 169, 177, 214
John Angelus (of Srem), 136, 161
John the Baptist, 6, 18, 82, 116, 218, 357, 375; (St. J's eve, 16, 17, 19, 37, 121).
John of Capistrano, 309, 326-27, 380
Jorga, N., 45
Jovan (Orthodox monk), 253
Jurijević (brothers), 251

Kačić family (pirates), 97, 135
Kačić, Bartholomaeus (17th century Jesuit visitor to Srem), 11
Kajmaković, Z., 15, 323

Kalocsa (Archbishopric of), 128, 135-39, 143-50, 154
Katarina (Daughter of Herceg Stefan; wife of King Stefan Tomaš), 106, 302, 305, 310, 330
Klaić, V., 202, 219
Klešić, Pavle (Bosnian nobleman), 171, 198, 220, 226-30, 249, 251, 269, 271, 272, 274, 281, 292, 321, 359
Klešić, Vladislav (Bosnian nobleman), 250-51, 269, 315
Klis, 179, 223
Ključ (in Hercegovina), 252, 253, 263, 365
Ključ (on the Sana), 189, 270, 304, 339, 345
Knezović family, 106
Kniewald, D., 2, 195, 366
Knin, 179, 210, 223
Konavli, 208, 210, 222, 238, 242, 267, 269, 278, 314, 316, 327
Konjic, 85, 191, 205, 253, 257, 282, 311, 327, 331, 332, 335, 363, 366, 372, 378; (bogus "Council of K.," 54, 108, 304)
Konstantin Filozof (Serbian author), 44, 86, 242
Kosača family, 220-21, 250, 251-54, 258, 263-65, 267, 270-71, 276, 278, 286, 300, 312, 323-24, 332. See also Vlatko Vuković, Sandalj Hranić, Stefan Vukčić.
Kosača, Beoka, 286, 293
Kosača, Stefan Vukčić. See Stefan Vukčić
Kosača, Stefan (Ahmed Hercegović, son of Herceg Stefan), 364, 373
Kosača, Vladislav (Son of Herceg Stefan), 253-54, 263, 270, 290, 314, 315, 317, 322, 339, 345, 349, 353, 373
Kosača, Vlatko (Son of Herceg Stefan), 254, 373, 375, 379
Kosače (village of), 286
Kosanović, S., 85
Košarići (village with medieval cemetery), 260
Kosman (Fortress of Sandalj and Herceg Stefan), 253
Kosovo, 198, 233, 379
Kotor, 42, 155, 156-57, 166, 255, 312, 373
Kotruljić, Benko (Ragusan merchant-author), 296, 329-30
Krajina, 170, 172, 202, 315, 345, 364, 380, 381
Krajinović, J., 289
Krbava, 179
Krčelić, B.A., 107
Kreševljani (medieval site), 91
Kreševo, 73, 85, 106, 114, 185, 285-86, 331, 332
Krmpota, 375
Krupa (on the Vrbas, also called Greben), 87-88, 92, 195, 207, 306
Kučinić, Djuraj (Heretical Bosnian nobleman), 335-38
Kujava (Second wife of King Ostoja), 282
Kulin, Ban (Ruler of Bosnia, ca. 1180-1204), 48, 113-15, 118, 121-29, 132, 133, 136, 140, 159, 160, 162, 174, 175, 205, 285
Kuripešić (16th century Austrian envoy), 80

Ladislas IV (Hungarian King, 1272-90), 153
Ladislas of Naples, 197-98, 220-25, 230, 232-34, 243, 300
Lampredius (Bishop of Trogir), 180
Languedoc, 99
Laonikos Chalkokondyles, 324, 325, 326
Lašva, 60, 106, 108, 183, 184, 216, 220, 227, 260
Lašvanin, Nikola (chronicler), 76, 377
Lazar, Knez (Serbian local ruler), 208, 233, 252, 283
Lea, Ch., 99
Leonardo (Archbishop of Dubrovnik), 134-35
Lepenica, 383
Lim, 44, 114, 158, 172, 191, 203, 208, 253, 277, 278, 281
Livno, See Hlivno

Ljubskovo (Village with Patarin hiža), 204, 256-57, 262, 270, 277, 287-88
Ljubomišlja (Village; possibly same as Ljubskovo), 288
Ljubušak, Mehmedbeg Kapetanović, 85
Ljubuški, 315, 331
Lockwood, W., 31
Lombardy, 118, 119
Lomnica (Orthodox monastery), 380
Louis (or Lajosh, King of Hungary, 1342-82), 188-89, 190, 207
Lubin (Bosnian Catholic monk), 128, 133, 216, 217
Luccari, 65, 73-74, 244
Lucius, 233, 235
Luke, Saint (relics of), 330-31, 339

Macedonia, 44, 75, 116, 117, 155, 171
Mačva, 43, 153, 154, 196, 201, 207, 289, 351
Makarska, 170, 171
Mandić, D., 61, 110-11
Manoilo Grk, 94
Mantua (Bagnolo), 118
Maravić, Marijan (17th century visitor), 384
Marianus, fra. of Florence, 110-11
Marica, battle of, 208
Marinus (Archdeacon of Dubrovnik), 48, 125, 126-28
Marinus (Catholic Bishop of Ston), 203
Marinus (Bosnian Franciscan, Custodian father), 314
Marko Kraljević, 282
Martić, fra., 85
Massarechi, Peter (17th century visitor), 65, 67-68, 69, 73, 167, 384, 389
Matasović, J., 61
Matheus (Dalmatian heretic), 118
Matyas Corvinus (Hungarian King, 1458-90), 328, 340, 341, 345, 353, 369
Medjedji (location of medieval cemetery), 261
Menhard of Ortenberg, 167
Miaša gost, 239, 243, 267, 291
Michael of Paris (chronicler), 74-75
Mihailović, Konstantin (medieval historian), 340
Miklosich, F., 283
Milac, pop, (Chaplain for Dabiša), 220
Milaja pop, 241
Milan, 71
Mileševo (monastery of), 19, 43, 106, 172, 191, 192, 208, 209, 277, 278, 310, 322, 324, 326, 327, 364, 365, 367
Milich Velimiseglich, (medieval chronicler), 41, 73
Miloje, djed, 303-04
Milorad Patarin (of Bradina), 257
Miloradović family, 175, 382
Milutin, gost, 263-64, 276
Milutin (Serbian King, 1282-1321), 155, 156, 171, 202
Mirohna, Abbot, 107
Miroslav, djed, 153, 155-57, 166, 215, 217, 373
Miroslav of Hum, 82-83, 114, 124, 158, 160
Mišljen, gost, 227, 262, 357
Modrića, 207
Moldavia, 59, 247, 301
Montenegro, 35, 89, 124, 161, 168
Mošin, V., 45, 56, 164, 213
Mostar, 13, 29, 70, 306, 331
Moštre (village with hiža), 174, 175, 176, 184, 204, 256, 290
Mrnavić, fra Ivan Tomko, 107-08
Muhašinovići (village; medieval site), 121, 159, 175, 176
Musa Kesedžija, 282

Nazarius (Cathar Bishop of Concorrezo), 84, 117
Nelipac family, 164, 170, 179, 206, 279
Nelipčić, Ivaniš, 223, 224, 279
Neretva, 85, 108, 114, 157, 160, 163, 172, 205, 253, 282, 306, 311, 315, 334, 345, 364, 366, 371, 372
Nevesinje, 91, 191, 282
Nicholas IV (pope), 154
Nicholas V (pope), 53, 54, 63, 302, 307-10, 314

Nicholas Modrussienses (Catholic Bishop), 340, 341, 353
Nicola (Cathar Bishop of Vicenza), 118
Nicopolis, 66, 72, 221, 377
Nikandar of Jerusalem (Orthodox elder), 253
Niketas (Bogomil bishop), 117
Nikšić (Onogost), 253, 255
Nin, Bishop of, 58, 296, 335, 336, 337
Ninoslav, Ban (mid 13th century Bosnian ruler), 48, 138-47, 149, 153, 161-65, 278
Niš, 300
Nodilo, S., 74
Novi Pazar, 29

Obižen, Patarin, 311, 371
Obradović, Dositej, 24-25
Ohmučević family, 108
Okiç, T., 4, 77-80, 261, 334, 369, 382
Olovo, 60, 183, 184, 220, 241, 282, 293, 327
Omiš, 97, 135, 171, 223
Orahovica (on the Neretva), 334
Orbini, M., 3, 41, 65-73, 101-02, 121-22, 123, 134, 150, 167, 181, 182, 187, 201, 202, 206, 208, 218, 276, 277, 330, 332, 352, 365, 377
Ostoja (King of Bosnia, 1398-1404, 1409-18), 221, 222, 225-31, 233-34, 236-38, 250, 255, 265, 269, 271, 274, 275, 281, 282, 291, 292, 298, 299, 302, 303, 359
Ostoja krstjanin (of Zgunje), 261, 290, 343
Ottocar of Bohemia, 153
Ozren (Orthodox monastery), 380

Padua, 331
Pagaminus (Bishop of Ulcinj), 316-17
Pal Chupor (Hungarian nobleman), 236
Palavestra, V., 92, 288
Papraća (Orthodox monastery), 380
Paul of Samosata, 66, 72, 73
Pavle Radenović, (Bosnian nobleman; from whom derived the Pavlovići), 220, 233, 234, 236, 237, 269, 278, 281, 282, 293, 350
Pavlović family, 220, 236, 237, 249, 258, 262, 264, 265, 267, 269-70, 271, 276, 278, 288, 293, 298, 311, 321-22, 329, 342, 347
Pavlović, Ivaniš, 74, 244, 255, 269, 300, 306, 309, 311, 322, 347, 353
Pavlović, Nikola, 292, 321-22
Pavlović, Peter I, 236-37, 269
Pavlović, Peter II, 292, 311, 321-22
Pavlović, Radoslav, 74, 237-44, 248, 249, 255, 256, 258, 265-67, 269, 272-74, 278, 283, 284, 291-93, 313
Peć, 19
Pelješac, 171
Peregrin Saxon (Vicar and then Bishop of Bosnia), 65, 67, 69, 95, 96, 148, 182-83, 185-88
Perojević, M., 340
Persia, 115, 119
Peter, Tsar of Bulgaria, 115
Peter (Bishop of Bosnia), 95, 148, 188, 189, 192
Peter, St. and St. Paul Church on the Lim, 114, 158, 172, 191, 277
Petko Krstjanin, 262, 323
Petrić, M., 92
Petrović, L., 2-3, 343
Philadelphia, 55, 119
Philippopolis (Plovdiv), 115, 117
Piedmont, 56, 57, 98
Pietro Livio of Verona, 3, 4, 41, 62, 63, 65-73, 101, 167, 202
Pišče (village on Pivska planina), 290, 317, 323, 350
Pius II (pope; Enea Silvio de Piccolomini), 1, 46, 62-65, 68-71, 75, 76, 111, 150, 300, 303, 330, 332-40, 385
Piva River, 220, 262, 323, 363
Pivska planina, 317, 323, 350
Plevlje, 385
Podbrežja (medieval site), 122
Poljice, 331

Poljska (stećci site), 260
Pomazanić, Bernard (16th century traveler), 385
Ponsa (Catholic Bishop of Bosnia), 140-44, 146-48, 163
Popovo polje, 191, 282
Požega, 136, 153
Prača, (medieval fortress), 220
Priezda (13th century Ban in north of Bosnia), 138, 154, 161, 162, 278
Prijepolje, 208
Prozor, 13, 345
Puhovac (site of important stećak), 227, 262
Rački, F., 2, 54, 56, 57, 108, 335, 339
Radak "Manichee", 65, 70, 339-40, 341, 353
Radašin Krstjanin (Patarin diplomat), 265, 272, 291
Radašin Krstjanin (referred to on stećak inscription), 261
Raden Jablanić (Bosnian nobleman; father of Pavle Radenović), 220
Radigost (Catholic Bishop of Bosnia), 115, 123, 285
Radimlja (Medieval cemetery site), 89, 175
Radin, gost (surname, Butković), 3, 4, 6, 45, 60, 70, 73, 79, 89, 153, 156, 176, 204, 239, 240, 244, 249, 253, 254, 256, 260, 264-65, 267, 272-73, 276, 283, 290, 291, 292, 293, 295-96, 300, 301, 307, 311, 314, 317-21, 334, 337, 344, 345, 357-60, 364-70, 372, 374, 378
Radin Seoničan (Gost Radin's nephew; also a gost), 205, 366-69, 372
Radin, starac (Patarin cleric), 81, 216
Radisav Patarin, 371-72
Radišić brothers (Pavle and one whose name unknown), 225, 228, 256, 272, 321
Radivoj gost (of Bijela, surname Priljubović), 369, 372
Radivoj (Brother of King Stefan Tomaš), 250, 255, 299, 306, 331, 338, 347
Radivojević, Djuraj (Bosnian nobleman), 233, 251
Radoe Rug(h)iza (Bosnian landlord and thief), 257-58, 259
Radohna Krstjanin, 175-76
Radomer (Radomir, djed), 81, 213-15, 219, 231
Radosav, starac (Patarin diplomat for Pavlovići), 292, 321-22
Radosav Bradiević, gost (Pavlović diplomat and a stroinik), 292, 321-22
Radosav, krstjanin (author of Radosav Ritual), 81, 83, 296, 304-05, 359
Radoslav, djed, 156, 173-74, 215, 217
Radoslav, gost, 156, 173-74, 176, 215, 217
Radoslav Pavlović, see under Pavlović
Radovac krstjanin (Pavlović official in Konavli), 238
Radoycho of Ljubskovo, 257
Ragnina, Nicolo di (Ragusan chronicler), 122, 123, 134
Ragusa, See Dubrovnik
Rainer (Archbishop of Split), 114
Rajković, Stjepan (recipient of grant in dubious charter), 106-09, 200
Rama, 85, 331, 345, 369
Ramberto (16th century traveler), 365
Ranzanus, Peter (Hungarian historian), 346, 359
Rastudije (Bosnian Church leader), 156, 213, 215-17
Raško gost (possibly Ratko, from Košarići stećak), 260, 288
Raštani, 306
Ratko (early Bosnian Church leader), 156, 157, 216, 217, 218
Ratko, djed, 304
Ratko (Presbyter and protovestijar for Tvrtko I), 200, 210
Ratko, pop, 222
Ratković, Radohna (nobleman in service of Sandalj), 252
Raynaldi, 1, 53, 309, 353
Rayner Sacconi (Inquisitor), 54, 55
Resa (Batalo's wife; Hrvoje's sister), 216, 220

Resti, Junius, 74, 115, 123, 134-35, 218, 222, 240, 252, 299, 303
Restoje, protovestijar, 306, 309
Roger of Wendover (Medieval chronicler), 74-75
Romanija (mountain), 311
Rome (excluding references to papacy), 66, 336
Rudnik, 191, 192, 208
Runciman, S., 108
Ruthenia, 59
Ruxin (Patarin of Ljubskovo), 257
Rycaut, Paul, 18-19

Sabellico, Marc Antonio, 1, 64, 65, 67, 68, 69, 71
Saint Felix-de-Caraman (France), 117
Salonika, 325
Samobor (on the Drina), 253
Sana, 216, 220, 270
Sandalj Hranić, (Bosnian nobleman; of Kosača family), 6, 172, 220-21, 222, 225, 230-31, 233-41, 243, 248-53, 262, 264, 267, 269, 270, 273, 274, 278, 281, 282, 283, 286, 291, 299, 321, 323, 324, 358, 365
Sanko Miltenović (Bosnian nobleman; from whom derived the Sankovići), 171, 191, 277
Sanković family, 203-04, 281, 282, 332
Sanković, Beljak, 171, 191
Sanković, Radača, 191
Sanković, Radič, 171, 191, 278, 321
Sarajevo, 29, 255, 311, 326, 379, 380, 384, 385 (Oriental Institute of, 77: Zemaljski muzej of, 88, 93, 121, 263)
Sardica (Sofia), 155, 300
Sava River, 114, 136, 139, 144, 196, 248, 260, 289
Sava, saint (including his cult and relics), 19, 39, 160, 191, 310-11, 324, 358, 388
"Sclavania" (Region in which a dualist "Church"), 54, 55, 117, 118-21, 125, 133, 139, 159, 162, 199-200
Senj (Bishop of, 147), (Prince of, 179)
Seonica, 205, 342, 366, 372
Serbia (including references to Serbs and to Raška), 48, 59, 75, 89-91, 106-08, 114, 117, 119, 124, 149, 153-57, 160, 168, 171-73, 181, 191-94, 196, 197, 203, 208, 214, 215, 232, 233-34, 241-42, 244, 248, 250, 254, 278, 279, 284, 300, 301, 310-12, 314, 322, 326-28, 330, 344, 379, 381, 388
Sibislav of Usora, 140, 162
Sigismund (King of Hungary, 1387-1437), 51, 197-99, 220, 221-26, 228-30, 232-35, 236, 242, 243, 249-50, 255, 307
Sigismund (Son of Stefan Tomaš), 374
Sinai, Mt., 324, 325, 388
Skarić, V., 263, 325
Skopje, 235
Skr'batno (monastery of), 81, 215
Slavonia, 46, 59, 95, 136-37, 139, 140, 144, 148, 150, 153, 154, 161, 163, 177, 178, 179, 180, 187, 223, 224, 225, 233, 305, 381
Smederevo, 328, 330, 352
Soko (Sokograd in Plivska župa), 106-07, 189
Sokolgrad (on Drina above Šćepan polje), 253, 262, 290, 323, 350, 363, 378
Sol (Tuzla), 43, 106, 135, 140, 154, 165, 170, 172, 185, 212, 233, 331
Solovjev, A., 2, 18, 82, 83, 203, 213, 217, 320, 321, 368
Split, 42, 45, 118, 123-24, 126, 128, 134, 135, 147, 163, 164, 170, 190, 198, 219, 223, 224, 225, 233
Srebrnica, 29, 44, 86, 189, 207, 220, 241-42, 257, 283-84, 287-88, 311, 312
Srebrnik (in Usora), 106, 189, 345
Srem, 11, 59, 61, 136, 161, 221, 248, 289
Stanko Kromirijenin, 216
Stefan Dragutin (Serbian King, later ruler Mačva etc.) 43, 153-54, 167, 170, 191, 196, 201, 278, 289, 351

Stefan Dušan (King — then Tsar — of Serbia, 1331-55), 66-67, 171, 181, 208, 278
Stefan Lazarević (Serbian Despot, 1389-1427), 44, 86, 233, 241-42, 282
Stefan Nemanja (Grand Župan of Serbia, 1168-96), 114, 117, 158, 160
Stefan Ostojić (King of Bosnia, 1418-21), 237-38, 279, 282
Stefan Prvovenčani (Ruler of Serbia, 1196-1227), 124, 160
Stefan Tomaš (King of Bosnia, 1443-61), 53, 54, 63, 70, 73, 76, 81, 106, 255, 256, 270, 273, 275, 278, 299-312, 314-15, 318, 328-35, 338, 342, 352
Stefan Tomašević (King of Bosnia, 1461-63), 70, 279, 300, 328, 330, 331, 338, 339, 341
Stefan Vukčić (Herceg Stefan), 70, 73, 153, 156, 166, 172, 238, 240, 244, 249, 252-56, 258-59, 262-65, 267, 270, 272, 273, 276, 283, 286, 290, 291, 292, 296, 298-302, 305-07, 309-24, 328-33, 335, 339, 340, 342, 344-46, 349-51, 353, 357, 358, 363-65, 369-71, 373, 378, 388
Stephen, Brother (a Franciscan used as a diplomat), 237, 238, 282
Stjepan (a 13th century ban), 140, 162
Stjepan Kotroman, 122, 154, 167, 191, 201, 278
Stjepan Kotromanić (Ban of Bosnia ca. 1315-53), 66-67, 70, 71, 73, 95, 122, 154, 167-71, 173-75, 177-88, 201-04, 206, 207, 268, 274, 278, 279, 290, 307
Stjepanići-Hrvatinići (Noble family of Donji kraji), 170, 173; see Hrvatinići
Stolac, 175, 382
Ston, 13, 67, 114, 136, 158, 171, 203, 327
Suibert (Peter Patak; Dominican chronicler), 121, 133, 139, 162
Sutjeska (Kraljeva S.), 60, 73, 75, 76, 183, 184, 198, 220, 236, 246, 277, 338, 389
Symeon (Archbishop of Salonika), 324-26
Symeon (Tsar of Bulgaria), 116
Symeon Semeonis (Franciscan visitor), 168, 359
Šabanović, H., 255, 383
Šćepan polje, 323-24
Šidak, J., 2, 82, 83, 107, 108, 183, 206
Šišić, F., 45, 46, 95, 107
Šunjić, M., 4, 370
Šubić family, 154, 156-57, 164, 167, 170, 173, 179, 201-02, 206
Šubić Djuraj, 170
Šubić, Jelena, 187
Šubić, Mladen I, 154, 201
Šubić, Mladen II, 76, 167, 169, 201, 202
Šubić, Pavle, 201, 202
Šurmin, Dj., 106

Tara River, 220, 323, 363
Tardislavić, Geo. (Hum nobleman), 306
Tatars, 144, 164
Tavan, 92
Tavna (Orthodox monastery), 380
Tephrice, 115
Terbipolensis bishop, 51, 52, 97, 195, 249-50
Tešanj, 331
Thalloczy, L., 173, 224
Theobald of Rouen (chronicler), 74-75
Theodora (Sister of Herceg Stefan; wife of Radoslav Pavlović), 238, 255
Theodore (Archdeacon of Scutari), 222, 233
Thomas (Bishop of Eger), 232
Thomas (Bishop of Hvar; papal legate to Bosnia; surname Tomasini), 53, 63, 299, 302, 303, 309, 310, 314, 316, 317, 329, 353
Thomas (Despot of the Morea), 263
Thomas (Archdeacon of Split), 44, 62, 118, 120-21, 124, 135, 161, 163, 164, 217
Thrace, 66, 72, 115, 117
Thuroczi, Janos (Hungarian chronicler), 235-36

Toričan, 216, 220
Torquemada, Johannes Cardinal of St. Sixtus, 3, 58, 335-37, 355
Toulouse, 117
Transylvania, 59, 66, 69, 72
Travnik, 20, 29, 220, 260
Trebinje, 62, 142, 208, 210, 282, 305, 327, 369, 376, 377, 384
Trnjačka (site of medieval stećak), 260
Trogir, 124, 164, 179, 180, 232
Truhelka, Ć., 13, 17, 85, 243, 251, 260, 263, 264, 289, 314, 320, 368, 370
Tudisić, S. (18th century visitor), 376
Turbe, 220
Turin, 56, 58, 98, 199
Turks, 60, 61, 66, 69, 76, 77-80, 86, 144, 151, 172, 198, 208, 221, 223, 234-37, 242, 250, 254-55, 261, 279, 287, 290, 293, 298-301, 307, 309-15, 321, 326, 328-29, 331, 333, 334, 338-45, 352, 353, 363, 364, 369, 370, 373, 376-84, 386, 387, 390
Tvrdisav krstjanin (Patarin diplomat), 292, 364-65
Tvrtko I (Ban, then King, of Bosnia, 1353-91), 43, 101, 106-07, 108, 154, 170, 184, 187-92, 196-98, 200-04, 207-12, 216, 219, 221, 255, 268, 274, 277, 278, 307
Tvrtko II (King of Bosnia, 1404-09, 1421-43), 53, 70, 107, 229, 230-33, 235-38, 241-51, 255, 256, 271, 274, 278, 281, 284, 285, 287, 291, 299, 307
Tvrtko Pripković (Copier of a gospel), 305
Tvrtković, Stojšan (Bosnian heretical nobleman), 335-38

Ugrin (Archbishop of Kalocsa), 135-36, 144
Ukrina River, 196
Ulcinj, 148, 316
Urban V (pope), 190
Uroš, Tsar (Tsar of Serbia 1355-71), 208
Uskoplje (in Bosnia), 205, 342, 344, 345, 369, 372, 375, 378, 387, 388
Uskoplje (near Trebinje), 369
Usora, 43, 67, 106, 108, 135, 140, 145, 154, 161, 162, 165, 170, 172, 185, 188, 189, 212, 232, 233, 351
Ustikolina, 220, 282
Užice, 208

Valentinus (Bishop of Makarska), 171
Valjevo, 89
Vardić, Radovan (secular diplomat), 239, 320
Varna, 301, 346-47
Večinić, Radmilo (heretical Bosnian nobleman), 335-38
Vego, M., 88, 286, 287, 288, 293, 311
Veličani (site of medieval cemetery), 191
Venanzio da Fabriano, fra (Franciscan author), 60
Venice (and Venetians), 42, 56, 57, 84, 186, 222, 231, 232, 240, 241, 281, 287, 299, 315-17, 322, 330, 331, 334, 340, 344, 364, 370, 373, 375, 378
Vesela straža, 345
Vicenza, 118
Vidin, 192
Vinodol, 375
Višegrad, 29, 91, 261, 287-88
Visoki Dečani (Serbian monastery), 19
Visoko, 29, 60, 114, 121, 141, 167, 174, 176, 183, 184, 187, 220, 233, 237, 246, 343-44, 375
Visoko-Srebrnica bishopric, 251, 338
Vitača (first wife of King Ostoja), 282
Vlachs, 89, 90, 91, 102, 256-57, 312, 375, 381-82, 388-89
Vladimirović, Radivoj, 108
Vladislav Kotromanić (brother of Stjepan Kotromanić), 187, 206, 207
Vlatko Tumarlić (Patarin diplomat), 226, 228, 236-37, 239, 264, 265, 269, 272, 274, 291

Vlatko Vuković (of Kosača family), 220, 278
Voje, I. 315-16
Vojsalić, Djuraj (Hrvoje's nephew and successor), 237, 251, 269, 285-86, 304, 306
Vojsalić, Peter, 269, 306, 315
Voislav Voinović (Serbian nobleman), 208
Vojslava (wife of Ban Kulin), 121, 124
Volaterrano, Raphael, 1, 64, 65, 67, 68, 69, 70, 71
Vranduk, 307, 331
Vrbas, 76, 161, 172, 196, 220, 377
Vrbica, 207
Vrhbosna, 114, 140-41, 143, 144, 220, 237, 241, 257, 269, 282, 311, 351, 379
Vuk (Brother of Tvrtko I), 106-07, 188, 190, 192, 203, 208, 277
Vuk, gost (of Uskoplje; surname Radivojević), 369, 372
Vukan of Zeta, 124, 126, 160
Vukašin (Regent for Tsar Uroš in Serbia with title King), 208
Vukovska župa in Slavonia, 153, 196
Wadding, L., 70, 76-77, 111, 245, 248, 309, 310
Wenzel, M., 88, 91, 102
Wolff, R.L., 235

Zadar, 42, 118, 178-79, 197, 222, 224, 231, 233, 240
Zagreb, 42, 56, 137
Završje, 170, 172, 202, 345
Zemljanik, 161
Zenica, 106, 113, 114, 122, 141, 167, 220, 227, 262
Zeno (16th century traveler), 374
Zeta, 119, 124, 155, 160, 191, 203, 208, 233, 244, 253, 255, 301, 379
Zgunje (village with medieval cemetery), 205, 261-62, 290, 343
Zvijezdović, Angelo (Bosnian Franciscan leader), 351, 379
Zvornik, 29, 250, 331, 384
Žepe (village with medieval cemetery), 92
Žitomislić (Hercegovinian Orthodox monastery), 382

EAST EUROPEAN MONOGRAPHS

1. Political Ideas and the Enlightenment in the Romanian Principalities, 1750-1831. By Vlad Georgescu. 1971.

2. America, Italy and the Birth of Yugoslavia, 1917-1919. By Dragan R. Zivojinovic. 1972.

3. Jewish Nobles and Geniuses in Modern Hungary. By William O. McCagg, Jr. 1972.

4. Mixail Soloxov in Yugoslavia: Reception and Literary Impact. By Robert F. Price. 1973.

5. The Historical and Nationalistic Thought of Nicolae Iorga. By William O. Oldson. 1973.

6. Guide to Polish Libraries and Archives. By Richard C. Lewanski. 1974.

7. Vienna Broadcasts to Slovakia, 1938-1939: A Case Study in Subversion. By Henry Delfiner. 1974.

8. The 1917 Revolution in Latvia. By Andrew Ezergailis. 1974.

9. The Ukraine in the United Nations Organization: A Study in Soviet Foreign Policy, 1944-1950. By Konstantyn Sawczuk. 1975.

10. The Bosnian Church: A New Interpretation. By John V. A. Fine, Jr. 1975.